DODGE | FULL SIZE VANS
1989-91 REPAIR MANUAL

President	Gary R. Ingersoll
Senior Vice President	Ronald A. Hoxter
Publisher	Kerry A. Freeman, S.A.E.
Editor-In-Chief	Dean F. Morgantini, S.A.E.
Managing Editor	David H. Lee, A.S.E., S.A.E.
Manager of Manufacturing	John J. Cantwell
Production Manager	W. Calvin Settle, Jr., S.A.E.
Senior Editor	Richard J. Rivele, S.A.E.
Senior Editor	Nick D'Andrea
Senior Editor	Ron Webb
Editor	Ron Webb

CHILTON BOOK COMPANY
ONE OF THE **ABC PUBLISHING COMPANIES,**
A PART OF **CAPITAL CITIES/ABC, INC.**

Manufactured in USA
© 1991 Chilton Book Company
Chilton Way, Radnor, PA 19089
ISBN 0–8019–8169–7
Library of Congress Catalog Card No. 90–056118
1234567890   0987654321

Contents

Contents

SAFETY NOTICE

Proper service and repair procedures are vital to the safe, reliable operation of all motor vehicles, as well as the personal safety of those performing repairs. This manual outlines procedures for servicing and repairing vehicles using safe, effective methods. The procedures contain many NOTES, CAUTIONS and WARNINGS which should be followed along with standard safety procedures to eliminate the possibility of personal injury or improper service which could damage the vehicle or compromise its safety.

It is important to note that the repair procedures and techniques, tools and parts for servicing motor vehicles, as well as the skill and experience of the individual performing the work vary widely. It is not possible to anticipate all of the conceivable ways or conditions under which vehicles may be serviced, or to provide cautions as to all of the possible hazards that may result. Standard and accepted safety precautions and equipment should be used when handling toxic or flammable fluids, and safety goggles or other protection should be used during cutting, grinding, chiseling, prying, or any other process that can cause material removal or projectiles.

Some procedures require the use of tools specially designed for a specific purpose. Before substituting another tool or procedure, you must be completely satisfied that neither your personal safety, nor the performance of the vehicle will be endangered

Although information in this manual is based on industry sources and is complete as possible at the time of publication, the possibility exists that some car manufacturers made later changes which could not be included here. While striving for total accuracy, Chilton Book Company cannot assume responsibility for any errors, changes or omissions that may occur in the compilation of this data.

PART NUMBERS

Part numbers listed in this reference are not recommendations by Chilton for any product by brand name. They are references that can be used with interchange manuals and aftermarket supplier catalogs to locate each brand supplier's discrete part number.

SPECIAL TOOLS

Special tools are recommended by the vehicle manufacturer to perform their specific job. Use has been kept to a minimum, but where absolutely necessary, they are referred to in the text by the part number of the tool manufacturer. These tools can be purchased under the appropriate part number, from your Dodge dealer or regional distributor or an equivalent tool can be purchased locally from a tool supplier or parts outlet. Before substituting any tool for the recommended one, read the SAFETY NOTICE at the top of this page.

ACKNOWLEDGMENTS

The Chilton Book Company expresses its appreciation to Chrysler Motor Corporation, Detroit, Michigan for their generous assistance.

General Information and Maintenance

1

QUICK REFERENCE INDEX

GENERAL INDEX

HOW TO USE THIS BOOK

Chilton's Total Car Care Repair Manual for Dodge Van/Wagons is intended to help you learn more about the inner workings of your vehicle and save you money on its upkeep and operation.

The first two sections will be the most used, since they contain maintenance, performance and tune-up information and procedures. Studies have shown that a properly maintained van can get at least 10% better gas mileage and better performance than an unmaintained van. The other sections deal with the more complex systems of your van. Operating systems from engine through brakes are covered to the extent that the average do-it-yourselfer becomes mechanically involved. This book will give you detailed instructions to help maintain and repair your van such as how to change your own brake pads and shoes, replace spark plugs, and how to do many more jobs that will save you money, give you personal satisfaction, and help you avoid expensive problems.

A secondary purpose of this book is a reference for owners who want to understand their van and/or their mechanics better.

Before starting any repair or maintenance job or removing any bolts, read through the entire procedure. This will give you the overall view of what is involved and what tools and supplies will be required. There is nothing more frustrating than having to walk to the bus stop on Monday morning because you were short one bolt on Sunday afternoon. So read ahead and plan ahead. Each operation should be approached logically and all procedures thoroughly understood before attempting any work.

All sections contain adjustments, maintenance, removal and installation procedures, and repair or overhaul procedures. When repair is not considered practical, we tell you how to remove the part and then how to install the new or rebuilt replacement. In this way, you at least save the labor costs. Do-it-yourself repair of some components is just not practical.

Two basic mechanic's rules should be mentioned here. One, whenever the left side of the van or engine is referred to, it is meant to specify the driver's side of the van. Conversely, the right side of the van means the passenger's side. Secondly, most screws and bolts are removed by turning counterclockwise, and tightened by turning clockwise.

Safety is always the most important rule. Constantly be aware of the dangers involved in working on an vehicle and take the proper precautions. (See the paragraphs in this section Servicing Your Vehicle Safely and the SAFETY NOTICE on the acknowledgement page.)

Pay attention to the instructions provided. There are 3 common mistakes in mechanical work:

1. Incorrect order of assembly, disassembly or adjustment. When taking something apart or putting it together, doing things in the wrong order usually just costs you extra time; however, it CAN break something. Read the entire procedure before beginning disassembly. Do everything in the order in which the instructions say you should do it, even if you can't immediately see a reason for it. When you're taking apart something that is very intricate (for example, a carburetor), you might want to draw a picture of how it looks when assembled in order to make sure you get everything back in its proper position. (We will supply an exploded view whenever possible). When making adjustments, especially tune-up adjustments, do them in order; often, one adjustment affects another, and you cannot expect even satisfactory results unless each adjustment is made only when it cannot be changed by any other.

2. Overtorquing (or undertorquing). While it is more common for over-torquing to cause damage, undertorquing can cause a fastener to vibrate loose causing serious damage. Especially when dealing with aluminum parts, pay attention to torque specifications and utilize a torque wrench in assembly. If a torque figure is not available, remember that if you are using the right tool to do the job, you will probably not have to strain yourself to get a fastener tight enough. The pitch of most threads is so slight that the tension you put on the wrench will be multiplied many, many times in actual force on what you are tightening. A good example of how critical torque is can be seen in the case of spark plug installation, especially where you are putting the plug into an aluminum cylinder head. Too little torque can fail to crush the gasket (if used), causing leakage of combustion gases and consequent overheating of the plug and engine parts. Too much torque can damage the threads, or distort the plug which may change the spark gap.

There are many commercial products available for ensuring that fasteners won't come loose, even if they are not torqued just right (a very common brand is Loctite®). If you're worried about getting something together tight enough to hold, but loose enough to avoid mechanical damage during assembly, one of these products might offer substantial insurance. Read the label on the package and make sure the product is compatible with the materials, fluids, etc. involved.

3. Crossthreading. This occurs when a part such as a bolt is screwed into a nut or casting at the wrong angle and forced. Cross threading is more likely to occur if access is difficult. It helps to clean and lubricate fasteners, and to start threading with the part to be installed going straight in. Then, start the bolt, spark plug, etc. with your fingers. If you encounter resistance, unscrew the part and start over again at a different angle until it can be inserted and turned several turns without much effort. Keep in mind that many parts, especially spark plugs, used tapered threads so that gentle turning will automatically bring the part you're treading to the proper angle if you don't force it or resist a change in angle. Don't put a wrench on the part until its been turned a couple of turns by hand. If you suddenly encounter resistance, and the part has not seated fully, don't force it. Turn it back out and make sure it's clean and threading properly.

Always take your time and be patient; once you have some experience, working on your van will become an enjoyable hobby.

TOOLS AND EQUIPMENT

Naturally, without the proper tools and equipment it is impossible to properly service you vehicle. It would be impossible to catalog each tool that you would need to perform each or any operation in this book. It would also be unwise to rush out and buy an expensive set of tools on the theory that one or more of them may be need at sometime.

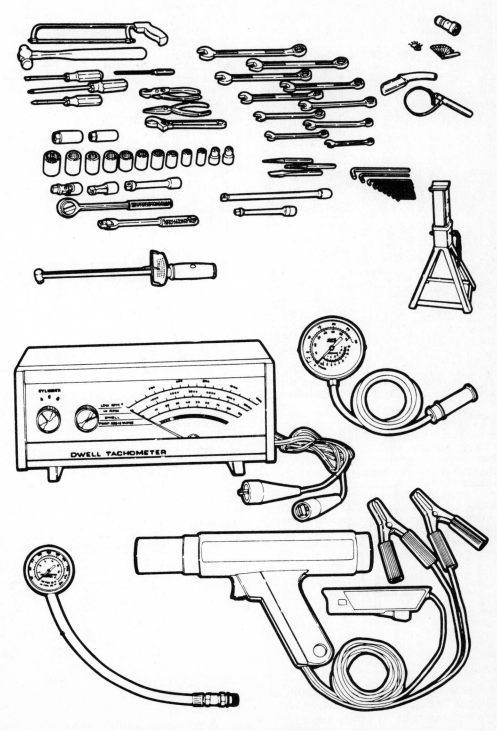

Necessary tool assortment needed for most jobs

The best approach is to proceed slowly gathering together a good quality set of those tools that are used most frequently. Don't be misled by the low cost of bargain tools. It is far better to spend a little more for better quality. Forged wrenches, 6- or 12-point sockets and fine tooth ratchets are by far preferable to their less expensive counterparts. As any good mechanic can tell you, there are few worse experiences than trying to work on a van with bad tools. Your monetary savings will be far out-weighed by frustration and mangled knuckles.

Begin accumulating those tools that are used most frequently; those associated with routine maintenance, tune-up and repairs.

Certain tools, such as a basic mechanic's tool set and a torque wrench, plus a basic ability to handle tools, are required to get started. In addition to the basic assortment of tools, including screwdrivers and pliers, you should have the following tools for routine maintenance and most repair jobs:

1. SAE and Metric, or SAE/Metric wrenches-sockets and combination open end-box end wrenches in sizes from 1/8 in. (3 mm) to 3/4 in. (19 mm) and a spark plug socket 13/16 in. or 5/8 in. depending on plug type.

If possible, buy various length socket drive extensions. One break in this department is that the metric sockets available in the U.S. will all fit the ratchet handles and extensions you may already have (1/4 in., 3/8 in., and 1/2 in. drive).

2. Jackstands for support and a floor jack to raise the vehicle.
3. Oil filter wrench.
4. Oil filler spout and a funnel for pouring oil.
5. Grease gun for chassis lubrication.
6. Hydrometer for checking the battery (non-sealed type).
7. A container for draining oil.
8. Many rags for wiping up the inevitable mess.

In addition to the above items there are several others that are not absolutely necessary, but handy to have around. these include oil dry (kitty litter works well and is usually less expensive) and the usual supply of lubricants, antifreeze and fluids, although these can be purchased as needed. This is a basic list for routine maintenance, but only your personal needs and desire can accurately determine your list of tools.

The second list of tools is for tune-ups. While the tools involved here are slightly more sophisticated, they need not be outrageously expensive. There are inexpensive tachometers on the market that are every bit as good for the average mechanic as a $100.00 plus professional model. Just be sure that it goes to at least 1,200-1,500 rpm on the tach scale and that it works on 4, 6, 8 cylinder engines. A basic list of tune-up equipment could include:

1. Tachometer.
2. Spark plug wrench.
3. Timing light (an induction model powered from the van's battery is best), and is required for use on most electronic ignition systems.
4. Wire spark plug gauge/adjusting tools.
5. Set of feeler blades.

In addition to these basic tools, there are several other tools and gauges you may find useful. These include:

1. A compression gauge. The screw-in type is slower to use, but eliminates the possibility of a faulty reading due to escaping pressure.
2. A manifold vacuum gauge.
3. A test light.
4. An volt/ohm meter. This is used for determining whether or not there is current in a wire. These are handy for use if a wire is broken somewhere in a wiring harness.

As a final note, you will probably find a torque wrench necessary for all but the most basic work. The beam type models are perfectly adequate, although the newer click type are more precise and easier to use when your sight is blocked when working in a restricted area.

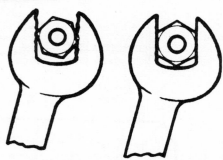

When using an open-end wrench, use the exact size needed and position it squarely on the flats of the bolt or nut

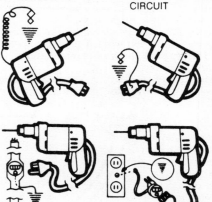

2-WIRE CONDUCTOR THIRD WIRE GROUNDING THE CASE

3-WIRE CONDUCTOR GROUNDING THROUGH A CIRCUIT

3-WIRE CONDUCTOR ONE WIRE TO A GROUND

3-WIRE CONDUCTOR GROUNDING THROUGH AN ADAPTER PLUG

When using electric tools, make sure they are properly grounded

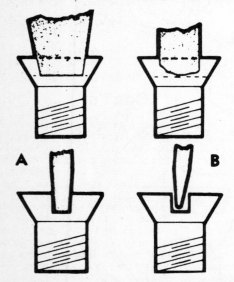

Keep screwdriver tips in good shape. They should fit in the screw head slots in the manner shown in "A". If they look like the tip shown in "B", they need grinding or replacing

Special Tools

Normally, the use of special factory tools is avoided for repair procedures, since these are not readily available for the do-it-yourself mechanic. When it is possible to preform the job with more commonly available tools, it will be pointed out, but occasionally, a special tool was designed to perform a specific function and should be used. Before substituting another tool, you should be convinced that neither your safety nor the performance of the vehicle will be compromised.

Some special tools are available commercially from major tool manufacturers. Others can be purchased from your Dodge or Plymouth Dealer or from:
Miller Special Tools
Utica Tool Co.
32615 Park La.
Garden City, MI 48135

SERVICING YOUR VEHICLE SAFELY

It is virtually impossible to anticipate all of the hazards involved with vehicle maintenance and service but care and common sense will prevent most accidents.

The rules of safety for mechanics range from "don't smoke around gasoline" to "use the proper tool for the job." The trick to avoiding injuries is to develop safe work habits and take every possible precaution.

Do's

• Do keep a fire extinguisher and first aid kit within easy reach.

• Do wear safety glasses or goggles when cutting, drilling, grinding, or prying, even if you have 20/20 vision. If you wear glasses for the sake of vision, then they should be made of hardened glass that can serve also as safety glasses, or wear safety glasses over your regular glasses.

• Do shield your eyes whenever you work around the battery. Batteries contain sulphuric acid; in case of contact with the eyes or skin, flush the area with water or a mixture of water and baking soda and get medical attention immediately.

• Do use safety stands for any under-van service. Jacks are for raising vehicles; safety stands are for making sure the vehicle stays raised until you want it to come down. Whenever the vehicle is raised, block the wheels remaining on the ground and set the parking brake.

• Do use adequate ventilation when working with any chemicals. Like carbon monoxide, the asbestos dust resulting from brake lining wear can be poisonous in sufficient quantities.

• Do disconnect the negative battery cable when working on the electrical system. The primary ignition system can contain up to 40,000 volts.

• Do follow manufacturer's directions whenever working with potentially hazardous materials. Both brake fluid and antifreeze are poisonous if taken internally.

• Do properly maintain your tools. Loose hammerheads, mushroomed punches and chisels, frayed or poorly grounded electrical cords, excessively worn screwdrivers, spread wrenches (open end), cracked sockets, slipping ratchets, or faulty droplight sockets can cause accidents.

• Do use the proper size and type of tool for the job being done.

• Do when possible, pull on a wrench handle rather than push on it, and adjust your stance to prevent a fall.

• Do be sure that adjustable wrenches are tightly adjusted on the nut or bolt and pulled so that the face is on the side of the fixed jaw.

• Do select a wrench or socket that fits the nut or bolt. The wrench or socket should sit straight, not cocked.

• Do strike squarely with a hammer. Avoid glancing blows.

• Do set the parking brake and block the drive wheels if the work requires that the engine be running.

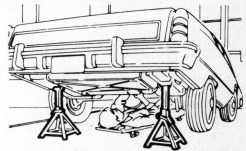

Always support the vehicle on jackstands

Don't's

• Don't run an engine in a garage or anywhere else without proper ventilation — EVER! Carbon monoxide is poisonous; it takes a long time to leave the human body and you can build up a deadly supply of it in your system by simply breathing in a little every day. You may not realize you are slowly poisoning yourself. Always use power vents, window, fans or open the garage door.

• Don't work around moving parts while wearing a necktie or other loose clothing. Short sleeves are much safer than long, loose sleeves and hard-toed shoes with neoprene soles protect your toes and give a better grip on slippery surfaces. Jewelry such as watches, fancy belt buckles, beads or body adornment of any kind is not safe working around a van. Long hair should be hidden under a hat or cap.

• Don't use pockets for toolboxes. A fall or bump can drive a screwdriver deep into your body. Even a wiping cloth hanging from the back pocket can wrap around a spinning shaft or fan.

• Don't smoke when working around gasoline, cleaning solvent or other flammable material.

• Don't smoke when working around the battery. When the battery is being charged, it gives off explosive hydrogen gas.

• Don't use gasoline to wash your hands or to clean parts; there are excellent soaps and safe part cleaning solvents available. Gasoline may contain lead, and lead can enter the body through a cut, accumulating in the body until you are very ill. Gasoline also removes all the natural oils from the skin so that bone dry hands will such up oil and grease.

• Don't service the air conditioning system unless you are equipped with the necessary tools and training. The refrigerant, R-12, is extremely cold and when exposed to the air, will instantly freeze any surface it comes in contact with, including your eyes. Although the refrigerant is normally non-toxic, R-12 becomes a deadly poisonous gas in the presence of an open flame. One good whiff of the vapors from burning refrigerant can be fatal.

• Don't ever use a bumper jack (the jack that comes with the vehicle) for anything other than changing tires! If you are serious about maintaining your van yourself, invest in a hydraulic floor jack of at least 1½ ton capacity. It will pay for itself many times over through the years.

SERIAL NUMBER IDENTIFICATION

Vehicle (V.I.N.)

The vehicle identification number plate is located on the upper left corner of the instrument panel, at the base of the windshield.

The vehicle identification number consists of a combination of 17 elements (numbers and letters). Use the following example as a key:

1 B 4 F B 1 3 X 1 L K 000001

Element or Position 1 - Country of Origin
• 1 = U.S.
• 2 = Canada

Element or Position 2 - Make
• B = Dodge
• E = Fargo

Element or Position 3 - Type of Vehicle
• 4 = Multipurpose Passenger
• 5 = Bus
• 6 = Incomplete
• 7 = Truck

Element or Position 4 - GVWR (lbs.) and Hydraulic Brakes
• F = 4001-5000
• G = 5001-6000
• H = 6001-7000
• J = 7001-8000
• K = 8001-9000
• W = Hydraulic Brakes

Element or Position 5 - Truck Line
• B = Wagon/Van/Bus

Element or Position 6 - Series
• 0 = 100
• 1 = 150
• 2 = 250
• 3 = 350

MFD BY	CHRYSLER CORPORATION	DATE OF MFR	GVWR
GAWR FRONT	WITH TIRES	RIMS AT	PSI COLD
GAWR REAR	WITH TIRES	RIMS AT	PSI COLD

THIS VEHICLE CONFORMS TO ALL APPLICABLE FEDERAL MOTOR VEHICLE SAFETY STANDARDS IN EFFECT ON THE DATE OF MANUFACTURE SHOWN ABOVE.

VIN:	TYPE:		SINGLE	DUAL

BAR CODE

MDH: VEHICLE MADE IN

Vehicle Safety Certification Label

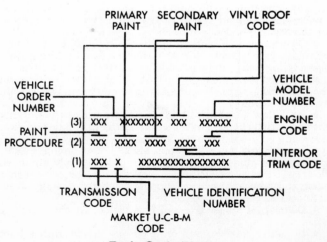

Body Code Plate

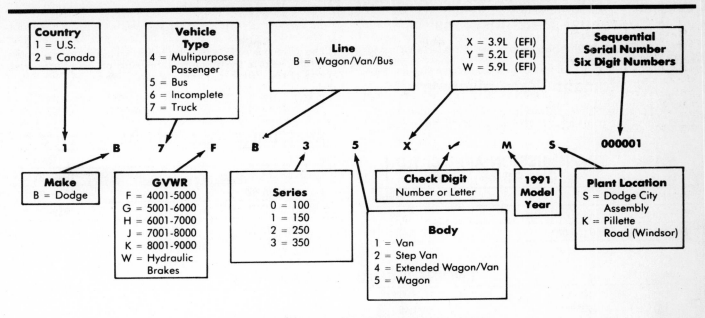

Vehicle Identification Number (VIN) Decoding

Element or Position 7 - Body
- 1 = Wagon(1989)
- 1 = Van(1990-91)
- 2 = Step Van(1990-91)
- 3 = Van(1989)
- 4 = Extended Wagon/Van(1990-91)
- 5 = Wagon(1990-91)
- 7 = Step Van(1989)

Element or Position 8 - Engines (CID/Liters)
- X = 238/3.9L EFI
- Y = 318/5.2L EFI
- W = 360/5.9L EFI

Element or Position 9
Check digit

Element or Position 10 - Model Year
- K = 1989
- L = 1990
- M = 1991

Element or Position 11 - Assembly Plant
- K = Pillette Road (Windsor)
- S = Dodge City

Element or Position 12 through 17
Sequence Number

Equipment Identification Plate

The Equipment Identification Plate is located on the inner surface of the hood or the front surface of the air conditioning or Heater housing. It contains the model, wheelbase, V.I.N., T.O.N. (Truck Order Number) and all production or special equipment on the vehicle when it was shipped from the factory. Always refer to this plate when ordering parts.

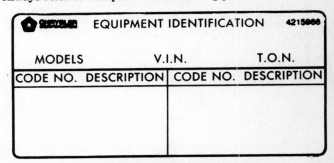

Equipment identification plate

Engine

The engine that the factory installed can be identified by the eight position on Vehicle Identification Number, as explained earlier. The engine itself can be identified, for parts replacement, by the engine serial number.

318 (5.2L) and 360 (5.9L) cu. in. V8s have the number on the front of the block, just below the left cylinder head.

The 238 (3.9L) cu. in. V6 has the serial number stamped on a pad on the right side of the block.

5.2L and 5.9L engine identification

Manual Transmissions

The transmission identification number is located on a tag secured by 2 bolts to cover.

Automatic Transmissions

The transmission identification number is located on the right side of the transmission case on a ground surface pad.

Axles

The drive axle code is found stamped on a flat surface on the axle tube, next to the differential housing, or, on a tag secured by one of the differential housing cover bolts.

REAR DRIVE AXLE APPLICATION CHART

Axle	Years
Chrysler 8³/₈ in.	1989–91 ①
Chrysler 9¹/₄ in HD	1989–91 ①
Spicer 60 and 60HD (9³/₄ in)	1989–91 ②

① Sure-grip available
② Trak-lok available

MANUAL TRANSMISSION APPLICATION CHART

Transmission Types	Years
New Process NP435 (4 speed OD)	1989
New Process NP2500 (5 speed)	1989–91

ENGINE IDENTIFICATION CHART

Years	No. of Cylinders and Cu. In. Displacement	Cu. In.	Actual Displacement CC	Liters	Fuel System	Type	Built by	Engine Code
1989	6-238	238.46	3,907.7	3.9	TBI	OHV	Chrysler	X
	8-318	317.95	5,210.3	5.2	TBI	OHV	Chrysler	Y
	8-360	359.90	5,897.7	5.9	TBI	OHV	Chrysler	W
1990	6-238	238.46	3,907.7	3.9	TBI	OHV	Chrysler	X
	8-318	317.95	5,210.3	5.2	TBI	OHV	Chrysler	Y
	8-360	359.90	5,897.7	5.9	TBI	OHV	Chrysler	W
1991	6-238	238.46	3,907.7	3.9	TBI	OHV	Chrysler	X
	8-318	317.95	5,210.3	5.2	TBI	OHV	Chrysler	Y
	8-360	359.90	5,897.7	5.9	TBI	OHV	Chrysler	W

AUTOMATIC TRANSMISSION APPLICATION CHART

Transmission	Years	Converter	Engine
Loadflite A998 (3 speed)	1989–90	Lockup	V6 & V8
Loadflite A999 (3 speed)	1989–91	Lockup	V6 & V8
Loadflite A727 (3 speed)	1989–91	Nonlockup	V8
Loadflite A500 (4 speed OD)	1989–91	Lockup	V6 & V8
Loadflite A518 (4 speed OD)	1990–91	Nonlockup	V8

CAPACITIES CHART

Years	Engine No. Cyl. Liters	Crankcase Includes Filter (qt)	Transmission (pts) 4-sp	5-sp	Auto	Drive Axle (pts)	Fuel Tank (gal)	Cooling System (qt) w/AC	wo/AC
1989	V6 3.9L	4	7.0	4.0	①	②	③	14.6	14.6
	V8 5.2L	5	—	4.0	①	②	③	16.5	16.5
	V8 5.9L	5	—	—	①	②	③	15.0	15.0④
1990	V6 3.9L	4	—	4.0	①	②	③	14.6	14.6
	V8 5.2L	5	—	4.0	①	②	③	16.5	16.5
	V8 5.9L	5	—	—	①	②	③	15.0	15.0④

CAPACITIES CHART

Years	Engine No. Cyl. Liters	Crankcase Includes Filter (qt)	Transmission (pts) 4-sp	5-sp	Auto	Drive Axle (pts)	Fuel Tank (gal)	Cooling System (qt) w/AC	wo/AC
1991	V6 3.9L	4	—	4.0	①	②	③	14.6	14.6
	V8 5.2L	5	—	4.0	①	②	③	16.5	16.5
	V8 5.9L	5	—	—	①	②	③	15.0	15.0

① Capacity w/Converter
A500/518 20.4
A727 16.8
A998 16.7
A999 16.7

② 8¼ 4.4
9¼ 4.5
Dana 60 6.0

③ Standard exc. 5.9L 22.0
Standard 5.9L 35.0
Optional exc. 5.9L 35.0
④ Add 1 qt. w/rear heater

VAN BODY CODE	WHEELBASE
11	109
12	127
13	127x

WAGON BODY CODE	WHEELBASE
52	109
52	127
53	127x

1989 POWER TEAM AVAILABILITY
Conventional Van/Wagon

Vehicle Line	Model Code*	GVWR	Engine Displacement & Sales Code: 3.9L 239in³ EFI EHB	5.2L 318in³ EFI ELG	5.9L 360in³ EFI EMG	5.9L 360in³ EFI EMJ	5.9L 360in³ Diesel ETA	Transmission: 5-Sp O.D. Man	4-Sp O.D. Man	4-Sp Man	3-Sp Auto	4-Sp O.D. Auto
VANS												
B150	L11, L12	5000, 5300	S	O				S			O	O
B250	L11, L12	6010	S	O				S			O	O
					O						S	
	L12, L13	6400	S	O				S			O	O
					O						S	O
B350	L12, L13	7500		S							S	O
					O						S	
		8510			O	O					S	
	L13	9000			O	O					S	
WAGONS												
B100, 150	E51, L51	5300,	S					S(a)			O	O
		6010		O							O	O
B150	L52	5300	S					S			O	O
		5500		O							O	O
B250	L52	6010	S					S			O	O
		6400		O							O	O
	L53	6400		S							S	O
	L52,L53	6400			O						O	
B350	L52	7500		S							S	O
					O						S	
	L53	8510			O	O					S	

(a) Not Available in California
* E 11 Long Range Van - 109.6
E 12 Long Range Van - 127.6
L 11 Van - 109.6
L 12 Van - 127.6
L 13 Maxivan - 127.6
E 51 Value Wagon - 109.6
L 51 Wagon - 109.6
L 52 Wagon - 127.6
L 53 Maxiwagon - 127.6

1989 Power Team Availability

RAM VAN

VEH. LINE	MODEL CODE	GVWR	ENGINE & SALES CODE				TRANSMISSION		
			3.9L EFI EHB	5.2L EFI ELG	5.9L EFI EMG	5.9L EFI EMJ	5-SP MAN. O.D.	3-SP AUTO	4-SP AUTO O.D.
B150 AB-1	L11	2268 kg (5000 lb)	S	O			S^bc	O	O^d
	L12	2404 kg (5300 lb)							
B250 AB-2	L11	2726 kg (6010 lb)	S	O			S^bc	O	O^d
	L12				O				S
	L12	2903 kg (5400 lb)	S	O				S	O^d
	L13				O				S
B350 AB-3	L12	3402 kg (7500 lb)	S						S
	L13				O				S
		3860 kg (8510 lb)			S	O			S
	L13	4082 kg (9000 lb)			S	O			S

RAM WAGON

VEH. LINE	MODEL CODE	GVWR	ENGINE & SALES CODE				TRANSMISSION		
			3.9L EFI EHB	5.2L EFI ELG	5.9L EFI EMG	5.9L EFI EMJ	5-SP MAN. O.D.	3-SP AUTO	4-SP AUTO O.D.
B150 AB-1	L51	2404 kg (5300 lb)	S				S^abc	O	
		2726 kg (6010 lb)		O				O	O^d
	L52	2404 kg (5300 lb)	S				S^bc	O	
		2495 kg (5500 lb)		O				O	O^d
B250 AB-2	L52	2726 kg (6010 lb)	S					S	
		2903 kg (6400 lb)		O				O	O^d
	L53	2903 kg (6400 lb)			S			S	O
	L52/53	2903 kg (6400 lb)			O				S
B350 AB-3	L52	3402 kg (7500 lb)			S				S
					O				
	L53	3860 kg (8510 lb)			O	O			S

(a) Not available in California.
(b) V-6 only.
(c) N.A. L12 and L52.
(d) V-8 only.
(e) Base engine for B350, L12 and L13 @ 7500 GVWR is 3.9L V-6.

L11 Van . . . . 2784 mm (109.6 in)
L12 Van . . . . 3190 mm (127.6 in)
L13 Maxivan . . . . 3190 mm (127.6 in)
L51 Wagon . . . . 2784 mm (109.6 in)
L52 Wagon . . . . 3190 mm (127.6 in)
L53 Maxiwagon . . . . 3190 mm (127.6 in)

1990-91 Power Team Availability

Model Name	Vehicle Family	Wheelbase	GVWR	Payload Allowance[1]	Curb Weight[2]
VANS					
B100	B-1	109.6	5300	1613	3687
		127.6	5300	1513	3787
B150	B-1	109.6	5000	1330	3670
			5300	1617	3683
		127.6	5000	1230	3770
			5300	1517	3783
B250	B-2	109.6	6010	2346	3664
		127.6	6010	2212	3798
			6400	2562	3838
		127.6 Maxi	6400	2468	3932
B350	B-3	127.6	7500	3336	4164
			8510	4275	4235
		127.6 Maxi	7500	3185	4312
			8510	4145	4365
			9000	4465	4535
WAGONS					
B100	B-1	109.6	5300	1315	3985
			6010	1970	4040
B150	B-1	109.6	5300	1310	3990
			6010	1965	4045
		127.6	5300	1161	4139
			5500	1230	4270
B250	B-2	127.6	6010	1835	4175
			6400	2178	4222
		127.6 Maxi	6400	1856	4544
B350	B-3	127.6	7500	2911	4889
			8510	3890	4620
		127.6 Maxi	7500	2657	4843
			8510	3642	4868

(1) Payload includes maximum weight of driver, passengers, cargo, and optional equipment not included in payload or GVW package
(2) Includes base engine

Vehicle Weights

STANDARD BODIES AVAILABLE						
VEHICLE FAMILY	WHEEL-BASE	BODY TYPE	LOAD SPACE DIMENSIONS			
			LENGTH (1)	WIDTH	HEIGHT	VOLUME
AB1, AB2	2784 mm 109.6 in	VAN/ WAGON	2360 mm 92.9 in	1834 mm 72.2 in	1351 mm 53.2 in	5.84 m³ 206.6 ft³
AB1, AB2 AB3	3190 mm 127.6 in	VAN/ WAGON	2817 mm 110.9 in	1834 mm 72.2 in	1351 mm 53.2 in	6.97 m³ 246.7 ft³
AB2, AB3 Maxi.	3190 mm 127.6 in	VAN/ WAGON	3477 mm 136.9 in	1834 mm 72.2 in	1351 mm 53.2 in	8.60 m³ 304.5 ft³

(1) Driver seat to rear door at belt line.

Body dimensions

VEHICLE	GVWR	MAX. TONGUE WEIGHT
B150 Van & Wagon	All	800
B250 Van & Wagon	All	800
B350 Van & Wagon	All	1000

GCWR—Gross Combined Weight Rating (Total vehicle, trailer, cargo, passengers & fluids)					
ENGINE—TRANSMISSION		AXLE RATIOS			MAX. TRAILER WEIGHT
		3.54	3.90	4.10	
B-150					
3.9L	Man. 5-spd.	8,000	8,000		4,100
	Auto. 3-spd.	8,800	9,200		4,900
	Auto. 4-spd. O/D	8,800	9,200		4,900
5.2L	Auto. 3-spd.	11,500	12,500		6,000
	Auto. 4-spd. O/D	10,500	10,500		6,400
B-250					
3.9L	Man. 5-spd.	8,000	8,000		4,100
	Auto. 3-spd.	8,800	9,200		4,900
	Auto. 4-spd. O/D	8,800	9,200		4,900
5.2L	Auto. 3-spd.	11,500	12,500		6,000
	Auto. 4-spd. O/D	10,500	10,500		6,500
	Auto. 4-spd. H. D.	11,500	12,500		7,500
5.9L	Auto. 4-spd. H. D.	12,000	13,000		8,000
B-350					
3.9L	Auto. 3-spd.	8,800	9,200		4,900
5.2L	Auto. 4-spd. O/D	10,500	10,500		6,100
	Auto. 4-spd. H. D.	11,500	12,500		8,100
5.9L	Auto. 4-spd. H. D.	12,000	13,000	13,500	8,500

Trailer towing specifications

ROUTINE MAINTENANCE

See the Maintenance Intervals Chart for the recommended intervals for the operations in this section.

Air Cleaner

The standard type is the traditional dry paper element.

REMOVAL AND INSTALLATION

1. Remove the top from the air cleaner and remove the element and wrapper (if equipped).
2. Clean the filter element by blowing it out with compressed air (from the inside out). Do not immerse the paper element in liquid. If the paper element is saturated for more than ½ of its circumference by oil from the wrapper, the element should be replaced and the rest of the crankcase ventilating system checked for proper functioning.
3. Wash the top and the air cleaner housing in solvent and wipe it dry.

NOTE: Do not immerse the temperature sensor, located in the housing, in cleaning solvent.

4. Replace the paper element and wrapper (if equipped) and replace the air cleaner cover.

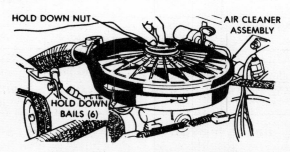

Removing the air cleaner element

Removing the air cleaner element

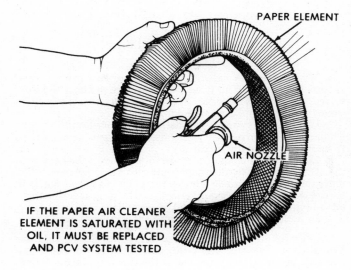

IF THE PAPER AIR CLEANER ELEMENT IS SATURATED WITH OIL, IT MUST BE REPLACED AND PCV SYSTEM TESTED

Cleaning the air cleaner with air pressure

Fuel Filter

— **CAUTION** —

Never smoke when working around gasoline! Avoid all sources of sparks or ignition. Gasoline vapors are EXTREMELY volatile!

Fuel Injected Gasoline Engines

The fuel filter is located inline on the inside of the left frame rail just ahead of the crossmember. To change the filter:

— **CAUTION** —

The TBI (Throttle Body Injection) system is under a constant pressure of approximately 100 kpa (14.5 psi). Before servicing the fuel system, the fuel pressure must be released.

1. Relieve the fuel system pressure as described in section 5.
2. Remove the filter retaining screw and remove the filter from the frame rail.
3. Loosen the fuel hose clamps, wrap a shop towel around the hoses and disconnect the hoses from the filter.
4. Installation is the reverse of removal. Install new fuel line clamps and tighten to 10 inch lbs. Tighten the filter mounting screw to 75 inch lbs.

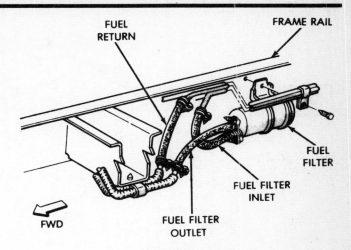

Fuel filter location

Positive Crankcase Ventilation System

Blow-by gases or crankcase vapors must be removed from the crankcase to prevent oil dilution and to prevent the formation of sludge. Traditionally, this was accomplished with a road draft tube. Air entered the rocker arm cover through an open oil filler cap and flowed down past the pushrods, mixing the blow-by gases in the crankcase. It was finally routed into the road draft tube where a partial vacuum was created, drawing the mixture into the road draft tube and out into the atmosphere.

The closed PCV system, on your van/wagon consists of an air inlet filter, a flow-control (PCV) valve and the necessary hose connection system.

The ventilation system operates by intake manifold vacuum as follows:

- fresh air flows from the air cleaner
- through the crankcase inlet air filter
- circulates through the crankcase
- mixes with the crankcase oil vapor
- exits via the PCV valve
- enters and flows through a passage at the base of the throttle body fuel injector
- becomes part of the calibrated air/fuel mixture
- enters the engine combustion chambers and is burned

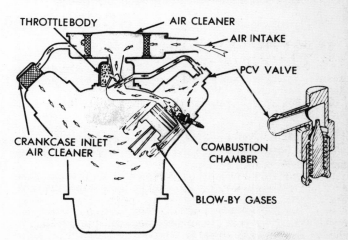

Operation of the crankcase ventilation system

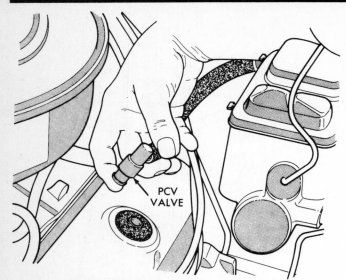

Check the vacuum at the PCV valve

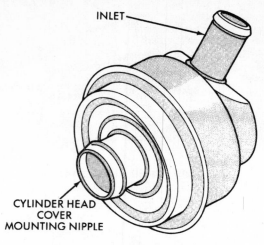

Crankcase air inlet filter

The PCV valve is used to control the rate at which crankcase vapors are returned to the intake manifold. The action of the valve plunger is controlled by intake manifold vacuum and the spring. During deceleration and idle, when manifold vacuum is high, it overcomes the tension of the valve spring and the plunger bottoms in the manifold end of the valve housing. Because of the valve construction, it reduces, but dies not stop, the passage of vapors to the intake manifold. When the engine is lightly accelerated or operated at constant speed, spring tension matches intake manifold vacuum pull and the plunger takes a mid-position in the valve body, allowing more vapors to flow into the manifold.

SERVICE

An inoperative PCV system will cause rough idling, sludge and oil dilution. In the event erratic idle, never attempt to compensate by disconnecting the PCV system. Disconnecting the PCV system will adversely affect engine ventilation. It could also shorten engine life through the buildup of sludge.

To inspect the PCV valve, proceed as follows:
1. With the engine idling, remove the PCV valve from the rocker cover. If the valve is not plugged, a hissing sound will be heard. A strong vacuum should be felt when you place your finger over the valve.
2. Install the PCV valve and allow about a minute for pressure to drop.
3. Remove the fresh air hose from the air cleaner housing and loosely hold as piece of paper at the end of the hose. The paper should be forced against the end of the hose opening by atmospheric pressure.
4. With the engine stopped, remove the PCV valve and shake it. A rattle or clicking should be heard to indicate that the valve is free.
5. If the system meets passes the above tests, no further service is required, unless replacement is specified in the Maintenance Intervals Chart. If the system does not meet the tests, the valve should be replaced with a new one.

NOTE: Do not attempt to clean a PCV valve.

6. With a new PCV valve installed, if the paper is not sucked against the fresh air in, it will be necessary to clean the PCV valve hose, the PCV hose port at the base of the throttle body injector, and the crankcase air inlet filter.
7. Clean the line with Combustion Chamber Conditioner or similar solvent. Do not leave the hoses in solvent for more than ½ hour. Allow the line to air dry.
8. Clean the crankcase air intake filter. Inspect the hose from the crankcase intake air cleaner and clean if it necessary. Remove the crankcase intake air cleaner and wash it thoroughly in kerosene or a similar solvent. Lubricate the filter by inverting it and filling with SAE 30 engine oil. Position the filter to allow excess oil to drain thoroughly through the vent nipple.

NOTE: After checking and/or servicing the Crankcase Ventilation System, any components that do not allow passage or air to the intake manifold should be replaced.

Evaporative Canister

The evaporative canister is part of the Evaporation Control System, the canister stores fuel vapors from the fuel tank. The only service associated with the canister is replacement should the canister become damaged. Should hose replacement be necessary, use only fuel resistant hose.

The vapor storage canister is located on the right side of the vehicle, in the engine compartment, behind the front wheelwell.

Battery

Loose, dirty, or corroded battery terminals are a major cause of "no-start." Every 3 months or so, remove the battery terminals and clean them, giving them a light coating of petroleum jelly when you are finished. This will help to retard corrosion.

Check the battery cables for signs of wear or chafing and replace any cable or terminal that looks marginal. Battery terminals can be easily cleaned and inexpensive terminal cleaning tools are an excellent investment that will pay for themselves many times over. They can usually be purchased from any well-equipped auto store or parts department. Side terminal batteries require a different tool to clean the threads in the battery case. The accumulated white powder and corrosion can be cleaned from the top of the battery with an old toothbrush and a solution of baking soda and water.

Unless you have a maintenance-free battery, check the electrolyte level (see Battery under Fluid Level Checks in this section) and check the specific gravity of each cell. Be sure that the vent holes in each cell cap are not blocked by grease or dirt. The vent holes allow hydrogen gas, formed by the chemical reaction in the battery, to escape safely.

REPLACEMENT BATTERIES

The cold power rating of a battery measures battery starting performance and provides an approximate relationship between battery size and engine size. The cold power rating of a replacement battery should match or exceed your engine size in cubic inches.

FLUID LEVEL (EXCEPT MAINTENANCE FREE BATTERIES)

Check the battery electrolyte level at least once a month, or more often in hot weather or during periods of extended van operation. The level can be checked through the case on translucent polypropylene batteries; the cell caps must be removed on other models. The electrolyte level in each cell should be kept filled to the split ring inside, or the line marked on the outside of the case.

If the level is low, add only distilled water, or colorless, odorless drinking water, through the opening until the level is correct. Each cell is completely separate from the others, so each must be checked and filled individually.

If water is added in freezing weather, the van should be driven several miles to allow the water to mix with the electrolyte. Otherwise, the battery could freeze.

SPECIFIC GRAVITY (EXCEPT MAINTENANCE FREE BATTERIES)

At least once a year, check the specific gravity of the battery. It should be between 1.20 in.Hg and 1.26 in.Hg at room temperature.

The specific gravity can be check with the use of an hydrometer, an inexpensive instrument available from many sources, including auto parts stores. The hydrometer has a squeeze bulb at one end and a nozzle at the other. Battery electrolyte is sucked into the hydrometer until the float is lifted from its seat. The specific gravity is then read by noting the position of the float. Generally, if after charging, the specific gravity between any two cells varies more than 50 points (0.50), the battery is bad and should be replaced.

It is not possible to check the specific gravity in this manner on sealed (maintenance free) batteries. Instead, the indicator built into the top of the case must be relied on to display any signs of battery deterioration. If the indicator is dark, the battery can be assumed to be OK. If the indicator is light, the specific gravity is low, and the battery should be charged or replaced.

Battery State of Charge at Room Temperature

Specific Gravity Reading	Charged Condition
1.260–1.280	Fully Charged
1.230–1.250	¾ Charged
1.200–1.220	½ Charged
1.170–1.190	¼ Charged
1.140–1.160	Almost no Charge
1.110–1.130	No Charge

Battery state of charge at room temperature

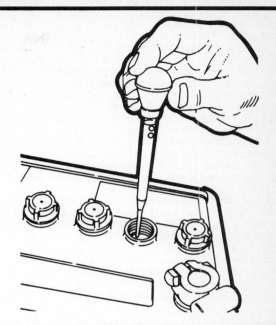

Testing the specific gravity of the battery

CABLES AND CLAMPS

Once a year, the battery terminals and the cable clamps should be cleaned. Loosen the clamps and remove the cables, negative cable first. On batteries with posts on top, the use of a puller specially made for the purpose is recommended. These are inexpensive, and available in auto parts stores. Side terminal battery cables are secured with a bolt.

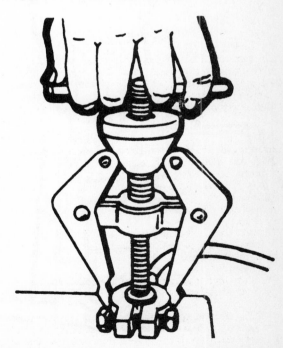

Special puller used to remove the cable end from the battery post

Clean the cable lamps and the battery terminal with a wire brush, until all corrosion, grease, etc., is removed and the metal is shiny. It is especially important to clean the inside of the clamp thoroughly, since a small deposit of foreign material or oxidation there will prevent a sound electrical connection and inhibit either starting or charging. Special tools are available for cleaning these parts, one type for conventional batteries and another type for side terminal batteries.

Before installing the cables, loosen the battery holddown clamp or strap, remove the battery and check the battery tray. Clear it of any debris, and check it for soundness. Rust should be wire brushed away, and the metal given a coat of anti-rust paint. Replace the battery and tighten the holddown clamp or strap securely, but be careful not to overtighten, which will crack the battery case.

After the clamps and terminals are clean, reinstall the cables, negative cable last; do not hammer on the clamps to install. Tighten the clamps securely, but do not distort them. Give the clamps and terminals a thin external coat of grease after installation, to retard corrosion.

Check the cables at the same time that the terminals are cleaned. If the cable insulation is cracked or broken, or if the ends are frayed, the cable should be replaced with a new cable of the same length and gauge.

CAUTION

Keep flame or sparks away from the battery; it gives off explosive hydrogen gas. Battery electrolyte contains sulphuric acid. If you should splash any on your skin or in your eyes, flush the affected area with plenty of clear water. If it lands in your eyes, get medical help immediately.

Clean the inside of the cable clamp with a wire brush

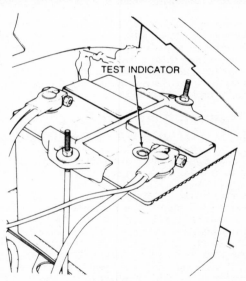

TEST INDICATOR

Indicator eye

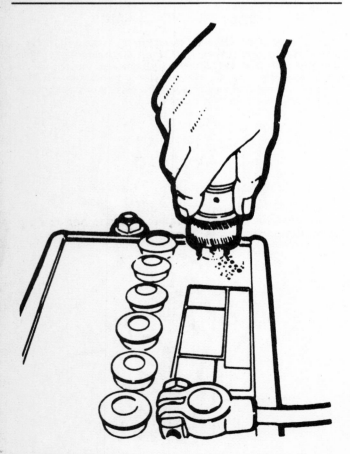

Clean the battery posts with a wire brush or the tool shown

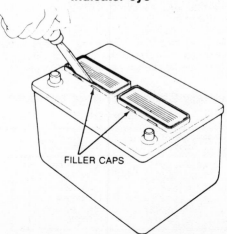

FILLER CAPS

Prying off the filler caps

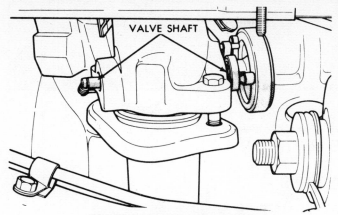

Manifold heat control valve

Heat Control Valve (Heat Riser)

The exhaust manifold heat valve should be serviced at the intervals as shown on the maintenance chart, or more frequently if the vehicle is operated under heavy duty conditions.

The exhaust manifold heat valve is located on the left exhaust manifold at the exhaust pipe connection. The valve should not have any restricted movement.

SERVICING

Observe the heat valve while accelerating the engine from idle. The counterweight on the opposite side of the spring should move clockwise. If no movement is observed, the shaft is binding or the spring is weak or broken. Allow the engine and manifold to cool. Apply MOPAR Manifold Heat Control Valve Solvent, or equivalent, to each end of the valve shaft. Allow the solvent to penetrate for a few minutes. Rotate the valve shaft (use the counterweight) back and forth until it moves freely. Replace the spring if it is weak or broken.

Belts

Once a year or at 12,000 mile intervals, the tension (and condition) of the alternator, power steering (if so equipped), air conditioning (if so equipped), and Thermactor air pump drive belts should be checked, and, if necessary, adjusted. Loose accessory drive belts can lead to poor engine cooling and diminish alternator, power steering pump, air conditioning compressor or Thermactor air pump output. A belt that is too tight places a severe strain on the water pump, alternator, power steering pump, compressor or air pump bearings.

INSPECTION

Replace any belt that is so glazed, worn or stretched that it cannot be tightened sufficiently.

NOTE: The material used in late model drive belts is such that the belts do not show wear. Replace belts at least every three years.

ADJUSTING

On vehicles with matched belts, replace both belts. New Alternator, Air Pump and Power Steering belts are to be adjusted to a tension of 140 lbs.; Used belts are adjusted to 80 lbs., measured on a belt tension gauge. Any belt that has been operating for a minimum of 10 minutes is considered a used belt. In the first 10 minutes, the belt should stretch to its maximum extent. After 10 minutes, stop the engine and recheck the belt tension. If a belt tension gauge is not available, the following procedures may be used.

Alternator/Air Conditioner Compressor Belts

1. Position the ruler perpendicular to the drive belt at its longest straight run. Test the tightness of the belt by pressing it firmly with your thumb. The deflection should not exceed ¼ in..
2. If the deflection exceeds ¼ in., loosen the alternator mounting bolts.
3. Carefully tighten the alternator belt adjustment jack screw to apply tension to the drive belts.
4. When the belt is properly tensioned, tighten the alternator mounting bolts.

Power Steering Drive Belt

1. Position a ruler perpendicular to the drive belt at its longest run. Test the tightness of the belt by pressing it firmly with your thumb. The deflection should be about ¼ in..
2. To adjust the belt tension, loosen the mounting bolts. Slots are provided in the pump front mounting bracket.
3. Position a ½ in. drive extension and ratchet into the square cutout provided in the pump housing bracket. Apply necessary tension to adjust the belt. While apply tension with the ratchet, tighten the mounting bolts and check belt tension. Adjust again, if necessary.

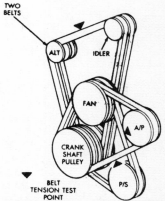

Alternator, air pump and power steering belt routing and tension testing location

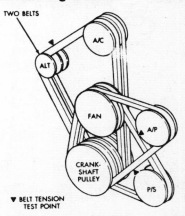

Alternator, air pump, A/C and power steering belt routing and tension testing location

HOW TO SPOT WORN V-BELTS

V–Belts are vital to efficient engine operation—they drive the fan, water pump and other accessories. They require little maintenance (occasional tightening) but they will not last forever. Slipping or failure of the V–belt will lead to overheating. If your V–belt looks like any of these, it should be replaced.

Cracking or Weathering

This belt has deep cracks, which cause it to flex. Too much flexing leads to heat build–up and premature failure. These cracks can be caused by using the belt on a pulley that is too small. Notched belts are available for small diameter pulleys.

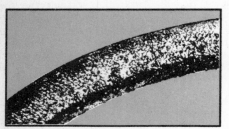

Softening (Grease and Oil)

Oil and grease on a belt can cause the belt's rubber compounds to soften and separate from the reinforcing cords that hold the belt together. The belt will first slip, then finally fail altogether.

Glazing

Glazing is caused by a belt that is slipping. A slipping belt can cause a run-down battery, erratic power steering, overheating or poor accessory performance. The more the belt slips, the more glazing will be built up on the surface of the belt. The more the belt is glazed, the more it will slip. If the glazing is light, tighten the belt.

Worn Cover

The cover of this belt is worn off and is peeling away. The reinforcing cords will begin to wear and the belt will shortly break. When the belt cover wears in spots or has a rough jagged appearance, check the pulley grooves for roughness.

Separation

This belt is on the verge of breaking and leaving you stranded. The layers of the belt are separating and the reinforcing cords are exposed. It's just a matter of time before it breaks completely.

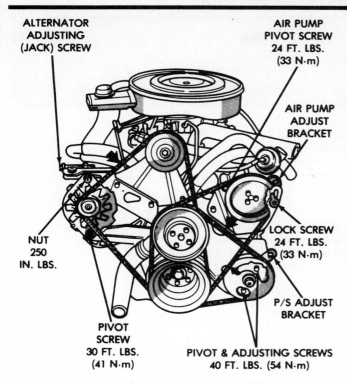

ALTERNATOR ADJUSTING (JACK) SCREW

AIR PUMP PIVOT SCREW 24 FT. LBS. (33 N·m)

AIR PUMP ADJUST BRACKET

NUT 250 IN. LBS.

LOCK SCREW 24 FT. LBS. (33 N·m)

P/S ADJUST BRACKET

PIVOT SCREW 30 FT. LBS. (41 N·m)

PIVOT & ADJUSTING SCREWS 40 FT. LBS. (54 N·m)

Belt drive system and mounting/adjusting bolt locations

Air Pump Drive Belt

1. Position a ruler perpendicular to the drive belt at its longest run. Test the tightness of the belt by pressing it firmly with your thumb. The deflection should be about ¼ in..

2. To adjust the belt tension, loosen the adjusting arm bolt slightly. If necessary, also loosen the mounting belt slightly.

3. Position a ½ in. drive extension and ratchet into the square cutout provided in the adjustment bracket, apply pressure with the ratchet to move the pump toward or away from the engine as necessary.

4. Hold pressure with the ratchet, tighten the adjusting arm bolt and check the tension. When the belt is properly tensioned, tighten the mounting bolt.

REMOVAL AND INSTALLATION

Refer to the previous Adjusting procedures. After loosening the mounting bolts, pivot the unit to relieve belt tension and remove the belt. Install the drive belt(s) and perform adjustment. Always replace dual drive belts with a new matched set. Tighten the mounting and adjustment bolts to the following torque specifications:

- Alternator Pivot Screw: 30 ft.lbs. (41 N.m)
- Alternator Upper Mounting Nut: 250 inch lbs.
- Air Pump Pivot Bolt: 24 ft.lbs. (33 N.m)
- Air Pump Lock Bolt: 24 ft.lbs. (33 N.m)
- Power Steering Pivot and Adjusting Bolts: 40 ft.lbs. (54 N.m)

Hoses

REMOVAL AND INSTALLATION

Inspect the condition of the radiator and heater hoses periodically. Early spring and at the beginning of the fall or winter, when you are performing other maintenance, are good times. Make sure the engine and cooling system are cold. Visually inspect for cracking, rotting or collapsed hoses, replace as necessary. Run your hand along the length of the hose. If a weak or swollen spot is noted when squeezing the hose wall, replace the hose.

1. Drain the cooling system into a suitable container (if the coolant is to be reused).

— **CAUTION** —

When draining the coolant, keep in mind that cats and dogs are attracted by the ethylene glycol antifreeze, and are quite likely to drink any that is left in an uncovered container or in puddles on the ground. This will prove fatal in sufficient quantity. Always drain the coolant into a sealable container. Coolant should be reused unless it is contaminated or several years old.

2. Loosen the hose clamps at each end of the hose that requires replacement.

3. Twist, pull and slide the hose off the radiator, water pump, thermostat or heater connection.

4. Clean the hose mounting connections. Position the hose clamps on the new hose.

5. Coat the connection surfaces with a water resistant sealer and slide the hose into position. Make sure the hose clamps are located beyond the raised bead of the connector (if equipped) and centered in the clamping area of the connection.

6. Tighten the clamps to 30-34 inch lbs. Do not overtighten.

7. Fill the cooling system.

8. Start the engine and allow it to reach normal operating temperature. Check for leaks.

Air Conditioning

— **CAUTION** —

The air conditioning system contains refrigerant under high pressure. Severe personal injury may result from improper service procedures. Wear safety goggles when servicing the refrigeration system.

SAFETY PRECAUTIONS

Because of the importance of the necessary safety precautions that must be exercised when working with air conditioning systems and R-12 refrigerant, a recap of the safety precautions are outlined.

1. Avoid contact with a charged refrigeration system, even when working on another part of the air conditioning system or vehicle. If a heavy tool comes into contact with a section of copper tubing or a heat exchanger, it can easily cause the relatively soft material to rupture.

2. When it is necessary to apply force to a fitting which contains refrigerant, as when checking that all system couplings are securely tightened, use a wrench on both parts of the fitting involved, if possible. This will avoid putting torque on the refrigerant tubing. (It is advisable, when possible, to use tube or line wrenches when tightening these flare nut fittings.)

3. Do not attempt to discharge the system by merely loosening a fitting, or removing the service valve caps and cracking these valves. Precise control is possibly only when using the service gauges. Place a rag under the open end of the center charging hose while discharging the system to catch any drops of liquid that might escape. Wear protective gloves when connecting or disconnecting service gauge hoses.

HOW TO SPOT BAD HOSES

Both the upper and lower radiator hoses are called upon to perform difficult jobs in an inhospitable environment. They are subject to nearly 18 psi at under hood temperatures often over 280°F, and must circulate nearly 7500 gallons of coolant an hour—3 good reasons to have good hoses.

Swollen Hose

A good test for any hose is to feel it for soft or spongy spots. Frequently these will appear as swollen areas of the hose. The most likely cause is oil soaking. This hose could burst at any time, when hot or under pressure.

Cracked Hose

Cracked hoses can usually be seen but feel the hoses to be sure they have not hardened; a prime cause of cracking. This hose has cracked down to the reinforcing cords and could split at any of the cracks.

Frayed Hose End (Due to Weak Clamp)

Weakened clamps frequently are the cause of hose and cooling system failure. The connection between the pipe and hose has deteriorated enough to allow coolant to escape when the engine is hot.

Debris In Cooling System

Debris, rust and scale in the cooling system can cause the inside of a hose to weaken. This can usually be felt on the outside of the hose as soft or thinner areas.

4. Discharge the system only in a well ventilated area, as high concentrations of the gas can exclude oxygen and act as an anesthetic. When leak testing or soldering this is particularly important, as toxic gas is formed when R-12 contacts any flame.

5. Never start a system without first verifying that both service valves are backseated, if equipped, and that all fittings are throughout the system are snugly connected.

6. Avoid applying heat to any refrigerant line or storage vessel. Charging may be aided by using water heated to less than 125°F (52°C) to warm the refrigerant container. Never allow a refrigerant storage container to sit out in the sun, or near any other source of heat, such as a radiator.

7. Always wear goggles when working on a system to protect the eyes. If refrigerant contacts the eye, it is advisable in all cases to see a physician as soon as possible.

8. Frostbite from liquid refrigerant should be treated by first gradually warming the area with cool water, and then gently applying petroleum jelly. A physician should be consulted.

9. Always keep refrigerant can fittings capped when not in use. Avoid sudden shock to the can which might occur from dropping it, or from banging a heavy tool against it. Never carry a refrigerant can in the passenger compartment of a van.

10. Always completely discharge the system before painting the vehicle (if the paint is to be baked on), or before welding anywhere near the refrigerant lines.

SYSTEM INSPECTION

The most important aspect of air conditioning service is the maintenance of pure and adequate charge of refrigerant in the system. A refrigeration system cannot function properly if a significant percentage of the charge is lost. Leaks are common because the severe vibration encountered in an automobile can easily cause a sufficient cracking or loosening of the air conditioning fittings. As a result, the extreme operating pressures of the system force refrigerant out.

The problem can be understood by considering what happens to the system as it is operated with a continuous leak. Because the expansion valve regulates the flow of refrigerant to the evaporator, the level of refrigerant there is fairly constant. The receiver/drier stores any excess of refrigerant, and so a loss will first appear there as a reduction in the level of liquid. As this level nears the bottom of the vessel, some refrigerant vapor bubbles will begin to appear in the stream of liquid supplied to the expansion valve. This vapor decreases the capacity of the expansion valve very little as the valve opens to compensate for its presence. As the quantity of liquid in the condenser decreases, the operating pressure will drop there and throughout the high side of the system. As the R-12 continues to be expelled, the pressure available to force the liquid through the expansion valve will continue to decrease, and, eventually, the valve's orifice will prove to be too much of a restriction for adequate flow even with the needle fully withdrawn.

At this point, low side pressure will start to drop, and severe reduction in cooling capacity, marked by freeze-up of the evaporator coil, will result. Eventually, the operating pressure of the evaporator will be lower than the pressure of the atmosphere surrounding it, and air will be drawn into the system wherever there are leaks in the low side.

Because all atmospheric air contains at least some moisture, water will enter the system and mix with the R-12 and the oil. Trace amounts of moisture will cause sludging of the oil, and corrosion of the system. Saturation and clogging of the filter/drier, and freezing of the expansion valve orifice will eventually result. As air fills the system to a greater and greater extend, it will interfere more and more with the normal flows of refrigerant and heat.

A list of general precautions that should be observed while doing this follows:

1. Keep all tools as clean and dry as possible.

2. Thoroughly purge the service gauges and hoses of air and moisture before connecting them to the system. Keep them capped when not in use.

3. Thoroughly clean any refrigerant fitting before disconnecting it, in order to minimize the entrance of dirt into the system.

4. Plan any operation that requires opening the system beforehand in order to minimize the length of time it will be exposed to open air. Cap or seal the open ends to minimize the entrance of foreign material.

5. When adding oil, pour it through an extremely clean and dry tube or funnel. Keep the oil capped whenever possible. Do not use oil that has not been kept tightly sealed.

6. Use only refrigerant 12. Purchase refrigerant intended for use in only automotive air conditioning system. Avoid the use of refrigerant 12 that may be packaged for another use, such as cleaning, or powering a horn, as it is impure.

7. Completely evacuate any system that has been opened to replace a component, other than when isolating the compressor, or that has leaked sufficiently to draw in moisture and air. This requires evacuating air and moisture with a good vacuum pump for at least one hour.

If a system has been open for a considerable length of time it

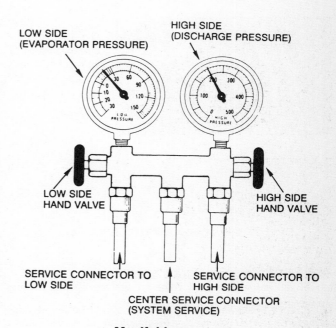

Manifold gauge set

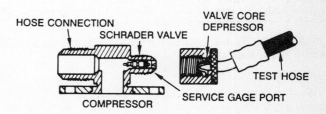

Schraeder valve cross-section

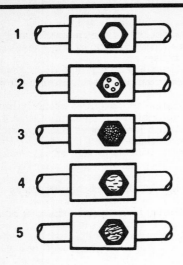

1 Clear sight glass — system correctly charged or overcharged

2 Occasional bubbles — refrigerant charge slightly low

3 Oil streaks on sight glass — total lack of refrigerant

4 Heavy stream of bubbles — serious shortage of refrigerant

5 Dark or clouded sight glass — contaminent present

Sight glass inspection

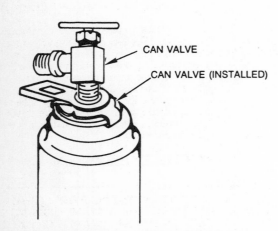

CAN VALVE

CAN VALVE (INSTALLED)

R-12 can opener/valve connection

may be advisable to evacuate the system for up to 12 hours
(overnight).

8. Use a wrench on both halves of a fitting that is to be disconnected, so as to avoid placing torque on any of the refrigerant
lines.

Antifreeze

In order to prevent heater core freeze-up during A/C operation, it is necessary to maintain permanent type antifreeze protection of +15°F (–9°C) or lower. A reading of –15°F (–26°C) is
ideal since this protection also supplies sufficient corrosion inhibitors for the protection of the engine cooling system.

**WARNING: Do not use antifreeze longer than speci-
fied by the manufacturer.**

Radiator Cap

For efficient operation of an air conditioned van's cooling system, the radiator cap should have a holding pressure which
meets manufacturer's specifications. A cap which fails to hold
these pressure should be replaced.

Condenser

Any obstruction of or damage to the condenser configuration
will restrict the air flow which is essential to its efficient operation. It is therefore, a good rule to keep this unit clean and in
proper physical shape.

NOTE: Bug screens are regarded as obstructions.

Condensation Drain Tube

This single molded drain tube expels the condensation, which
accumulates on the bottom of the evaporator housing, into the
engine compartment.

If this tube is obstructed, the air conditioning performance
can be restricted and condensation buildup can spill over onto
the vehicle's floor.

REFRIGERANT LEVEL CHECK

—————————— **CAUTION** ——————————
*The compressed refrigerant used in the air conditioning system expands
into the atmosphere at a temperature of –21.7°F (–30°C) or lower. This
will freeze any surface, including your eyes, that it contacts. In addition,
the refrigerant decomposes into a poisonous gas in the presence of a
flame. Do not open or disconnect any part of the air conditioning system.*

Sight Glass Check

You can safely make a few simple check to determine if your
air conditioning system needs service. The tests work best if the
temperature is warm (about 70°F [21.1°C]).

**NOTE: If your vehicle is equipped with an aftermar-
ket air conditioner, the following system check may not
apply. You should contact the manufacturer of the unit
for instructions on systems checks.**

1. Place the automatic transmission in Park or the manual
transmission in Neutral. Set the parking brake.
2. Run the engine at a fast idle (about 1,500 rpm) either with
the help of a friend or by temporarily readjusting the idle speed
screw.
3. Set the controls for maximum cold with the blower on
High.
4. Locate the sight glass at the top of the receiver/dryer.
5. If you see bubbles, the system must be recharged. Very
likely there is a leak at some point.

6. If there are no bubbles, there is either no refrigerant at all or the system is fully charged. Feel the two hoses going to the belt driven compressor. If they are both at the same temperature, the system is empty and must be recharged.

7. If one hose (high pressure) is warm and the other (low pressure) is cold, the system may be all right. However, you are probably making these tests because you think there is something wrong, so proceed to the next step.

8. Have an assistant in the van turn the fan control on and off to operate the compressor clutch. Watch the sight glass.

9. If bubbles appear when the clutch is disengaged and disappear when it is engaged, the system is properly charged.

10. If the refrigerant takes more than 45 seconds to bubble when the clutch is disengaged, the system is overcharged. This usually causes poor cooling at low speeds.

NOTE: If it is determined that the system has a leak, it should be corrected as soon as possible. Leaks may allow moisture to enter and cause a very expensive rust problem.

Exercise the air conditioner for a few minutes, every two weeks or so, during the cold months. This avoids the possibility of the compressor seals drying out from lack of lubrication.

TEST GAUGES

Most of the service work performed on the air conditioning system requires the use of a set of test gauges, one for the high (head) pressure side of the system, the other for the low (suction) side.

The low side gauge records both pressure and vacuum. Vacuum readings are calibrated from 0 to 30 inches Hg and the pressure graduations read from 0 to no less than 150 psi.

The high side gauge measures pressure from 0 to at least 300 psi.

With the gauge set you can perform the following procedures:
1. Test high and low side pressures.
2. Remove air, moisture, and contaminated refrigerant.
3. Purge the system of refrigerant.
4. Charge the system with refrigerant.

Gauge sets used on the C-171 compressor use 2 gauges: 1 compound gauge for suction and 1 discharge pressure gauge.

All gauge sets must have 3 hoses, with 1 being for center manifold outlet.

WARNING: When connecting the hoses to the compressor service ports, the manifold gauge valves must be closed!

The suction gauge valve is opened to provide a passage between the suction gauge and the center manifold outlet. The discharge gauge valve is opened to provide a passage between the discharge pressure gauge and the center manifold outlet.

Connecting the Gauge Set

The left side Suction gauge is connected to the suction service port of the air conditioner compressor. A special service port adapter may be required. The Discharge Pressure Gauge is connected to the discharge service port located on the discharge line from the compressor. The center located manifold outlet provides the necessary connection for a long service hose used when discharging the system with a vacuum pump, and charging the system from the refrigerant supply.

DISCHARGING THE SYSTEM

1. Remove the caps from the high and low pressure charging valves in the high and low pressure lines.
2. Turn both manifold gauge set hand valves to the fully closed (clockwise) position.

3. Connect the manifold gauge set.
4. If the van does not have a service access gauge port valve, connect the gauge set low pressure hose to the evaporator service access gauge port valve.
5. Place the end of the center hose away from you and the van, preferably into a container such as an old coffee can to catch the refrigerant oil.
6. Open the low pressure gauge valve slightly and allow the system pressure to bleed off.
7. When the system is just about empty, open the high pressure valve very slowly to avoid losing an excessive amount of refrigerant oil. Allow any remaining refrigerant to escape.

EVACUATING THE SYSTEM

NOTE: This procedure requires the use of a vacuum pump.

1. Connect the manifold gauge set.
2. Discharge the system.
3. Make sure that the low pressure gauge set hose is connected to the low pressure service gauge port and the high pressure hose connected to the high pressure service gauge port on the compressor discharge line.
4. Connect the center service hose to the inlet fitting of the vacuum pump.
5. Turn both gauge set valves to the wide open position.
6. Start the pump and note the low side gauge reading.
7. Operate the pump until the low pressure gauge reads at least 26 in.Hg. Continue running the vacuum pump for 5 to 10 minutes more. If you've replaced some component in the system, run the pump for an additional 20-30 minutes.
8. Close both gauge set valves. Turn off the pump. The needle should remain stationary at the point at which the pump was turned off. If the needle drops to zero rapidly, there is a leak in the system which must be repaired.

CHARGING THE SYSTEM

------------------------ **CAUTION** ------------------------
NEVER OPEN THE HIGH PRESSURE SIDE WITH A CAN OF REFRIGERANT CONNECTED TO THE SYSTEM! OPENING THE HIGH PRESSURE SIDE WILL OVER PRESSURIZE THE CAN, CAUSING IT TO EXPLODE!

NOTE: The following procedure is for charging the system with small 1lb. cans of R-12 refrigerant.

1. Connect the gauge set.
2. Close (clockwise) both gauge set valves.
3. Connect the center hose to the refrigerant dispenser can opener valve.
4. Make sure the can opener valve is closed, that is, the needle is raised, and connect the valve to the can. Open can dispenser valve to puncture the top of the can with the needle. Turn the dispenser valve back to the start position after the can has been punctured.
5. Loosen the center hose fitting at the pressure gauge, allowing refrigerant to purge the hose of air. When the air is bled, tighten the fitting.

------------------------ **CAUTION** ------------------------
CHARGE THE A/C SYSTEM THROUGH THE LOW PRESSURE (SUCTION) SIDE ONLY. KEEP THE CAN IN AN UPRIGHT POSITION!

6. Place the can of R-12 refrigerant in a pan containing water heated to 125 degrees F. Start the engine and move the air conditioning controls to the low blower position.
7. Open the low side gauge set valve.

8. Allow refrigerant to be drawn into the system. Adjust the valve so that charging pressure does not exceed 50 psi. Keep the water temperature in the pan at the required temperature by adding hot water as necessary.

NOTE: The low pressure (cycling) cut-out switch will prevent the compressor clutch from energizing until refrigerant is added to the system. If the clutch does not engage, replace the switch.

9. Raise the engine speed to about 1,300 rpm. The compressor will operate and pull refrigerant gas into the system.

10. If more than one can of refrigerant is needed, close the can valve and gauge set low side valve when the can is empty and connect a new can to the opener. Repeat the charging process until the sight glass indicates a full charge. The frost line on the outside of the can will indicate what portion of the can has been used.

—————————— CAUTION ——————————
NEVER ALLOW THE HIGH PRESSURE SIDE READING TO EXCEED 240 psi!
————————————————————————————

11. When the charging process has been completed, close the gauge set valve and can valve. Run the system for at least five minutes to allow it to normalize. Low pressure side reading should be 18-30 psi; high pressure reading should be 120-210 psi at an ambient temperature of 70-90°F (21-32°C).

12. Loosen both service hoses at the gauges to allow any refrigerant to escape. Remove the gauge set and install the dust caps on the service valves.

NOTE: Multi-can dispensers are available which allow a simultaneous hook-up of up to four 1 lb. cans of R-12.

—————————— CAUTION ——————————
Never exceed the recommended maximum charge for the system. The maximum charge for the system is 40 oz.
————————————————————————————

LEAK TESTING

Some leak tests can be performed with a soapy water solution. There must be at least a ½ lb. charge in the system for a leak to be detected. The most extensive leak tests are performed with either a Halide flame type leak tester or the more preferable electronic leak tester.

In either case, the equipment is expensive, and the use of a Halide detector can be **extremely** hazardous!

When using either method of leak detection, follow the manufacturer's instructions as the design and function of the detection may vary significantly.

Windshield Wipers

Intense heat from the sun, snow, and ice, road oils and the chemicals used in windshield washer solvent combine to deteriorate the rubber wiper refills. The refills should be replaced about twice a year or whenever the blades begin to streak or chatter.

WIPER REFILL REPLACEMENT

Normally, if the wipers are not cleaning the windshield properly, only the refill has to be replaced. The blade and arm usually require replacement only in the event of damage. It is not necessary (except on new Tridon® refills) to remove the arm or the blade to replace the refill (rubber part), though you may have to position the arm higher on the glass. You can do this turning the ignition switch on and operating the wipers. When they are positioned where they are accessible, turn the ignition switch off.

There are several types of refills and your vehicle could have any kind, since aftermarket blades and arms may not use exactly the same type refill as the original equipment.

Most Anco® styles use a release button that is pushed down to allow the refill to slide out of the yoke jaws. The new refill slides in and locks in place.

Some Trico® refills are removed by locating where the metal backing strip or the refill is wider. Insert a small screwdriver blade between the frame and metal backing strip. Press down to release the refill from the retaining tab.

Other Trico® blades are unlocked at one end by squeezing 2 metal tabs, and the refill is slid out of the frame jaws. When the new refill is installed, the tabs will click into place, locking the refill.

The polycarbonate type is held in place by a locking lever that is pushed downward out of the groove in the arm to free the refill. When the new refill is installed, it will lock in place automatically.

The Tridon® refill has a plastic backing strip with a notch about 1 in. (25mm) from the end. Hold the blade (frame) on a hard surface so that the frame is tightly bowed. Grip the tip of the backing strip and pull up while twisting counterclockwise. The backing strip will snap out of the retaining tab. Do this for the remaining tabs until the refill is free of the arm. The length of these refills is molded into the end and they should be replaced with identical types.

No matter which type of refill you use, be sure that all of the frame claws engage the refill. Before operating the wipers, be sure that no part of the metal frame is contacting the windshield.

Tires and Wheels

The tires should be rotated as specified in the Maintenance Intervals Chart.

The tires on your van should have built-in tread wear indicators, which appear as ½ in. (12.7mm) bands when the tread depth gets as low as $\frac{1}{16}$ in. (1.6mm). When the indicators appear in 2 or more adjacent grooves, it's time for new tires.

TIRE ROTATION

It is recommended that you have the tires rotated and the balance checked every 6,000 miles. There is no way to give a tire rotation diagram for every combination of tires and vehicles, but the accompanying diagrams are a general rule to follow. Radial tires should not be cross-switched; they last longer if their direction of rotation is not changed. Van tires and some high-performance tires sometimes have directional tread, indicated by arrows on the sidewalls; the arrow shows the direction of rotation. They will wear very rapidly if reversed. Studded snow tires will lose their studs if their direction of rotation is reversed.

NOTE: Mark the wheel position or direction of rotation on radial tires or studded snow tires before removing them.

If your van is equipped with tires having different load ratings on the front and the rear, the tires should not be rotated front to rear. Rotating these tires could affect tire life (the tires with the lower rating will wear faster, and could become overloaded), and upset the handling of the van.

When installing the wheels on the vehicle, tighten the lug nuts in a criss-cross pattern. Lug nuts should be torqued to the following figures:
- 150-250 Models; ½-20 stud: 85-110 ft.lbs.
- 350 Model; 90 degree cone; ½-20 stud: 85-110 ft.lbs.
- 350 Model w/H.D. axle (90 degree cone ⅝-18 stud: 175-225 ft.lbs.
- 350 Model w/H.D. axle (Flanged nut ⅝-18 stud): 300-350 ft.lbs.

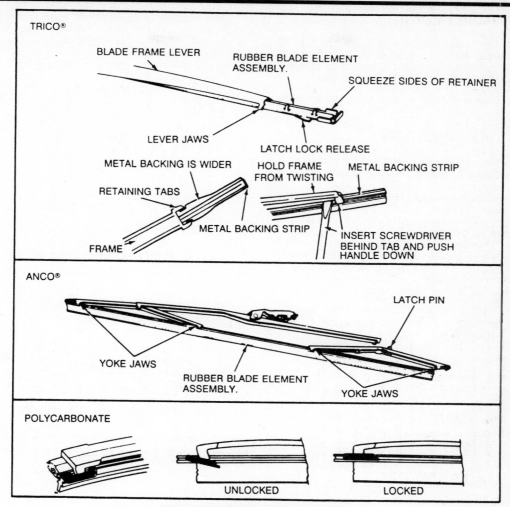

Windshield wiper blade refills

capacity of the vehicle, although they will provide an extra margin of tread life. Be sure to check overall height before using larger size tires which may cause interference with suspension components or wheel wells. When replacing conventional tire sizes with other tire size designations, be sure to check the manufacturer's recommendations. Interchangeability is not always possible because of differences in load ratings, tire dimensions, wheel well clearances, and rim size. Also due to differences in handling characteristics, 70 Series and 60 Series tires should be used only in pairs on the same axle; radial tires should be used only in sets of four.

NOTE: Many states have vehicle height restrictions; some states prohibit the lifting of vehicles beyond their design limits.

The wheels must be the correct width for the tire. Tire dealers have charts of tire and rim compatibility. A mismatch can cause sloppy handling and rapid tread wear. The old rule of thumb is that the tread width should match the rim width (inside bead to inside bead) within an inch. For radial tires, the rim width should be 80% or less of the tire (not tread) width.

The height (mounted diameter) of the new tires can greatly change speedometer accuracy, engine speed at a given road speed, fuel mileage, acceleration, and ground clearance. Tire manufacturers furnish full measurement specifications. Speedometer drive gears are available for correction.

NOTE: Dimensions of tires marked the same size may vary significantly, even among tires from the same manufacturer.

The spare tire should be of the same size, construction and design as the tires on the vehicle. It's not a good idea to carry a spare of a different construction.

For maximum satisfaction, tires should be used in sets of five. Mixing or different types (radial, bias-belted, fiberglass belted) should be avoided. Conventional bias tires are constructed so that the cords run bead-to-bead at an angle. Alternate plies run at an opposite angle. This type of construction gives rigidity to both tread and sidewall. Bias-belted tires are similar in construction to conventional bias ply tires. Belts run at an angle and also at a 90° angle to the bead, as in the radial tire. Tread life is improved considerably over the conventional bias tire. The radial tire differs in construction, but instead of the carcass plies running at an angle of 90° to each other, they run at an angle of 90° to the bead. This gives the tread a great deal of rigidity and the sidewall a great deal of flexibility and accounts for the characteristic bulge associated with radial tires.

Radial tire are recommended for use on most Dodge Vans.

When radial tires are used, tire sizes and wheel diameters should be selected to maintain ground clearance and tire load capacity equivalent to the minimum specified tire. Radial tires should always be used in sets of five, but in an emergency, radial tires can be used with caution on the rear axle only. If this is

TIRE DESIGN

The tires on your van were selected to provide the best all around performance for normal operation when inflated as specified. Oversize tires will not increase the maximum carrying done, both tires on the rear should be of radial design.

WARNING: Radial tires should never be used on only the front axle!

TIRE INFLATION

For optimum tire life, you should keep the tires properly inflated, rotate them often and have the wheel alignment checked periodically.

A tire inflation pressure guide decal is usually mounted on the left door frame. In general, pressure of 32-35 psi would be suitable for highway use with moderate loads and passenger van type tires (load range B, non-flotation) of original equipment size. Pressures should be checked before driving, since pressure can increase as much as 6 psi due to heat. It is a good idea to have an accurate gauge and to check pressures weekly. Not all gauges on service station air pumps are to be trusted. In general, van type tires require higher pressures and flotation type tires, lower pressures.

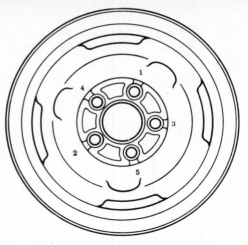

Wheel lug tightening sequence

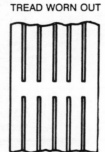

TREAD STILL GOOD   TREAD WORN OUT

Tread wear indicators appear as solid bands when the tread is worn

Tread depth can be checked with a penny; when the top of Lincoln's head is visible, it's time for new tires

Tread depth can also be checked with an inexpensive gauge made for the purpose

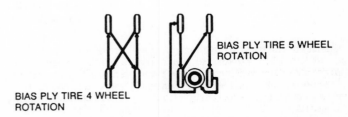

BIAS PLY TIRE 4 WHEEL ROTATION

BIAS PLY TIRE 5 WHEEL ROTATION

This rotation is for bias-belted tires only

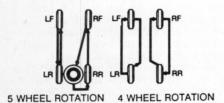

5 WHEEL ROTATION   4 WHEEL ROTATION

This rotation is for radial tires

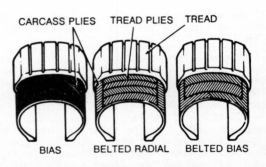

CARCASS PLIES   TREAD PLIES   TREAD

BIAS   BELTED RADIAL   BELTED BIAS

Types of tire construction

Troubleshooting Basic Wheel Problems

Problem	Cause	Solution
The car's front end vibrates at high speed	• The wheels are out of balance • Wheels are out of alignment	• Have wheels balanced • Have wheel alignment checked/adjusted
Car pulls to either side	• Wheels are out of alignment • Unequal tire pressure • Different size tires or wheels	• Have wheel alignment checked/adjusted • Check/adjust tire pressure • Change tires or wheels to same size
The car's wheel(s) wobbles	• Loose wheel lug nuts • Wheels out of balance • Damaged wheel • Wheels are out of alignment • Worn or damaged ball joint • Excessive play in the steering linkage (usually due to worn parts) • Defective shock absorber	• Tighten wheel lug nuts • Have tires balanced • Raise car and spin the wheel. If the wheel is bent, it should be replaced • Have wheel alignment checked/adjusted • Check ball joints • Check steering linkage • Check shock absorbers
Tires wear unevenly or prematurely	• Incorrect wheel size • Wheels are out of balance • Wheels are out of alignment	• Check if wheel and tire size are compatible • Have wheels balanced • Have wheel alignment checked/adjusted

Troubleshooting Basic Tire Problems

Problem	Cause	Solution
The car's front end vibrates at high speeds and the steering wheel shakes	• Wheels out of balance • Front end needs aligning	• Have wheels balanced • Have front end alignment checked
The car pulls to one side while cruising	• Unequal tire pressure (car will usually pull to the low side) • Mismatched tires • Front end needs aligning	• Check/adjust tire pressure • Be sure tires are of the same type and size • Have front end alignment checked
Abnormal, excessive or uneven tire wear See "How to Read Tire Wear"	• Infrequent tire rotation • Improper tire pressure • Sudden stops/starts or high speed on curves	• Rotate tires more frequently to equalize wear • Check/adjust pressure • Correct driving habits
Tire squeals	• Improper tire pressure • Front end needs aligning	• Check/adjust tire pressure • Have front end alignment checked

Tire Size Comparison Chart

"Letter" sizes			Inch Sizes	Metric-inch Sizes		
"60 Series"	"70 Series"	"78 Series"	1965–77	"60 Series"	"70 Series"	"80 Series"
			5.50-12, 5.60-12	165/60-12	165/70-12	155-12
		Y78-12	6.00-12			
		W78-13	5.20-13	165/60-13	145/70-13	135-13
		Y78-13	5.60-13	175/60-13	155/70-13	145-13
			6.15-13	185/60-13	165/70-13	155-13, P155/80-13
A60-13	A70-13	A78-13	6.40-13	195/60-13	175/70-13	165-13
B60-13	B70-13	B78-13	6.70-13	205/60-13	185/70-13	175-13
			6.90-13			
C60-13	C70-13	C78-13	7.00-13	215/60-13	195/70-13	185-13
D60-13	D70-13	D78-13	7.25-13			
E60-13	E70-13	E78-13	7.75-13			195-13
			5.20-14	165/60-14	145/70-14	135-14
			5.60-14	175/60-14	155/70-14	145-14
			5.90-14			
A60-14	A70-14	A78-14	6.15-14	185/60-14	165/70-14	155-14
	B70-14	B78-14	6.45-14	195/60-14	175/70-14	165-14
	C70-14	C78-14	6.95-14	205/60-14	185/70-14	175-14
D60-14	D70-14	D78-14				
E60-14	E70-14	E78-14	7.35-14	215/60-14	195/70-14	185-14
F60-14	F70-14	F78-14, F83-14	7.75-14	225/60-14	200/70-14	195-14
G60-14	G70-14	G77-14, G78-14	8.25-14	235/60-14	205/70-14	205-14
H60-14	H70-14	H78-14	8.55-14	245/60-14	215/70-14	215-14
J60-14	J70-14	J78-14	8.85-14	255/60-14	225/70-14	225-14
L60-14	L70-14		9.15-14	265/60-14	235/70-14	
	A70-15	A78-15	5.60-15	185/60-15	165/70-15	155-15
B60-15	B70-15	B78-15	6.35-15	195/60-15	175/70-15	165-15
C60-15	C70-15	C78-15	6.85-15	205/60-15	185/70-15	175-15
	D70-15	D78-15				
E60-15	E70-15	E78-15	7.35-15	215/60-15	195/70-15	185-15
F60-15	F70-15	F78-15	7.75-15	225/60-15	205/70-15	195-15
G60-15	G70-15	G78-15	8.15-15/8.25-15	235/60-15	215/70-15	205-15
H60-15	H70-15	H78-15	8.45-15/8.55-15	245/60-15	225/70-15	215-15
J60-15	J70-15	J78-15	8.85-15/8.90-15	255/60-15	235/70-15	225-15
	K70-15		9.00-15	265/60-15	245/70-15	230-15
L60-15	L70-15	L78-15, L84-15	9.15-15			235-15
	M70-15	M78-15				255-15
		N78-15				

NOTE: Every size tire is not listed and many size comaprisons are approximate, based on load ratings. Wider tires than those supplied new with the vehicle should always be checked for clearance

HOW TO READ TIRE WEAR

The way your tires wear is a good indicator of other parts of your car. Abnormal wear patterns are often caused by the need for simple tire maintenance, or for front end alignment.

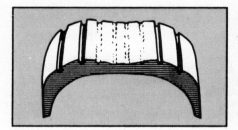

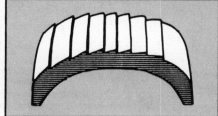

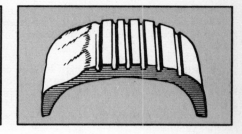

Over-Inflation

Excessive wear at the center of the tread indicates that the air pressure in the tire is consistently too high. The tire is riding on the center of the tread and wearing it prematurely. Occasionally, this wear pattern can result from outrageously wide tires on narrow rims. The cure for this is to replace either the tires or the wheels.

Feathering

Feathering is a condition when the edge of each tread rib develops a slightly rounded edge on one side and a sharp edge on the other. By running your hand over the tire, you can usually feel the sharper edges before you'll be able to see them. The most common causes of feathering are incorrect toe-in setting or deteriorated bushings in the front suspension.

Cupping

Cups or scalloped dips appearing around the edge of the tread almost always indicate worn (sometimes bent) suspension parts. Adjustment of wheel alignment alone will seldom cure the problem. Any worn component that connects the wheel to the vehicle can cause this type of wear. Occasionally, wheels that are out of balance will wear like this, but wheel imbalance usually shows up as bald spots between the outside edges and center of the tread.

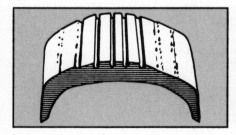

Under-Inflation

This type of wear usually results from consistent under-inflation. When a tire is under inflated, there is too much contact with the road by the outer threads, which wear prematurely. When this type of wear occurs, and the tire pressure is known to be consistently correct, a bent or worn steering component or the need for wheel alignment could be indicated.

One Side Wear

When an inner or outer rib wears faster than the rest of the tire, the need for wheel alignment is indicated. There is excessive camber in the front suspension, causing the wheel to lean too much, putting excessive load on one side of the tire. Misalignment could also be due to sagging springs, worn ball joints, or worn control arm bushings. Be sure the vehicle is loaded the way it's normally driven when you have the wheels aligned.

Second-Rib Wear

Second-rib wear is normally found only in radial tires, and appears where the steel belts end in relation to the tread. Normally, it can be kept to a minimum by paying careful attention to tire pressure and frequently rotation the tires. This is often considered normal wear but excessive amounts indicate that the tires are too wide for the wheels.

FLUIDS AND LUBRICANTS

Fuel and Oil Recommendations

Any van originally equipped with a catalyic converter must use unleaded gasoline. All vehicles so equipped with catalyst emission control systems have labels located on the instrument panel and adjacent to the fuel filler cap stating Unleaded Fuel Only. Vehicles requiring unleaded gasoline are equipped with a restrictor in the fuel filler tube especially designed to accept the smaller unleaded fuel dispensing nozzles only.

Unleaded gasoline having a minimum octane rating of 87 (R+M)/2 should be used. Engines may respond differently to gasolines having the same octane rating. Should the engine in your van develop spark knock (ping), trying purchasing your gasoline from a different source or try a different brand.

Use gasolines containing a high level of detergent additives. The use of a detergent type gasoline will reduce fuel injector and intake system deposit build-up and help maintain an excellent degree of vehicle driveability.

The recommended oil viscosities for sustained temperatures ranging from below 0°F (−18°C) to above 32°F (0°C) are listed in this section. They are broken down into multi-viscosities and single viscosities. Multi Viscosity oils are recommended because of their wider range of acceptable temperatures and driving conditions.

When adding oil to the crankcase or changing the oil or filter, it is important that oil of an equal quality to original equipment be used in your van. The use of inferior oils may void the warranty, damage your engine, or both.

The SAE (Society of Automotive Engineers) grade number of oil indicates the viscosity of the oil (its ability to lubricate at a given temperature). The lower the SAE number, the lighter the oil; the lower the viscosity, the easier it is to crank the engine in cold weather but the less the oil will lubricate and protect the engine in high temperatures. This number is marked on every oil container.

Oil viscosities should be chosen from those oils recommended for the lowest anticipated temperatures during the oil change interval. Due to the need for an oil that embodies both good lubrication at high temperatures and easy cranking in cold weather, multigrade oils have been developed. Basically, a multigrade oil is thinner at low temperatures and thicker at high tempera-

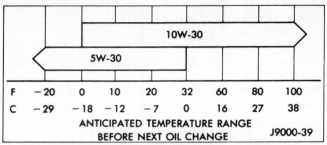
Temperature/Oil viscosity recommendations

tures. For example, a 10W-30 oil (the W stands for winter) exhibits the characteristics of a 10 weight (SAE 10) oil when the van is first started and the oil is cold. Its lighter weight allows it to travel to the lubricating surfaces quicker and offer less resistance to starter motor cranking than, say, a straight 30 weight (SAE 30) oil. But after the engine reaches operating temperature, the 10W-30 oil begins acting like straight 30 weight (SAE 30) oil, its heavier weight providing greater lubrication with less chance of foaming than a straight 30 weight oil.

The API (American Petroleum Institute) designations, also found on the oil container, indicates the classification of engine oil used under certain given operating conditions. Only oils designated for use Service SG or SG/CD heavy duty detergent should be used in your van. Oils of the SG type perform may functions inside the engine besides their basic lubrication. Through a balanced system of metallic detergents and polymeric dispersants, the oil prevents high and low temperature deposits and also keeps sludge and dirt particles in suspension. Acids, particularly sulphuric acid, as well as other by-products of engine combustion are neutralized by the oil. If these acids are allowed to concentrate, they can cause corrosion and rapid wear of the internal engine parts.

Oils, currently available, marked Energy Conserving, Fuel Saving, Fuel Efficient, Gas Saving, etc. on the lower part of the container logo, that meet the viscosity grade requirements are recommended.

WARNING: Non-detergent motor oils or straight mineral oils should not be used in your gasoline engine.

Engine

OIL LEVEL CHECK

Check the engine oil level every time you fill the gas tank. The oil level dipstick is located on the right hand front of the engine near the alternator. The oil level should be above the ADD mark and not above the FULL mark on the dipstick. Make sure that the dipstick is inserted into the crankcase as far as possible and that the vehicle is resting on level ground. Also, allow a few minutes after turning off the engine for the oil to drain into the pan or an inaccurate reading will result.

1. Open the hood and remove the engine oil dipstick.
2. Wipe the dipstick with a clean, lint-free rag and reinsert it. Be sure to insert it all the way.
3. Pull out the dipstick and note the oil level. It should be between the **SAFE (MAX)** mark and the **ADD (MIN)** mark.
4. If the level is below the lower mark, replace the dipstick and add fresh oil to bring the level within the proper range. Do not overfill.
5. Recheck the oil level and close the hood.

NOTE: Use a multi-grade oil with API classification SG or SG/CD.

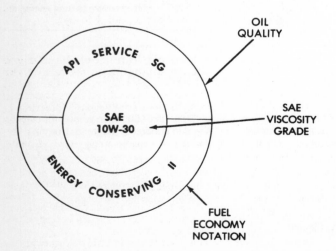
Engine oil container standard notations

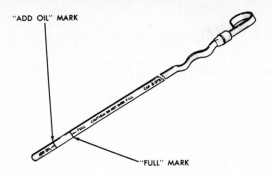

"ADD OIL" MARK

"FULL" MARK

Engine oil dipstick

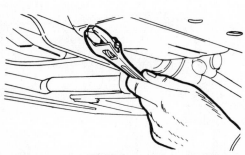

Loosen, but do not remove the drain plug

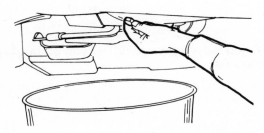

Remove the plug by hand, keeping an inward pressure on it

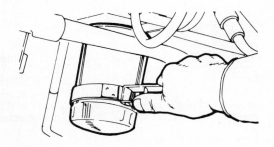

Loosening the oil filter with an oil filter wrench

Unscrew the filter by hand

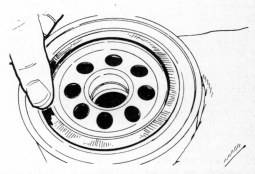

Coat the gasket on the new filter with fresh engine oil

OIL AND FILTER CHANGE

───── **CAUTION** ─────

The EPA warns that prolonged contact with used engine oil may cause a number of skin disorders, including cancer! You should make every effort to minimize your exposure to used engine oil. Protective gloves should be worn when changing the oil. Wash your hands and any other exposed skin areas as soon as possible after exposure to used engine oil. Soap and water, or waterless hand cleaner should be used.

NOTE: The engine oil and oil filter should be changed at the same time, at the recommended intervals on the maintenance schedule chart. However, for the small price of an oil filter, it's cheap insurance to replace the filter at every oil change. One of the larger filter manufacturers points out in its advertising that not changing the filter leaves about one quart of dirty oil in the engine. This claim is true and should be kept in mind when changing your oil.

The oil should be changed more frequently if the vehicle is being operated in very dusty areas. Before draining the oil, make sure that the engine is at operating temperature. Hot oil will hold more impurities in suspension and will flow better, allowing the removal of more oil and dirt.

You should have available a container to hold a minimum of 7 quarts of oil, a wrench for the drain plug, a spout (depending on what type of container the oil comes in), or a funnel, and a rag or two.

1. Position the vehicle on a level surface and set the parking brake and block the rear wheels.

2. If additional clearance is necessary to work underneath the vehicle, raise and safely support the front.

3. Position a suitable drain pan under the engine oil pan drain plug. Loosen the drain plug with a wrench, then, unscrew the plug with your fingers, using a rag to shield your fingers from the heat. Push in on the plug as you unscrew it so you can feel when all of the screw threads are out of the hole. You can then remove the plug quickly with the minimum amount of oil running down your arm and you will also have the plug in your hand and not in the bottom of a pan of hot oil. Drain the oil into the drain pan.

CAUTION

The engine oil will be HOT! Keep your arms, face and hands clear of the oil as it drains out!

---

4. When all of the oil has drained, clean off the drain plug and screw it back into the engine oil pan. Tighten the drain plug to 20 ft.lbs.

5. Remove the drain pan from under the vehicle. Lower the vehicle and pour the correct quantity of fresh oil into the engine. Start the engine and allow it warm up. Check for leaks. Shut off the engine and check the oil level. Add additional oil if necessary. Do not overfill the crankcase. Overfilling will cause oil aeration and loss of oil pressure.

NOTE: Used oil, indiscriminately discarded, can present a problem to the environment. Ask the supplier of the oil, your service station or car/truck dealer where used oil can be safely discarded.

OIL FILTER CHANGE

To remove the filter, you may need an oil filter wrench since the filter may have been fitted too tightly and the heat from the engine may have made it even tighter. A filter wrench can be obtained at an auto parts store and is well worth the investment.

1. Raise and safely support the front of the vehicle to gain more working room, if necessary.

2. Drain the engine oil.

3. Loosen the filter with the filter wrench. With a rag wrapped around the filter, unscrew the filter from the boss on the side of the engine. Be careful of hot oil that will run down the side of the filter. Make sure that you have a pan under the filter before you start to remove it from the engine; should some of the hot oil happen to get on you, you will have a place to dump the filter in a hurry.

4. Wipe the base of the mounting boss with a clean, dry cloth. When you install the new filter, smear a small amount of oil on the gasket with your finger, just enough to coat the entire surface, where it comes in contact with the mounting plate.

5. Place the filter in position over the mounting stud and tighten according to the instructions from the filter manufacturer (usually hand tight from the point when the filter contacts the engine filter boss-about a half a turn).

6. Lower the vehicle. Fill the crankcase with the proper amount of fresh oil. Start the engine and check for leaks.

NOTE: When you have finished changing the oil and oil filter pour the oil from the drain pan into suitable containers. Plastic gallon milk or antifreeze containers will work fine. Find a service station or garage which accepts waste oil for recycling and dispose of it there.

Manual Transmission

FLUID RECOMMENDATIONS

Chrysler Corporation has used a variety of transmission lubricants through the years covered in this manual. Please refer to the owner's manual for the proper Lubricant.

- Overdrive-4 NP435 (New Process) 4-sp: SAE 10W-30
- New Process 2500 5-sp: SAE 10W-30

NOTE: Dexron®II ATF may be used if a high shift effort is experience during cold weather operation. Should a high mileage/hot weather operation situation occur, the use of GL-5 SAE 90 to 140 is allowed in continued above 90 degree F temperatures.

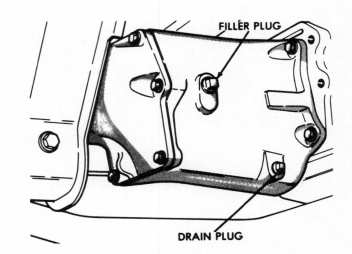

NP-435 4-speed overdrive fill and drain plug locations

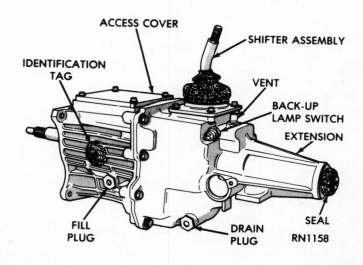

NP2500 5-speed fill and drain plug locations

LEVEL CHECK

The fluid level should be checked every 6 months/6,000 miles, whichever comes first.

1. Park the van on a level surface, turn off the engine, apply the parking brake and block the wheels.
2. Remove the filler plug from the side of the transmission case with a proper size wrench. The fluid level should be even with the bottom of the filler hole.
3. If additional fluid is necessary, add it through the filler hole using a siphon pump or squeeze bottle.
4. Replace the filler plug; do not overtighten.

DRAIN AND REFILL

1. Position the van on a level surface.
2. Place a pan of sufficient capacity under the transmission and remove the upper (fill) plug to provide a vent opening.
3. Remove the drain plug on the lower side of the transmission case and drain the fluid into the catch pan. Removing the upper fill/level check plug will help to vent the transmission and permit the fluid to drain faster.
4. Wipe the drain plug clean and install it into the transmission case.
5. Pump in sufficient lubricant to bring the level to the bottom of the filler plug opening.

Automatic Transmissions

FLUID RECOMMENDATIONS
- All Types: MOPAR ATF PLUS or Dexron®II ATF

LEVEL CHECK

It is very important to maintain the proper fluid level in an automatic transmission. If the level is either too high or too low, poor shifting operation and internal damage are likely to occur. For this reason a regular check of the fluid level is essential.

1. Drive the vehicle for 15-20 minutes to allow the transmission to reach operating temperature.
2. Park the van on a level surface, apply the parking brake (block the wheels, if necessary) and leave the engine idling. Shift the transmission and engage each gear, then place the gear selector in N(NEUTRAL).
3. Wipe away any dirt in the areas of the transmission dipstick to prevent it from falling into the filler tube. Withdraw the dipstick, wipe it with a clean, lint-free rag and reinsert it until it seats.
4. Withdraw the dipstick and note the fluid level. It should be between the upper (FULL) mark and the lower (ADD) mark.
5. If the level is below the lower mark, use a funnel and add fluid in small quantities through the dipstick filler neck. Keep the engine running while adding fluid and check the level after each small amount. Do not overfill.

DRAIN AND REFILL

1. Raise the front of the van and support it on jackstands. Place a large drain pan under the transmission.
2. Loosen the pan attaching bolts and tap the pan at one corner to break it loose.
3. Allow the fluid to drain into the drain pan.
4. After most of the fluid has drained, carefully remove the attaching bolts, lower the pan and drain the rest of the fluid.
5. Remove the filter attaching screws and remove the filter.
6. Install a new filter. Tighten the screws to 35 inch lbs. (4 N.m).
7. Thoroughly clean the fluid pan with safe solvent and allow it to dry.

8. Using a new gasket, install the pan to the transmission. Tighten the attaching bolts to 150 inch lbs. (17 N.m).
9. Pour in two quarts of the recommended ATF fluid through the dipstick tube.
10. Start the engine and allow it to run for a few minutes. With the parking brake set, slowly move the gear selector to each position. Return it to the Neutral position.
11. Check the fluid level. Add more fluid as necessary to bring it up to the "ADD " level.
12. Drive the van to bring the temperature up to normal operating temperature. Check the level again. It should be between the "Add" and "Full" marks.

PAN AND FILTER SERVICE

The automatic transmission, used on your van, circulates transmission fluid through a filter mounted on the valve body. When a transmission fluid change is required, always replace the filter. Refer to the DRAIN AND REFILL service procedure for instructions.

Drive Axles

FLUID RECOMMENDATIONS

All axles use SAE 80-90 multi-purpose type hypoid gear lubricant (GL-5).
However, if your vehicle is operated in 90 degree F temperatures, SAE 80W-140 or SAE 140 mat be used. Models that are equipped with a limited slip differential must have 4 ounces of friction additive (modifier) added when a fluid change is performed.

FLUID LEVEL CHECK

To check the axle lubricant level, remove the axle filler plug with the van level.

NOTE: The rear axle cover is equipped with a pressed in rubber filler plug instead of the usual screw in plug.

Remove the filler plug. Use your finger for a dipstick.(Being care of sharp edges). Add lubricant with a suction gun or squeeze bottle. The lubricant level should be ⅜ in. below the filler plug hole.

Automatic transmission dipstick markings

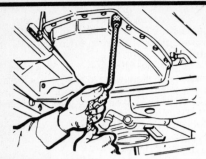

Loosen but do not remove the pan attaching bolts

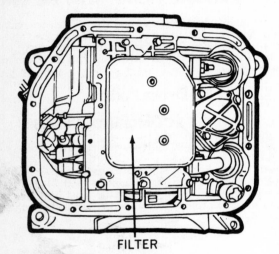

FILTER

Transmission fluid filter location

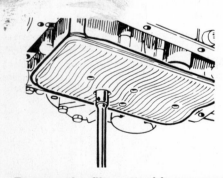

Remove the filter attaching screws

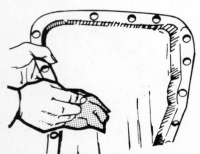

Clean the pan with a safe solvent and a rag. Be sure the pan is completely dry before installation

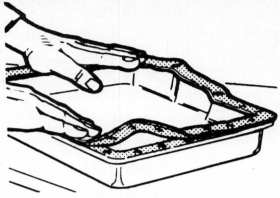

Install a new gasket

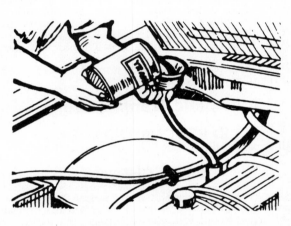

Fill the transmission with a funnel through the transmission dipstick tube

DRAIN AND REFILL

NOTE: Axles on Dodge vans are not equipped with drain plugs. The old lubricant must be drained by removing the differential housing cover. Most models no longer use a paper gasket under the rear axle cover. Instead of the paper gasket, a bead of RTV silicone sealant is now used in production. The sealant is available for service. The sealer should be applied as follows:

1. Loosen the cover mounting bolts after placing a suitable catch container under the rear axle. After the bolts have all been loosened, carefully pry the cover away from the housing and allow the fluid to start draining. Remove the cover. Flush the old fluid from the axle housing, EXCEPT ON MODELS EQUIPPED WITH LIMITED SLIP DIFFERENTIALS. Scrape away any remains of the gasket/silicone material.
2. Clean the cover surface with mineral spirits. Any axle lubricant on the cover or axle housing will prevent the sealant from taking.
3. Apply a 1/8 in. bead of sealant to the clean, dry cover flange. Apply the bead in a continuous bead along the bolt circle of the cover, looping inside the bolt holes as shown.
4. Allow the sealant to air dry.
5. Clean the carrier gasket flange and air dry. Install the cover. If, for any reason, the cover is not installed within 20 minutes of applying the sealant, remove the sealant and start over.

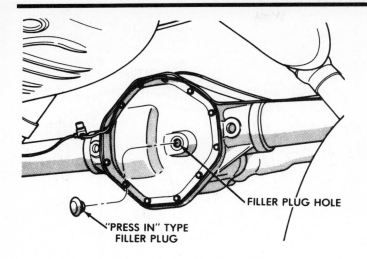

Rear axle fill plug location

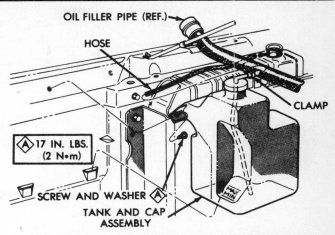

Coolant recovery system-level marks shown

Cooling System

─── **CAUTION** ───

Never remove the radiator cap under any conditions while the engine is running! Failure to follow these instructions could result in damage to the cooling system or engine and/or personal injury. To avoid having scalding hot coolant or steam blow out of the radiator, use extreme care when removing the radiator cap from a hot radiator. Wait until the engine has cooled, then wrap a thick cloth around the radiator cap and turn it slowly to the first stop. Step back while the pressure is released from the cooling system. When you are sure the pressure has been released, press down on the radiator cap (still have the cloth in position) turn and remove the radiator cap.

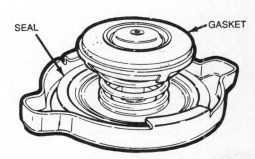

Check the condition of the radiator cap gasket and seal

At least once every 2 years, the engine cooling system should be inspected, flushed, and refilled with fresh coolant. If the coolant is left in the system too long, it loses its ability to prevent rust and corrosion. If the coolant has been diluted with too much water, it won't protect against freezing.

The radiator cap should be looked at for signs of age or deterioration. Fan belt and other drive belts should be inspected and adjusted to the proper tension (See Belt Tension Adjustment).

Hose clamps should be tightened, and soft or cracked hoses replaced. Damp spots, or accumulations of rust or dye near hoses, water pump or other areas, indicate possible leakage. This must be corrected before filling the system with fresh coolant.

FLUID RECOMMENDATIONS

A 50/50 mixture of anti-freeze (ethylene glycol) and water (distilled water if possible) should be used, in most temperature zones. A greater ratio of anti-freeze may be required for extremely cold zones. Coolant should be tested at least once a year for freeze protection.

LEVEL CHECK

A coolant reservoir is provided for fluid expansion and also a quick visual method for checking coolant level. When the engine is idling (gear selector in proper position and parking brake applied), and at normal operating temperature, the coolant level should be between the minimum and maximum marks on the side of the reservoir.

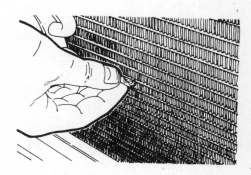

Remove any debris from the radiator's cooling fins

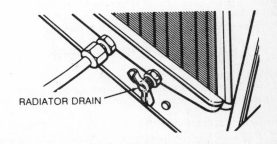

Radiator drain cock

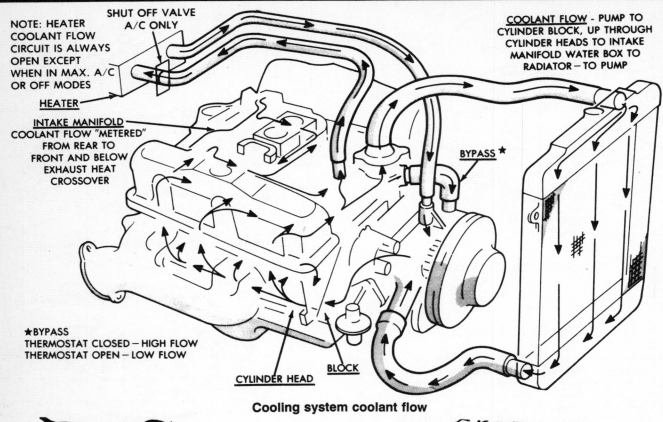

NOTE: HEATER COOLANT FLOW CIRCUIT IS ALWAYS OPEN EXCEPT WHEN IN MAX. A/C OR OFF MODES

SHUT OFF VALVE A/C ONLY

HEATER

INTAKE MANIFOLD COOLANT FLOW "METERED" FROM REAR TO FRONT AND BELOW EXHAUST HEAT CROSSOVER

COOLANT FLOW - PUMP TO CYLINDER BLOCK, UP THROUGH CYLINDER HEADS TO INTAKE MANIFOLD WATER BOX TO RADIATOR – TO PUMP

BYPASS ★

★BYPASS
THERMOSTAT CLOSED – HIGH FLOW
THERMOSTAT OPEN – LOW FLOW

CYLINDER HEAD

BLOCK

Cooling system coolant flow

The system should be pressure tested once a year

CHECK THE RADIATOR CAP

While you are checking the coolant level, check the radiator cap for a worn or cracked gasket. If the cap doesn't seal properly, fluid will be lost in the form of steam and the engine will overheat. Replace the cap with a new one, if necessary.

CLEAN RADIATOR OF DEBRIS

Periodically clean any debris—leaves, paper, insects, etc.—from the radiator fins. Pick the large pieces off by hand. The smaller pieces can be washed away with water pressure from a hose.

Carefully straighten any bent radiator fins with a pair of needle nose pliers. Be careful—the fins are very soft! Don't wiggle the fins back and forth too much. Straighten them once and try not to move them again.

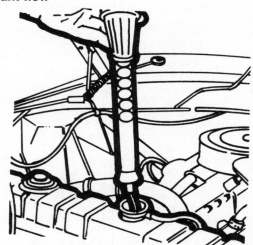

Coolant protection can be checked with an inexpensive tester

DRAIN AND REFILL

Completely draining and refilling the cooling system every two years at least will remove the accumulated rust, scale and other deposits. Coolant in late model vans is a 50/50 mixture of ethylene glycol and water for year round use. Use a good quality antifreeze with water pump lubricants, rust inhibitors and other corrosion inhibitors along with acid neutralizers.

1. Remove the radiator cap. Drain the existing coolant by opening the radiator and engine drain petcocks, or disconnecting the bottom radiator hose at the radiator outlet.

NOTE: Before opening the radiator petcock, spray it with some penetrating lubricant.

2. Close the petcock or re-connect the lower hose and fill the system with water.

3. Add a can of quality radiator flush.

4. Idle the engine until the upper radiator hose gets hot.

5. Drain the system again.

6. Repeat this process until the drained water is clear and free of scale.

7. Close all petcocks and connect all the hoses.

8. If equipped with a coolant recovery system, flush the reservoir with water and leave empty.

9. Determine the capacity of your cooling system (see the Capacities Chart). Add a 50/50 mix of quality antifreeze (ethylene glycol) and water to provide the desired protection.

10. Run the engine to operating temperature.

11. Stop the engine and check the coolant level.

12. Check the level of protection with an antifreeze tester, replace the cap and check for leaks.

FLUSHING AND CLEANING THE SYSTEM

The method of chemical flushing the cooling system is given in the previous DRAIN AND REFILL procedure. Reverse flushing of the cooling system is the forcing of water through the cooling system, using air pressure, in a direction opposite of the normal circulation flow. Reverse flushing is usually only necessary with very dirty systems that show some evidence of partial blockage. Reverse flushing can be used to clean both the radiator and the engine block separately. However, if just the radiator is suspect, It made be wise, in these extreme cases, to remove the radiator and have it cleaned by a radiator repair shop.

No more than 20 psi air pressure should be used when reverse flushing. Drain the cooling system and remove the upper and lower hoses. For radiator flushing, insert the flushing gun into the lower radiator hose connection. For servicing the engine block, remove the thermostat and install the thermostat housing back onto the engine. Place a suitable flushing gun into the housing.

When the flushing gun in properly in position, turn on the water and when the radiator or engine is full, apply air pressure. Allow the system to fill between blasts of air. Continue the procedure until the water runs clear.

After flushing is completed, install the thermostat (if removed),connect the hoses and fill the cooling system

Master Cylinder (Brake & Clutch)

FLUID RECOMMENDATIONS

MOPAR Heavy Duty Brake Fluid, or the equivalent with FMVSS No. 116, DOT 3 and SAE J-1703 Standard designations.

LEVEL CHECK

Brake

The master cylinder reservoir is located under the hood, on the left side firewall. Before removing the master cylinder reservoir caps, make sure the vehicle is resting on level ground and clean all dirt away from the top of the master cylinder. The reservoir level should be filled to the bottom of the filler hole ring. Unscrew the and remove the cap. Add fluid as necessary and replace the cap.

If the level of the brake fluid is less than half the volume of the reservoir, it is advised that you check the brake system for leaks. Leaks in the hydraulic brake system most commonly occur at the wheel cylinder.

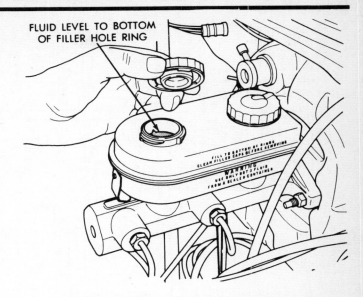

Brake master cylinder fluid level check

Hydraulic Clutch

The hydraulic fluid reservoirs on these systems are mounted on the firewall. Unscrew the cap. The fluid level should be even with the lid retaining collar. Keep the reservoir topped up with DOT-3; do not overfill.

Carefully clean the top and sides of the reservoir before opening, to prevent contamination of the system with dirt, etc. Remove the reservoir diaphragm before adding fluid, and replace after filling.

Power Steering Reservoir

FLUID RECOMMENDATIONS

Use only MOPAR Power Steering Fluid, or the equivalent. Only power steering fluid that is specially formulated for minimum effect on rubber hoses should be used.

LEVEL CHECK

1. Clean the outside of the reservoir cover before removing it.

2. Remove the cover from the reservoir. Wipe the cap mounted dipstick clean, and insert the cap and stick. Once again remove the cap and dipstick and observe the fluid level indicated on the dipstick.

3. When the fluid is HOT, the level will be approximately ½-1 in. below the top of the filler next or at the level indicated on the dipstick.

4. If the fluid is at ROOM TEMPERATURE (approx. 70°F), the level will be about 1½-2½ in. below the top of the filler neck.

5. Add fluid as required and replace the cap.

Chassis Greasing

The lubrication chart indicates where the grease fittings are located. The vehicle should be greased according to the intervals shown on the Maintenance Interval Recommendations at the end of this section.

Water resistant NGL grade 2 EP Multi-mileage grease) should be used for all chassis grease points.

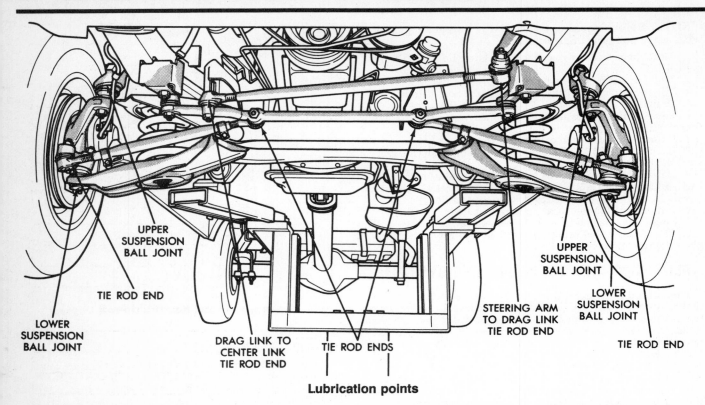

UPPER
SUSPENSION
BALL JOINT

TIE ROD END

LOWER
SUSPENSION
BALL JOINT

DRAG LINK TO
CENTER LINK
TIE ROD END

TIE ROD ENDS

STEERING ARM
TO DRAG LINK
TIE ROD END

UPPER
SUSPENSION
BALL JOINT

LOWER
SUSPENSION
BALL JOINT

TIE ROD END

Lubrication points

Every year or 7,500 miles (factory recommendations are for longer intervals between lubes. See maintenance chart) the front suspension ball points, both upper and lower on each side of the van, must be greased. Most vans covered in this guide should be equipped with grease nipples on the ball joints, although some may have plugs which must be removed and nipples fitted.

NOTE: Do not pump so much grease into the ball joint that excess grease squeezes out of the rubber boot. This destroys the watertight seal.

Jack up the front end of the van and safely support it with jackstands. Block the rear wheels and firmly apply the parking brake. If the van has been parked in temperatures below 20°F for any length of time, park it in a heated garage for an hour or so until the ball joints loosen up enough to accept the grease.

Depending on which front wheel you work on first, turn the wheel and tire outward, either full-lock right or full-lock left. You now have the ends of the upper and lower suspension control arms in front of you; the grease nipples are visible pointing up (top ball joint) and down (lower ball joint) through the end of each control arm. If the nipples are not accessible enough, remove the wheel and tire. Wipe all dirt and crud from the nipples or from around the plugs (if installed). If plugs are on the van, remove them and install grease nipples in the holes (nipples are available in various thread sizes at most auto parts stores). Using a hand operated, low pressure grease gun loaded with a quality chassis grease, grease the ball joint only until the rubber joint boot begins to swell out.

Steering Linkage

The steering linkage should be greased at the same interval as the ball joints. Grease nipples are installed on the steering tie rod ends on most models. Wipe all dirt and hardened grease from around the nipples at each tie rod end. Using a hand operated, low pressure grease gun loaded with the recommended chassis grease, grease the linkage until the old grease begins to squeeze out around the tie rod ends. Wipe off the nipples and any excess grease. Also grease the nipples on the steering drag link.

Parking Brake Linkage

Use chassis grease on the parking brake cable where it contacts the cable guides, levers and linkage.

Automatic Transmission Linkage

Apply a small amount of clean engine oil to the kickdown and shift linkage points at 7,500 mile intervals.

Body Lubrication and Maintenance

Hood Latch And Hinges

Clean the latch surfaces and apply clean engine oil to the latch pilot bolts and the spring anchor. Also lubricate the hood hinges with engine oil. Use a chassis grease to lubricate all the pivot points in the latch release mechanism.

Door Hinges

The gas tank filler door and van doors should be wiped clean and lubricated with clean engine oil once a year. The door lock cylinders and latch mechanisms should be lubricated periodically with a few drops of graphite lock lubricant or a few shots of silicone spray.

Front Wheel Bearings

The front wheels each rotate on a set of opposed, tapered roller bearings. A grease retainer at the inside of the hub prevents lubricant from leaking out of the disc brake rotor hub.

COMPONENT	SERVICE INTERVAL	LUBRICANT	
Door Hinges	As Required	Engine Oil	
Cargo Door Hinges (Van)	As Required	Engine Oil	
Door Latches	As Required	Multi-Purpose Grease (Water Resistant)	1
Hood Hinges	As Required	Engine Oil	
Seat Regulator & Track Adjuster	As Required	Mult-Purpose Grease NLGI Grade 2 EP	2
Sliding Door — Upper Hinge & Swing Lock Striker	As Required	Multi-Purpose Grease NLGI Grade 2 EP	2
Sliding Door — Open Position Striker	As Required	Multi-Purpose Grease NLGI Grade 2 EP	2
Sliding Door — Upper Hinge Assist Spring Ends & Casting Contact Surface	As Required	Multi-Purpose Grease NLGI Grade 2 EP	2
Sliding Door — Lower Track	As Required	Multi-Purpose Grease NLGI Grade 2 EP	2
Sliding Door — Open Position Catch Pivot	As Required	Light Engine Oil	
Sliding Door — Rear Latch Striker Shaft & Wedge	As Required	Stainless Wax Lubricant	
Window System Components (Regulators, Tracks, Links, Channel Areas — Except Glass Run Weatherstrips and Felt Lubricator, if Equipped)	As Required	Smooth White Body Hardware Lubricant	3
Lock Cylinders	Twice/Year	Lock Cylinder Lubricant	4
Window Regulator Felt Tube Lubricator, if so Equipped	As Required	Engine Oil	
Parking Brake Mechanism	As Required	Multi-Purpose Grease	1

1 Mopar Wheel Bearing Grease (High Temperature)
2 Mopar Multi-Mileage Lubricant
3 Mopar Spray White Lube
4 Mopar Lock Cylinder Lubricant

Body lubricant recommendations

For light duty cycle vehicles GVW rating of 3 855 Kg (8,500 lbs.) and below.

Component	Truck Models	Fittings	Service Interval	Lubricant
Drag Link Ball Joints	AB-1, -2, -3	2	22,500 Miles (36 000 km) or 2 Years	Multi-purpose grease—NGLI grade 2 EP (Multi-mileage lubricant)
Suspension Ball Joints	AB-1, -2, -3	4	22,500 Miles (36 000 km) or 2 Years. Every Engine Oil Change for Off-Highway Operation	Multi-purpose grease—NGLI grade 2 EP (Multi-mileage lubricant)
Tie Rod Ball Joints	AB-1, -2, -3	4	22,500 Miles (36 000 km) or 2 Years	Multi-purpose grease—NGLI grade 2 EP (Multi-mileage lubricant)
Wheel Stop	AB-1, -2, -3	—	22,500 Miles (36 000 km) or as Required	Wax type lubricant
Brake Booster Bellcrank Pivot	AB-3	—	15,000 Miles (24 000 km) or 12 Months	Light engine oil

For heavy duty cycle vehicles GVW rating of 3 856 Kg (8,501 lbs.) and above.

Component	Truck Models	Fittings	Service Interval	Lubricant
Brake Booster Bellcrank Pivot	AB-3	—	12,000 Miles (19 000 km) or 1 Year	Light engine oil
Drag Link Ball Joints	AB-3	2	24,000 Miles (39 000 km) or 2 Years	Multi-purpose grease—NGLI grade 2 EP (Multi-mileage lubricant)
Suspension Ball Joints	AB-3	4	24,000 Miles (39 000 km)	Multi-purpose grease—NGLI grade 2 EP (Multi-mileage lubricant)
Tie Rod Ball Joints	AB-3	4	24,000 Miles (39 000 km) or 2 Years	Multi-purpose grease—NGLI grade 2 EP (Multi-mileage lubricant)
Wheel Stop	AB-3	—	24,000 Miles (39 000 km) or as Required	Wax type lubricant

Chassis lubricants

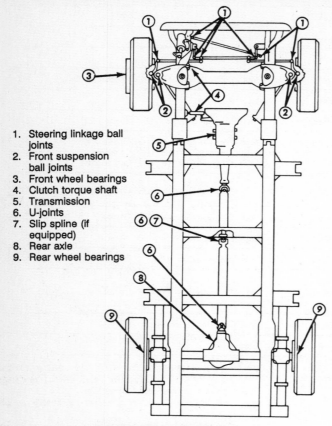

1. Steering linkage ball joints
2. Front suspension ball joints
3. Front wheel bearings
4. Clutch torque shaft
5. Transmission
6. U-joints
7. Slip spline (if equipped)
8. Rear axle
9. Rear wheel bearings

Chassis lubrication points

REMOVAL, REPACKING, INSTALLATION AND ADJUSTMENT

Before handling the bearings, there are a few things that you should remember to do and not to do.

Remember to DO the following:

• Remove all outside dirt from the hub housing before exposing the bearing.
• Treat a used bearing as gently as you would a new one.
• Work with clean tools in clean surroundings.
• Use clean, dry canvas gloves, or at least clean, dry hands.
• Clean solvents and flushing fluids are a must.
• Use clean paper when laying out the bearings to dry.
• Protect disassembled bearings from rust and dirt. Cover them up.
• Use clean rags to wipe bearings.
• Keep the bearings in oil-proof paper when they are to be stored or are not in use.
• Clean the inside of the housing before replacing the bearing.

Do NOT do the following:

• Don't work in dirty surroundings.
• Don't use dirty, chipped or damaged tools.
• Try not to work on wooden work benches or use wooden mallets.
• Don't handle bearings with dirty or moist hands.
• Do not use gasoline for cleaning; use a safe solvent.
• Do not spin-dry bearings with compressed air. They will be damaged.
• Do not spin dirty bearings.
• Avoid using cotton waste or dirty cloths to wipe bearings.
• Try not to scratch or nick bearing surfaces.

• Do not allow the bearing to come in contact with dirt or rust at any time.

1. Raise and support the front end on jackstands.
2. Remove the wheel cover. Remove the wheel.
3. Remove the caliper from the disc rotor and suspend it with wire to the underbody to prevent damage to the brake hose.
4. Remove the grease cap from the hub. Then, remove the cotter pin, nut lock, adjusting nut and flat washer from the spindle. Remove the outer bearing assembly from the hub.
5. Pull the hub and disc assembly off the wheel spindle.
6. Remove and discard the old grease retainer. Remove the inner bearing cone and roller assembly from the hub.
7. Clean all grease from the inner and outer bearing cups with solvent. Inspect the cups for pits, scratches, or excessive wear. If the cups are damaged, remove them with a drift.
8. Clean the inner and outer cone and roller assemblies with solvent and shake them dry. If the cone and roller assemblies show excessive wear or damage, replace them with the bearing cups as a unit.
9. Clean the spindle and the inside of the hub with solvent to thoroughly remove all old grease.
10. Covering the spindle with a clean cloth, brush all loose dirt and dust from the brake assembly. Remove the cloth carefully so as to not get dirt on the spindle.
11. If the inner and/or outer bearing cups were removed, install the replacement cups on the hub. Be sure that the cups seat properly in the hub.
12. It is imperative that all old grease be removed from the bearings and surrounding surfaces before repacking. The new EP High Temperature Grease is not compatible with the sodium base grease used in the past. Use a bearing packer, if available. If not, work the grease in between the rollers and the outer and inner races of the bearing, using the palm of your hand. Work from the larger diameter of the bearing. Make sure the grease goes through the width of the bearing to the small diameter. Pack and install the inner bearing first. Install the inner bearing into the hub and install the grease seal.
13. Install the hub and disc on the wheel spindle. To prevent damage to the grease retainer and spindle threads, keep the hub centered on the spindle.
14. Install the outer bearing cone and roller assembly and the flat washer on the spindle. Install the adjusting nut.
15. Adjust the wheel bearings. Rotate the wheel, hub and rotor assembly while tightening the adjusting nut to 240–300 inch lbs. (27–34 N.m) in order to seat the bearings. Back off the adjusting nut ¼ turn (90°), then tighten the adjusting nut finger tight. Locate the nut lock on the adjusting nut so that the castellations on the lock are lined up with the cotter pin hole in the spindle. Install the new cotter pin, bending the ends of the cotter pin around the castellated flange of the nut lock. Check the wheel for proper rotation, then install the grease cap. If the wheel still does not rotate properly, inspect and clean or replace the wheel bearings and cups.

NOTE: Bend the ends of the cotter pin around the castellations of the locknut to prevent interference with the radio static collector in the grease cap. Install the grease cap.

16. Install the wheels.
17. Install the wheel cover. Lower the vehicle.

Packing the wheel bearing by hand

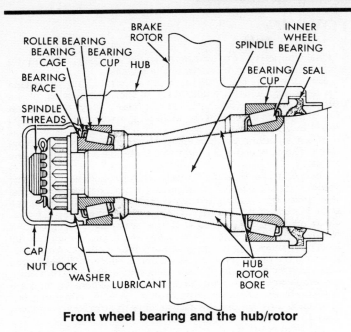

Front wheel bearing and the hub/rotor

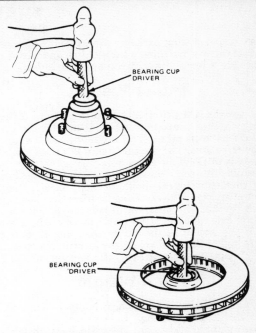

Bearing cup removal/installation

Dana 60 and 60HD Rear Axle Bearings

REMOVAL, REPACKING, INSTALLATION AND ADJUSTMENT

The wheel bearings used on the Dana/Spicer 60 and 60HD rear axle assemblies use full floating rear axles that are packed with wheel bearing grease. Axle lubricant can also flow into the wheel hubs and bearings, however, wheel bearing grease is the primary lubricant. The wheel bearing grease provides lubrication until the axle lubricant reaches the bearings during normal operation.

1. Set the parking brake and loosen the axle shaft bolts.
2. Raise the rear wheels off the floor and place jackstands under the rear axle housing so that the axle is parallel with the floor.
3. Remove the axle shaft bolts.
4. Remove the axle shaft and gaskets.
5. With the axle shaft removed, remove the gasket from the axle shaft flange studs.
6. Bend the lockwasher tab away from the locknut, and then remove the locknut, lockwasher, and the adjusting nut.
7. Remove the outer bearing cone and pull the wheel straight off the axle.
8. With a piece of hardwood or a brass drift which will just

13. Pack each bearing cone and roller with a bearing packer or in the manner previously outlined for the front wheel bearings. Use a EP multi-purpose wheel bearing grease.
14. Place the inner bearing cone and roller assembly in the wheel hub.
15. Install a new inner seal in the hub with a seal installation tool.
16. Install the adjusting nut.
17. While rotating the wheel, tighten the adjusting nut to 120-140 ft. lbs.
18. Back off the nut ⅓ turn (120°). This will provide 0.001-0.008 in. endplay.
19. Install the lock ring onto the spindle keyway.
20. Install the axle shaft and new gasket.

clear the outer bearing cup, drive the inner bearing cone and inner seal out of the wheel hub.
9. Wash all the old grease or axle lubricant out of the wheel hub, using a suitable solvent.
10. Wash the bearing cups and rollers and inspect them for pitting, galling, and uneven wear patterns. Inspect the roller for end wear.
11. If the bearing cups are to be replaced, drive them out with a brass drift. Install the new cups with a block of wood and hammer or press them in.
12. if the bearing cups are properly seated, a 0.0015 in. (0.038mm) feeler gauge will not fit between the cup and the wheel hub. The gauge should not fit beneath the cup. Check several places to make sure the cups are squarely seated.

TRAILER TOWING

Factory trailer towing packages are available on most vans. However, if you are installing a trailer hitch and wiring on your van, there are a few thing that you ought to know.

Trailer Weight

Trailer weight is the first, and most important, factor in determining whether or not your vehicle is suitable for towing the trailer you have in mind. The horsepower-to-weight ratio should be calculated. The basic standard is a ratio of 35:1. That is, 35 pounds of GVW for every horsepower.

To calculate this ratio, multiply you engine's rated horsepower by 35, then subtract the weight of the vehicle, including passengers and luggage. The resulting figure is the ideal maximum trailer weight that you can tow. One point to consider: a numerically higher axle ratio can offset what appears to be a low trailer weight. If the weight of the trailer that you have in mind is somewhat higher than the weight you just calculated, you might consider changing your rear axle ratio to compensate.

Hitch Weight

There are three kinds of hitches: bumper mounted, frame mounted, and load equalizing.

Bumper mounted hitches are those which attach solely to the vehicle's bumper. Many states prohibit towing with this type of hitch, when it attaches to the vehicle's stock bumper, since it subjects the bumper to stresses for which it was not designed. Aftermarket rear step bumpers, designed for trailer towing, are acceptable for use with bumper mounted hitches.

Frame mounted hitches can be of the type which bolts to two or more points on the frame, plus the bumper, or just to several points on the frame. Frame mounted hitches can also be of the tongue type, for Class I towing, or, of the receiver type, for Classes II and III.

Load equalizing hitches are usually used for large trailers. Most equalizing hitches are welded in place and use equalizing bars and chains to level the vehicle after the trailer is hooked up.

The bolt-on hitches are the most common, since they are relatively easy to install.

Check the gross weight rating of your trailer. Tongue weight is usually figured as 10% of gross trailer weight. Therefore, a trailer with a maximum gross weight of 2,000 lb. will have a maximum tongue weight of 200 lb. Class I trailers fall into this category. Class II trailers are those with a gross weight rating of 2,000-3,500 lb., while Class III trailers fall into the 3,500-6,000 lb. category. Class IV trailers are those over 6,000 lb. and are for use with fifth wheel vans, only.

When you've determined the hitch that you'll need, follow the manufacturer's installation instructions, exactly, especially when it comes to fastener torques. The hitch will subjected to a lot of stress and good hitches come with hardened bolts. Never substitute an inferior bolt for a hardened bolt.

Wiring

Wiring the van for towing is fairly easy. There are a number of good wiring kits available and these should be used, rather than trying to design your own. All trailers will need brake lights and turn signals as well as tail lights and side marker lights. Most states require extra marker lights for overly wide trailers. Also, most states have recently required back-up lights for trailers, and most trailer manufacturers have been building trailers with back-up lights for several years.

Additionally, some Class I, most Class II and just about all Class III trailers will have electric brakes.

Add to this number an accessories wire, to operate trailer internal equipment or to charge the trailer's battery, and you can have as many as seven wires in the harness.

Determine the equipment on your trailer and buy the wiring kit necessary. The kit will contain all the wires needed, plus a plug adapter set which included the female plug, mounted on the bumper or hitch, and the male plug, wired into, or plugged into the trailer harness.

When installing the kit, follow the manufacturer's instructions. The color coding of the wires is standard throughout the industry.

One point to note, some domestic vehicles, and most imported vehicles, have separate turn signals. On most domestic vehicles, the brake lights and rear turn signals operate with the same bulb. For those vehicles with separate turn signals, you can purchase an isolation unit so that the brake lights won't blink whenever the turn signals are operated, or, you can go to your local electronics supply house and buy four diodes to wire in series with the brake and turn signal bulbs. Diodes will isolate the brake and turn signals. The choice is yours. The isolation units are simple and quick to install, but far more expensive than the diodes. The diodes, however, require more work to install properly, since they require the cutting of each bulb's wire and soldering in place of the diode.

One final point, the best kits are those with a spring loaded cover on the vehicle mounted socket. This cover prevents dirt and moisture from corroding the terminals. Never let the vehicle socket hang loosely. Always mount it securely to the bumper or hitch.

Cooling

ENGINE

One of the most common, if not THE most common, problem associated with trailer towing is engine overheating.

With factory installed trailer towing packages, a heavy duty cooling system is usually included. Heavy duty cooling systems are available as optional equipment on most vans, with or without a trailer package. If you have one of these extra-capacity systems, you shouldn't have any overheating problems.

If you have a standard cooling system, without an expansion tank, you'll definitely need to get an aftermarket expansion tank kit, preferably one with at least a 2 quart capacity. These kits are easily installed on the radiator's overflow hose, and come with a pressure cap designed for expansion tanks.

Another helpful accessory is a Flex Fan. These fan are large diameter units are designed to provide more airflow at low speeds, with blades that have deeply cupped surfaces. The blades then flex, or flatten out, at high speed, when less cooling air is needed. These fans are far lighter in weight than stock fans, requiring less horsepower to drive them. Also, they are far quieter than stock fans.

If you do decide to replace your stock fan with a flex fan, note that if your van has a fan clutch, a spacer between the flex fan and water pump hub will be needed.

Aftermarket engine oil coolers are helpful for prolonging engine oil life and reducing overall engine temperatures. Both of these factors increase engine life.

While not absolutely necessary in towing Class I and some Class II trailers, they are recommended for heavier Class II and all Class III towing.

Engine oil cooler systems consist of an adapter, screwed on in place of the oil filter, a remote filter mounting and a multi-tube, finned heat exchanger, which is mounted in front of the radiator or air conditioning condenser.

TRANSMISSION

An automatic transmission is usually recommended for trail-

Recommended Equipment Checklist

Equipment	Class I Trailers Under 2,000 pounds	Class II Trailers 2,000-3,500 pounds	Class III Trailers 3,500-6,000 pounds	Class IV Trailers 6,000 pounds and up
Hitch	Frame or Equalizing	Equalizing	Equalizing	Fifth wheel Pick-up truck only
Tongue Load Limit**	Up to 200 pounds	200-350 pounds	350-600 pounds	600 pounds and up
Trailer Brakes	Not Required	Required	Required	Required
Safety Chain	3/16" diameter links	1/4" diameter links	5/16" diameter links	—
Fender Mounted Mirrors	Useful, but not necessary	Recommended	Recommended	Recommended
Turn Signal Flasher	Standard	Constant Rate or heavy duty	Constant Rate or heavy duty	Constant Rate or heavy duty
Coolant Recovery System	Recommended	Required	Required	Required
Transmission Oil Cooler	Recommended	Recommended	Recommended	Recommended
Engine Oil Cooler	Recommended	Recommended	Recommended	Recommended
Air Adjustable Shock Absorbers	Recommended	Recommended	Recommended	Recommended
Flex or Clutch Fan	Recommended	Recommended	Recommended	Recommended
Tires	•••	•••	•••	•••

NOTE The information in this chart is a guide Check the manufacturer's recommendations for your car if in doubt

*Local laws may require specific equipment such as trailer brakes or fender mounted mirrors .Check your local laws
Hitch weight is usually 10-15% of trailer gross weight and should be measured with trailer loaded
**Most manufacturer's do not recommend towing trailers of over 1,000 pounds with compacts Some intermediates cannot tow Class III trailers
***Check manufacturer s recommendations for your specific car trailer combination
—Does not apply

Trailer towing equipment checklist

er towing. Modern automatics have proven reliable and, of course, easy to operate, in trailer towing.

The increased load of a trailer, however, causes an increase in the temperature of the automatic transmission fluid. Heat is the worst enemy of an automatic transmission. As the temperature of the fluid increases, the life of the fluid decreases.

It is essential, therefore, that you install an automatic transmission cooler.

The cooler, which consists of a multi-tube, finned heat exchanger, is usually installed in front of the radiator or air conditioning compressor, and hooked inline with the transmission cooler tank inlet line. Follow the cooler manufacturer's installation instructions.

Select a cooler of at least adequate capacity, based upon the combined gross weights of the van and trailer.

Cooler manufacturers recommend that you use an aftermarket cooler in addition to, and not instead of, the present cooling tank in your van's radiator. If you do want to use it in place of the radiator cooling tank, get a cooler at least two sizes larger than normally necessary.

NOTE: A transmission cooler can, sometimes, cause slow or harsh shifting in the transmission during cold weather, until the fluid has a chance to come up to normal operating temperature. Some coolers can be purchased with or retrofitted with a temperature bypass valve which will allow fluid flow through the cooler only when the fluid has reached operating temperature, or above.

PUSHING AND TOWING

Pushing

Dodge vans equipped with manual transmission can be push started, although this is not recommended if you value the appearance of your van.

To push start, make sure that the bumpers of both vehicles are in reasonable alignment. Bent sheet metal and inflamed tempers are both predictable results from misaligned bumpers when push starting. Turn the ignition key to ON and engage high gear. Depress the clutch pedal. When a speed of about 10 mph is reached, slightly depress the gas pedal and slowly release the clutch. The engine should start. Never get an assist by having the vehicle towed. There is too much risk of the towed vehicle ramming the towing vehicle once it starts.

Vehicles equipped with automatic transmission cannot be started by pushing or towing.

Towing

Tow only in Neutral and at speeds not over 30 mph and distances not exceeding 15 miles. If either the transmission or rear axle is not functioning properly, or if the vehicle is to be towed more than 15 miles, the driveshaft should be disconnected or the van towed with the rear wheels off the ground.

JACKING

A jack is a tire change tool only. If it is necessary to work under the vehicle, place the vehicle on jackstands. Do not operate the engine when the vehicle is raised on a jack.

A jack may used under the rear axle at the spring U-bolts or under the front suspension crossmember in the reinforced area inboard and next to the lower control arm pivot.

— CAUTION —

Never use a floor jack under any part of the underbody. Do not attempt to raise one entire side of the vehicle by placing a jack midway between the front and rear wheels. This may result in permanent damage to the body!

For models supplied with a bumper jack, notches are provided in the bumper for raising the vehicle.

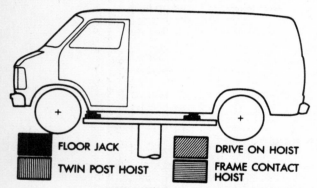

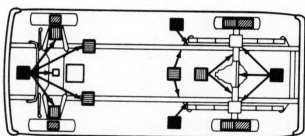

Vehicle lifting points. Places for positioning floor jack are the dark squares

TORQUE SPECIFICATIONS

Component	English	Metric
Fuel line clamps:	10 inch lbs.	1.12 Nm
Fuel filter mounting screw:	75 inch lbs.	8.4 Nm
Alternator pivot screw:	30 ft.lbs.	41 Nm
Alternator upper mounting nut:	250 inch lbs.	28 Nm
Air pump pivot bolt:	24 ft.lbs.	33 Nm
Air Pump Lock Bolt:	24 ft.lbs.	33 Nm
Power Steering Pivot and Adjusting Bolts:	40 ft.lbs.	54 Nm
Cooling system hose clamps:	30-34 inch lbs.	3.4-3.8 Nm
Lug nuts		
150-250 Models; ½-20 stud:	85-110 ft. lbs.	116–150 Nm
350 Model; 90 degree cone; ½-20 stud:	85-110 ft. lbs.	116–150 Nm
350 Model w/H.D. axle; 90 degree cone; ⅝-18 stud:	75-225 ft. lbs.	102–306 Nm
350 Model w/H.D. axle; Flanged nut; ⅝-18 stud:	300-350 ft. lbs.	408–476 Nm
Engine oil pan drain plug:	20 ft. lbs.	27 Nm
Automatic transmission filter screws:	35 inch lbs.	4 Nm
Automatic transmission pan bolts:	150 inch lbs.	17 Nm

HEAVY DUTY CYCLE — GASOLINE ENGINES

Inspection and service should be performed anytime a malfunction is observed or suspected.

Where both time and mileage are shown, follow the interval which occurs first. Miles (Thousand)	6	12	18	24	30	36	42	48	54	60	66	72	78	82½	84	90	96	102	108
Kilometers (Thousand) (12 months)	9.6	19	29	38	48	58	67	77	85	96	106	116	125	132	135	145	154	164	174
Coolant Condition, Coolant Hoses/Clamps (12 months)	X	X	X	X	X	X	X	X	X	X	X	X	X		X	X	X	X	X
Exhaust System — Check	X	X	X	X	X	X	X	X	X	X	X	X	X		X	X	X	X	X
Oil — Change (12 Months)	X	X	X	X	X	X	X	X	X	X	X	X	X		X	X	X	X	X
Oil Filter — Change (2nd Oil Change)*		X		X		X		X		X		X			X		X		X
Drive Belt Tension — Inspect & Adjust			X¹			X			X¹			X				X¹			X
Drive Belt (V-Type) — Replace									X										
Air Filter/Air Pump Air Filter — Replace				X				X				X					X		
Crankcase Inlet Air Filter — Clean				X				X				X							
Spark Plug — Replace					X					X						X			
Fuel Filter — Replace as necessary																			
Coolant — Flush/Replace (36 months) & 24 months/48 000 km (30,000 miles) thereafter									X										
EGR Valve & Tube — Replace										X²									
EGR Tube — Clean Passengers										X²									
PCV Valve — Replace										X²									
Vacuum Emission Components — Replace										X									
Ignition Timing — Adjust to Specs, as necessary										X									
Ignition Cables, Distributor Cap & Rotor — Replace										X									
Manifold Heat Control Valve — Lubricate										X									
Battery — Replace										X									
Oxygen Sensor — Replace														X²					

* If accumulated mileage is less than 9 600 km (6,000 miles) for 12 months, replace the filter at each oil change.

¹ For California vehicles, this maintenance is recommended by Chrysler Motors to the owner but, is not required to maintain the warranty on the air pump drive belt.

² Requires Emission Maintenance Reminder Light. If so equipped, these parts are to be replaced at the indicated mileage, or when the emissions maintenance reminded light remains on continuously with the key in the "on" position, whichever occurs first.

LIGHT DUTY CYCLE

Inspection and service is also necessary any time a malfunction is observed or suspected.

X = Scheduled maintenance for all vehicles.
O = Scheduled maintenance for all vehicles (except California). Recommended for proper vehicle performance for vehicles built for sale in California.

Where both time and mileage are shown, follow the interval which occurs first. Miles (Thousand) Kilometers (Thousand)	7½ 12	15 24	22½ 36	30 48	37½ 60	45 72	52½ 84	60 96	67½ 108	75 120	82½ 132	90 144	97½ 156	105 168	112½ 180	120 192
Coolant Condition, Coolant Hoses/Clamps	X	X	X	X	X	X	O	O	O	O	O	O	O	O	O	O
Exhaust System — Check	X	X	X	X	X	X	O	O	O	O	O	O	O	O	O	O
Oil — Change (6 months)	X	X	X	X	X	X	O	O	O	O	O	O	O	O	O	O
Oil Filter — Change (2nd Oil Change)		X		X		X		O		O		O		O		O
Drive Belt Tension — Inspect & Adjust		O¹		X		O¹		O		O		O		O		O
Drive Belt (V-Type) — Replace								O				O				O
Spark Plug — Replace				X				O				O				O
Air Filter — Replace				X				O				O				O
Fuel Filter — Replace as necessary																
Coolant — Flush/Replace (36 months) & 24 months/48 000 km (30,000 miles) thereafter							O									
EGR Valve & Tube — Replace (clean passages at 60 months)								O²								O²
PCV Breather — Clean								O								O
PCV Valve — Replace (60 months)								O²								O²
Vacuum Emission Components — Replace (60 months)								O								O
Ignition Timing — Adjust to Specs as necessary								O								O
Ignition Cables, Distributor Cap & Rotor — Replace								O								O
Manifold Heat Control Valve — Lubricate								O								O
Battery — Replace								O								
Oxygen Sensor — Replace											O²					

¹ For California vehicles, this maintenance is recommended by Chrysler Motors to the owner, but is not required to maintain the warranty on the air pump drive belt. This maintenance is not allowed on California Durability Vehicles.

² Requires Emission Maintenance Reminder Light. If so equipped, these parts are to be replaced at the indicated mileage, or when the emissions maintenance reminder light remains on continuously with the key in the "on" position, whichever occurs first.

Engine Performance and Tune-Up

2

QUICK REFERENCE INDEX

GENERAL INDEX

TUNE-UP PROCEDURES

In order to extract the full measure of performance and economy from your engine it is essential that it be properly tuned at regular intervals. A regular tune-up will keep your vehicle's engine running smoothly and will prevent the annoying minor breakdowns and poor performance associated with an untuned engine.

With today's sophisticated electronic engine controls and longer-life spark plugs, the manufacturer recommends a complete tune-up be performed every 50,000 miles (80,000km). This interval should be halved if the vehicle is operated under severe conditions, such as trailer towing, prolonged idling, continual stop and start driving, or if starting or running problems are noticed. It is assumed that the routine maintenance described in Section 1 has been kept up, as this will have a decided effect on the results of a tune-up. All of the applicable steps of a tune-up should be followed in order, as the result is a cumulative one.

If the specifications on the tune-up sticker in the engine compartment disagree with the Tune-Up Specifications chart in this section, the figures on the sticker must be used. The sticker often reflects changes made during the production run.

Spark Plugs

A typical spark plug consists of a metal shell surrounding a ceramic insulator. A metal electrode extends downward through the center of the insulator and protrudes a small distance. Located at the end of the plug and attached to the side of the outer metal shell is the side electrode. The side electrode bends in at a 90° angle so that its tip is even with, and parallel to, the tip of the center electrode. The distance between these two electrodes (measured in thousandths of an inch) is called the spark plug gap. The spark plug in no way produces a spark but merely provides a gap across which the current can arc. The coil produces anywhere from 20,000 to 40,000 volts which travels to the distributor where it is distributed through the spark plug wires to the spark plugs. The current passes along the center electrode and jumps the gap to the side electrode, igniting the air/fuel mixture in the combustion chamber.

SPARK PLUG HEAT RANGE

Spark plug heat range is the ability of the plug to dissipate heat. The longer the insulator (or the farther it extends into the engine), the hotter the plug will operate; the shorter the insulator the cooler it will operate. A plug that absorbs little heat and remains too cool will quickly accumulate deposits of oil and carbon since it is not hot enough to burn them off. This leads to plug fouling and consequently to misfiring. A plug that absorbs too much heat will have no deposits, but, due to the excessive heat, the electrodes will burn away quickly and in some instances, preignition may result. Preignition takes place when plug tips get so hot that they glow sufficiently to ignite the fuel/air mixture before the actual spark occurs. This early ignition will usually cause a pinging during low speeds and heavy loads.

The general rule of thumb for choosing the correct heat range when picking a spark plug is: if most of your driving is long distance, high speed travel, use a colder plug; if most of your driving is stop and go, use a hotter plug. Original equipment plugs are compromise plugs, but most people never have occasion to change their plugs from the factory-recommended heat range.

REPLACING SPARK PLUGS

A set of spark plugs usually requires replacement after about 20,000 to 30,000 miles (32,000–48,000km), depending on your style of driving. In normal operation, plug gap increases about 0.001 in. (0.025mm) for every 1,000-2,500 miles (1600–

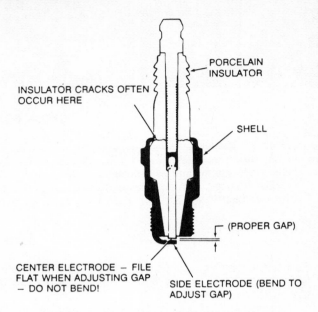

PORCELAIN INSULATOR

INSULATOR CRACKS OFTEN OCCUR HERE

SHELL

(PROPER GAP)

CENTER ELECTRODE — FILE FLAT WHEN ADJUSTING GAP — DO NOT BEND!

SIDE ELECTRODE (BEND TO ADJUST GAP)

Cross section of a typical spark plug

4000km). As the gap increases, the plug's voltage requirement also increases. It requires a greater voltage to jump the wider gap and about two to three times as much voltage to fire a plug at high speeds than at idle.

When you're removing spark plugs, you should work on one at a time. Don't start by removing the plug wires all at once, because unless you number them, they may become mixed up. Take a minute before you begin and number the wires with tape. The best location for numbering is near where the wires come out of the cap. You can buy adhesive-backed numbering tags at most auto parts stores to make this job easier.

NOTE: Apply a small amount of silicone dielectric compound to the inside of the terminal boots whenever an ignition wire is disconnected from the plug, or coil/distributor cap connection.

1. Twist the spark plug boot and remove the boot and wire from the plug. Do not pull on the wire itself as this will ruin the wire.

2. If possible, use a brush or shop rag to clean the area around the spark plug. Make sure that all the dirt is removed so that none will enter the cylinder after the plug is removed.

3. Remove the spark plug using the proper size socket. Truck models use either a ⅝ in. or ¹³⁄₁₆ in. size socket depending on the engine. Turn the socket counterclockwise to remove the plug. Be sure to hold the socket straight on the plug to avoid breaking the plug ceramic insulator, or rounding off the hex on the plug.

4. Once the plug is out, examine the tip since plug readings are vital signs of engine condition.

5. Use a round wire feeler gauge to check the plug gap. The correct size gauge should pass through the electrode gap with a slight drag. If you're in doubt, try one size smaller and one larger. The smaller gauge should go through easily while the larger one shouldn't go through at all. If the gap is incorrect, use the electrode bending tool on the end of the gauge to adjust the gap. When adjusting the gap, always bend the side electrode. The center electrode is non-adjustable.

6. Squirt a drop of penetrating oil on the threads of the new plug and install it. Don't oil the threads too heavily. Turn the plug in clockwise by hand until it is snug.

7. When the plug is finger tight, tighten it with a wrench. If you don't have a torque wrench, tighten the plug about ⅛ turn after it bottoms out. A torque wrench is strongly recommended, especially on aluminum cylinder heads.

8. Install the plug boot firmly over the plug. Proceed to the next plug.

CHECKING AND REPLACING SPARK PLUG CABLES

Visually inspect the spark plug cables for burns, cuts, or breaks in the insulation. Check the spark plug boots and the nipples on the distributor cap and coil. Replace any damaged wiring. If no physical damage is obvious, the wires can be checked with an ohmmeter for excessive resistance.

When installing a new set of spark plug cables, replace the cables one at a time so there will be no mixup. Start by replacing the longest cable first. Install the boot firmly over the spark plug. Route the wire exactly the same as the original. Insert the nipple firmly into the tower on the distributor cap. Repeat the process for each cable.

Make sure you route the plug wires carefully and install them into any wire holders present. A plug wire laying against a metal component or the engine can wear or misfire. If it falls against a hot exhaust manifold, it will burn through and cause a miss.

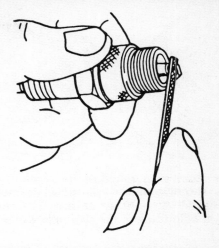

Plugs which are in good condition can be filed and reused

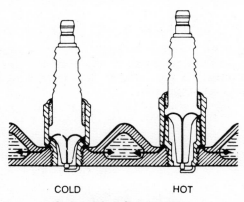

Spark plug heat range

COLD   HOT

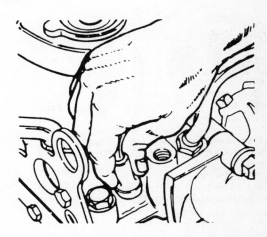

Twist and pull on the rubber boot to remove the spark plug wires. Never pull on the wire itself

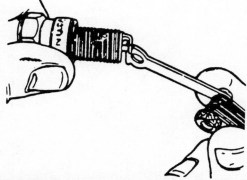

Adjust the electrode gap by bending the side electrode

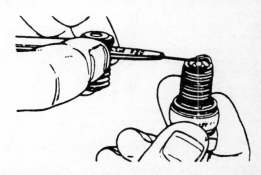

Always use a wire gauge to check the electrode gap; a flat feeler gauge may not give the proper reading

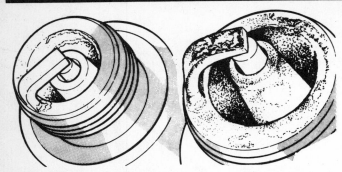

Check the spark plug condition for signs of abnormal operation. The plug on the left shows normal firing; the deposits will be light tan or gray. The plug on the right shows cold fouling; the deposits will be dry and black

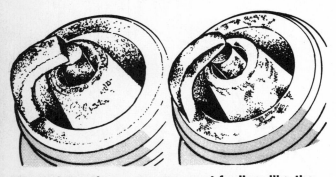

Worn piston rings can cause wet fouling, like the plug on the left. Overheating is indicated by a white or light gray insulator which appears blistered, like the plug on the right

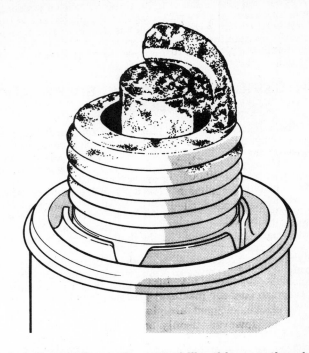

If the plug is oil or ash crusted like this one, there's oil getting into the combustion chamber. Further engine diagnosis is indicated to determine where the excess oil is coming from

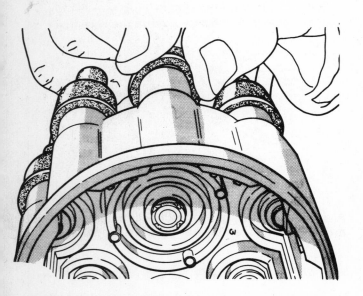

Installing the secondary cable and nipple at the distributor cap

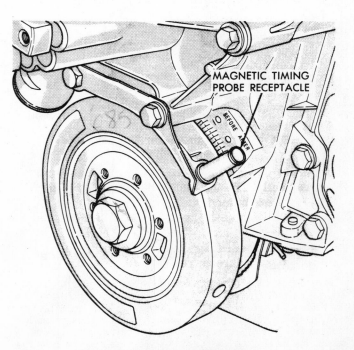

MAGNETIC TIMING PROBE RECEPTACLE

Crank damper timing marks

FIRING ORDERS

To avoid confusion, replace spark plug wires one at a time.

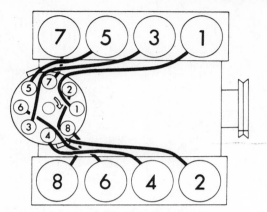

V8 engine firing order

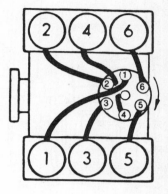

V6 engine firing order

ELECTRONIC IGNITION

The Single Module Engine Controller (SMEC) on 1989 models and the Single Board Engine Controller (SBEC) on 1990-91 models, are the on-board computers which control the entire ignition system. Both automatically regulate the spark advance to fire the spark plugs according to different engine operating conditions. Both also have a built in microprocessor which continually receives input from various engine sensors. These input signals are then used to compute the optimum ignition timing for the best driveability during operation. During the crank-start period, the SMEC or SBEC provides a set timing advance to insure quick and efficient starting.

The amount of electronic spark advance provided is determined by three input factors: coolant temperature, engine rpm and available manifold vacuum. The computer also receives signals from the oxygen sensor and electronically adjusts the air/fuel mixture to insure the most efficient combustion with the lowest possible engine emissions.

Components

COOLANT SENSOR

The coolant temperature sensor, located in the intake manifold near the thermostat housing, informs the computer when the coolant temperature reaches a predetermined operating level. This information is required to prevent changing of the air/fuel ratio with the engine in a non-operating temperature mode. Its signals to the computer also help to control the amount of spark advance with a cold engine.

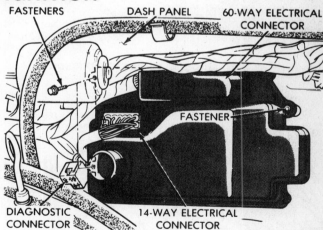

Single Module Engine Controller (1989 SMEC) showing typical mounting location and wire connectors, including the 14-way connector

THROTTLE POSITION SENSOR (TPS)

The throttle position sensor is located on the throttle body and tells the computer when the engine is at idle, off-idle, or at wide-open throttle (WOT). At idle, the computer cancels the spark advance. Idle speed is computer-controlled through the automatic idle speed motor.

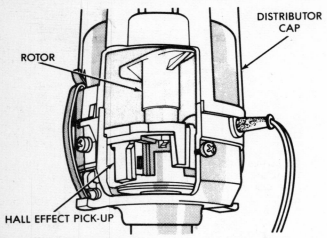

Hall Effect distributor used on the V6 and V8 engines

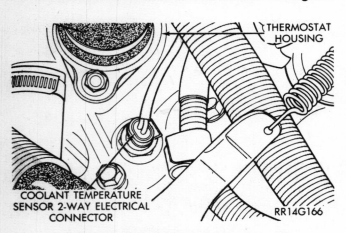

Coolant temperature sensor location

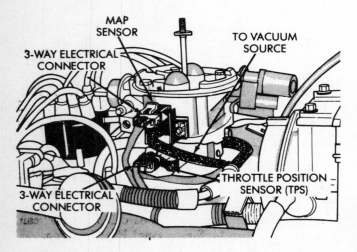

The Manifold Absolute Pressure (MAP) sensor is used to measure the amount of available manifold vacuum

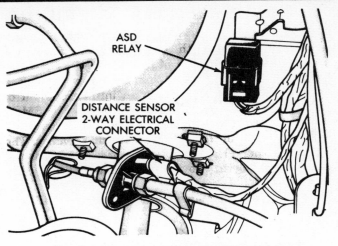

Relay identification on V6 and V8 engines

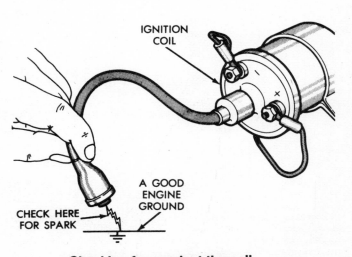

Checking for spark at the coil

MANIFOLD ABSOLUTE PRESSURE (MAP) SENSOR

The manifold absolute pressure (MAP) sensor is a device which monitors manifold vacuum. It is either mounted on the firewall or throttle body and is linked electrically to the on-board computer. The MAP transmits information on manifold vacuum conditions to the computer. This data determines engine load and is used with data from other sensors to adjust the air/fuel mixture.

OXYGEN SENSOR

The oxygen sensor is located in the exhaust manifold and through the use of a self-produced electrical current, signals the computer as to the oxygen content within the exhaust gases flowing past it. Since the electrical output of the oxygen sensor reflects the amount of oxygen in the exhaust, the results are proportional to the rich and lean mixture of the air/fuel ratio. The computer then adjusts the air/fuel ratio to a level that maintains the operating efficiency of the three-way catalytic converter and the engine.

HALL EFFECT PICKUP ASSEMBLY

The Hall Effect pickup is located in the distributor assembly and supplies the engine speed and ignition timing data to the SMEC or SBEC to advance or retard the ignition spark as required by running conditions.

AUTO SHUTDOWN (ASD) RELAY

When there is no ignition signal present with the ignition key in the RUN position, there is no need for fuel delivery. When this condition occurs, the auto shutdown relay interrupts power to the electric fuel pump, fuel injectors and ignition coil. The relays are located on the fender well in the engine compartment. Late models have a power distribution center installed behind the washer reservoir which encases all the relays in one location.

IGNITION COIL

The SMEC or SBEC computer energizes the ignition coil through the ASD relay. When the relay is energized by the computer, battery voltage is supplied to the ignition coil positive terminal. The computer will not energize the ASD relay until it receives input from the distributor pickup. The ignition coil is designed to operate without an external ballast resistor.

If the ignition coil is replaced because of a burned tower, carbon tracking or arcing at either the tower, nipple or boot at the coil end of the secondary cable, replace the cable too. Any arcing at the tower will carbonize the nipple. Installing the cable on a new coil will cause the coil to fail. If the ignition wires show any signs of damage, they should be replaced. Damaged ignition wires can cause arcing and failure of the coil.

V6 and V8 engines use an oil filled coil. It's mounted to the firewall at the rear of the engine compartment.

TROUBLESHOOTING

NOTE: The electronic ignition system can be tested with either special ignition testers or a voltmeter with a 20,000 ohm/volt rating and an ohmmeter using a 9 volt battery as a power source. Since the special ignition system testers have manufacturer's instructions accompanying the units, the technician can refer to the procedural steps necessary to operate them. The following outline will cover the ohm/volt meter unit.

SECONDARY CIRCUIT TEST

1. Remove the coil wire from the distributor cap and hold it cautiously about ¼ in. (6mm) away from an engine ground, then crank the engine while checking for spark.
2. If a good spark is present, slowly move the coil wire away from the engine and check for arcing at the coil while cranking.
3. If good spark is present and it is not arcing at the coil, check the rest of the parts of the ignition system.

IGNITION SYSTEM STARTING TEST

1. Visually inspect all secondary cables at the coil, distributor and spark plugs for cracks and tightness.
2. Check the primary wire at the coil for tightness.

─────────────── CAUTION ───────────────
Whenever removing or installing the wiring harness connector to the control unit, the ignition switch must be in the OFF position.
──

3. With a voltmeter, measure the voltage at the battery and to ascertain that enough current is available to operate the cranking and ignition systems.

4. Crank the engine for 5 seconds while monitoring the voltage at the coil positive (+) terminal. If the voltage remains near zero for the entire period of cranking, see the tests under Electronic Engine Controls for diagnosis of the computer and auto shutdown relay.

5. If the voltage is near 12V and drops to zero after 1-2 seconds of cranking, see the tests outlined in Section 4 under Electronic Engine Controls for diagnosis of the distributor reference pickup circuit to the computer.

6. If the voltage remains near 12V during the entire 5 seconds, remove the 14-way connector on 1989 models, or the 60-way connector on 1990-91 models, and check for any spread terminals.

7. Remove the test lead from the coil positive terminal and connect an 18 gauge jumper wire between the battery positive terminal and the coil positive terminal.

8. Construct a special test jumper as shown in the illustration. Using the jumper, momentarily ground terminal 12 on the 14-way connector (1989 models), or terminal 19 on the 60-way connector (1990-91 models). A a spark should be generated at the coil cable when the ground is removed.

9. If a spark is generated, replace the engine control computer (SMEC or SBEC).

10. If no spark is generated, use the special jumper to ground

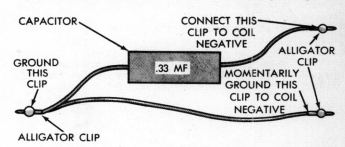

Special jumper necessary for testing computer controlled ignition systems

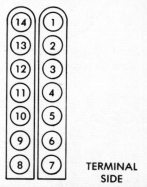

Terminal locations on 14-way connector used on 1989 models

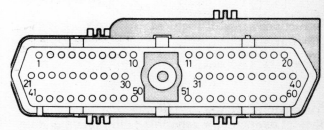

Terminal locations on 60-way connector

the coil negative (−) terminal directly. If a spark is produced, look for an open wire in the harness. If there's still no spark, replace the ignition coil.

COOLANT TEMPERATURE SENSOR TEST

1. Turn the ignition key OFF and disconnect the wire connector from the coolant temperature sensor.
2. Connect an ohmmeter to the terminals on the sensor.
3. With the engine/sensor at normal operating temperature (about 200°F/93°C), sensor resistance should read approximately 700-1000Ω. At room temperature (about 70°F/21°C), sensor resistance should be about 7000-13000Ω.
4. If your test results differ, replace the coolant temperature sensor.

TESTING FOR POOR PERFORMANCE

Basic Timing

Ignition timing is the measurement, in degrees of crankshaft rotation, of the point at which the spark plugs fire in each of the cylinders. It is measured in degrees before or after Top Dead Center (TDC) of the compression stroke.

Ideally, the air/fuel mixture in the cylinder will be ignited by the spark plug just as the piston passes TDC of the compression stroke. If this happens, the piston will be beginning the power stroke just as the compressed and ignited air/fuel mixture starts to expand. The expansion of the air/fuel mixture then forces the piston down on the power stroke and turns the crankshaft.

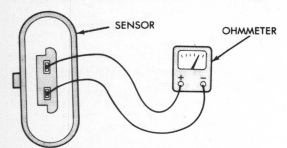

Coolant temperature sensor test

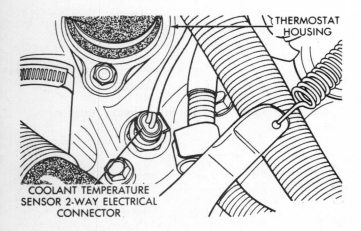

Coolant temperature sensor test
Coolant temperature sensor location on V6 and V8 engines

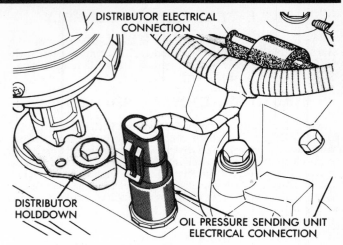

Distributor hold-down bolt location on V6 and V8 engines

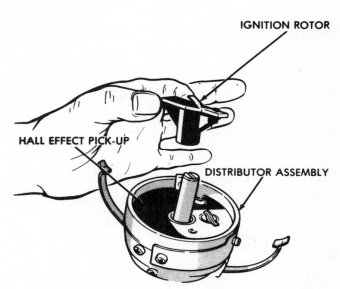

Ignition rotor removal on V6 and V8 engine distributor

Because it takes a fraction of a second for the spark plug to ignite the mixture in the cylinder, the spark plug must fire a little before the piston reaches TDC. Otherwise, the mixture will not be completely ignited as the piston passes TDC and the full power of the explosion will not be used by the engine.

The timing measurement is given in degrees of crankshaft rotation before the piston reaches TDC (BTDC, or Before Top Dead Center). If the setting for the ignition timing is 5°BTDC, each spark plug must fire 5° before each piston reaches TDC. This only holds true, however, when the engine is at idle speed.

As the engine speed increases, the piston go faster. The spark plugs have to ignite the fuel even sooner if it is to be completely ignited when the piston reaches TDC.

If the ignition is set too far advanced (BTDC), the ignition and expansion of the fuel in the cylinder will occur too soon and tend to force the piston down while it is still traveling up. This causes engine ping. If the ignition spark is set too far retarded after TDC (ATDC), the piston will have already passed TDC and started on its way down when the fuel is ignited. This will cause the piston to be forced down for only a portion of its travel. This

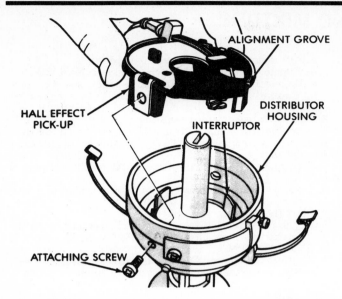

Removing the Hall Effect pickup on V6 and V8 engine distributor

will result in poor engine performance and lack of power.

The timing is best checked with a timing light. This device is connected in series with the No. 1 spark plug. The current that fires the spark plug also causes the timing light to flash.

It is a good idea to paint the timing mark with day-glow or white paint to make it quickly and easily visible. Be sure that all wires, hands and arms are out of the way of the fan. Do not wear any loose clothing when reaching anywhere near the fan.

Correct basic timing is essential for optimum engine performance. Before any testing and service is begun on a poor performance complaint, the basic timing must be checked and adjusted as required. Refer to the underhood specifications label for timing adjustment specifications.

IGNITION TIMING ADJUSTMENT

NOTE: On engines with computer-controlled electronic ignition, your timing light may or may not work, depending on the construction of the light. Consult the manufacturer of the light if in doubt.

ADJUSTMENT

1. Set the gearshift selector in Park or Neutral and apply the parking brake firmly. All lights and accessories must be OFF.
2. Insert the pickup probe of a magnetic timing light into the tube near the timing marks on V6 and V8 engines. If a magnetic timing light is not available, use a conventional power timing light connected to the No. 1 spark plug wire.

NOTE: DO NOT puncture spark plug wires, boots or nipples with test probes. Always use proper adapters. Puncturing the spark plug cables with probes will damage them. Breaking the rubber insulator may permit a secondary current arc which can ruin the coil.

3. Connect a tachometer to the engine and turn selector to the proper cylinder position. Start engine and run until operating temperature is reached.

4. Disconnect coolant temperature sensor connector. The instrument panel warning lights should come on. Engine rpm should be within emission label specifications. If not, refer to the tests in Section 4 for the throttle body minimum air flow check procedure before continuing.
5. Aim the power timing light at the timing marks or the hole in the bell housing, or read the magnetic timing unit.
6. Loosen the distributor hold-down bolt and adjust timing to emission label specifications if the timing isn't within ±2° of the value given.
7. Shut the engine off. Reconnect coolant temperature sensor. Disconnect and reconnect positive battery quick disconnect, or erase the fault codes as described in Section 4, then start the vehicle. The loss of power (check engine) lamp should be off.
8. Shut engine off, then disconnect all test equipment.

DISTRIBUTOR REMOVAL & INSTALLATION

1. Make sure the ignition switch is OFF.
2. Disconnect the distributor pickup lead wire at the wiring harness connector.
3. Matchmark the distributor base to engine block for an installation reference.
4. On V6 and V8 engines, unfasten the distributor cap retaining clips and lift off the distributor cap.
5. Scribe a mark on the edge of the distributor housing to indicate the position of the rotor to use as a reference when installing the distributor.
6. Remove the distributor hold-down screw and clamp, then carefully lift out the distributor.
To Install:
7. Position the distributor in the engine. Make sure the rubber O-ring seal is in the groove in the distributor housing. If the O-ring is cracked or nicked, replace it with a new one.
8. Clean the top of the cylinder block to insure a good seal between the base of the distributor and the block.
9. Carefully insert the distributor and engage the distributor drive gear with the slot in the oil pump drive gear (V6 and V8)
10. If the engine was turned while the distributor was removed, it's necessary to establish the proper relationship between the distributor drive and the No. 1 piston position. Rotate the crankshaft until the No. 1 piston is at the top of its compression stroke. The mark on the crankshaft damper or flywheel should be in line with the 0/TDC timing mark. Rotate the rotor to the position of the No. 1 distributor cap terminal (see the Firing Order diagrams if you're not sure). Once done, install the distributor as described in the previous step.
11. Connect the pickup coil leads and the distributor cap. Make sure all high tension wires snap firmly into the cap towers. Install the distributor hold-down screw finger tight.
12. Connect the distributor pickup lead wire(s) at the wiring harness connectors.
13. Adjust ignition timing to specifications as previously described.

HALL EFFECT PICKUP REPLACEMENT

1. Remove the distributor cap. Disconnect the pickup wire connector.
2. Remove the ignition rotor from the distributor shaft.
3. Remove the Hall Effect pickup attaching screws.
4. Carefully lift the Hall Effect pickup assembly out of the distributor housing.
5. Install the replacement pickup assembly into the distributor. Be careful. The pickup leads may be damaged if not properly installed.
6. Install the attaching screws. Install the rotor.
7. Install the distributor cap and reconnect the Hall Effect wire connector.

IDLE SPEED AND MIXTURE

Various sensors, switches and mechanical linkage positions provide information to the Engine Controller (SMEC or SBEC), which determines the correct air/fuel ratio during varying conditions like acceleration, deceleration, wide open throttle and idle.

A TPS (Throttle Position Sensor) is mounted on the throttle body and senses the angle of the throttle blade opening. A 5 volt signal is supplied to the TPS and a portion of the voltage is sent back to the controller based on the position of the throttle blade. The signal received assists in making drivability adjustments such as mixture control.

An ISC (Idle Speed Control) actuator adjusts the idle speed by physically moving the throttle lever. The moving of the throttle lever will change the air flow which results in the controller changing the amount of fuel supplied. The ISC compensates for varying engine loads and ambient temperatures.

Idle Speed
ADJUSTMENT

1. Before adjusting the idle on an electronic fuel injected vehicle the following items must be checked.

a. AIS motor has been checked for operation.
b. Engine has been checked for vacuum or EGR leaks.
c. Engine timing has been checked and set to specifications.
d. Coolant temperature sensor has been checked for operation.
2. Connect a tachometer to the engine.
3. Start the engine allow it to run for about two minutes.
4. Shut off the engine, allow 60 seconds for the throttle kicker to fully engage.
5. Disconnect the wiring connector to the idle speed control actuator.
6. Disconnect the wiring connector to the coolant temperature sensor.
7. Start the engine.
8. Adjust the extension screw on the ISC actuator until the rpm is within specifications: 3.9L 2500-2600 RPM. 5.2L and 5.9L 2750-2850 RPM.
9. Shut off the engine. Connect the ISC actuator. Connect the coolant temperature sensor. Disconnect the tach.

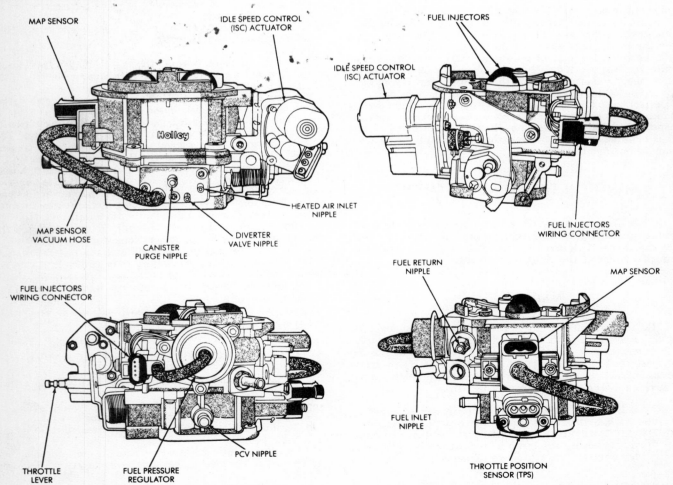

V6 engine throttle body assembly

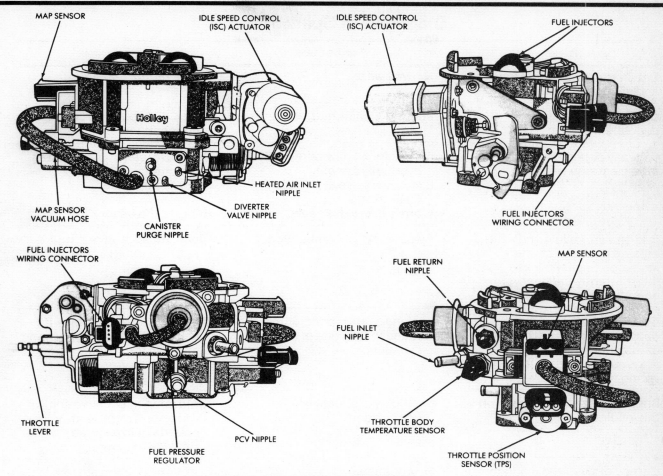

V8 engine throttle body assembly

TUNE-UP SPECIFICATIONS

| Years | Engine No. Cyl. Liters | Spark Plugs | | Distributor | | Ignition Timing (deg.) | | Idle Speed | | Valve Clearance (in.) | |
		Type	Gap (in.)	Point Gap (in.)	Dwell (deg.)	Man. Trans.	Auto. Trans.	Man. Trans.	Auto. Trans.	In.	Exh.
1989	6-3.9	RN12YC	0.035	Electronic		10B	10B	750	750	0.060–0.210②	
	8-5.2	RN12YC	0.035	Electronic		10B	10B	700	700	0.060–0.210②	
	8-5.9	RN12YC	0.035	Electronic		10B	10B	①	①	0.060–0.210②	
1990	6-3.9	RN12YC	0.035	Electronic		10B	10B	①	①	0.060–0.210②	
	8-5.2	RN12YC	0.035	Electronic		10B	10B	①	①	0.060–0.210②	
	8-5.9	RN12YC	0.035	Electronic		10B	10B	①	①	0.060–0.210②	
1991	6-3.9	RN12YC	0.035	Electronic		10B	10B	①	①	0.060–0.210②	
	8-5.2	RN12YC	0.035	Electronic		10B	10B	①	①	0.060–0.210②	
	8-5.9	RN12YC	0.035	Electronic		10B	10B	①	①	0.060–0.210②	

Always check emission control label specifications
before adjustment. If a difference is noted, use the
label specifications.
B—Before Top Dead Center
① See Emission Label for idle speed specifications
② Hydraulic roller tappets
   Adjustment is not required

Diagnosis of Spark Plugs

Problem	Possible Cause	Correction
Brown to grayish-tan deposits and slight electrode wear.	• Normal wear.	• Clean, regap, reinstall.
Dry, fluffy black carbon deposits.	• Poor ignition output.	• Check distributor to coil connections.
Wet, oily deposits with very little electrode wear.	• "Break-in" of new or recently overhauled engine. • Excessive valve stem guide clearances. • Worn intake valve seals.	• Degrease, clean and reinstall the plugs. • Refer to Section 3. • Replace the seals.
Red, brown, yellow and white colored coatings on the insulator. Engine misses intermittently under severe operating conditions.	• By-products of combustion.	• Clean, regap, and reinstall. If heavily coated, replace.
Colored coatings heavily deposited on the portion of the plug projecting into the chamber and on the side facing the intake valve.	• Leaking seals if condition is found in only one or two cylinders.	• Check the seals. Replace if necessary. Clean, regap, and reinstall the plugs.
Shiny yellow glaze coating on the insulator.	• Melted by-products of combustion.	• Avoid sudden acceleration with wide-open throttle after long periods of low speed driving. Replace the plugs.
Burned or blistered insulator tips and badly eroded electrodes.	• Overheating.	• Check the cooling system. • Check for sticking heat riser valves. Refer to Section 1. • Lean air-fuel mixture. • Check the heat range of the plugs. May be too hot. • Check ignition timing. May be over-advanced. • Check the torque value of the plugs to ensure good plug-engine seat contact.
Broken or cracked insulator tips.	• Heat shock from sudden rise in tip temperature under severe operating conditions. Improper gapping of plugs.	• Replace the plugs. Gap correctly.

Engine and Engine Overhaul 3

ENGINE ELECTRICAL

Understanding the Engine Electrical System

The engine electrical system can be broken down into three separate and distinct systems:
1. The starting system.
2. The charging system.
3. The ignition system.

BATTERY AND STARTING SYSTEM

Basic Operating Principles

The battery is the first link in the chain of mechanisms which work together to provide cranking of the automobile engine. In most modern vehicles, the battery is a lead/acid electrochemical device consisting of six 2v subsections connected in series so the unit is capable of producing approximately 12v of electrical pressure. Each subsection, or cell, consists of a series of positive and negative plates held a short distance apart in a solution of sulfuric acid and water. The two types of plates are of dissimilar metals. This causes a chemical reaction to be set up, and it is this reaction which produces current flow from the battery when its positive and negative terminals are connected to an electrical appliance such as a lamp or motor. The continued transfer of electrons would eventually convert the sulfuric acid in the electrolyte to water, and make the two plates identical in chemical composition. As electrical energy is removed from the battery, its voltage output tends to drop. Thus, measuring battery voltage and battery electrolyte composition are two ways of checking the ability of the unit to supply power. During the starting of the engine, electrical energy is removed from the battery. However, if the charging circuit is in good condition and the operating conditions are normal, the power removed from the battery will be replaced by the generator (or alternator) which will force electrons back through the battery, reversing the normal flow, and restoring the battery to its original chemical state.

The battery and starting motor are linked by very heavy electrical cables designed to minimize resistance to the flow of current. Generally, the major power supply cable that leaves the battery goes directly to the starter, while other electrical system needs are supplied by a smaller cable. During starter operation, power flows from the battery to the starter and is grounded through the vehicle's frame and the battery's negative ground strap.

The starting motor is a specially designed, direct current electric motor capable of producing a very great amount of power for its size. One thing that allows the motor to produce a great deal of power is its tremendous rotating speed. It drives the engine through a tiny pinion gear (attached to the starter's armature), which drives the very large flywheel ring gear at a greatly reduced speed. Another factor allowing it to produce so much power is that only intermittent operation is required of it. This, little allowance for air circulation is required, and the windings can be built into a very small space.

The starter solenoid is a magnetic device which employs the small current supplied by the starting switch circuit of the ignition switch. This magnetic action moves a plunger which mechanically engages the starter and electrically closes the heavy switch which connects it to the battery. The starting switch circuit consists of the starting switch contained within the ignition switch, a transmission neutral safety switch or clutch pedal switch, and the wiring necessary to connect these in series with the starter solenoid or relay.

A pinion, which is a small gear, is mounted to a one-way drive clutch. This clutch is splined to the starter armature shaft. When the ignition switch is moved to the **start** position, the solenoid plunger slides the pinion toward the flywheel ring gear via a collar and spring. If the teeth on the pinion and flywheel match properly, the pinion will engage the flywheel immediately. If the gear teeth butt one another, the spring will be compressed and will force the gears to mesh as soon as the starter turns far enough to allow them to do so. As the solenoid plunger reaches the end of its travel, it closes the contacts that connect the battery and starter and then the engine is cranked.

As soon as the engine starts, the flywheel ring gear begins turning fast enough to drive the pinion at an extremely high rate of speed. At this point, the one-way clutch begins allowing the pinion to spin faster than the starter shaft so that the starter will not operate at excessive speed. When the ignition switch is released from the starter position, the solenoid is de-energized, and a spring contained within the solenoid assembly pulls the gear out of mesh and interrupts the current flow to the starter.

Some starter employ a separate relay, mounted away from the starter, to switch the motor and solenoid current on and off. The relay thus replaces the solenoid electrical switch, buy does not eliminate the need for a solenoid mounted on the starter used to mechanically engage the starter drive gears. The relay is used to reduce the amount of current the starting switch must carry.

THE CHARGING SYSTEM

Basic Operating Principles

The automobile charging system provides electrical power for operation of the vehicle's ignition and starting systems and all the electrical accessories. The battery services as an electrical surge or storage tank, storing (in chemical form) the energy originally produced by the engine driven generator. The system also provides a means of regulating generator output to protect the battery from being overcharged and to avoid excessive voltage to the accessories.

The storage battery is a chemical device incorporating parallel lead plates in a tank containing a sulfuric acid/water solution. Adjacent plates are slightly dissimilar, and the chemical reaction of the two dissimilar plates produces electrical energy when the battery is connected to a load such as the starter motor. The chemical reaction is reversible, so that when the generator is producing a voltage (electrical pressure) greater than that produced by the battery, electricity is forced into the battery, and the battery is returned to its fully charged state.

The vehicle's generator is driven mechanically, through V-belts, by the engine crankshaft. It consists of two coils of fine wire, one stationary (the stator), and one movable (the rotor). The rotor may also be known as the armature, and consists of fine wire wrapped around an iron core which is mounted on a shaft. The electricity which flows through the two coils of wire (provided initially by the battery in some cases) creates an intense magnetic field around both rotor and stator, and the interaction between the two fields creates voltage, allowing the generator to power the accessories and charge the battery.

There are two types of generators: the earlier is the direct current (DC) type. The current produced by the DC generator is generated in the armature and carried off the spinning armature by stationary brushes contacting the commutator. The commutator is a series of smooth metal contact plates on the end of the armature. The commutator is a series of smooth metal contact plates on the end of the armature. The commutator plates, which are separated from one another by a very short gap, are connected to the armature circuits so that current will flow in one directions only in the wires carrying the generator output. The generator stator consists of two stationary coils of wire which draw some of the output current of the generator to form a powerful magnetic field and create the interaction of fields which generates the voltage. The generator field is wired in series with the regulator.

Newer automobiles use alternating current generators or alternators, because they are more efficient, can be rotated at higher speeds, and have fewer brush problems. In an alternator, the field rotates while all the current produced passes only through the stator winding. The brushes bear against continuous slip rings rather than a commutator. This causes the current produced to periodically reverse the direction of its flow. Diodes (electrical one-way switches) block the flow of current from traveling in the wrong direction. A series of diodes is wired together to permit the alternating flow of the stator to be converted to a pulsating, but unidirectional flow at the alternator output. The alternator's field is wired in series with the voltage regulator.

The regulator consists of several circuits. Each circuit has a core, or magnetic coil of wire, which operates a switch. Each switch is connected to ground through one or more resistors. The coil of wire responds directly to system voltage. When the voltage reaches the required level, the magnetic field created by the winding of wire closes the switch and inserts a resistance into the generator field circuit, thus reducing the output. The contacts of the switch cycle open and close many times each second to precisely control voltage.

While alternators are self-limiting as far as maximum current is concerned, DC generators employ a current regulating circuit which responds directly to the total amount of current flowing through the generator circuit rather than to the output voltage. The current regulator is similar to the voltage regulator except that all system current must flow through the energizing coil on its way to the various accessories.

Ignition Coil

TESTING

1. Remove the secondary cable from the distributor.
2. Hold the end of the cable (use insulated pliers) about ¼ in. away from a good engine ground.
3. Have a helper crank the engine. Check for spark from the cable end to the engine ground point.
4. If a spark occurs, it must be constant. If the spark is constant, continue to crank the engine and while slowly moving the secondary cable away from the ground, check for spark arching at the coil tower.
5. If arcing occurs replace the coil.
6. If spark is not constant or spark is not present, proceed to the failure to start test.

FAILURE TO START TEST

Perform the previous coil testing procedure. If constant spark or no spark is present proceed with the following test.

Set the parking brake and block the drive wheels before proceeding with this test.

1. Determine that sufficient battery voltage of 12.4 volts is present for the starting and ignition systems.
2. Crank the engine for five seconds while checking the voltage at the positive terminal of the ignition coil.
3. If the voltage remains near zero during the entire period of cranking, check the engine controller and the auto shutdown relay.
4. If the voltage is at near battery voltage and drops to zero after one or two seconds of cranking, check the distributor reference pickup to controller circuit.
5. If voltage remains at near battery reading during the entire five seconds of cranking, turn the key off. Remove the sixty pin connector from the engine controller. Check the sixty pin connection for spread terminals.

6. Remove the positive lead from the coil. Connect an 18 gauge jumper wire between the positive post of the batter and the positive terminal of the coil.
7. Make a special jumper as shown in the illustration. On 1989 models; Using the special jumper, momentarily ground terminal twelve (12) of the fourteen way connector (The connector is plugged into the front of the engine controller, unplug it). On 1990-91 models; Using the special jumper, momentarily ground terminal nineteen (19) of the sixty way connector to the engine controller.
8. A spark should be generated when the special jumper is removed from the terminal.
9. If a spark is generated, replace the engine controller.
10. If no spark is seen, use the special jumper to ground the negative terminal of the coil directly.
11. If a spark is produced, the wiring harness should be checked for an open condition.
12. If no spark is produced, replace the ignition coil.

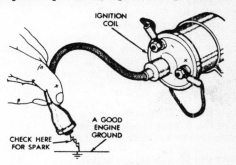

Checking for spark

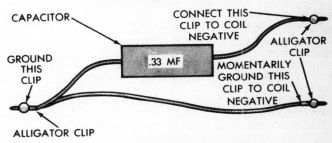

Make a special jumper as shown

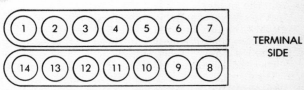

14-Way electrical connector pin identification

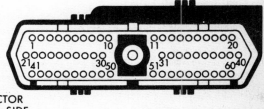

60-Way electrical connector pin identification

REMOVAL AND INSTALLATION

1. Disconnect the negative battery ground cable.
2. Disconnect the two small wires (note the position and color of the wires, as to which is negative and which is positive. The wires must be connected to their proper terminals when installing the coil) and secondary ignition cable.
3. Disconnect the condenser connector from the coil, if equipped.
4. Unbolt and remove the coil.
5. Place the coil into its mounting position and secure.
6. Connect the ignition cable and two lead wires.

Engine/Ignition Controller Module

REMOVAL AND INSTALLATION

NOTE: DO NOT remove the grease from the sixty way connector, or the fourteen way connector (1989), or from the connector cavity in the controller. The grease is used in order to prevent moisture from contacting the connector terminals. Make sure that at least a ⅛ in. thick coating of grease covers the bottom of connector cavities. Apply MOPAR Multi-Purpose grease, or equivalent to the connector prior to connecting it to the controller.

1. Remove the air cleaner duct from the controller.
2. Remove the 3 module retaining screws.
3. Remove the 14 way (1989) and 60 way connectors from the controller.
4. Connect the 14 way (1989) and 60 way connectors to the module.
5. Place the controller in position and secure it with the mounting screws.
6. Install the air cleaner duct to the controller.

Distributor

REMOVAL AND INSTALLATION

1. Disconnect the distributor pickup lead at the wiring harness connector.
2. Unfasten the clips or screws that retain the distributor cap and lift off the cap.
4. Bump the engine around until the rotor is pointing at No. 1 cylinder firing position and the timing marks on the front case and crank pulley are aligned. Disconnect the negative battery cable from the battery.
5. Mark the distributor body and the engine block to indicate the position of the distributor in the block. Mark the distributor body to indicate the rotor position. These marks are used as guides when installing the distributor.
6. Remove the distributor hold down bolt and bracket. Carefully lift the distributor from the engine. The shaft may rotate slightly as the distributor is removed. Make a note of where the movement stops. That point is where the rotor must point when the distributor is reinstalled into the block.
7. If the crankshaft has not been rotated while the distributor was removed from the engine, use the reference marks made before removal to correctly position the distributor in the block. The shaft may have to be rotated slightly to engage the intermediate shaft gear.
If the crankshaft was rotated or otherwise distributed (e.g. during engine rebuilding) after the distributor was removed, proceed as follows:

INSTALLATION - ENGINE DISTURBED

1. Rotate the crankshaft until No. 1 cylinder is at top dead center (TDC) of the compression stroke. To do this, remove the spark plug from cylinder No. 1 and place your thumb over the hole. Slowly turn the engine by hand in the normal direction of rotation until compression is felt at the hole. The **0** mark on the crankshaft pulley should be aligned with the pointer on the timing case cover.
2. Hold the distributor over the mounting pad on the cylinder block so that the distributor body flange coincides with the mounting pad and the rotor points to the No. 1 cylinder firing position.
3. Install the distributor while holding the rotor in position, allowing it to move only enough to engage the slot in the drive gear.
4. Install the cap, tighten the hold-down bracket bolt. Connect the wiring and vacuum hose. Check the timing with a timing light. Adjust if necessary.

Alternator

The alternator is belt-driven from the engine. Energy is supplied from the alternator system to the rotating field through two brushes to two slip-rings. The slip-rings are mounted on the rotor shaft and are connected to the field coil. This energy supplied to the rotating field from the battery is called excitation current and is used to initially energize the field to begin the generation of electricity. Once the alternator starts to generate electricity, the excitation current comes from its own output rather than the battery.

The alternator produces power in the form of alternating current. The alternating current is rectified by diodes into direct current. The direct current is used to charge the battery and power the rest of the electrical system.

When the ignition key is turned on, current flows from the battery, through the charging system indicator light on the instrument panel, to the voltage regulator section of the engine controller and to the alternator. Since the alternator is not producing any current, the alternator warning light comes on. When the engine is started, the alternator begins to produce current and turns the alternator light off. As the alternator turns and produces current, the current is divided in two ways: part to the battery to charge the battery and power the electrical components of the vehicle, and part is returned to the alternator to enable it to increase its output. In this situation, the alternator is receiving current from the battery and from itself. The voltage regulator section of the engine controller controls the current supply to the alternator to prevent it from receiving too much current which would cause it to put out too much current. Conversely, if the voltage regulator does not allow the alternator to receive enough current, the battery will not be fully charged and will eventually go dead.

The battery is connected to the alternator at all times, whether the ignition key is turned on or not. If the battery were shorted to ground, the alternator would also be shorted. This would damage the alternator. To prevent this, a fuse link is installed in the wiring between the battery and the alternator. If the battery is shorted, the fuse link is melted, protecting the alternator.

PRECAUTIONS

To prevent damage to the alternator and regulator, the following precautions should be taken when working with the electrical system.
1. Never reverse the battery connections.
2. Booster batteries for starting must be connected properly: positive-to-positive and negative-to-ground.
3. Disconnect the battery cables before using a fast charger; the charger has a tendency to force current through the diodes in the opposite direction for which they were designed. This burns out the diodes.

Troubleshooting Basic Charging System Problems

Problem	Cause	Solution
Noisy alternator	• Loose mountings • Loose drive pulley • Worn bearings • Brush noise • Internal circuits shorted (High pitched whine)	• Tighten mounting bolts • Tighten pulley • Replace alternator • Replace alternator • Replace alternator
Squeal when starting engine or accelerating	• Glazed or loose belt	• Replace or adjust belt
Indicator light remains on or ammeter indicates discharge (engine running)	• Broken fan belt • Broken or disconnected wires • Internal alternator problems • Defective voltage regulator	• Install belt • Repair or connect wiring • Replace alternator • Replace voltage regulator
Car light bulbs continually burn out— battery needs water continually	• Alternator/regulator overcharging	• Replace voltage regulator/alternator
Car lights flare on acceleration	• Battery low • Internal alternator/regulator problems	• Charge or replace battery • Replace alternator/regulator
Low voltage output (alternator light flickers continually or ammeter needle wanders)	• Loose or worn belt • Dirty or corroded connections • Internal alternator/regulator problems	• Replace or adjust belt • Clean or replace connections • Replace alternator or regulator

4. Never use a fast charger as a booster for starting the vehicle.

5. Never disconnect the voltage regulator while the engine is running.

6. Avoid long soldering times when replacing diodes or transistors. Prolonged heat is damaging to AC generators.

7. Do not use test lamps of more than 12 volts (V) for checking diode continuity.

8. Do not short across or ground any of the terminals on the AC generator.

9. The polarity of the battery, generator, and regulator must be matched and considered before making any electrical connections within the system.

10. Never operate the alternator on an open circuit. make sure that all connections within the circuit are clean and tight.

11. Disconnect the battery terminals when performing any service on the electrical system. This will eliminate the possibility of accidental reversal of polarity.

12. Disconnect the battery ground cable if arc welding is to be done on any part of the vehicle.

CHARGING SYSTEM TROUBLESHOOTING

There are many possible ways in which the charging system can malfunction. Often the source of a problem is difficult to diagnose, requiring special equipment and a good deal of experience. This is usually not the case, however, where the charging system fails completely and causes the dash board warning light to come on or the battery to become dead. To get an idea of what the problem might be, only two pieces of equipment are needed: a test light, to determine that current is reaching a certain point; and a current indicator (milliampmeter), to determine the direction of the current flow and its measurement in amps.

NOTE: In order for the current indicator to give a valid reading, the vehicle must be equipped with battery cables which are of the same gauge size and quality as original equipment battery cables. A normal vehicle electrical system will draw 5 to 30 milliamperes from the battery, with the ignition in the OFF position and all nonignition controlled circuits working properly. A vehicle that has not been operated for an extended period of time may discharge the battery to an inadequate level. If this is the case, the main fusible link (located to the rear of the battery on the engine wiring harness) should be disconnected. If the ignition off draw (IOD) is over 30 milliamperes, the defect must be found and corrected.

Ignition Off Draw (IOD) Tests

1. Turn off all electrical components on the vehicle. Make sure the doors of the vehicle are closed. If the vehicle is equipped with electronic accessories that stay on for a period of time after the doors are closed, allow the systems to automatically shut off, up to three minutes.

2. Raise the hood and disconnect the battery cables, negative cable first. If the vehicle is equipped with an under the hood light, disconnect the light.

3. Connect the ground wire on a test light to the disconnected positive battery cable. Hold the probe end of the test light to the positive battery post, and connect the negative battery cable. If the test light does not light, proceed to Step 4. If the test light does come on, go to Step 5.

NOTE: If the test light comes on an IOD greater than 3 amps is indicated. IOD greater than 3 amps may damage the milliampmeter.

4. With the test light connected and not lit, connect an amme-

ter (millampere scale) between the positive cable clamp and the positive post of the battery. Disconnect the test lamp. A reading of 30 milliamperes or less indicates normal electrical draw. If the reading is more than 30 milliamperes, excessive IOD must be corrected.

5. Remove the fuses and circuit breakers from the fuse panel, one at a time. If the test light goes out, and the reading drops below 30 milliamperes when a certain fuses or circuit breaker is removed, that circuit should be suspect.

6. If IOD is detected after all of the fuses and circuit breakers have been removed, disconnect the 60 way connector from the engine controller.

7. If excessive IOD is detected after all fused circuits and the 60 way connector have been disconnected, disconnect the B+ terminal from the alternator. If the reading drops below 30 milliamperes, the alternator should be suspect. Install all the fuses, circuit breakers, the 60 way connector, the B+ terminal, the battery cables and the under hood lamp.

REMOVAL AND INSTALLATION

While internal alternator repairs are possible, they require specialized tools and training. Therefore, it is advisable to replace a defective alternator, or have it repaired by a qualified shop.

1. Disconnect the battery ground cable at the battery.
2. Disconnect and label the alternator output leads.
3. Loosen the alternator adjusting bolt to relieve belt tension. Remove the drive belts.
4. Remove the alternator mounting bolts and remove the alternator from the vehicle.
5. Position the alternator and install the mounting bolts loosely. Place the drive belts over the alternator pulley. Tighten the adjusting screw until proper belt tension is maintained. Tighten the mounting bolts. Connect the electrical wiring.

Regulator

The voltage regulator is incorporated with the engine electronics and is not serviced separately. See Section 2 for using the Check Engine Lamp for alternator and charging system fault code reading.

Battery

REMOVAL AND INSTALLATION

1. Make sure that the ignition switch and all battery fed accessories are turned OFF.
2. Disconnect the battery cables from the battery, negative cable first.
3. Remove the battery hold down and remove the battery from the vehicle.
4. Clean the battery tray and battery cable ends.
5. Position the battery on to the support tray. Install the battery hold down.
6. Connect the battery cables, positive first then the negative.
7. Apply a thin coating of grease to the outer parts of the cable ends.

Starter

All models are equipped with a reduction gear type starter motor.

NOTE: While internal starter repairs are possible, they require specialized tools and training. Therefore, it is advisable to replace a defective starter, or have it repaired by a qualified shop.

REMOVAL AND INSTALLATION

1. Disconnect the negative ground cable at the battery.
2. Remove the cable from the starter.
3. Disconnect the solenoid leads at their solenoid terminals.
4. Remove the starter attachment bolts and withdraw the starter from the engine flywheel housing. On some models with automatic transmissions, the oil cooler tube bracket will interfere with the starter removal. In this case, remove the starter attachment bolts, slide the cooler tube bracket off the stud, and then withdraw the starter.
5. Position the starter into the bell housing and install the mounting bolts. Be sure that the starter and flywheel housing mating surfaces are free of dirt and oil to make a good electrical contact. Connect the solenoid and starter cable to the starter. Connect the negative battery cable.

ALTERNATOR AND REGULATOR SPECIFICATIONS

| Years | Engine No. Cyl. Liters | Alternator | | | Regulator | | |
		Field Current @ 12v (amps)	Output (amps)	Regulated Volts @ 75°F	Air Gap (in.)	Point Gap (in.)	Back Gap (in.)
1989	6-3.9	2.5–5.0	①	12		Electronic	
	8-5.2	2.5–5.0	①	12		Electronic	
	8-5.9	2.5–5.0	①	12		Electronic	
1990–91	6-3.9	NA	②	12		Electronic	
	8-5.2	NA	②	12		Electronic	
	8-5.9	NA	②	12		Electronic	

NA—Not available
① Nippondenso: 75HS=68
           90HS=87
           120HS=98
Chrysler: 90RS=87
        120RS=98

② Nippondenso: 53005984=73
           5234026=75
           5234028=90
           5234199=120
Bosch 5235028=90

Troubleshooting Basic Starting System Problems

Problem	Cause	Solution
Starter motor rotates engine slowly	• Battery charge low or battery defective • Defective circuit between battery and starter motor • Low load current • High load current	• Charge or replace battery • Clean and tighten, or replace cables • Bench-test starter motor. Inspect for worn brushes and weak brush springs. • Bench-test starter motor. Check engine for friction, drag or coolant in cylinders. Check ring gear-to-pinion gear clearance.
Starter motor will not rotate engine	• Battery charge low or battery defective • Faulty solenoid • Damage drive pinion gear or ring gear • Starter motor engagement weak • Starter motor rotates slowly with high load current • Engine seized	• Charge or replace battery • Check solenoid ground. Repair or replace as necessary. • Replace damaged gear(s) • Bench-test starter motor • Inspect drive yoke pull-down and point gap, check for worn end bushings, check ring gear clearance • Repair engine
Starter motor drive will not engage (solenoid known to be good)	• Defective contact point assembly • Inadequate contact point assembly ground • Defective hold-in coil	• Repair or replace contact point assembly • Repair connection at ground screw • Replace field winding assembly
Starter motor drive will not disengage	• Starter motor loose on flywheel housing • Worn drive end busing • Damaged ring gear teeth • Drive yoke return spring broken or missing	• Tighten mounting bolts • Replace bushing • Replace ring gear or driveplate • Replace spring
Starter motor drive disengages prematurely	• Weak drive assembly thrust spring • Hold-in coil defective	• Replace drive mechanism • Replace field winding assembly
Low load current	• Worn brushes • Weak brush springs	• Replace brushes • Replace springs

STARTER SPECIFICATIONS

Years	Engine No. Cyl. Liters	Cranking Draw Test (Amps)	Solenoid Closing Voltage	No-Load Test			Brush Spring Tension (oz.)
				Amps	Volts	RPM	
1989	6-3.9	150–220	7.5	82	11	3625	NA
	8-5.2	150–220	7.5	82	11	3625	NA
	8-5.9	150–220	7.5	82	11	3625	NA

STARTER SPECIFICATIONS

Years	Engine No. Cyl. Liters	Cranking Draw Test (Amps)	Solenoid Closing Voltage	No-Load Test			Brush Spring Tension (oz.)
				Amps	Volts	RPM	
1990	6-3.9	150–220	7.5	82	11	3625	NA
	8-5.2	150–220	7.5	82	11	3625	NA
	8-5.9	150–220	7.5	82	11	3625	NA
1991	6-3.9	150–220	7.5	73	11	3601	NA
	8-5.2	150–220	7.5	73	11	3601	NA
	8-5.9	150–220	7.5	73	11	3601	NA

Engine should be at operating temperature for cranking draw test. Heavy oil or tight engine will increase starter amperage draw.

ENGINE MECHANICAL

Engine Overhaul Tips

Most engine overhaul procedures are fairly standard. In addition to specific parts replacement procedures and complete specifications for your individual engine, this section also is a guide to accept rebuilding procedures. Examples of standard rebuilding practice are shown and should be used along with specific details concerning your particular engine.

Competent and accurate machine shop services will ensure maximum performance, reliability and engine life.

In most instances it is more profitable to remove, clean and inspect the component, buy the necessary parts and deliver these to a shop for actual machine work. On the other hand, much of the rebuilding work is well within the scope of the do-it-yourself mechanic.

TOOLS

The tools required for an engine overhaul or parts replacement will depend on the depth of your involvement. With a few exceptions, they will be the tools found in a mechanic's tool kit. More in-depth work will require any or all of the following:
- a dial indicator (reading in thousandths) mounted on a universal base
- micrometers and telescope gauges
- jaw and screw-type pullers
- scraper
- valve spring compressor
- ring groove cleaner
- piston ring expander and compressor
- ridge reamer
- cylinder hone or glaze breaker
- Plastigage®
- engine stand

The use of most of these tools is illustrated in this section. Many can be rented for a one-time use from a local parts jobber or tool supply house specializing in automotive work.

Occasionally, the use of special tools is called for. See the information on Special Tools and Safety Notice in the front of this book before substituting another tool.

INSPECTION TECHNIQUES

Procedures and specifications are given in this section for inspecting, cleaning and assessing the wear limits of most major components. Other procedures such as Magnaflux® and Zyglo® can be used to locate material flaws and stress cracks. Magnaflux® is a magnetic process applicable only to ferrous materials. The Zyglo® process coats the material with a fluorescent dye penetrant and can be used on any material Check for suspected surface cracks can be more readily made using spot check dye. The dye is sprayed onto the suspected area, wiped off and the area sprayed with a developer. Cracks will show up brightly.

OVERHAUL TIPS

Aluminum has become extremely popular for use in engines, due to its low weight. Observe the following precautions when handling aluminum parts:
- Never hot tank aluminum parts (the caustic hot tank solution will eat the aluminum.
- Remove all aluminum parts (identification tag, etc.) from engine parts prior to the tanking.
- Always coat threads lightly with engine oil or anti-seize compounds before installation, to prevent seizure.
- Never overtorque bolts or spark plugs especially in aluminum threads.

Stripped threads in any component can be repaired using any of several commercial repair kits (Heli-Coil®, Microdot®, Keenserts®, etc.).

When assembling the engine, any parts that will be frictional contact must be prelubed to provide lubrication at initial start-up. Any product specifically formulated for this purpose can be used, but engine oil is not recommended as a prelube.

When semi-permanent (locked, but removable) installation of bolts or nuts is desired, threads should be cleaned and coated with Loctite® or other similar, commercial non-hardening sealant.

REPAIRING DAMAGED THREADS

Several methods of repairing damaged threads are available. Heli-Coil® (shown here), Keenserts® and Microdot® are among the most widely used. All involve basically the same principle—drilling out stripped threads, tapping the hole and installing a prewound insert—making welding, plugging and oversize fasteners unnecessary.

Two types of thread repair inserts are usually supplied: a standard type for most Inch Coarse, Inch Fine, Metric Course and Metric Fine thread sizes and a spark lug type to fit most spark plug port sizes. Consult the individual manufacturer's catalog to determine exact applications. Typical thread repair kits will contain a selection of prewound threaded inserts, a tap (corresponding to the outside diameter threads of the insert) and an installation tool. Spark plug inserts usually differ because they require a tap equipped with pilot threads and a combined reamer/tap section. Most manufacturers also supply blister-packed thread repair inserts separately in addition to a master kit containing a variety of taps and inserts plus installation tools.

Before effecting a repair to a threaded hole, remove any snapped, broken or damaged bolts or studs. Penetrating oil can be used to free frozen threads. The offending item can be removed with locking pliers or with a screw or stud extractor. After the hole is clear, the thread can be repaired, as shown in the series of accompanying illustrations.

Checking Engine Compression

A noticeable lack of engine power, excessive oil consumption and/or poor fuel mileage measured over an extended period are all indicators of internal engine war. Worn piston rings, scored or worn cylinder bores, blown head gaskets, sticking or burnt valves and worn valve seats are all possible culprits here. A check of each cylinder's compression will help you locate the problems.

As mentioned earlier, a screw-in type compression gauge is more accurate that the type you simply hold against the spark plug hole, although it takes slightly longer to use. It's worth it to obtain a more accurate reading. Follow the procedures below.

1. Warm up the engine to normal operating temperature.
2. Remove all the spark plugs.
3. Disconnect the high tension lead from the ignition coil.
4. Fully open the throttle either by operating the carburetor throttle linkage by hand or by having an assistant floor the accelerator pedal.
5. Screw the compression gauge into the No.1 spark plug hole until the fitting is snug.

WARNING: Be careful not to crossthread the plug hole. On aluminum cylinder heads use extra care, as the threads in these heads are easily ruined.

6. Ask an assistant to depress the accelerator pedal fully. Then, while you read the compression gauge, ask the assistant to crank the engine two or three times in short bursts using the ignition switch.
7. Read the compression gauge at the end of each series of cranks, and record the highest of these readings. Repeat this procedure for each of the engine's cylinders. Compare the highest reading to the lowest, a ten percent difference is acceptable. The difference between any two cylinders should be no more than 12–14 pounds.
8. If a cylinder is unusually low, pour a tablespoon of clean engine oil into the cylinder through the spark plug hole and repeat the compression test. If the compression comes up after adding the oil, it appears that the cylinder's piston rings or bore are damaged or worn. If the pressure remains low, the valves may not be seating properly (a valve job is needed), or the head gasket may be blown near that cylinder. If compression in any two adjacent cylinders is low, and if the addition of oil doesn't help the compression, there is leakage past the head gasket. Oil and coolant water in the combustion chamber can result from this problem. There may be evidence of water droplets on the engine dipstick when a head gasket has blown.

Troubleshooting Engine Mechanical Problems

Problem	Cause	Solution
External oil leaks	• Fuel pump gasket broken or improperly seated	• Replace gasket
	• Cylinder head cover RTV sealant broken or improperly seated	• Replace sealant; inspect cylinder head cover sealant flange and cylinder head sealant surface for distortion and cracks
	• Oil filler cap leaking or missing	• Replace cap
External oil leaks	• Oil filter gasket broken or improperly seated	• Replace oil filter
	• Oil pan side gasket broken, improperly seated or opening in RTV sealant	• Replace gasket or repair opening in sealant; inspect oil pan gasket flange for distortion
	• Oil pan front oil seal broken or improperly seated	• Replace seal; inspect timing case cover and oil pan seal flange for distortion

Troubleshooting Engine Mechanical Problems (cont.)

Problem	Cause	Solution
External oil leaks	• Oil pan rear oil seal broken or improperly seated	• Replace seal; inspect oil pan rear oil seal flange; inspect rear main bearing cap for cracks, plugged oil return channels, or distortion in seal groove
	• Timing case cover oil seal broken or improperly seated	• Replace seal
	• Excess oil pressure because of restricted PCV valve	• Replace PCV valve
	• Oil pan drain plug loose or has stripped threads	• Repair as necessary and tighten
	• Rear oil gallery plug loose	• Use appropriate sealant on gallery plug and tighten
	• Rear camshaft plug loose or improperly seated	• Seat camshaft plug or replace and seal, as necessary
	• Distributor base gasket damaged	• Replace gasket
Excessive oil consumption	• Oil level too high	• Drain oil to specified level
	• Oil with wrong viscosity being used	• Replace with specified oil
	• PCV valve stuck closed	• Replace PCV valve
	• Valve stem oil deflectors (or seals) are damaged, missing, or incorrect type	• Replace valve stem oil deflectors
	• Valve stems or valve guides worn	• Measure stem-to-guide clearance and repair as necessary
	• Poorly fitted or missing valve cover baffles	• Replace valve cover
	• Piston rings broken or missing	• Replace broken or missing rings
	• Scuffed piston	• Replace piston
	• Incorrect piston ring gap	• Measure ring gap, repair as necessary
	• Piston rings sticking or excessively loose in grooves	• Measure ring side clearance, repair as necessary
	• Compression rings installed upside down	• Repair as necessary
	• Cylinder walls worn, scored, or glazed	• Repair as necessary
	• Piston ring gaps not properly staggered	• Repair as necessary
	• Excessive main or connecting rod bearing clearance	• Measure bearing clearance, repair as necessary
No oil pressure	• Low oil level	• Add oil to correct level
	• Oil pressure gauge, warning lamp or sending unit inaccurate	• Replace oil pressure gauge or warning lamp
	• Oil pump malfunction	• Replace oil pump
	• Oil pressure relief valve sticking	• Remove and inspect oil pressure relief valve assembly
	• Oil passages on pressure side of pump obstructed	• Inspect oil passages for obstruction
	• Oil pickup screen or tube obstructed	• Inspect oil pickup for obstruction
	• Loose oil inlet tube	• Tighten or seal inlet tube

Troubleshooting Engine Mechanical Problems (cont.)

Problem	Cause	Solution
Low oil pressure	• Low oil level	• Add oil to correct level
	• Inaccurate gauge, warning lamp or sending unit	• Replace oil pressure gauge or warning lamp
	• Oil excessively thin because of dilution, poor quality, or improper grade	• Drain and refill crankcase with recommended oil
	• Excessive oil temperature	• Correct cause of overheating engine
	• Oil pressure relief spring weak or sticking	• Remove and inspect oil pressure relief valve assembly
	• Oil inlet tube and screen assembly has restriction or air leak	• Remove and inspect oil inlet tube and screen assembly. (Fill inlet tube with lacquer thinner to locate leaks.)
	• Excessive oil pump clearance	• Measure clearances
	• Excessive main, rod, or camshaft bearing clearance	• Measure bearing clearances, repair as necessary
High oil pressure	• Improper oil viscosity	• Drain and refill crankcase with correct viscosity oil
	• Oil pressure gauge or sending unit inaccurate	• Replace oil pressure gauge
	• Oil pressure relief valve sticking closed	• Remove and inspect oil pressure relief valve assembly
Main bearing noise	• Insufficient oil supply	• Inspect for low oil level and low oil pressure
	• Main bearing clearance excessive	• Measure main bearing clearance, repair as necessary
	• Bearing insert missing	• Replace missing insert
	• Crankshaft end play excessive	• Measure end play, repair as necessary
	• Improperly tightened main bearing cap bolts	• Tighten bolts with specified torque
	• Loose flywheel or drive plate	• Tighten flywheel or drive plate attaching bolts
	• Loose or damaged vibration damper	• Repair as necessary
Connecting rod bearing noise	• Insufficient oil supply	• Inspect for low oil level and low oil pressure
	• Carbon build-up on piston	• Remove carbon from piston crown
	• Bearing clearance excessive or bearing missing	• Measure clearance, repair as necessary
	• Crankshaft connecting rod journal out-of-round	• Measure journal dimensions, repair or replace as necessary
	• Misaligned connecting rod or cap	• Repair as necessary
	• Connecting rod bolts tightened improperly	• Tighten bolts with specified torque
Piston noise	• Piston-to-cylinder wall clearance excessive (scuffed piston)	• Measure clearance and examine piston
	• Cylinder walls excessively tapered or out-of-round	• Measure cylinder wall dimensions, rebore cylinder
	• Piston ring broken	• Replace all rings on piston

Troubleshooting Engine Mechanical Problems (cont.)

Problem	Cause	Solution
Piston noise	• Loose or seized piston pin	• Measure piston-to-pin clearance, repair as necessary
	• Connecting rods misaligned	• Measure rod alignment, straighten or replace
	• Piston ring side clearance excessively loose or tight	• Measure ring side clearance, repair as necessary
	• Carbon build-up on piston is excessive	• Remove carbon from piston
Valve actuating component noise	• Insufficient oil supply	• Check for: (a) Low oil level (b) Low oil pressure (c) Plugged push rods (d) Wrong hydraulic tappets (e) Restricted oil gallery (f) Excessive tappet to bore clearance
	• Push rods worn or bent	• Replace worn or bent push rods
	• Rocker arms or pivots worn	• Replace worn rocker arms or pivots
	• Foreign objects or chips in hydraulic tappets	• Clean tappets
	• Excessive tappet leak-down	• Replace valve tappet
	• Tappet face worn	• Replace tappet; inspect corresponding cam lobe for wear
	• Broken or cocked valve springs	• Properly seat cocked springs; replace broken springs
	• Stem-to-guide clearance excessive	• Measure stem-to-guide clearance, repair as required
	• Valve bent	• Replace valve
	• Loose rocker arms	• Tighten bolts with specified torque
	• Valve seat runout excessive	• Regrind valve seat/valves
	• Missing valve lock	• Install valve lock
	• Push rod rubbing or contacting cylinder head	• Remove cylinder head and remove obstruction in head
	• Excessive engine oil (four-cylinder engine)	• Correct oil level

Troubleshooting the Cooling System

Problem	Cause	Solution
High temperature gauge indication—overheating	• Coolant level low • Fan belt loose • Radiator hose(s) collapsed • Radiator airflow blocked	• Replenish coolant • Adjust fan belt tension • Replace hose(s) • Remove restriction (bug screen, fog lamps, etc.)
	• Faulty radiator cap • Ignition timing incorrect • Idle speed low • Air trapped in cooling system • Heavy traffic driving	• Replace radiator cap • Adjust ignition timing • Adjust idle speed • Purge air • Operate at fast idle in neutral intermittently to cool engine
	• Incorrect cooling system component(s) installed • Faulty thermostat • Water pump shaft broken or impeller loose • Radiator tubes clogged • Cooling system clogged • Casting flash in cooling passages	• Install proper component(s) • Replace thermostat • Replace water pump • Flush radiator • Flush system • Repair or replace as necessary. Flash may be visible by removing cooling system components or removing core plugs.
	• Brakes dragging • Excessive engine friction • Antifreeze concentration over 68%	• Repair brakes • Repair engine • Lower antifreeze concentration percentage
	• Missing air seals • Faulty gauge or sending unit	• Replace air seals • Repair or replace faulty component
	• Loss of coolant flow caused by leakage or foaming • Viscous fan drive failed	• Repair or replace leaking component, replace coolant • Replace unit
Low temperature indication—undercooling	• Thermostat stuck open • Faulty gauge or sending unit	• Replace thermostat • Repair or replace faulty component
Coolant loss—boilover	• Overfilled cooling system • Quick shutdown after hard (hot) run • Air in system resulting in occasional "burping" of coolant • Insufficient antifreeze allowing coolant boiling point to be too low • Antifreeze deteriorated because of age or contamination • Leaks due to loose hose clamps, loose nuts, bolts, drain plugs, faulty hoses, or defective radiator • Faulty head gasket • Cracked head, manifold, or block	• Reduce coolant level to proper specification • Allow engine to run at fast idle prior to shutdown • Purge system • Add antifreeze to raise boiling point • Replace coolant • Pressure test system to locate source of leak(s) then repair as necessary • Replace head gasket • Replace as necessary

Troubleshooting the Cooling System (cont.)

Problem	Cause	Solution
Coolant loss—boilover	• Faulty radiator cap	• Replace cap
Coolant entry into crankcase or cylinder(s)	• Faulty head gasket • Crack in head, manifold or block	• Replace head gasket • Replace as necessary
Coolant recovery system inoperative	• Coolant level low • Leak in system • Pressure cap not tight or seal missing, or leaking • Pressure cap defective • Overflow tube clogged or leaking • Recovery bottle vent restricted	• Replenish coolant to FULL mark • Pressure test to isolate leak and repair as necessary • Repair as necessary • Replace cap • Repair as necessary • Remove restriction
Noise	• Fan contacting shroud • Loose water pump impeller • Glazed fan belt • Loose fan belt • Rough surface on drive pulley • Water pump bearing worn • Belt alignment	• Reposition shroud and inspect engine mounts • Replace pump • Apply silicone or replace belt • Adjust fan belt tension • Replace pulley • Remove belt to isolate. Replace pump. • Check pulley alignment. Repair as necessary.
No coolant flow through heater core	• Restricted return inlet in water pump • Heater hose collapsed or restricted • Restricted heater core • Restricted outlet in thermostat housing • Intake manifold bypass hole in cylinder head restricted • Faulty heater control valve • Intake manifold coolant passage restricted	• Remove restriction • Remove restriction or replace hose • Remove restriction or replace core • Remove flash or restriction • Remove restriction • Replace valve • Remove restriction or replace intake manifold

NOTE: *Immediately after shutdown, the engine enters a condition known as heat soak. This is caused by the cooling system being inoperative while engine temperature is still high. If coolant temperature rises above boiling point, expansion and pressure may push some coolant out of the radiator overflow tube. If this does not occur frequently it is considered normal.*

Troubleshooting the Serpentine Drive Belt

Problem	Cause	Solution
Tension sheeting fabric failure (woven fabric on outside circumference of belt has cracked or separated from body of belt)	• Grooved or backside idler pulley diameters are less than minimum recommended • Tension sheeting contacting (rubbing) stationary object • Excessive heat causing woven fabric to age • Tension sheeting splice has fractured	• Replace pulley(s) not conforming to specification • Correct rubbing condition • Replace belt • Replace belt

Troubleshooting the Serpentine Drive Belt (cont.)

Problem	Cause	Solution
Noise (objectional squeal, squeak, or rumble is heard or felt while drive belt is in operation)	• Belt slippage • Bearing noise • Belt misalignment • Belt-to-pulley mismatch • Driven component inducing vibration • System resonant frequency inducing vibration	• Adjust belt • Locate and repair • Align belt/pulley(s) • Install correct belt • Locate defective driven component and repair • Vary belt tension within specifications. Replace belt.
Rib chunking (one or more ribs has separated from belt body)	• Foreign objects imbedded in pulley grooves • Installation damage • Drive loads in excess of design specifications • Insufficient internal belt adhesion	• Remove foreign objects from pulley grooves • Replace belt • Adjust belt tension • Replace belt
Rib or belt wear (belt ribs contact bottom of pulley grooves)	• Pulley(s) misaligned • Mismatch of belt and pulley groove widths • Abrasive environment • Rusted pulley(s) • Sharp or jagged pulley groove tips • Rubber deteriorated	• Align pulley(s) • Replace belt • Replace belt • Clean rust from pulley(s) • Replace pulley • Replace belt
Longitudinal belt cracking (cracks between two ribs)	• Belt has mistracked from pulley groove • Pulley groove tip has worn away rubber-to-tensile member	• Replace belt • Replace belt
Belt slips	• Belt slipping because of insufficient tension • Belt or pulley subjected to substance (belt dressing, oil, ethylene glycol) that has reduced friction • Driven component bearing failure • Belt glazed and hardened from heat and excessive slippage	• Adjust tension • Replace belt and clean pulleys • Replace faulty component bearing • Replace belt
"Groove jumping" (belt does not maintain correct position on pulley, or turns over and/or runs off pulleys)	• Insufficient belt tension • Pulley(s) not within design tolerance • Foreign object(s) in grooves • Excessive belt speed • Pulley misalignment • Belt-to-pulley profile mismatched • Belt cordline is distorted	• Adjust belt tension • Replace pulley(s) • Remove foreign objects from grooves • Avoid excessive engine acceleration • Align pulley(s) • Install correct belt • Replace belt
Belt broken (Note: identify and correct problem before replacement belt is installed)	• Excessive tension • Tensile members damaged during belt installation • Belt turnover	• Replace belt and adjust tension to specification • Replace belt • Replace belt

Troubleshooting the Serpentine Drive Belt (cont.)

Problem	Cause	Solution
Belt broken (Note: identify and correct problem before replacement belt is installed)	• Severe pulley misalignment • Bracket, pulley, or bearing failure	• Align pulley(s) • Replace defective component and belt
Cord edge failure (tensile member exposed at edges of belt or separated from belt body)	• Excessive tension • Drive pulley misalignment • Belt contacting stationary object • Pulley irregularities • Improper pulley construction • Insufficient adhesion between tensile member and rubber matrix	• Adjust belt tension • Align pulley • Correct as necessary • Replace pulley • Replace pulley • Replace belt and adjust tension to specifications
Sporadic rib cracking (multiple cracks in belt ribs at random intervals)	• Ribbed pulley(s) diameter less than minimum specification • Backside bend flat pulley(s) diameter less than minimum • Excessive heat condition causing rubber to harden • Excessive belt thickness • Belt overcured • Excessive tension	• Replace pulley(s) • Replace pulley(s) • Correct heat condition as necessary • Replace belt • Replace belt • Adjust belt tension

GENERAL ENGINE SPECIFICATIONS

Years	Engine No. Cyl. Liters	Fuel System Type	SAE net Horsepower @ rpm	SAE net Torque ft. lb. @ rpm	Bore × Stroke	Comp. Ratio	Oil Press. (psi.) @ 2000 rpm
1989	6-3.9L	TBI	125 @ 4000	195 @ 2000	3.91 × 3.31	9.0:1	30–80 ①
	8-5.2L	TBI	170 @ 4000	260 @ 2000	3.91 × 3.31	9.0:1	30–80 ①
	8-5.9L	TBI	185 @ 4000	283 @ 1600	4.00 × 3.58	8.1:1	30–80 ①
1990	6-3.9L	TBI	125 @ 4000	195 @ 2000	3.91 × 3.31	9.0:1	30–80 ①
	8-5.2L	TBI	170 @ 4000	260 @ 2000	3.91 × 3.31	9.0:1	30–80 ①
	8-5.9L	TBI	185 @ 4000	283 @ 1600	4.00 × 3.58	8.1:1	30–80 ①
1991	6-3.9L	TBI	125 @ 4000	195 @ 2000	3.91 × 3.31	9.0:1	30–80 ①
	8-5.2L	TBI	170 @ 4000	260 @ 2000	3.91 × 3.31	9.0:1	30–80 ①
	8-5.9L	TBI	185 @ 4000	283 @ 1600	4.00 × 3.58	8.1:1	30–80 ①

① 3000 rpm; at least 6 psi at carb idle. If pressure is 0 at curb idle do not run engine to 3000 rpm.

VALVE SPECIFICATIONS

Year	Engine No. Cyl. Liters	Seat Angle (deg.)	Face Angle (deg.)	Spring Test Pressure (lbs. @ in.)	Spring Installed Height (in.)	Stem-to-Guide Clearance (in.) Intake	Stem-to-Guide Clearance (in.) Exhaust	Stem Diameter (in.) Intake	Stem Diameter (in.) Exhaust
1989	6-3.9	45	45	①	②	0.001–0.003	0.002–0.004	0.372	0.371
	8-5.2	45	45	①	②	0.001–0.003	0.002–0.004	0.372	0.371
	8-5.9	45	45	①	②	0.001–0.003	0.002–0.004	0.372	0.371

VALVE SPECIFICATIONS

Year	Engine No. Cyl. Liters	Seat Angle (deg.)	Face Angle (deg.)	Spring Test Pressure (lbs. @ in.)	Spring Installed Height (in.)	Stem-to-Guide Clearance (in.)		Stem Diameter (in.)	
						Intake	Exhaust	Intake	Exhaust
1990	6-3.9	45	45	①	②	0.001–0.003	0.002–0.004	0.372	0.371
	8-5.2	45	45	①	②	0.001–0.003	0.002–0.004	0.372	0.371
	8-5.9	45	45	①	②	0.001–0.003	0.002–0.004	0.372	0.371
1991	6-3.9	45	45	①	②	0.001–0.003	0.002–0.004	0.372	0.371
	8-5.2	45	45	①	②	0.001–0.003	0.002–0.004	0.372	0.371
	8-5.9	45	45	①	②	0.001–0.003	0.002–0.004	0.372	0.371

① Intake = 78–88 @ $1^{11}/_{16}$
Exhaust = 80–90 @ $1^{13}/_{64}$

② Intake = $1^5/_8$–$1^{11}/_{16}$
Exhaust = $1^{29}/_{64}$–$1^{33}/_{64}$

CAMSHAFT SPECIFICATIONS

(All specifications in inches)

Years	Engine No. Cyl. Liters	Journal Diameter					Bearing Clearance	Elevation		End Play
		1	2	3	4	5		Int.	Exh.	
1989	6-3.9	1.998–1.999	1.967–1.968	1.951–1.952	1.5605–1.5615	—	0.002	0.373	0.400	0.002–0.010
	8-5.2	1.998–1.999	1.982–1.983	1.967–1.968	1.951–1.952	1.5605–1.5615	0.002	0.373	0.410	0.010
	8-5.9	1.998–1.999	1.982–1.983	1.967–1.968	1.951–1.952	1.5605–1.5615	0.002	0.400	0.410	0.010
1990	6-3.9	1.998–1.999	1.967–1.968	1.951–1.952	1.5605–1.5615	—	0.002	0.373	0.400	0.002–0.010
	8-5.2	1.998–1.999	1.982–1.983	1.967–1.968	1.951–1.952	1.5605–1.5615	0.002	0.373	0.410	0.010
	8-5.9	1.998–1.999	1.982–1.983	1.967–1.968	1.951–1.952	1.5605–1.5615	0.002	0.400	0.410	0.010
1991	6-3.9	1.998–1.999	1.967–1.968	1.951–1.952	1.5605–1.5615	—	0.002	0.373	0.400	0.002–0.010
	8-5.2	1.998–1.999	1.982–1.983	1.967–1.968	1.951–1.952	1.5605–1.5615	0.002	0.373	0.410	0.010
	8-5.9	1.998–1.999	1.982–1.983	1.967–1.968	1.951–1.952	1.5605–1.5615	0.002	0.400	0.410	0.010

CRANKSHAFT AND CONNECTING ROD SPECIFICATIONS

(All specifications in inches)

Years	Engine No. Cyl. Liters	Crankshaft				Connecting Rod		
		Main Bearing Journal Dia.	Main Bearing Oil Clearance	Shaft End Play	Thrust on No.	Journal Dia.	Oil Clearance	Side Clearance
1989	6.39L	2.4995–2.5005	①	0.002–0.007	2	2.1240–2.1250	0.0005–0.0022	0.006–② 0.014

CRANKSHAFT AND CONNECTING ROD SPECIFICATIONS
(All specifications in inches)

Years	Engine No. Cyl. Liters	Crankshaft Main Bearing Journal Dia.	Main Bearing Oil Clearance	Shaft End Play	Thrust on No.	Connecting Rod Journal Dia.	Oil Clearance	Side Clearance
1989								
	8-5.2L	2.4995–2.5005	③	0.002–0.007	3	2.1240–2.1250	0.0005–0.0022	0.006–② 0.014
	8-5.9L	2.8095–2.8105	③	0.002–0.009	3	2.1240–2.1250	0.0005–0.0022	0.006–② 0.014
1990	6-3.9L	2.4995–2.5005	①	0.002–0.007	2	2.1240–2.1250	0.0005–0.0022	0.006–② 0.014
	8-5.2L	2.4995–2.5005	③	0.002 0.007	3	2.1240–2.1250	0.0005–0.0022	0.006–② 0.014
	8-5.9L	2.8095–2.8105	③	0.002–0.009	3	2.1240–2.1250	0.0005–0.0022	0.006–② 0.014
1991	6-3.9L	2.4995–2.5005	①	0.002–0.007	2	2.1240–2.1250	0.0005–0.0022	0.006–② 0.014
	6-5.2L	2.4995–2.5005	③	0.002–0.007	3	2.1240–2.1250	0.0005–0.0022	0.006–② 0.014
	6-5.9L	2.8095–2.8105	③	0.002–0.009	3	2.1240–2.1250	0.0005–0.0022	0.006–② 0.014

① No. 1—.0005–.0015
Nos. 2, 3 & 4—.0005–.0020
② Two rods
③ No. 1—.0005–.0015
Nos. 2, 3, 4 & 5—.0005–.0020

PISTON AND RING SPECIFICATIONS
(All specifications in inches)

Years	Engine No. Cyl. Liters	Ring Gap #1 Compr.	#2 Compr.	Oil Control	Ring Side Clearance #1 Compr.	#2 Compr.	Oil Control	Piston Clearance
1989	6-3.9	0.010–0.020	0.010–0.020	0.015–0.055	0.015–0.030	0.015–0.030	0.002–0.005	0.0005–0.0015
	8-5.2	0.010–0.020	0.010–0.020	0.015–0.055	0.015–0.030	0.015–0.030	0.002–0.005	0.0005–0.0015
	8-5.9	0.010–0.020	0.010–0.020	0.015–0.055	0.015–0.030	0.015–0.030	0.002–0.005	0.0005–0.0015
1990	6-3.9	0.010–0.020	0.010–0.020	0.015–0.055	0.015–0.030	0.015–0.030	0.002–0.005	0.0005–0.0015
	8-5.2	0.010–0.020	0.010–0.020	0.015–0.055	0.015–0.030	0.015–0.030	0.002–0.005	0.0005–0.0015
	8-5.9	0.010–0.020	0.010–0.020	0.015–0.055	0.015–0.030	0.015–0.030	0.002–0.005	0.0005–0.0015
1991	6-3.9	0.010–0.020	0.010–0.020	0.015–0.055	0.015–0.030	0.015–0.030	0.002–0.005	0.0005–0.0015
	8-5.2	0.010–0.020	0.010–0.020	0.015–0.055	0.015–0.030	0.015–0.030	0.002–0.005	0.0005–0.0015
	8-5.9	0.010–0.020	0.010–0.020	0.015–0.055	0.015–0.030	0.015–0.030	0.002–0.005	0.0005–0.0015

TORQUE SPECIFICATIONS

(All specifications in ft. lbs.)

Years	Engine No. Cyl. Liters	Cyl. Head	Conn. Rod	Main Bearing	Crankshaft Damper	Flywheel	Manifold Intake	Manifold Exhaust
1989	6-3.9	105	45	85	135	55	45	①
	8-5.2	105	45	85	100	55	45	①
	8-5.9	105	45	85	100	55	45	①
1990-91	6-3.9	105	45	85	135	55	45	①
	8-5.2	105	45	85	100	55	45	①
	8-5.9	105	45	85	100	55	45	①

① Screw = 20
  Nut = 15

Engine Design

All V6 and V8 engines used are valve-in head (overhead head valve) type engines with wedge shaped combustion chambers. All are equipped with hydraulic roller tappets. The lubrication system is a rotor type oil pump mounted on the rear main bearing cap and a full-flow, throwaway element filter located on the lower right-hand side of the block.

Engine

REMOVAL AND INSTALLATION

1. Drain the cooling system.

——————— CAUTION ———————

When draining the coolant, keep in mind that cats and dogs are attracted by the ethylene glycol antifreeze, and are quite likely to drink any that is left in an uncovered container or in puddles on the ground. This will prove fatal in sufficient quantity. Always drain the coolant into a sealable container. Coolant should be reused unless it is contaminated or several years old.

2. Disconnect the battery.
3. Drain the engine oil.

——————— CAUTION ———————

The EPA warns that prolonged contact with used engine oil may cause a number of skin disorders, including cancer! You should make every effort to minimize your exposure to used engine oil. Protective gloves should be worn when changing the oil. Wash your hands and any other exposed skin areas as soon as possible after exposure to used engine oil. Soap and water, or waterless hand cleaner should be used.

4. Remove the oil filter and oil dipstick.
5. Remove the engine-to-transmission strut.
6. Remove the engine cover.
7. Remove the air cleaner.
8. Remove the throttle body.
9. Remove the starter.
10. If so equipped, discharge the air conditioning system. See Section 1.
11. Disconnect the refrigerant lines at the condenser. Cap all openings at once.
12. Remove the front bumper.
13. Remove the grille and support brace.
14. Remove the fan shroud.
15. Disconnect the upper radiator hose.
16. Disconnect the lower radiator hose. Disconnect the transmission cooler lines (if equipped).

17. Remove the radiator and condenser as an assembly.
18. Remove the air conditioning compressor.
19. Remove the power steering pump from its bracket and position it out of the way, WITHOUT DISCONNECTING THE HOSES!
20. Disconnect the heater hoses.
21. Tag and disconnect all vacuum lines at the engine.
22. Disconnect the coil.
23. Disconnect the alternator and air pump.
24. Disconnect the temperature sending unit.
25. Disconnect the oil pressure sending unit.
26. Disconnect the engine-to-body ground strap.
27. Disconnect the throttle linkage.
28. Remove the alternator.
29. Remove the fan, pulley and drive belts.
30. Disconnect the fuel line(s) at the pump.
31. Remove the oil dipstick and tube.
32. Remove the heater blower motor (if necessary to gain clearance).
33. Check for any lines, vacuum hoses, or wiring harnesses that have not been disconnected. Label and disconnect.
34. Remove the left exhaust manifold.
35. Raise and support the van on jackstands.
36. Remove the entire exhaust system.
37. Matchmark and remove the driveshaft.
38. Disconnect the shift rods.
39. Remove the clutch torque shaft, or the shifter torque shaft and oil cooler lines.
40. Disconnect the speedometer cable at the transmission.
41. Disconnect any electrical connections at the transmission.
42. Remove the inspection plate or converter cover plate from the bellhousing (automatic transmission). Remove the converter to flexplate bolts/nuts.
43. Support the transmission on a transmission jack.
44. Remove the rear engine support.
45. Install an engine lifting fixture to the heads or intake manifold.
46. Attach a shop crane and take up the weight of the engine.
47. Remove the front engine mount insulator top stud nuts and washers.
48. Remove the transmission-to-bellhousing bolts, or the converter housing-to-engine bolts.
49. Roll the transmission back until it is clear of the engine or clutch, then lower it and move it away.
50. On vans with manual transmission, remove the clutch.
51. Remove the flywheel or flex plate.
52. Raise the rear of the engine about 2 in.
53. Turn the crankshaft until the cutout portion of the crankshaft flange is at the 3 o'clock position.

54. Remove the oil pan bolts, lower the pan just enough to reach the oil pick-up tube and turn the tube and strainer slightly to the right side to clear the pan. Remove the pan.

55. Lower the van to the ground.

56. Raise the engine and maneuver it through the front of the van. It may be necessary to raise the van slightly to keep the crane arm horizontal.

To install:

58. Lower the engine into the van.

59. Raise and support the front end on jackstands.

60. Install the oil pan. Torque the bolts to 15 ft.lbs.

61. Install the flywheel or flex plate. Torque the bolts to 55 ft.lbs.

62. On vans with manual transmission, install the clutch.

63. Position the transmission and install.

64. Install the transmission-to-bellhousing bolts, or the converter housing-to-engine bolts. Torque the ⅜-16 bolts to 30 ft.lbs.; the ⁷/₁₆-14 bolts to 50 ft.lbs.

65. Install the front engine mount insulator top stud nuts and washers. Torque the nuts to 75 ft.lbs.

66. Remove the shop crane.

67. Remove the engine lifting fixture.

68. Install the rear engine support. Torque the bolts to 50 ft.lbs.

69. Remove the transmission jack.

70. Install the inspection plate or converter cover plate on the bellhousing.

71. Connect any electrical connections at the transmission.

72. Connect the speedometer cable at the transmission.

73. Install the clutch torque shaft, or the shifter torque shaft and oil cooler lines.

74. Connect the shift rods.

75. Install the transmission-to-engine strut.

76. Install the driveshaft.

77. Install the exhaust system.

78. Lower the van.

80. Install the left exhaust manifold.

81. Install the intake manifold.

82. Install the heater blower motor.

83. Install the oil dipstick and tube.

84. Connect the fuel line(s) at the pump.

85. Install the fan, pulley and drive belts.

86. Install the alternator.

87. Connect the throttle linkage and install the operating lever from the accelerator bellcrank.

88. Connect the engine-to-body ground strap.

89. Connect the oil pressure sending unit.

90. Connect the temperature sending unit.

91. Connect the alternator.

92. Connect the coil.

93. Connect all vacuum lines at the engine.

94. Connect the heater hoses.

95. Install the power steering pump.

96. Install the air conditioning compressor.

97. Install the throttle body.

98. Install the radiator and condenser assembly.

99. Connect the lower radiator hose.

100. Connect the upper radiator hose.

101. Install the fan shroud.

102. Install the grille and support brace.

103. Install the front bumper.

104. Connect the refrigerant lines at the condenser.

105. Evacuate, charge and leak test the system.

106. Install the starter.

107. Install the air cleaner.

108. Install the engine cover.

109. Install the oil filter.

110. Fill the crankcase.

111. Connect the battery.

112. Fill the cooling system.

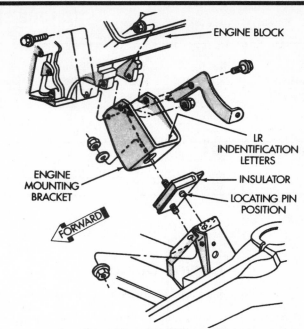

V8 right side front engine mount

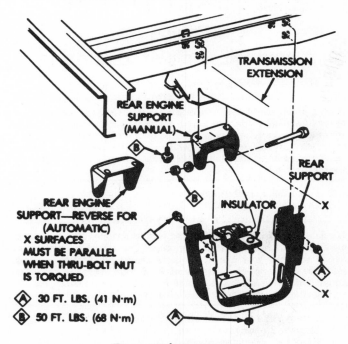

Rear engine support

Rocker Arm Cover

REMOVAL AND INSTALLATION

1. Disconnect the negative battery cable. Disconnect all wires, cable and hoses crossing the rocker cover.

2. Remove the rocker cover bolts.

3. Using a soft mallet, tap loose the rocker cover.

WARNING: Never pry the cover loose. You may damage the cover or head surface!

4. Lift the cover off the engine.

5. Thoroughly clean all gasket material from the mating surfaces of the head and cover. On models which use RTV silicone sealant instead of a gasket, make sure that all material is removed.

6. On models using a gasket, place a new gasket on the head, coated with sealer. Coat the rocker cover mating surface with sealer and install the cover. Tighten the bolts to 80 in.lbs. on models with a gasket.

On models with RTV silicone material, torque the nuts to 80 in.lbs.; the studs to 115 in.lbs. Connect all wires and hoses etc. removed. Connect the negative battery cable.

Rocker Arms/Shaft Assemblies
REMOVAL AND INSTALLATION

The stamped steel rocker arms are arranged on one rocker arm shaft per cylinder head. Because the angle of the pushrods tend to force the rocker arm pairs toward each other, oilite spacers are fitted to absorb the side thrust at each rocker arm. The

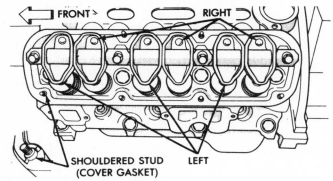

V6 rocker arm location — left bank shown

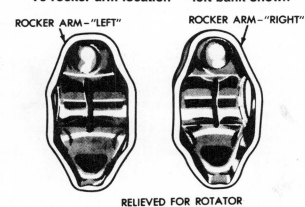

RELIEVED FOR ROTATOR CLEARANCE

INTAKE ROCKER ARM

EXHAUST ROCKER ARM

Intake and exhaust rocker arm identification

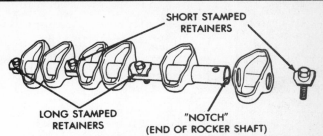

V6 rocker arm retainers and notch location

shaft is secured by bolts and steel retainers attached to the brackets on the cylinder head. To remove the arm and shaft from each cylinder head:

1. Disconnect the negative battery cable. Disconnect the spark plug wires.

2. Disconnect the closed ventilation system and evaporative control system from the valve cover.

3. Remove each valve cover and gasket.

4. Remove the rocker shaft bolts and retainer.

5. Remove each rocker arm and shaft assembly. Service as necessary. If the rocker arms are removed from the shaft, keep everything in order for installation in the original position. The exhaust rocker arm is backcut for rotator clearance.

6. Clean the cylinder head cover and cylinder head of old gasket or RTV material.

7. Inspect the cylinder head cover for any distortion. Straighten the cylinder head cover, if necessary.

8. Assembly the rocker arms on the shafts. Position the rocker arm assemblies on the cylinder heads.

9. The notch on the end of the rocker shafts should point to the engine centerline and toward the front of the engine on the left cylinder head and toward the rear on the right side. Install the mounting retainers and bolts. Install the long stamped retainers in the No. 2 and No. 4 positions.

10. Torque the rocker shaft bolts to 200 in.lbs (23 N.m). The rocker arm shaft should be torqued down slowly, starting with the centermost bolts. After installing the rocker arm shaft assemblies, allow 20 minutes of lifter bleed down time before starting the engine.

11. Install the engine rocker arm cover and new gasket. Tighten to 80 in. lbs. (9 N.m). Connect the spark plug wires and the negative battery cable.

Thermostat

REMOVAL AND INSTALLATION

1. Disconnect the negative battery cable. Drain the cooling system to below the level of the thermostat.

―――――――――― **CAUTION** ――――――――――

When draining the coolant, keep in mind that cats and dogs are attracted by the ethylene glycol antifreeze, and are quite likely to drink any that is left in an uncovered container or in puddles on the ground. This will prove fatal in sufficient quantity. Always drain the coolant into a sealable container. Coolant should be reused unless it is contaminated or several years old.

――――――――――――――――――――――

2. Remove the upper radiator hose from the thermostat housing.

3. Loosen and remove the mounting bolts and the thermostat housing.

4. Check to make sure that the thermostat valve closes tightly. If the valve does not close completely due to foreign material, carefully clean the sealing edge of the valve while being careful not to damage the sealing edge. If the valve does not close tightly after it has been cleaned, a new thermostat must be installed.

5. Immerse the thermostat in a container of warm water so

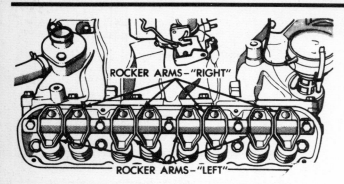

V8 rocker arm location — left bank shown

that its pellet is completely covered and does not touch the bottom or sides of the container.

6. Heat the water and, while stirring the water continuously (to ensure uniform temperature), check the water temperature with a thermometer at the point when a 0.001 in. feeler gauge can be inserted in the valve opening at a water temperature with ±5° of the standard thermostat temperature. If the thermostat does not open within the temperature range, replace it with a new thermostat.

7. Continue heating the water to a temperature of approximately 20° higher than the standard thermostat opening temperature. At this point, the thermostat should be fully open. If it is not, install a new thermostat.

8. To install: Clean all old gasket material from the gasket mounting surfaces of the thermostat housing and the intake manifold. On 3.9L and 5.2L engines, place the gasket on the intake manifold. Insert the thermostat, spring side down, through the gasket opening and center it on the manifold. Place the housing over the thermostat, make sure the thermostat is positioned in the housing recess and install the mounting bolts. On the 5.9L engine, install the thermostat into the intake manifold, spring side down. Place a new gasket over the thermostat. Install the thermostat housing making sure the thermostat fits into the recess provided. Install the mounting bolts.

9. Tighten the mounting bolts to 200 in.lbs. (23 N.m). Connect the upper radiator hose. Fill the cooling system. Connect the negative battery cable.

NOTE: Poor heater output and slow engine warm-ups is often caused by a thermostat stuck in the open position; occasionally one sticks shut causing immediate overheating. Do not attempt to correct an overheating condition by permanently removing the thermostat. Thermostat flow restriction is designed into the system; without it, localized overheating due to turbulence may occur.

Intake Manifold

REMOVAL AND INSTALLATION

1. Release the fuel system pressure. Drain the cooling system and disconnect the negative battery cable.

-------- CAUTION --------

When draining the coolant, keep in mind that cats and dogs are attracted by the ethylene glycol antifreeze, and are quite likely to drink any that is left in an uncovered container or in puddles on the ground. This will prove fatal in sufficient quantity. Always drain the coolant into a sealable container. Coolant should be reused unless it is contaminated or several years old.

2. Remove the engine cover. Remove the alternator, air cleaner and fuel lines (intake and return).

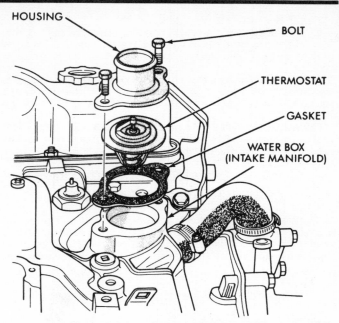

3.9L and 5.2L thermostat installation

3. Disconnect accelerator linkage. If equipped, disconnect the speed control and transmission kickdown cables.

4. Remove the return spring.

5. Remove the distributor cap and wires.

6. Disconnect the coil wires, temperature sending unit wire, heater hoses and bypass hose. Check for, disconnect or remove anything else that would interfere with the intake manifold removal.

7. Remove the right side engine valve cover. Remove the intake manifold mounting bolts. Remove the intake manifold with the throttle body attached. Remove the throttle body from the intake manifold.

8. Clean all gasket mounting surfaces on the intake manifold, both the head side and throttle body side. Take care when cleaning the aluminum manifold on the 3.9L V6, avoid gouging and scratching of the surfaces. Clean the cylinder head mounting surfaces being careful not to drop material into the opened ports. Inspect the crossover passages of the manifold, make sure they are not plugged up.

9. Use a new gasket and mount the throttle body onto the intake manifold. Tighten the bolts to 175 in.lbs. (20 N.m).

10. Install the intake manifold gaskets onto the cylinder heads. The gaskets are marked RT for the right side and LT for the left. They MUST be installed as designated to provide proper coolant flow through the water cooled intake manifold.

11. Place a small spot (about ¼ in.) of RTV material at each block to cylinder head corner. Apply a thin uniformed coating of quick dry cement to the front and rear manifold gaskets (block side) and to the cylinder block mounting surfaces. Allow about 4 or 5 minutes of drying time for the cement.

12. Carefully install the front and rear manifold gaskets. The center hole in the gasket MUST engage the dowels in the block. The end holes in the front and back gaskets MUST lock into the tangs of the manifold gaskets.

13. Carefully lower the intake manifold into position. Loosely install the near center mounting bolts, two on each side. Check the manifold end seals to make sure they did not slide out of position. Install the rest of the mounting bolts finger tight.

14. Tighten the bolts in sequence, starting from the middle of the manifold and work towards the ends. Tighten the bolts in two stages; first to 25 ft.lbs (34 N.m), then to 40 ft.lbs. (54 N.m).

15. Install the engine valve cover, the disconnected lines, link-

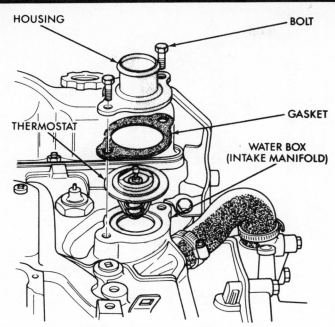

5.9L thermostat installation

age, hoses, alternator and air cleaner. Install the distributor cap and wires. Fill the cooling system and connect the battery cable.

Exhaust Manifold

REMOVAL AND INSTALLATION

1. Disconnect the negative battery cable. Raise and safely support the vehicle. Disconnect the exhaust manifold at the flange where it mates to the exhaust pipe.

2. Remove the engine cover.

3. Remove the exhaust manifold by removing the securing bolts and washers. To reach these bolts, it may be necessary to jack the engine slightly off its front mounts. When the exhaust manifold is removed, sometimes the securing studs will screw out with the nuts. If this occurs, the studs must be replaced with the aid of sealing compound on the coarse thread ends. If this is not done, water leaks may develop at the studs.

4. Place the manifolds in position on the cylinder heads and install the mounting nuts and bolts. On the center branch of 318 and the 360 exhaust manifold, no conical washers are used.

5. Tighten the bolts/nuts working from the center towards the ends of the manifold. Tighten the bolts to 20 ft.lbs. (27 N.m) and the nuts to 15 ft.lbs. (20 N.m).

6. Connect the exhaust pipes to the manifold. Tighten to 24 ft.lbs. (33 N.m). Lower the vehicle and install the engine cover. Connect the battery cable.

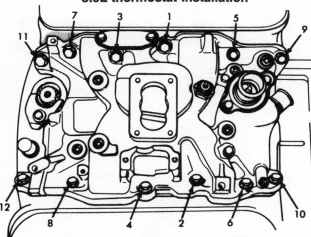

Tightening sequence on the V6 intake manifold

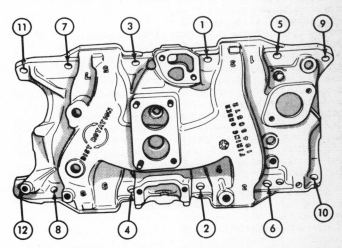

Tightening sequence on the V8 intake manifold

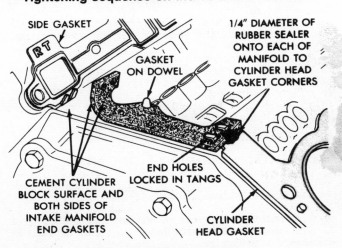

Intake manifold end seals

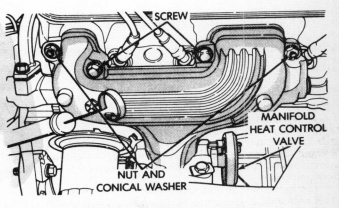

V6 exhaust manifold installation

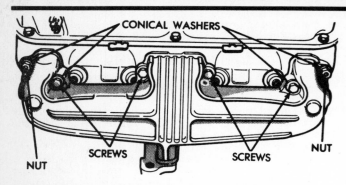

V8 5.2L exhaust manifold installation

Air Conditioning Compressor

REMOVAL AND INSTALLATION

C-171 Compressor

1. Disconnect the negative battery cable. Discharge the re-frigerant system. See Section 1 for the proper procedure.
2. Disconnect the two refrigerant lines from the compressor. Cap the openings immediately!
3. Remove tension from the drive belt. Remove the belt
4. Disconnect the clutch wire at the connector.
5. Remove the compressor-to-bracket bolts. It may be neces-sary to loosen the bracket-to-engine bolts.
6. Remove the compressor.
7. Service as required. Place the compressor in position and secure it with the mounting bolts, after properly adjusting the drive belt. Use new O-rings coated with clean refrigerant oil at all fittings. New, replacement compressors contain 10 oz. of re-frigerant oil. Prior to installation, pour off 4 oz. of oil. This will maintain the oil charge in the system. Connect the battery ca-ble. Evacuate, charge and leak test the A/C system.

Radiator

REMOVAL AND INSTALLATION

1. Drain the cooling system.

CAUTION

When draining the coolant, keep in mind that cats and dogs are attracted by the ethylene glycol antifreeze, and are quite likely to drink any that is left in an uncovered container or in puddles on the ground. This will prove fatal in sufficient quantity. Always drain the coolant into a sealable container. Coolant should be reused unless it is contaminated or several years old.

---

2. Disconnect the negative battery ground cable.
3. Detach the upper hose from the radiator. Disconnect the coolant reservoir hose from the radiator.
4. Remove the shroud mounting nuts and position it out of the way, back over the front of the engine.
5. Remove the radiator top mounting screws. If equipped with air conditioning, remove the condenser attaching screws, accessible through the grille. Do not disconnect any air condi-tioning lines.
6. Raise the vehicle and support it. Disconnect and plug the automatic transmission cooler lines and cap the openings in the cooler. Disconnect the lower radiator hose.

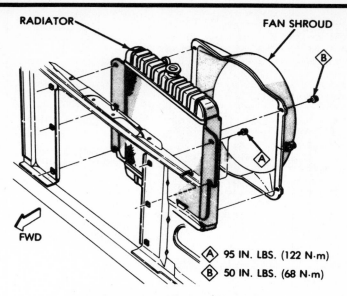

Ⓐ 95 IN. LBS. (122 N·m)
Ⓑ 50 IN. LBS. (68 N·m)

Radiator and shroud mounting

Radiator removal/installation

7. Hold the radiator in place and remove the lower mounting screws. Carefully slide it down and out of the truck.
8. Hold the radiator in place and install the lower mounting screws.
9. Connect the automatic transmission cooler lines. Connect the lower radiator hose.
10. Install the radiator top mounting screws. If equipped with air conditioning, install the condenser attaching screws, accessi-ble through the grille.
11. Install the shroud mounting nuts.
12. Connect the upper hose from the radiator.
13. Connect the battery ground cable.
14. Fill the cooling system.
15. Check all fluid levels and run the engine, making sure there are no leaks.

Air Conditioning Condenser

REMOVAL AND INSTALLATION

1. Discharge the refrigerant system. See Section 1.
2. Disconnect the refrigerant lines from the condenser. Cap all openings immediately!
3. Drain the cooling system.

CAUTION

When draining the coolant, keep in mind that cats and dogs are attracted by the ethylene glycol antifreeze, and are quite likely to drink any that is left in an uncovered container or in puddles on the ground. This will prove fatal in sufficient quantity. Always drain the coolant into a sealable container. Coolant should be reused unless it is contaminated or several years old.

4. Disconnect the upper radiator hose.
5. Remove the bolts retaining the ends of the radiator upper support to the side supports.
6. Carefully pull the top edge of the radiator rearward and remove the condenser upper support.
7. Lift out the condenser.
8. If a new condenser is being installed, add 1 fl.oz. of new refrigerant oil to the new condenser. Installation is the reverse of removal. Always use new O-rings coated with clean refrigerant oil on the line fittings. Evacuate, charge and leak test the system.

Engine Fan

REMOVAL AND INSTALLATION

Your van is usually equipped with a thermal controlled engine cooling fan. The thermal control fan consists of the viscous fan drive clutch and a thermostatic spring mounted on the front face. The viscous fan drive clutch is a silicone-fluid-filled coupling connecting the fan assembly to the water pump pulley. The coupling allows the fan to be driven in a normal manner at low engine speeds while limiting the top speed of the fan to a predetermined maximum level at higher engine speeds.

The thermostatic spring coil reacts to the temperature of the air passing through the radiator fins. If the air temperature rises above a certain point the drive clutch engages to operate the fan at a higher cooling speed. Until additional engine cooling is necessary, the fan will remain at a reduced RPM regardless of engine speed.

Regardless of increased engine speed, once the fan has reached its maximum operating speed it will not rotate any faster. When the necessary engine cooling has been accomplished causing a reduction in the temperature of the air flowing through the radiator fins, the spring coil again reacts and the fan speed is reduced to the previous disengaged speed.

NOTE: Do not place in an upright (shaft pointing straight up) position. The silicone fluid could drain into the bearing assembly and contaminate the lubricant.

1. Disconnect the negative battery ground cable.
2. Loosen the drive belts and remove them from the water pump/fan drive pulley.
3. Remove the fan or viscous drive assembly to water pump pulley hub mounting bolts.
4. Remove the fan shroud mounting screws.
5. Remove the fan assembly and shroud from the engine compartment.
6. Service as required. Remove the fan from the viscous drive unit if necessary.
7. If removed, install the fan to the drive unit. Tighten the mounting bolts to 17 ft.lbs. (23 N.m).

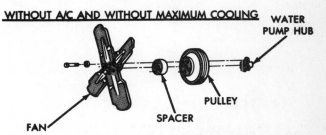

WITHOUT A/C AND WITHOUT MAXIMUM COOLING

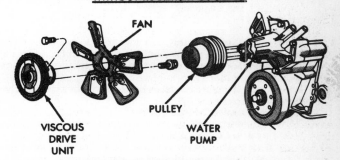

ALL 5.2L/5.9L AND ALL 3.9L EXCEPT WITHOUT A/C AND WITHOUT MAXIMUM COOLING

Fan/fan assembly installation

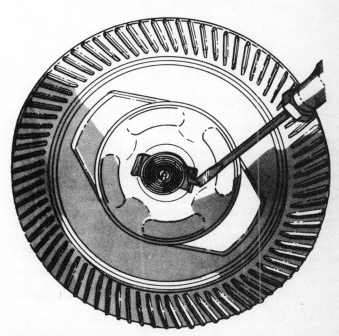

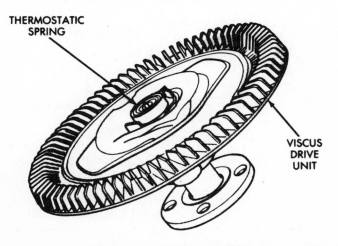

Thermal control fan drive

Disconnecting the end of the thermostatic spring

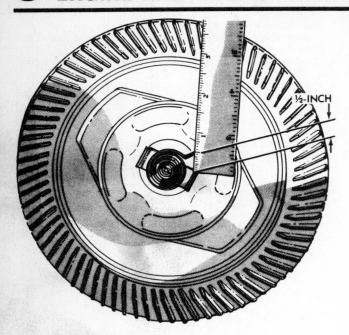

Testing the spring and shaft rotation

8. Place the fan or assembly and shroud into position in the engine compartment. Tighten the fan or assembly mounting bolts to 17 ft.lbs. (23 N.m). Tighten the shroud mounting screws to 95 in.lbs. (11 N.m).

9. Install and tension the drive belts. Connect the negative battery cable.

In case of engine overheating, inspect the thermal control fan drive for proper operation by observing the movement of the thermostatic spring coil and shaft. Lift the end of the thermostatic spring up and out of the retaining slot. (Fan assembly has been removed from van). Rotate the spring counterclockwise until a stop is felt. The gap between the end of the coil and the retaining clip should be approximately 12mm (0.5 ins.). Replace the unit if the shaft does not rotate with the coil. After testing install the end of the coil back into the retaining clip slot.

Water Pump

REMOVAL AND INSTALLATION

1. Disconnect the negative battery ground cable. Drain the cooling system. Remove the radiator. Although radiator removal is not absolutely necessary, in some cases, it will make the job of water pump replacement much easier.

--- CAUTION ---

When draining the coolant, keep in mind that cats and dogs are attracted by the ethylene glycol antifreeze, and are quite likely to drink any that is left in an uncovered container or in puddles on the ground. This will prove fatal in sufficient quantity. Always drain the coolant into a sealable container. Coolant should be reused unless it is contaminated or several years old.

2. Loosen all the accessories that are belt driven and remove the drive belts.

3. On engines without air conditioning, remove the alternator bracket attaching bolts and tie the alternator and bracket out of the way.

4. On engines with air conditioning, remove the front bracket that supports the alternator and air conditioner compressor. The A/C compressor will be supported by the back mounting bracket. Do not disconnect any A/C refrigerant lines. Remove the alternator adjusting bracket and A/C idler pulley.

5. Remove the fan blade, spacer (or fluid unit), pulley, and bolts as an assembly.

NOTE: To prevent silicone fluid from draining into the drive bearing and ruining the lubricant, do not place the thermostatic fan drive unit with the shaft pointing downward.

6. Remove the air pump.

7. Remove the power steering pump front bracket. Support the pump with mechanic's wire. Do not allow a strain to be put on any power steering hoses.

8. Disconnect all hoses (radiator and heater) from the water pump.

9. Remove the water pump mounting bolts and the water pump.

10. Clean all gasket mounting surfaces. Use a new gasket coated with sealer. Place the water pump into position and install the mounting bolts. Torque the bolts to 30 ft.lbs. (41 N.m).

11. Connect the hoses (radiator and heater) to the water pump.

12. Install the power steering pump bracket, the air pump, alternator and A/C compressor brackets. Install the idler pulley.

13. Install the cooling fan or assembly. Install and tension the drive belts.

14. Install the radiator and fill the cooling system. Start the engine, after connecting the negative battery cable, and check for coolant leaks.

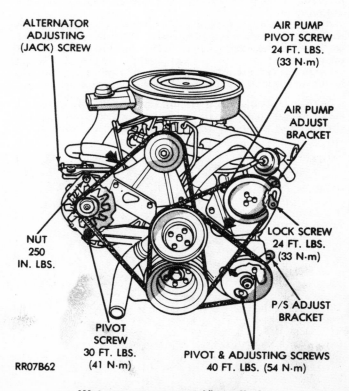

Water pump removal/installation

Cylinder Head
REMOVAL AND INSTALLATION

1. Discharge the fuel system pressure. Drain the cooling system and disconnect the negative battery ground cable. Remove the engine cover.

------ **CAUTION** ------

When draining the coolant, keep in mind that cats and dogs are attracted by the ethylene glycol antifreeze, and are quite likely to drink any that is left in an uncovered container or in puddles on the ground. This will prove fatal in sufficient quantity. Always drain the coolant into a sealable container. Coolant should be reused unless it is contaminated or several years old.

2. Remove the alternator, air cleaner, and fuel line.
3. Disconnect the accelerator linkage.
4. Remove the vacuum hose from the throttle body.
5. Remove the distributor cap and wires as an assembly.
6. Disconnect the coil wires, water temperature sending unit, heater hoses, and bypass hose.
7. Remove the closed ventilation system, the evaporative control system from the engine valve covers. Remove the valve covers.
8. Remove the intake manifold, ignition coil, and throttle body as an assembly.
9. Remove the exhaust manifold mounting bolts.
10. Remove the rocker and shaft assemblies.
11. Remove the pushrods and keep them in order to ensure installation in their original locations.
12. Remove the head bolts from each cylinder head and remove the cylinder heads. Remove the spark plugs.
13. Clean all the gasket surfaces of the engine block and the cylinder heads.
14. Inspect and check the cylinder head surface with a straight edge. If warpage is indicated, measure the amount. This amount must not exceed 0.00075 times the span length in any direction. For example, if a 12 in. span is 0.004 in. warped, the maximum allowable difference is $12 \times 0.00075 = 0.009$ in. In this case, the head is within limits. If the warpage exceeds the specified limits, either replace the head or lightly machine the head gasket surface.
15. Service the cylinder heads and valve system as required. Check the instructions that came with the gasket set you are using. Some head gaskets do not require the use of a sealer for installation. If required, coat the cylinder block and head mounting surfaces with sealer. Place the gaskets in position on the cylinder block and install the cylinder heads.
16. Install the cylinder head bolts. Tighten the bolts in two steps. Torque the cylinder head bolts in the proper sequence (working for the center of the head outward). Tighten the head bolts to 50 ft.lbs (68 N.m). Repeat the procedure and tighten to 105 ft.lbs. (143 N.m).
17. Inspect the pushrods. Replace any that are bent or worn.
18. Install the pushrods into the original position they were removed from.
19. Install the rocker and shaft assemblies.
20. Install the exhaust manifolds.
21. Install the intake manifold, ignition coil, and throttle body as an assembly.
22. Install the valve covers. Install the closed ventilation system and evaporative control system.
23. Connect the coil wires, water temperature sending unit, heater hoses, and bypass hose.
24. Install the distributor cap and wires as an assembly.
25. Install the vacuum hose to the throttle body.
26. Connect the accelerator linkage.
27. Install the alternator, air cleaner, and fuel line.
28. Fill the cooling system and connect the battery ground cable.

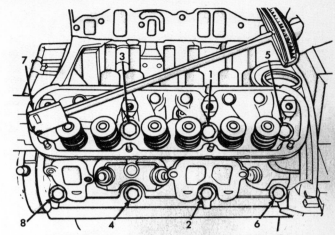

V6 cylinder head bolt tightening sequence

V8 cylinder head bolt tightening sequence

CLEANING AND INSPECTION

1. With the valves installed to protect the valve seats, remove deposits from the combustion chambers and valve heads with a scraper and a wire brush. Be careful not to damage the cylinder head gasket surface. After the valves are removed, clean the valve guide bores with a valve guide cleaning tool. Using cleaning solvent to remove dirt, grease and other deposits, clean all bolts holes; be sure the oil passage is clean (V8 engines).
2. Remove all deposits from the valves with a fine wire brush or buffing wheel.
3. Inspect the cylinder heads for cracks or excessively burned areas in the exhaust outlet ports.
4. Check the cylinder head for cracks and inspect the gasket surface for burrs and nicks. Replace the head if it is cracked.
5. On cylinder heads that incorporate valve seat inserts, check the inserts for excessive wear, cracks, or looseness.

RESURFACING

Cylinder Head Flatness

When the cylinder head is removed, check the flatness of the cylinder head gasket surfaces.

1. Place a straightedge across the gasket surface of the cylinder head. Using feeler gauges, determine the clearance at the center of the straightedge.
2. If warpage exceeds 0.003 in. in a 6 in. span, the cylinder head must be resurfaced.
3. If necessary to refinish the cylinder head gasket surface, do not plane or grind off more than 0.010 in. from the original gasket surface.

NOTE: When milling the cylinder heads of V6 and V8 engines, the intake manifold mounting position is altered, and must be corrected by milling the manifold flange a proportionate amount. Consult an experienced machinist about this.

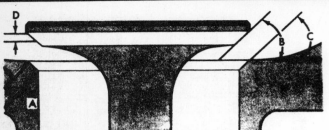

A-SEAT WIDTH (INTAKE 1/16 (1.587 mm) TO 3/32 (2.381 mm) INCH
EXHAUST: (.080-.100 INCH) – (2-2.5 mm)
B-FACE ANGLE (INTAKE & EXHAUST: 44½°-45°)
C-SEAT ANGLE (INTAKE & EXHAUST: 45°-45½°)
D-CONTACT SURFACE

Valve face and seat angles

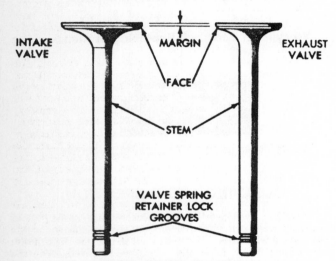

Intake and exhaust valves

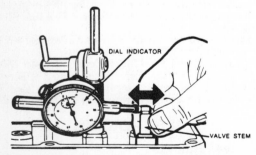

Check the valve stem to guide clearance

Reaming a valve seat with a hand reamer

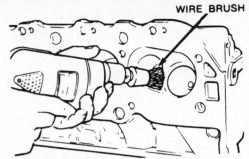

Remove the carbon from the cylinder head with a wire brush and electric drill

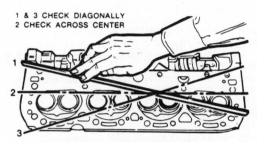

Check the cylinder head for warpage

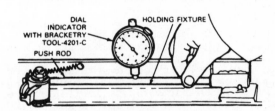

Checking pushrod runout

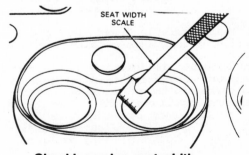

Checking valve seat width

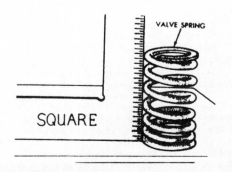

Checking the valve spring for squareness

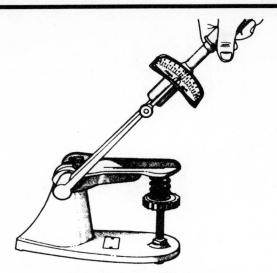

Testing valve spring pressure

Valves

REMOVAL AND INSTALLATION

1. Block the head on its side, or install a pair of head-holding brackets made especially for valve removal.

2. Use a socket slightly larger than the valve stem and keepers, place the socket over the valve stem and gently hit the socket with a plastic hammer to break loose any varnish buildup.

3. Remove the valve keepers, retainer, spring shield and valve spring using a valve spring compressor (the locking C-clamp type is the easiest kind to use).

4. Put the parts in a separate container numbered for the cylinder being worked on; do not mix them with other parts removed.

5. Remove and discard the valve stem oil seals. A new seal will be used at assembly time.

6. Remove the valves from the cylinder head and place them, in order, through numbered holes punched in a stiff piece of cardboard or wood valve holding stick.

NOTE: The exhaust valve stems, on some engines, are equipped with small metal caps. Take care not to lose the caps. Make sure to reinstall them at assembly time. Replace any caps that are worn.

7. Use an electric drill and rotary wire brush to clean the intake and exhaust valve ports, combustion chamber and valve seats. In some cases, the carbon will need to be chipped away. Use a blunt pointed drift for carbon chipping. Be careful around the valve seat areas.

8. Use a wire valve guide cleaning brush and safe solvent to clean the valve guides.

9. Clean the valves with a revolving wires brush. Heavy carbon deposits may be removed with the blunt drift.

NOTE: When using a wire brush to clean carbon on the valve ports, valves etc., be sure that the deposits are actually removed, rather than burnished.

10. Wash and clean all valve springs, keepers, retaining caps etc., in safe solvent.

11. Clean the head with a brush and some safe solvent and wipe dry.

12. Check the head for cracks. Cracks in the cylinder head usually start around an exhaust valve seat because it is the hottest part of the combustion chamber. If a crack is suspected but cannot be detected visually have the area checked with dye penetrant or other method by the machine shop.

13. After all cylinder head parts are reasonably clean, check the valve stem-to-guide clearance. If a dial indicator is not on hand, a visual inspection can give you a fairly good idea if the guide, valve stem or both are worn.

14. Insert the valve into the guide until slight away from the valve seat. Wiggle the valve sideways. A small amount of wobble is normal, excessive wobble means a worn guide or valve stem. If a dial indicator is on hand, mount the indicator so that the stem of the valve is at 90° to the valve stem, as close to the valve guide as possible. Move the valve off the seat, and measure the valve guide-to-stem clearance by rocking the stem back and forth to actuate the dial indicator. Measure the valve stem using a micrometer and compare to specifications to determine whether stem or guide wear is causing excessive clearance.

15. The valve guide, if worn, must be repaired before the valve seats can be resurfaced. Chrysler supplies valves with oversize stems to fit valve guides that are reamed to oversize for repair. The machine shop will be able to handle the guide reaming for you. In some cases, if the guide is not too badly worn, knurling may be all that is required.

16. Reface, or have the valves and valve seats refaced. The valve seats should be a true 45° angle. Remove only enough material to clean up any pits or grooves. Be sure the valve seat is not too wide or narrow. Use a 60° grinding wheel to remove material from the bottom of the seat for raising and a 30° grinding wheel to remove material from the top of the seat to narrow.

17. After the valves are refaced by machine, hand lap them to the valve seat. Clean the grinding compound off and check the position of face-to-seat contact. Contact should be close to the center of the valve face. If contact is close to the top edge of the valve, narrow the seat; if too close to the bottom edge, raise the seat.

18. Valves should be refaced to a true angle of 45°. Remove only enough metal to clean up the valve face or to correct runout. If the edge of a valve head, after machining, is $\frac{1}{32}$ in. (0.8mm) or less replace the valve. The tip of the valve stem should also be dressed on the valve grinding machine, however, do not remove more than 0.010 in. (0.254mm).

19. After all valve and valve seats have been machined, check the remaining valve train parts (springs, retainers, keepers, etc.) for wear. Check the valve springs for straightness and tension.

20. Install the valves in the cylinder head and metal caps.

21. Install new valve stem oil seals.

22. Install the valve keepers, retainer, spring shield and valve spring using a valve spring compressor (the locking C-clamp type is the easiest kind to use).

23. Check the valve spring installed height, shim or replace as necessary.

Valve Springs

REMOVAL AND INSTALLATION

Refer to the previous procedure concerning servicing the valves for instructions on valve spring removal and installation.

INSPECTION

Place the valve spring on a flat surface next to a carpenter's square. Measure the height of the spring, and rotate the spring against the edge of the square to measure distortion. If the spring height varies (by comparison) by more than $\frac{1}{16}$ in. (1.6mm) or if the distortion exceeds $\frac{1}{16}$ in. (1.6mm), replace the spring.

Have the valve springs tested for spring pressure at the installed and compressed (installed height minus valve lift) height using a valve spring tester. Springs should be within one pound, plus or minus each other. Replace springs as necessary.

VALVE SPRING INSTALLED HEIGHT

After installing the valve spring, measure the distance between the spring mounting pad and the lower edge of the spring retainer. Compare the measurement to specifications. If the installed height is incorrect, add shim washers between the spring mounting pad and the spring. Use only washers designed for valve springs, available at most parts houses.

Valve Seats

If the valve seat is damaged or burnt and cannot be serviced by refacing, it may be possible to have the seat machined and an insert installed. Consult an automotive machine shop for their advice.

Valve Guides

Worn valve guides can, in most cases, be reamed to accept a valve with an oversized stem. Valve guides that are not excessively worn or distorted may, in some cases, be knurled rather than reamed. However, if the valve stem is worn reaming for an oversized valve stem is one answer to the problem. Another solution to the problem of worn guides is having the guide reamed to accept an insert bushing. The bushing will permit the use of a standard valve.

Knurling is a process in which metal is displaced and raised, thereby reducing clearance. Knurling also produces excellent oil control. The possibility of knurling instead of reaming the valve guides should be discussed with a machinist.

Valve Stem Oil Seal

REPLACEMENT WITH THE CYLINDER HEAD INSTALLED

If valve stem oil seals are found to be the cause of excessive oil consumption, they may be replaced without removing the cylinder block.

1. Remove the air cleaner.
2. Remove rocker arm covers and spark plugs.
3. Detach the coil wire from the distributor.
4. Turn the engine so that No. 1 cylinder is at Top Dead Center on the compression stroke. Both Valves for No. 1 cylinder should be fully closed and the crankshaft damper timing mark at TDC. The distributor rotor will point at the No. 1 spark plug wire location in the cap.
5. Remove the rocker shaft and install a dummy shaft.
6. Apply 90-100 psi air pressure to No. 1 cylinder, using a spark plug hole air hose adaptor.
7. Use a valve spring compressor to compress each No. 1 cylinder valve spring and remove the retainer locks and the spring. Remove the old seals.
8. Install a cup shield on the exhaust valve stem. Position it down against the valve guide.
9. Push the intake valve stem seal firmly and squarely over the valve guide.
10. Compress the valve spring only enough to install the lock.
11. Repeat the operation on each successive cylinder in the firing order, making sure that the crankshaft is exactly on TDC for each cylinder. See the Firing Order and Distributor Rotation illustrations in the Specifications section of this section for cylinder numbering.
12. Replace the rocker arms, covers, spark plugs and coil wire.

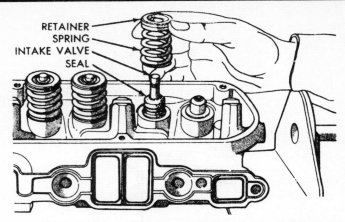

Installing the valve, cup seal, spring and retainer

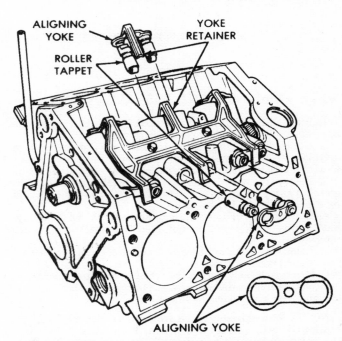

V6 roller tappets, aligning yoke and retainer

Valve Lifters

REMOVAL AND INSTALLATION

NOTE: Special tool C-4129 (V6) or C-4129A (V8), or the equivalent Hydraulic Lifter Puller is required.

V6 3.9L

The V6 3.9L engine is equipped with roller tappets (lifters). The system consists of the roller tappet assemblies, aligning yokes and yoke retainers. Roller tappet alignment is maintained by machined flats on the tappet body being fitted (in pairs) into six alignment yokes. The yokes are secured by an alignment yoke retainer.

1. Refer to the necessary procedures in the section. Remove the valve cover and rocker arm assemblies. Remove and identify the pushrods (they must be returned to their original installed positions).

2. Remove the tappet yoke retainer and aligning yokes. Slide the special tool through the opening in the cylinder head and seat the tool firmly into the head of the tappet.

3. Pull the tappet out of the mounting bore using a twisting motion. Keep the tappets in order, they must be installed in the same hole they were removed from.

4. Service as required. Lubricate the tappets and install them in their original positions.

5. Install the alignment yokes with the marking arrow toward the camshaft. Install the yoke retainer. Tighten the mounting bolts to 200 in.lbs. (23 N.m).

6. Install the intake manifold, pushrods, rocker arm assemblies and valve covers.

V8 5.2L & 5.9L

1. Refer to the necessary procedures in the section. Remove the valve cover and rocker arm assemblies. Remove and identify the pushrods (they must be returned to their original installed positions).

2. Slide the special tool through the opening in the cylinder head and seat the tool firmly into the head of the tappet.

3. Pull the tappet out of the mounting bore using a twisting motion. Keep the tappets in order, they must be installed in the same hole they were removed from.

4. Service as required. Lubricate the tappets and install them in their original positions.

5. Install the intake manifold, pushrods, rocker arm assemblies and valve covers.

Oil Pan

REMOVAL AND INSTALLATION

1. Disconnect the battery ground cable.
2. Remove the oil dipstick.
3. Raise and support the front end on jackstands.
4. Drain the oil.

CAUTION

The EPA warns that prolonged contact with used engine oil may cause a number of skin disorders, including cancer! You should make every effort to minimize your exposure to used engine oil. Protective gloves should be worn when changing the oil. Wash your hands and any other exposed skin areas as soon as possible after exposure to used engine oil. Soap and water, or waterless hand cleaner should be used.

---

5. Remove the exhaust cross-over pipe.
6. Remove the left engine-to-transmission strut.
7. Remove the bolts and lower the oil pan.
8. Thoroughly clean the gasket mating surfaces.
9. When installing the pan, always use new gaskets coated with sealer. Notches on the gaskets (if equipped) are at the rear of the engine, pointing outward. Apply a drop of RTV silicone sealer where the cork and rubber gaskets meet. Torque the oil pan bolts to 200 in.lbs. (23 N.m). Tighten the cross-over pipe to 24 ft.lbs. (33 N.m).

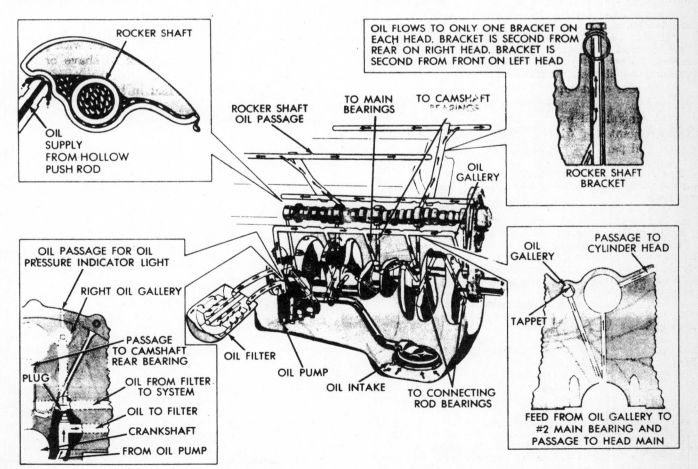

Engine oiling system

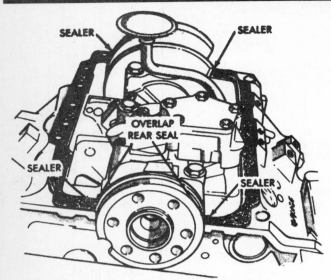

V6 oil pan gasket and seal installation

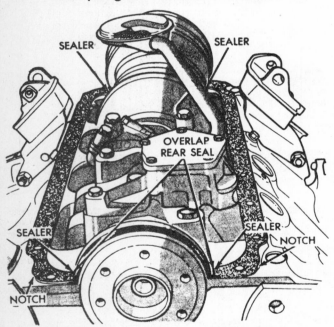

V8 oil pan gasket and seal installation

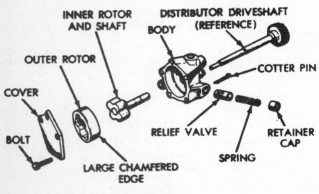

Oil pump

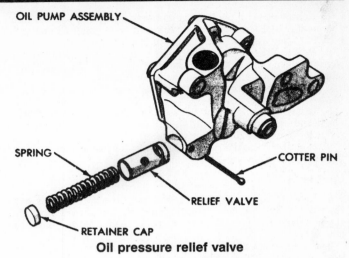

Oil pressure relief valve

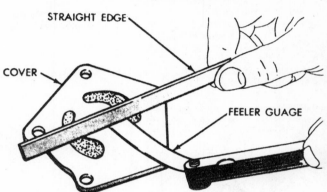

Checking the oil pump cover flatness

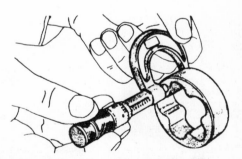

Measuring outer rotor thickness

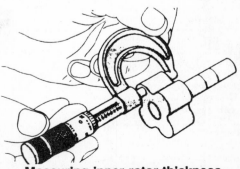

Measuring inner rotor thickness

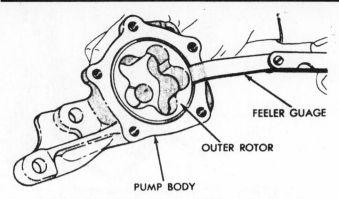

Measuring outer rotor clearance in housing

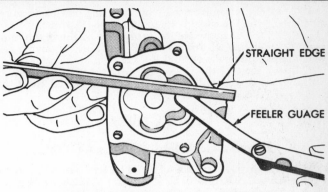

Measuring clearance over rotors

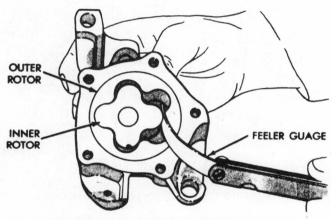

Measuring clearance between rotors

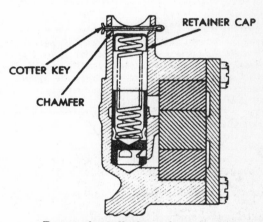

Proper installation of retainer cap

Oil Pump

REMOVAL AND INSTALLATION

NOTE: It is necessary to remove the oil pan, and to remove the oil pump from the rear main bearing cap to service the oil pump.

1. Drain the engine oil and remove the oil pan.

CAUTION

The EPA warns that prolonged contact with used engine oil may cause a number of skin disorders, including cancer! You should make every effort to minimize your exposure to used engine oil. Protective gloves should be worn when changing the oil. Wash your hands and any other exposed skin areas as soon as possible after exposure to used engine oil. Soap and water, or waterless hand cleaner should be used.

2. Remove the oil pump mounting bolts and remove the oil pump from the rear main bearing cap.
3. To remove the relief valve, drill a ⅛ in. hole into the relief valve retainer cap and insert a self-threading sheet metal screw into the cap. Clamp the screw into a vise and while supporting the oil pump, clamp the cap by tapping the pump body using a soft hammer. Discard the retainer cap and remove the spring and the relief valve.
4. Remove the oil pump cover and lockwashers, and lift off the cover. Discard the oil ring seal. Remove the pump rotor and shaft, and lift out the outer rotor.

NOTE: Wash all parts in solvent and inspect for damage or wear. The mating surfaces of the oil pump cover should be smooth. Replace the pump assembly if this is not the case.

5. Lay a straight edge across the pump cover surface and if a 0.0015 in. feeler gauge can be inserted between the cover and the straight edge, the pump assembly should be replaced. Measure the thickness and the diameter of the outer rotor. If the outer rotor thickness measures 0.825 in. or less, or if the diameter is 2.469 in. or less, replace the outer rotor. If the inner rotor measures 0.825 in. or less, then the inner rotor and shaft assembly must be replaced.
6. Slide the outer rotor into the pump body, do this by pressing it to one side with your fingers and measure the clearance between the rotor and the pump body. If the measurement is 0.014 in. or more, replace the oil pump assembly. Install the inner rotor and shaft into the pump body. If the clearance between the inner and outer rotors is 0.008 in. or more, replace the shaft and both rotors.
7. Place a straightedge across the face of the pump, between the bolt holes. If a feeler gauge of 0.004 in. or more can be inserted between the rotors and the straight edge, replace the pump assembly.
8. Inspect the oil pressure relief valve plunger for scoring and free operation in its bore. Small marks may be removed with 400-grit wet or dry sandpaper.
9. The relief valve spring has a free length of 1.95 in. and should test between 19.5 and 20.5 lbs. when compressed to 1¹¹⁄₃₂ in. Replace the spring if it fails to meet this specification.

10. To install, assemble the oil pump, using new parts as required. Tighten the cover bolts to 95 in.lbs. (11 N.m).

11. Prime the oil pump before installation by filling the rotor cavity with engine oil. Install the oil pump on the engine and tighten attaching bolts to 30 ft.lbs. (41 N.m).

12. Install the oil pan.

13. Fill the engine with the proper grade motor oil. Start the engine and check for leaks.

Front Cover

REMOVAL AND INSTALLATION

NOTE: A Crankshaft Vibration Damper Puller/Installer Kit, Tool C-3688 and C-3732A, or equivalent is required.

1. Discharge the fuel system pressure. Disconnect the battery and drain the cooling system. Remove the radiator. Remove the engine cover.

--- CAUTION ---

When draining the coolant, keep in mind that cats and dogs are attracted by the ethylene glycol antifreeze, and are quite likely to drink any that is left in an uncovered container or in puddles on the ground. This will prove fatal in sufficient quantity. Always drain the coolant into a sealable container. Coolant should be reused unless it is contaminated or several years old.

2. Remove the water pump.

3. Remove the power steering pump.

4. Remove the crankshaft pulley.

5. Remove the vibration damper with a puller. Remove the fuel lines. Loosen the oil pan bolts and remove the front bolt on each side.

6. Remove the timing gear cover and the crankshaft oil slinger.

7. Clean all gasket surfaces thoroughly.

8. Install a new seal in the case.

9. Install the timing case cover with a new gasket. A ⅛ in. bead of RTV sealer is recommended on the oil pan gasket. Install, but do not tighten the cover mounting bolts. The crankshaft damper will act as an alignment tool when it is installed.

10. Lubricate the seal lip with lithium white grease and slide the damper back into position. Using tool C-3688, or equivalent, press the damper onto the shaft. Torque the bolt to 135 ft.lbs. (183 N.m).

11. Torque the cover bolts to 30 ft.lbs. (41 N.m). Retighten the engine oil pan to 200 in.lbs. (23 N.m). Install the fuel pump and lines.

12. Install the water pump.

13. Install the power steering pump.

14. Replace the radiator and hoses.

15. Tighten all drive belts. Fill the cooling system and connect the battery cable.

Front Cover Seal

REPLACEMENT

NOTE: With the timing case still mounted on the engine, a seal remover and installer tool is required to prevent seal damage. With the timing case unmounted from the engine the seal can be carefully press into the case. Take care not to damage the sealing lips.

1. Refer to the previous procedures. Remove the crankshaft damper. Using a seal puller, remove the seal from the timing cover.

2. Using a seal puller, remove the seal from the timing cover. Clean the case mounting groove.

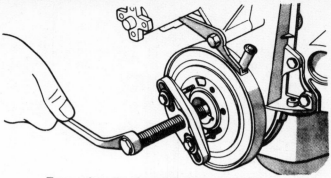

Removing the front damper assembly

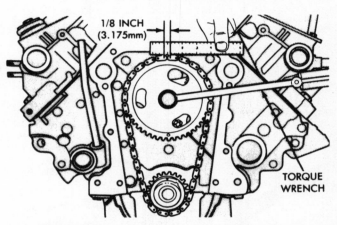

Measuring timing chain wear and stretch

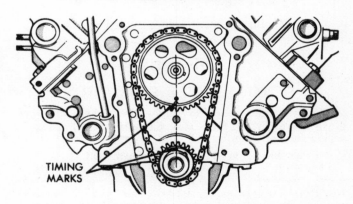

Alignment of timing marks

3. Place the seal into position on the timing case. Using the seal installation tool press the seal into the case.

4. Seat the seal tightly against the cover face. There should be a maximum clearance of 0.0014 in. between the seal and the cover. Be careful not to over-compress the seal.

Timing Chain

REMOVAL AND INSTALLATION

1. Rotate the engine until No. 1 cylinder is at TDC (top dead center) on its compression stroke. Refer to the previous procedures. Remove the front cover.

2. Remove the camshaft sprocket attaching bolt and cup

washer. Remove the timing chain with both the cam and crankshaft sprockets.

3. To begin the installation procedure, place the camshaft and crankshaft sprockets on a flat surface with the timing indicator marks on an imaginary centerline through both sprocket bosses. Place the timing chain around both sprockets. Be sure that the timing marks are in alignment.

4. If necessary, turn the crankshaft and camshaft to align them with the keyway location in the crankshaft sprocket and the keyway or dowel hole in the camshaft sprocket.

5. Lift the sprockets and timing chain while keeping the sprockets tight against the chain in the correct position. Slide both sprockets evenly onto their respective shafts.

6. Use a straightedge to measure the alignment of the sprocket timing marks. They must be perfectly aligned.

7. Install the cup washer and camshaft sprocket bolt. Torque to 35 ft.lbs. (47 N.m). If camshaft end play exceeds 0.010 in., install a new thrust plate. It should be 0.002-0.006 in. with the new plate.

CHECKING TIMING CHAIN SLACK

1. Position a scale (ruler or straightedge) next to the timing chain to detect any movement in the chain.

2. Place a torque wrench and socket on the camshaft sprocket attaching bolt. Apply either 30 ft.lbs. (if the cylinder heads are installed on the engine) or 15 ft. lbs. (cylinder heads removed) of force to the bolt and rotate the bolt in the direction of crankshaft rotation in order to remove all slack from the chain.

3. While applying torque to the camshaft sprocket bolt, the crankshaft should not be allowed to rotate. It may be necessary to block the crankshaft to prevent rotation.

4. Position the scale over the edge of a timing chain link and apply an equal amount of torque in the opposite direction. If the movement of the chain exceeds ⅛ in., replace the chain.

VALVE TIMING

1. Turn the crankshaft until the No.6 exhaust valve is closing and the intake valve is opening.

2. Insert a ¼ in. spacer between the rocker arm pad and stem tip of No.1 intake valve. Allow the spring load to bleed the tappet down.

3. Install a dial indicator so that the plunger contacts the valve spring retainer and nearly perpendicular as possible. Zero the indicator.

4. Rotate the crankshaft clockwise until the valve has lifted 0.010 in. on the 3.9L and 5.2L; 0.034 in. on the 5.9L.

WARNING: Do not rotate the crankshaft any further, as serious damage will result!

5. The ignition timing marks should now read anywhere from 10°BTDC to 2°ATDC.

6. If the reading is not within specifications, check for timing mark alignment, timing chain wear and the accuracy of the ignition timing indicator.

Camshaft

REMOVAL AND INSTALLATION

1. Engine has been removed from the van. Remove the intake manifold, cylinder head covers, rocker arm assemblies, push rods, and valve tappets, keeping them in order to insure the installation in their original locations.

2. Remove the timing gear cover, the camshaft and the crankshaft sprockets, and the timing chain.

3. Remove the distributor and lift out the oil pump and distributor driveshaft.

4. Remove the camshaft thrust plate (on 318 and 360).

5. Install a long bolt into the front of the camshaft and remove the camshaft, being careful not to damage the cam bearings with the cam lobes.

6. Prior to installation, lubricate the camshaft lobes and bearings journals. It is recommended that 1 pt. of Crankcase Conditioner be added to the initial crankcase oil fill. Insert the camshaft into the engine block within 2 in. of its final position in the block.

7. Have an assistant support the camshaft with a suitable tool to prevent the camshaft from contacting the plug in the rear of the engine block. Position the suitable tool against the rear side of the cam gear and be careful not to damage the cam lobes.

8. Replace the camshaft thrust plate. If camshaft end play exceeds 0.010 in., install a new thrust plate. It should be 0.002-0.006 in. with the new plate.

9. Install the timing chain and sprockets, timing gear cover, and pulley.

10. Install the tappets, pushrods, rocker arms, and cylinder head covers. Install fuel pump, if removed.

11. Install the distributor and oil pump driveshaft, Install the distributor.

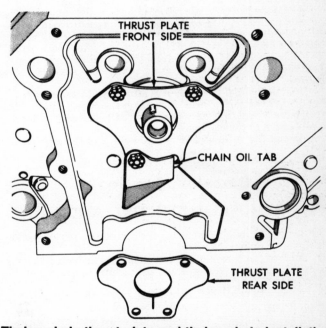

Timing chain thrust plate and timing chain installation

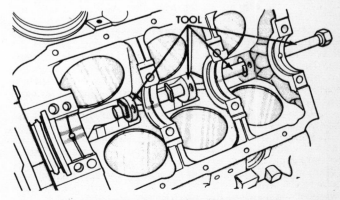

Removal/installation of cam bearings

CHECKING CAMSHAFT

Camshaft Lobe Lift

Check the lift of each lobe in consecutive order and make a note of the reading.

1. Remove the fresh air inlet tube and the air cleaner. Remove the heater hose and crankcase ventilation hoses. Remove valve rocker arm cover(s).
2. Remove the rocker arm stud nut or fulcrum bolts, fulcrum seat and rocker arm.
3. Make sure the pushrod is in the valve tappet socket. Install a dial indicator so that the actuating point of the indicator is in the push rod socket (or the indicator ball socket adaptor is on the end of the push rod) and in the same plane as the push rod movement.
4. Disconnect the I terminal and the S terminal at the starter relay. Install an auxiliary starter switch between the battery and S terminals of the start relay. Crank the engine with the ignition switch off. Turn the crankshaft over until the tappet is on the base circle of the camshaft lobe. At this position, the push rod will be in its lowest position.
5. Zero the dial indicator. Continue to rotate the crankshaft slowly until the push rod is in the fully raised position.
6. Compare the total lift recorded on the dial indicator with the specification shown on the Camshaft Specification chart.

To check the accuracy of the original indicator reading, continue to rotate the crankshaft until the indicator reads zero. If the left on any lobe is below specified wear limits listed, the camshaft and the valve tappet operating on the worn lobe(s) must be replaced.

7. Install the dial indicator and auxiliary starter switch.
8. Install the rocker arm, fulcrum seat and stud nut or fulcrum bolts. Check the valve clearance. Adjust if required (refer to procedure in this section).
9. Install the valve rocker arm cover(s) and the air cleaner.

Camshaft Bearings

REMOVAL AND INSTALLATION

1. Remove the engine following the procedures in this section and install it on a work stand.
2. Remove the camshaft, flywheel and crankshaft, following the appropriate procedures. Push the pistons to the top of the cylinder.
3. Remove the camshaft rear bearing bore plug. Remove the camshaft bearings with a bearing removal tool.
4. Select the proper size expanding collet and back-up nut and assemble on the mandrel. With the expanding collet collapsed, install the collet assembly in the camshaft bearing and tighten the back-up nut on the expanding mandrel until the collet fits the camshaft bearing.
5. Assemble the puller screw and extension (if necessary) and install on the expanding mandrel. Wrap a cloth around the threads of the puller screw to protect the front bearing or journal. Tighten the pulling nut against the thrust bearing and pulling plate to remove the camshaft bearing. Be sure to hold a wrench on the end of the puller screw to prevent it from turning.
6. To remove the front bearing, install the puller from the rear of the cylinder block.
7. Position the new bearings at the bearing bores, and press them in place. Be sure to center the pulling plate and puller screw to avoid damage to the bearing. Failure to use the correct expanding collet can cause severe bearing damage. Align the oil holes in the bearings with the oil holes in the cylinder block before pressing bearings into place.
8. Install the camshaft rear bearing bore plug.

9. Install the camshaft, crankshaft, flywheel and related parts, following the appropriate procedures.
10. Install the engine.

Pistons and Connecting Rods

REMOVAL AND INSTALLATION

1. Remove the engine from the vehicle. Remove the cylinder heads.
2. Remove the timing chain/gears.
3. Remove the oil pan.
4. Pistons should be removed following the firing order of the engine. Turn the crankshaft until the piston to be removed is at the bottom of its stroke.
5. Place a cloth on the head of the piston to be removed and, using a ridge reamer, remove the ridge and deposits from the upper end of the cylinder bore.

WARNING: Never remove more than $\frac{1}{32}$ in. from the ring travel area when removing the ridges!

6. Mark all connecting rod bearing caps so that they may be returned to their original locations in the engine. The connecting rod caps are usually marked. The marks must be matched when re-assembling the engine. Mark all pistons so they can be returned to their original cylinders. Remove the rod cap nuts and the cap. Push the piston and rod assembly up and out of the cylinder bore. Install the rod cap and nuts. Proceed to the next cylinder.

NOTE: After removing the connecting rod cap and bearing, place a short length of rubber hose over the rod bolts to prevent cylinder wall and crank journal scoring when removing or installing the piston and rod assembly.

7. After all the piston assemblies have been removed. Using an internal micrometer, measure the bores across the thrust faces of the cylinder and parallel to the axis of the crankshaft at a minimum of four equally spaced locations. The bore must not be out-of-round by more than 0.005 in. and it must not taper more than 0.010 in. Taper is the difference in wear between two bore measurements in any cylinder.
8. If the cylinder bore is in satisfactory condition, place each ring in the bore in turn and square it in the bore with the head of the piston. Measure the ring gap. If the ring gap is greater than the limit, get a new ring. If the ring gap is less than the limit, file the end of the ring to obtain the correct gap.
9. Check the ring side clearance by installing rings on the piston, and inserting a feeler gauge of the correct dimension between the ring and the lower land. The gauge should slide freely around the ring circumference without binding. Any wear will form a step on the lower land. Remove any pistons having high steps. Before checking the ring side clearance, be sure that the ring grooves are clean and free of carbon, sludge, or grit.
10. Piston rings should be installed so that their ends are at three equal spacings. Avoid installing rings with their ends in line with the piston pin bosses and the thrust direction.
11. Install the pistons in their original bores, if you are reusing the same pistons. A notch or v mark indicate the front of the piston. The ID mark should face toward the front of the engine when the piston is installed. Install short lengths of rubber hose over the connecting rod bolts to prevent damage to the cylinder walls or rod journal.
12. Turn the crank so that the rod journal is in the lowest position. Install a connecting rod bearing in position on the connecting rod. Lubricate the bearing. Install a ring compressor over the rings on the piston. Lower the piston and rod assembly into the bore until the ring compressor contacts the block. Using a wooden hammer handle, push the rod into the bore while guiding the rod onto the journal. Install the rod cap with a lubricated

bearing insert in position. Torque the rod nuts to required specification. Proceed to the next cylinder.

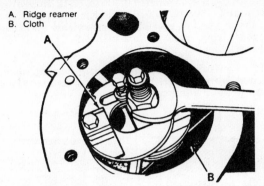

A. Ridge reamer
B. Cloth

Removing the ridge at the top of the cylinder

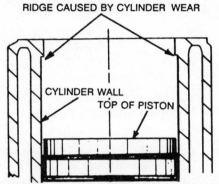

RIDGE CAUSED BY CYLINDER WEAR
CYLINDER WALL
TOP OF PISTON

Ridge formed by the pistons at the top of their travel

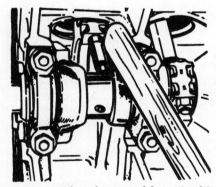

Push the piston and rod assembly out with a hammer handle

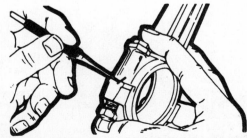

Match the connecting rods to their caps with a scribe mark

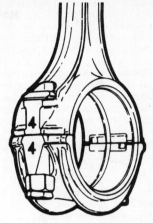

Mark the rod and cap for location position

CLEANING AND INSPECTION

All of the engines utilize pressed-in wrist pins, which can only be removed by an arbor press with proper fixtures attached.

A piston ring expander is necessary for removing the piston rings without damaging them; any other method (screwdriver blades, pliers, etc.) usually results in the rings being bent, scratched or distorted, or the piston itself being damaged. When the rings are removed, clean the ring grooves using an appropriate ring groove cleaning tool, using care not to cut too deeply. Thoroughly clean all carbon and varnish from the piston with solvent.

WARNING: Do not use a wire brush or caustic solvent (acids, etc.) on pistons.

Inspect the pistons for scuffing, scoring, cracks, pitting, or excessive ring groove wear. If these are evident, the piston must be replaced.

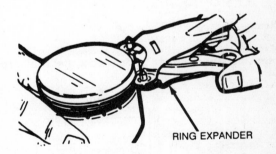

RING EXPANDER

Removing/installing the piston rings

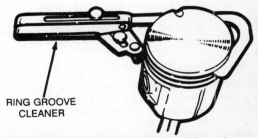

RING GROOVE
CLEANER

Cleaning the piston ring grooves with a ring groove cleaning tool

The piston should also be checked in relation to the cylinder diameter. Using a telescoping gauge and micrometer, or a dial gauge, measure the cylinder bore diameter perpendicular (90%) to the piston pin, 2½ in. (64mm) below the cylinder block deck (surface where the block mates with the heads). Then, with the micrometer, measure the piston, perpendicular to its wrist pin on the skirt. the difference between the two measurements is the piston clearance. If the clearance is within specifications or slightly below (after the cylinders have been bored or honed), finish honing is all that is necessary. If the clearance is excessive, try to obtain a slightly larger piston to bring clearance to within specifications. If this is not possible, obtain the first oversize piston and machine hone/bore the cylinder to size. Generally, if the cylinder bore is tapered 0.005 in. (0.127mm) or more or is out-of-round 0.003 in. (0.076mm) or more, it is advisable to rebore for the smallest possible oversize piston and rings.

After measuring, mark pistons with a felt tip pen for reference and for assembly.

NOTE: Cylinder machine honing/boring should be performed by a reputable, automotive machine shop with the proper equipment.

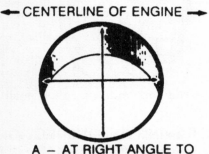

A – AT RIGHT ANGLE TO
CENTERLINE OF ENGINE
B – PARALLEL TO
CENTERLINE OF ENGINE

Cylinder bore measuring points

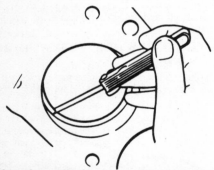

Measuring the cylinder bore with a dial indicator

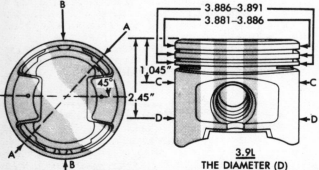

ELLIPTICAL SHAPE OF THE PISTON SKIRT SHOULD BE .010 (.254mm) TO .012 (.304mm) IN. LESS AT DIAMETER (A) THAN ACROSS THE THRUST FACES AT DIAMETER (B)

3.9L
THE DIAMETER (D) SHOULD BE .000 TO .0006 (.0152mm) IN. LARGER THAN (C)

3.9L piston measurements

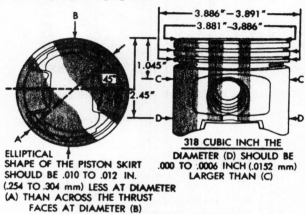

ELLIPTICAL SHAPE OF THE PISTON SKIRT SHOULD BE .010 TO .012 IN. (.254 TO .304 mm) LESS AT DIAMETER (A) THAN ACROSS THE THRUST FACES AT DIAMETER (B)

318 CUBIC INCH THE
DIAMETER (D) SHOULD BE .000 TO .0006 INCH (.0152 mm) LARGER THAN (C)

5.2L piston measuring points

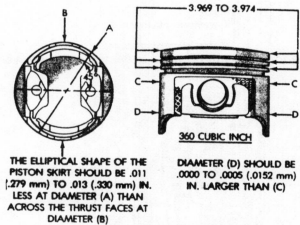

THE ELLIPTICAL SHAPE OF THE PISTON SKIRT SHOULD BE .011 (.279 mm) TO .013 (.330 mm) IN. LESS AT DIAMETER (A) THAN ACROSS THE THRUST FACES AT DIAMETER (B)

360 CUBIC INCH
DIAMETER (D) SHOULD BE .0000 TO .0005 (.0152 mm) IN. LARGER THAN (C)

5.9L piston measuring points

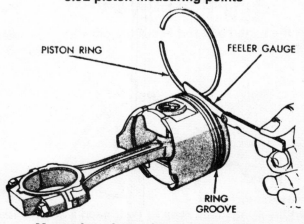

PISTON RING

FEELER GAUGE

RING GROOVE

Measuring piston ring side clearance

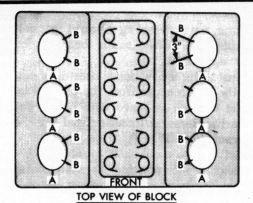

TOP VIEW OF BLOCK

A-EXPANDER GAPS   B-RAIL GAPS

IF YOU HAVE FOLLOWED THE
INSTRUCTIONS, THE RING WILL BE
IN THIS POSITION ON THE PISTON.

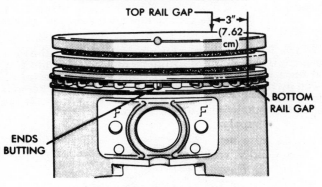

Proper oil ring installation

Measuring The Old Pistons

Check used piston-to-cylinder bore clearance as follows:
1. Measure the cylinder bore diameter with a telescope gauge.
2. Measure the piston diameter. When measuring the pistons for size or taper, measurements must be made with the piston pin removed.
3. Subtract the piston diameter from the cylinder bore diameter to determine piston-to-bore clearance.
4. Compare the piston-to-bore clearances obtained with those clearances recommended. Determine if the piston-to-bore clearance is in the acceptable range.
5. When measuring taper, the largest reading must be at the bottom of the skirt.

Selecting New Pistons

1. If the used piston is not acceptable, check the service piston size and determine if a new piston can be selected. (Service pistons are available in standard, high limit and standard oversize.
2. If the cylinder bore must be reconditioned, measure the new piston diameter, then machine hone/bore the cylinder bore to obtain the preferred clearance.
3. Select a new piston and mark the piston to identify the cylinder for which it was fitted.

Cylinder Honing

1. When cylinders are being honed to break the cylinder wall glaze for new piston ring installation, follow the vehicle manu-facturer's recommendations for the required cross hatch cylinder pattern.
2. Honing with a hand held hone should only be done to provide a cross-hatch pattern, not for increasing the bore size.
3. When finish-honing a cylinder bore, the hone should be moved up and down at a sufficient speed to obtain a very fine uniform surface finish in a cross-hatch pattern of approximately 50-60° included angle. The finish marks should be clean but not sharp, free from imbedded particles and torn or folded metal.
4. Thoroughly clean the bores with hot water and detergent. Scrub well with a stiff bristle brush and rinse thoroughly with hot water. It is extremely essential that a good cleaning operation be performed. If any of the abrasive material is allowed to remain in the cylinder bores, it will rapidly wear the new rings and cylinder bores. The bores should be swabbed several times with light engine oil and a clean cloth and then wiped with a clean dry cloth. CYLINDERS SHOULD NOT BE CLEANED WITH KEROSENE OR GASOLINE! Clean the remainder of the cylinder block to remove the excess material spread during the honing operation.

Piston Ring End Gap

Piston ring end gap should be checked while the rings are removed from the pistons. Incorrect end gap indicates that the wrong size rings are being used; ring breakage could occur.

Compress the piston rings to be used in a cylinder, one at a time, into that cylinder. Squirt clean oil into the cylinder, so that the rings and the top 2 in. (51mm) of cylinder wall are coated. Using an inverted piston, press the rings approximately 1 in. (25mm) below the deck of the block (on diesels, measure ring gap clearance with the ring positioned at the bottom of ring travel in the bore). Measure the ring end gap with the feeler gauge, and compare to the Ring Gap chart in this chapter. Carefully pull the ring out of the cylinder and file the ends squarely with a fine file to obtain the proper clearance.

Piston Ring Side Clearance Check

Check the pistons to see that the ring grooves and oil return holes have been properly cleaned. Slide a piston ring into its groove, and check the side clearance with a feeler gauge. Insert the gauge between the ring and its lower land (lower edge of the groove), because any wear that occurs forms a step at the inner portion of the lower land. If the piston grooves have worn to the extent that relatively high steps exist on the lower land, the piston should be replaced, because these will interfere with the operation of the new rings and ring clearance will be excessive. Piston rings are not furnished in oversize widths to compensate for ring groove wear.

Install the rings on the piston, lowest ring first, using a piston ring expander. There is a high risk of breaking or distorting the rings, or scratching the piston, if the rings are installed by hand or other means.

Position the rings on the piston as illustrated; spacing of the various piston ring gaps is crucial to proper oil retention and even cylinder wear. When installing new rings, refer to the installation diagram furnished with the new parts.

Connecting Rod Bearings

Connecting rod bearings consist of two halves or shells which are interchangeable in the rod and cap. when the shells are placed in position, the ends extend slightly beyond the rod and cap surfaces so that when the rod bolts are torqued the shells will be clamped tightly in place to insure positive seating and to prevent turning. A tang holds the shells in place.

NOTE: The ends of the bearing shells must never be filed flush with the mating surfaces of the rod and cap.

If a rod bearing becomes noisy or is worn so that its clearance on the crank journal is sloppy, a new bearing of the correct undersize must be selected and installed.

WARNING: Under no circumstances should the rod end or cap be filed to adjust the bearing clearance, nor should shims of any kind be used.

Inspect the rod bearings while the rod assemblies are out of the engine. If the shells are scored or show flaking, they should be replaced. If they are in good shape, check for proper clearance on the crank journal. Any scoring or ridges on the crank journal means the crankshaft must be reground and fitted with undersized bearings, or replaced.

Checking Rod Bearing Clearance

NOTE: Make sure connecting rods and their caps are kept together, and that the caps are installed in the proper direction.

Replacement bearings are available in standard size, and in undersizes for reground crankshaft. Connecting rod-to-crankshaft bearing clearance is checked using Plastigage® at either the top or bottom of each crank journal. the Plastigage® has a range of 0 to 0.003 in. (0.076mm).

1. Remove the rod cap with the bearing shell. Completely clean the bearing shell and the crank journal, and blow any oil from the oil hole in the crankshaft.

NOTE: The journal surfaces and bearing shells must be completely free of oil, because Plastigage® is soluble in oil.

2. Place a strip of Plastigage® lengthwise along the bottom

center of the lower bearing shell, then install the cap with shell and torque the bolt or nuts to specification. DO NOT TURN the crankshaft with the Plastigage® installed in the bearing.

3. Remove the bearing cap with the shell. The flattened Plastigage® will be found sticking to either the bearing shell or crank journal. Do not remove it yet.

4. Use the printed scale on the Plastigage® envelope to measure the flattened material at its widest point. The number within the scale which most closely corresponds to the width of the Plastigage® indicated bearing clearance in thousandths of an inch.

5. Check the specifications chart in this chapter for the desired clearance. It is advisable to install a new bearing if clearance exceeds 0.003 in. (0.076mm); however, if the bearing is in good condition and is not being checked because of bearing noise, bearing replacement is not necessary.

6. If you are installing new bearings, try a standard size, then each undersize in order until one is found that is within the specified limits when checked for clearance with Plastigage®. Each under size has its size stamped on it.

7. When the proper size shell is found, clean off the Plastigage® material from the shell, oil the bearing thoroughly, reinstall the cap with its shell and torque the rod bolt nuts to specification.

NOTE: With the proper bearing selected and the nuts torqued, it should be possible to move the connecting rod back and forth freely on the crank journal as allowed by the specified connecting rod end clearance. If the rod cannot be moved, either the rod bearing is too far undersize or the rod is misaligned.

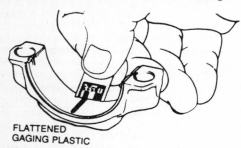

FLATTENED
GAGING PLASTIC

Checking rod bearing clearance with Plastigage®

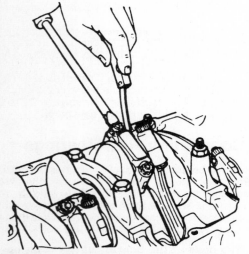

Checking the connecting rod side clearance. Use a small prybar to carefully spread the rods

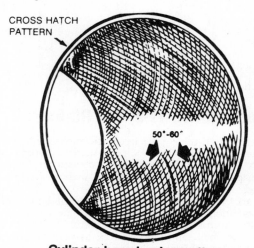

CROSS HATCH PATTERN

50°-60°

Cylinder bore honing pattern

ROD BEARING REPLACEMENT

WITH THE ENGINE OUT OF THE VAN: Install the connecting rod to the piston, the use of an arbor press and special fixtures are required. Make sure the piston installation notches and any marks on the rod are in proper relation to one another. Lubricate the wrist pin with clean engine oil and install the pin into the rod and piston assembly by using an arbor press as required. To install the piston and rod assemblies:

1. Make sure the connecting rod bearings (including end cap) are of the correct size and properly installed.

2. Fit rubber hoses over the connecting rod bolt to protect the crankshaft journals, as in the Piston Removal procedure. Coat the rod bearings with clean oil.

3. Using the proper ring compressor, insert the piston assembly into the cylinder so that the notch in the top of the piston

faces the front of the engine (this assumes that the dimple(s) or other markings on the connecting rods are in correct relation to the piston notch(s)).

4. From beneath the engine, coat each crank journal with clean oil. Pull the connecting rod, with the bearing shell in place, into position against the crank journal.

5. Remove the rubber hoses. Install the bearing cap and cap nuts and torque to specification.

6. Check the clearance between the sides of the connecting rods and the crankshaft using a feeler gauge. Spread the rods slightly with a screwdriver to insert the gauge. If clearance is below the minimum tolerance, the rod may be machined to provide adequate clearance. If clearance is excessive, substitute an unworn rod, and recheck. If clearance is still outside specifications, the crankshaft must be welded and reground, or replaced.

7. WITH THE ENGINE IN THE VAN: Remove the spark plugs. Remove the oil pan.

8. Turn the crank to its lowest position for the connecting rod bearing that is to be replaced.

9. Remove the rod cap retaining nuts and the cap. Slide the bearing shell from the cap. Push the rod assembly slowly and carefully up into the cylinder until enough clearance is gained to remove the bearing shell. Take care not to scratch the crankshaft journal with the connecting rod bolts.

WARNING: Do not push the rod and piston assembly up to far or the top ring make come out of the cylinder bore. Engine removal and disassembly will then be necessary.

10. Remove the upper bearing shell from the connecting rod. Clean the rod bearing mounting surface and crankshaft journal. Place the new shell in position with the locking tab fully mounted in the connecting rod notch.

11. Lubricate the crankshaft journal and pull the rod assembly back down over the crank journal.

12. Install the new bearing shell into the rod cap. Tab indexed in the notch provided. Install the rod cap, in the correct position with locking notches butted. Torque the retaining nuts to specification. Proceed to the next connecting rod.

Freeze Plugs
REMOVAL AND INSTALLATION

—————————— **CAUTION** ——————————

Removing the block heater of freeze plug may cause personal injury if the engine is not completely cooled down. Even after the radiator has been drained, there will engine coolant still in the block. Use care when removing the assembly or freeze plug. When draining the coolant, keep in mind that cats and dogs are attracted by the ethylene glycol antifreeze, and are quite likely to drink any left in an uncovered container or in puddles on the ground. This will prove fatal in sufficient quantity. Always drain the coolant into a suitable container. Coolant should be reused unless it is contaminated or several years old.

1. Use a blunt tool such as a drift or a dull chisel and a hammer. Strike the lower edge of the freeze plug. The freeze plug should rotate.

2. Grasp the upper edge of the rotated freeze plug with a pair of pliers and pull the freeze plug out of its mounting hole.

3. Clean the freeze plug mounting hole thoroughly, removing all the built up scale, oil and old sealer.

4. Clean the new freeze plug with a non-filming safe solvent to remove any oil or grease.

5. Lightly coat the inside of the freeze plug mounting hole with Loctite® Stud 'N' Bearing Mount, or the equivalent.

6. Place the freeze plug into position in the mounting hole. Use the proper size driver and drive the plug into the hole until the sharp edge of the plug is at least 0.020 in. (0.50mm) inside the lead-in chamfer.

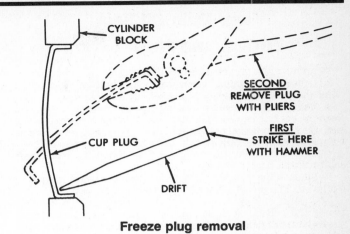

Freeze plug removal

7. It is not necessary to wait for the curing of the sealant. The cooling system can be filled and the van placed into service immediately.

Engine Block Heater
REMOVAL AND INSTALLATION

The engine block heater may be of the freeze plug mounted design, dipstick or heater hose mounted type.

—————————— **CAUTION** ——————————

Removing the block heater of freeze plug may cause personal injury if the engine is not completely cooled down. Even after the radiator has been drained, there will engine coolant still in the block. Use care when removing the assembly or freeze plug

1. To remove the engine freeze plug mounted block heater: Depending on the heaters location, accessories may have to be removed, such as the starter motor, motor mount, etc. Remove the obstruction before attempting to remove the assembly.

2. Disconnect the negative battery ground cable. Drain the cooling system.

—————————— **CAUTION** ——————————

When draining the coolant, keep in mind that cats and dogs are attracted by the ethylene glycol antifreeze, and are quite likely to drink any left in an uncovered container or in puddles on the ground. This will prove fatal in sufficient quantity. Always drain the coolant into a suitable container. Coolant should be reused unless it is contaminated or several years old.

3. Disconnect the electrical connector, loosen the retaining screw and remove the heater from the block.

4. Clean the mounting hole. Coat the heater mounting O-ring with engine oil. Install the heater and tighten the retaining screw. Connect the wiring harness and the battery cable.

5. On heater hose mounted heaters: After draining the cooling system (see Cautions), loosen the mounting hose clamps and slide them back down the hose. Twist and pull the hose from the heater hose connections and remove the heater.

6. Replace any hoses as necessary. Push the hoses over the mounting nipples of the heater and secure the hose clamps.

7. Fill the cooling system and check for leaks.

Rear Main Oil Seal

REMOVAL AND INSTALLATION

1. Remove the oil pan and the oil pump.
2. Loosen all the main bearing cap bolts, thereby lowering the crankshaft slightly but not to exceed 1/32 in. (0.8mm).
3. Remove the rear main bearing cap, and remove the oil seal from the bearing cap and cylinder block. On the block half of the seal use a seal removal tool, or install a small metal screw in one end of the seal, and pull on the screw to remove the seal. Exercise caution to prevent scratching or damaging the crankshaft seal surfaces.
4. Remove the oil seal retaining pin from the bearing cap if so equipped. The pin is not used with the split-lip seal.
5. Carefully clean the seal groove in the cap and block with a brush and solvent such as lacquer thinner, spot remover, or equivalent, or trichlorethylene. Also, clean the area thoroughly, so that no solvent touches the seal.
6. Dip the split lip-type seal halves in clean engine oil.
7. Carefully install the upper seal (cylinder block) into its groove with undercut side of the seal (rubber type) toward the FRONT of the engine, by rotating it on the seal journal of the crankshaft until approximately 3/8 in. (9.5mm) protrudes below the parting surface. On rope type, pull into position with the seal installing tool. Be sure no rubber has been shaved from the outside diameter of the seal by the bottom edge of the groove. Do not allow oil to get on the sealer area.
8. Tighten the main bearing cap bolts to the specifications listed in the Torque chart at the beginning of this chapter.
9. On the rope type seal, trim the upper seal ends flush with the block surface. Install the lower seal (rubber type) in the rear main bearing cap under undercut side of seal toward the FRONT of the engine, allow the seal to protrude approximately 3/8 in. (9.5mm) above the parting surface to mate with the upper seal when the cap is installed. With rope type seals, press the seal full and firmly into the cap groove. Trim the ends flush with the cap.
10. Install the side seals into the bearing cap. Apply an even 1/16 in. (1.6mm) bead of RTV silicone sealer at the bearing cap to block joint to provide oil pan end sealing.

NOTE: This sealer sets up in 15 minutes.

11. Install the rear main bearing cap. Tighten the cap bolts to specifications.
12. Install the oil pump and oil pan.

Undersize Journal	Identification Stamp
.001 inch (Rod)	R1-R2-R3-Etc.
.010 inch (Rod)	RX
.001 inch (Main)	M1-M2-M3 or M4
.010 inch (Main)	MX

STEEL STAMP IDENTIFICATION
R (ROD) AND/OR M (MAIN) FOLLOWED
BY THE ROD OR MAIN NUMBER

EXAMPLE
R2 M4
OR R3
OR M2

V6 crankshaft journal size identification

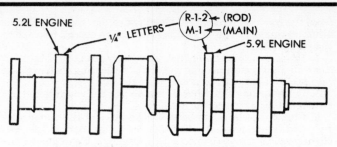

V8 crankshaft journal size identification

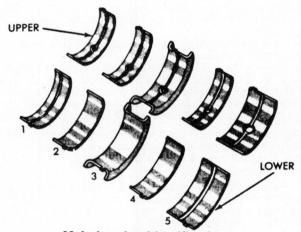

Main bearing identification

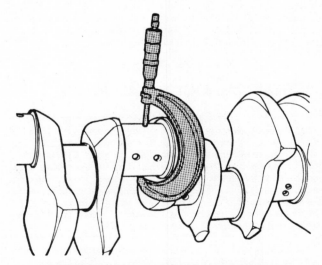

Measuring crankshaft bearing journals

Homemade roll-out pin

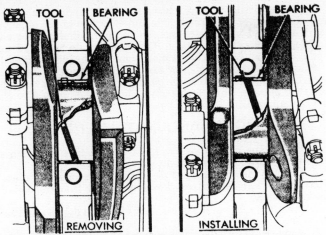

Removing and installing the main bearings using a roll-out pin

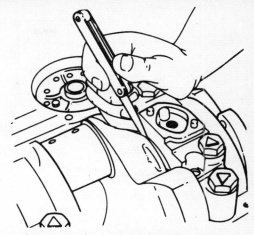

Measuring the crankshaft endplay

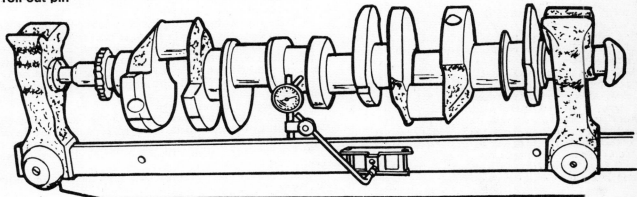

Measuring the crankshaft runout

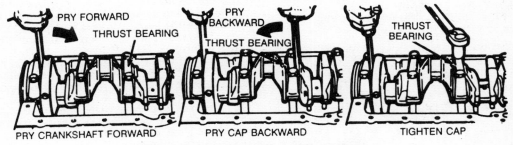

Aligning the crankshaft thrust bearing

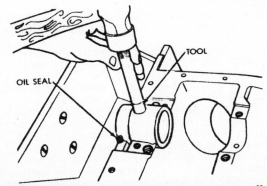

Installing the rear main bearing rope type oil seal

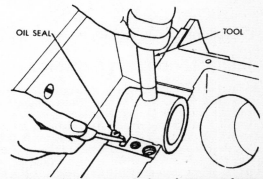

Trimming the rear main bearing rope type seal

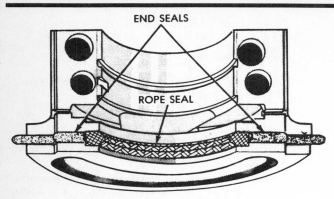

Rear main bearing cap

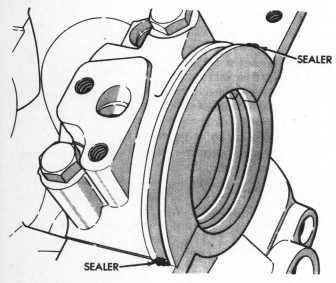

Oil pan end main bearing cap sealer points

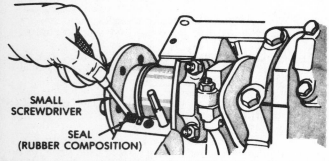

Removing the split rubber oil seal on the 5.9L engine

Crankshaft and Main Bearings

If the crankshaft has been fitted with an undersize bearing or bearings from the factory, a ¼ set of letters will be stamped on the milled flat of the No. 8 crankshaft counterweight on 5.2L engines. On the No. 3 counterweight for 5.9L engines, and on the No. 6 counterweight for the 3.9L engine. The letter **R** or **M** signifies whether the undersize journal is a rod or main, and the number following the letter indicates which one it is. The letters RX indicate that all rod journals are 0.010 in. undersize. MX indicates that all main journals are 0.010 in. undersize.

Upper and lower bearing inserts are not interchangeable on any of the engines due to oil hole and V-groove in the uppers. Mark all main bearing as to location and front of engine facing position before removing them. Remove the main bearing caps one at a time and check the condition and existing oil clearance.

The rear main bearing lower seal is held in place by the rear main bearing cap. Note that the oil pump is mounted on this cap and that there is a hollow dowel which must be in place when the cap is installed.

REMOVAL AND INSTALLATION

1. Follow the previous procedures and remove the engine from the vehicle. Attach it to an engine stand.
2. Remove the crankshaft pulley from the crankshaft vibration damper. Remove the capscrew and washer from the end of the crankshaft.
3. Install a puller on the crankshaft vibration damper and remove the damper.
4. Remove the cylinder front cover and crankshaft gear, refer to the necessary procedure in this section.
5. Invert the engine on the work stand. Remove the oil pan and gasket. Remove the oil pump.
6. Make sure all bearing caps (main and connecting rod) are marked so that they can be installed in their original locations. Turn the crankshaft until the connecting rod from which the cap is being removed is down, and remove the bearing cap. Push the connecting rod and piston assembly up into the cylinder. Repeat this procedure until all the connecting rod bearing caps are removed.
7. Remove the main bearings caps.
8. Carefully lift the crankshaft out of the block so that the thrust bearing surfaces are not damaged. Handle the crankshaft with care to avoid possible fracture to the finished surfaces.
9. Remove the rear journal seal from the block and rear main bearing cap.
10. Remove the main bearing inserts from the block and bearing caps.
11. Remove the connecting rod bearing inserts from the connecting rods and caps.
12. If the crankshaft main bearing journals have been refinished to a definite undersize, install the correct undersize bearings. Be sure the bearing inserts and bearing bores are clean. Foreign material under the inserts will distort the bearing and cause a failure.
13. Place the upper main bearing inserts in position in the bores with the tang fitting in the slot. Be sure the oil holes in the bearing inserts are aligned with the oil holes in the cylinder block.
14. Install the lower main bearing inserts in the bearing caps.
15. Clean the rear journal oil seal groove and the mating surfaces of the block and rear main bearing cap.
16. Install the rear main oil seals.
17. Carefully lower the crankshaft into place. Be careful not to damage the bearing surfaces.
Checking Main Bearing Clearance:
18. Check the clearance of each main bearing by using the following procedure:
   a. Place a piece of Plastigage® or its equivalent, on bearing surface across full width of bearing cap and about ¼ in. (6mm) off center.
   b. Install cap and tighten bolts to specifications. Do not turn crankshaft while Plastigage® is in place.
   c. Remove the cap. Using Plastigage® scale, check width of Plastigage® at widest point to get the minimum clearance. Check at narrowest point to get maximum clearance. Difference between readings is taper of journal.
   d. If clearance exceeds specified limits, try a 0.001 in. (0.0254mm) or 0.002 in. (0.051mm) undersize bearing in combination with the standard bearing. Bearing clearance must be within specified limits. If standard and 0.002 in.

(0.051mm) undersize bearing does not bring clearance within desired limits, refinish crankshaft journal, then install undersize bearings.

NOTE: Refer to Rear Main Oil Seal removal and installation, for special instructions in applying RTV sealer to rear main bearing cup.

19. Install all the bearing caps except the thrust bearing cap. Be sure the main bearing caps are installed in their original locations. Tighten the bearing cap bolts to specifications.
20. Install the thrust bearing cap with the bolts finger tight.
21. Pry the crankshaft forward against the thrust surface of the upper half of the bearing.
22. Hold the crankshaft forward and pry the thrust bearing cap to the rear. This will align the thrust surfaces of both halves of the bearing.
23. Retain the forward pressure on the crankshaft. Tighten the cap bolts to specifications.
24. Check the crankshaft end play using the following procedures:
    a. Force the crankshaft toward the rear of the engine.
    b. Install a dial indicator so that the contact point rests against the crankshaft flange and the indicator axis is parallel to the crankshaft axis.
    c. Zero the dial indicator. Push the crankshaft forward and note the reading on the dial.
    d. If the end play exceeds the wear limit listed in the Crankshaft and Connecting Rod Specifications chart, replace the thrust bearing. If the end play is less than the minimum limit, inspect the thrust bearing faces for scratches, burrs, nicks, or dirt. If the thrust faces are not damaged or dirty, then they probably were not aligned properly.
25. Lubricate and install the new thrust bearing and align the faces.
26. Install new bearing inserts in the connecting rods and caps. Check the clearance of each bearing.
27. After the connecting rod bearings have been fitted, apply a light coat of engine oil to the journals and bearings.
28. Turn the crankshaft throw to the bottom of its stroke. Push the piston all the way down until the rod bearing seats on the crankshaft journal.
29. Install the connecting rod cap. Tighten the nuts to specification.
30. After the piston and connecting rod assemblies have been installed, check the side clearance with a feeler gauge between the connecting rods on each connecting rod crankshaft journal. Refer to Crankshaft and Connecting Rod specifications chart.
31. Install the timing chain and gears, cylinder front cover, crankshaft pulley and oil pan. Install the engine.

Engine in the Van

1. With the oil pan, oil pump and spark plugs removed, remove the cap from one main bearing at a time. Remove the bearing from the cap.
2. Use a special roll out pin or make one, using a bent cotter pin as shown in the illustration. Install the end of the pin in the oil hole in the crankshaft journal.
3. Rotate the crankshaft clockwise as viewed from the front of the engine. This will roll the upper bearing out of the block.
4. Lube the new upper bearing with clean engine oil and insert the plain (unnotched) end between the crankshaft and the indented or notched side of the block. Roll the bearing into place, making sure that the oil holes are aligned. Remove the roll pin from the oil hole.
5. Lube the new lower bearing and install it in the main bearing cap. Install the main bearing cap onto the block, making sure it is positioned in proper direction with the matchmarks in alignment.
6. Torque the main bearing cap to specification. Repeat the procedure for the other main bearings.

NOTE: Refer to the previous Crankshaft Installation for thrust bearing alignment.

CLEANING AND INSPECTION

NOTE: handle the crankshaft carefully to avoid damage to the finish surfaces.

1. Clean the crankshaft with solvent, and blow out all oil passages with compressed air.
2. Use crocus cloth to remove any sharp edges, burrs or other imperfections which might damage the oil seal during installation or cause premature seal wear.

NOTE: Do not use crocus cloth to polish the seal surfaces. A finely polished surface may produce poor sealing or cause premature seal wear.

3. Inspect the main and connecting rod journals for cracks, scratches, grooves or scores.
4. Measure the diameter of each journal at least four places to determine out-of-round, taper or undersize condition.
5. On an engine with a manual transmission, check the fit of the clutch pilot bearing in the bore of the crankshaft. A needle roller bearing and adapter assembly is used as a clutch pilot bearing. It is inserted directly into the engine crank shaft. The bearing and adapter assembly cannot be serviced separately. A new bearing must be installed whenever a bearing is removed.
6. Inspect the pilot bearing, when used, for roughness, evidence of overheating or loss of lubricant. Replace if any of these conditions are found.
7. If you expect to use the old main bearings: clean the bearing inserts and caps thoroughly in solvent, and dry them with compressed air.

NOTE: Do not scrape varnish or gum deposits from the bearing shells.

8. Inspect each bearing carefully. Bearings that have a scored, chipped, or worn surface should be replaced.
9. The copper-lead bearing base may be visible through the bearing overlay in small localized areas. This may not mean that the bearing is excessively worn. It is not necessary to replace the bearing if the bearing clearance is within recommended specifications.
10. Check the clearance of bearings that appear to be satisfactory with Plastigage® or its equivalent. Fit the new bearings following the procedure Crankshaft and Main Bearings removal and installation, they should be reground to size for the next undersize bearing.
11. Regrind the journals to give the proper clearance with the next undersize bearing. If the journal will not clean up to maximum undersize bearing available, replace the crankshaft.
12. Always reproduce the same journal shoulder radius that existed originally. Too small a radius will result in fatigue failure of the crankshaft. Too large a radius will result in bearing failure due to radius ride of the bearing.
13. After regrinding the journals, chamfer the oil holes, then polish the journals with a #320 grit polishing cloth and engine oil. Crocus cloth may also be used as a polishing agent.

Flywheel/Flex Plate and Ring Gear

NOTE: Flex plate is the term for a flywheel mated with an automatic transmission.

REMOVAL AND INSTALLATION

NOTE: The ring gear is replaceable only on engines mated with a manual transmission. Engines with automatic transmissions have ring gears which are welded to the flex plate.

1. Remove the transmission.
2. Remove the clutch. The flywheel or flexplate bolts should be loosened a little at a time in a cross pattern to avoid warping the flywheel. On models with manual transmissions, replace the pilot bearing in the end of the crankshaft after removing the flywheel.
3. The flywheel should be checked for cracks and glazing. It can be resurfaced by a machine shop.
4. If the ring gear is to be replaced, drill a hole in the gear between two teeth, being careful not to contact the flywheel surface. Using a cold chisel at this point, crack the ring gear and remove it.

5. Polish the inner surface of the new ring gear and heat it in an oven to about 600°F (316°C). Quickly place the ring gear on the flywheel and tap it into place, making sure that it is fully seated.

WARNING: Never heat the ring gear past 800°F (426°C), or the tempering will be destroyed.

6. Position the flywheel or flexplate on the end of the crankshaft. Torque the bolts a little at a time, in a cross pattern, to the torque figure shown in the Torque Specifications Chart.
7. Install the clutch.
8. Install the transmission.

EXHAUST SYSTEM

The basic exhaust system consists of exhaust manifolds, exhaust pipe, catalytic converter, extension pipe, heat shields, muffler and tail pipe.

All engines have a thermostatic controlled heat valve in the exhaust manifold for faster warmup and improved driveability after cold start.

The exhaust manifolds are equipped with ball flange outlets to assure a tight seal and strain free connections. The exhaust system must be properly aligned to prevent stress, leakage and body contact. If the system contacts any body panel, it may amplify objectionable noises originating from the engine or body.

When inspecting an exhaust system, critically inspect for cracked or loose joints, stripped screw or bolt threads, corrosion damage and worn, cracked or broken hangers. Replace all components that are badly corroded or damaged. DO NOT attempt to repair.

When replacement is required, use original equipment parts (or their equivalent) to assure proper alignment and to provide acceptable exhaust noise levels.

— CAUTION —

Avoid application of rust prevention compounds or undercoating materials to exhaust system floor pan heat shields. Light overspray near the edges is permitted. Application of coating will result in excessive floor pan temperatures and objectionable fumes.

Catalytic Converter

The stainless steel catalytic converter body is designed to last the life of the vehicle. Excessive heat can result in bulging or other distortion, but excessive heat will not be the fault of the converter. A fuel system, air injection or ignition system malfunction that permits unburned fuel to enter the converter will usually cause overheating. If a converter is heat damaged, correct the cause of the damage at the same time the converter is replaced. Also, inspect all other components of the exhaust system for heat damage. Unleaded gasoline must be used to avoid contaminating the catalyst core.

Heat Shields

Heat shields are needed to protect both the vehicle and the environment from the high temperatures developed in the vicinity of the catalytic converter. The combustion reaction facilitated by the catalyst releases additional heat in the exhaust system. Under severe operating conditions, the temperature increases in the area of the reactor. Such conditions can exist when the engine misfires or otherwise does not operate at peak efficiency. Do Not remove spark plugs wires form plugs or by any other means short out cylinders. Failure of the catalytic converter can occur due to a temperature increase caused by unburned fuel passing through the converter.

Do not allow the engine to operate at fast idle for extended periods (over 5 minutes). This condition may result in excessive temperatures in the exhaust system and on the floor pan.

Exhaust Gas Recirculation (EGR)

To assist in the control of oxides of nitrogen (NOx) in engine exhaust, all engines are equipped with an exhaust gas recirculation system. The use of exhaust gas to dilute incoming air/fuel mixtures lowers peak flame temperatures during combustion, thus limiting the formation of NOx.
Exhaust gases are taken from openings in the exhaust gas crossover passage in the intake manifold.

Manifold Heat Control Valve

A manifold heat control valve is located in the right exhaust manifold. The thermostatic controlled valve directs heated exhaust gases to the heat chamber in the intake manifold beneath the throttle body. This helps vaporize the fuel mixture during the engine warm-up period. When the valve is closed, the exhaust gases are directed to the heat chamber through the right side of the exhaust crossover passage. After circulating through the heat chamber, the gases are returned to the exhaust manifold through the left side of the passage.

Safety Precautions

For a number of reasons, exhaust system work can be the most dangerous type of work you can do on your vehicle. Always observe the following precautions:

● Support the vehicle extra securely. Not only will you often be working directly under it, but you'll frequently be using a lot of force, say, heavy hammer blows, to dislodge rusted parts. This can cause a vehicle that's improperly supported to shift and possibly fall.

● Wear goggles. Exhaust system parts are always rusty. Metal chips can be dislodged, even when you're only turning rusted bolts. Attempting to pry pipes apart with a chisel makes the chips fly even more frequently.

● If you're using a cutting torch, keep it a great distance from either the fuel tank or lines. Stop what you're doing and feel the temperature of the fuel bearing pipes on the tank frequently. Even slight heat can expand and/or vaporize fuel, resulting in accumulated vapor, or even a liquid leak, near your torch.

● Watch where your hammer blows fall and make sure you hit squarely. You could easily tap a brake or fuel line when you hit an exhaust system part with a glancing blow. Inspect all lines and hoses in the area where you've been working.

─────────── CAUTION ───────────

Be very careful when working on or near the catalytic converter. External temperatures can reach 1,500°F (816°C) and more, causing severe burns. Removal or installation should be performed only on a cold exhaust system.

Special Tools

A number of special exhaust system tools can be rented from auto supply houses or local stores that rent special equipment. A common one is a tail pipe expander, designed to enable you to join pipes of identical diameter.

It may also be quite helpful to use solvents designed to loosen rusted bolts or flanges. Soaking rusted parts the night before you do the job can speed the work of freeing rusted parts considerably. Remember that these solvents are often flammable. Apply only to parts after they are cool!

COMPONENT REPLACEMENT

System components may be welded or clamped together. The system consists of a head pipe, catalytic converter, intermediate pipe, muffler and tail pipe, in that order from the engine to the back of the car.

The head pipe is bolted to the exhaust manifold, except on turbocharged engines, in which case it is bolted to the turbocharger outlet elbow. Various hangers suspend the system from the floor pan. When assembling exhaust system parts, the relative clearances around all system parts is extremely critical. See the accompanying illustration and observe all clearances during assembly. In the event that the system is welded, the various parts will have to be cut apart for removal. In these cases, the cut parts may not be reused. To cut the parts, a hacksaw is the best choice. An oxy-acetylene cutting torch may be faster but the sparks are DANGEROUS near the fuel tank and, at the very least, accidents could happen, resulting in damage to other under-car parts, not to mention yourself!

The following replacement steps relate to clamped parts:

1. Raise and support the vehicle on jackstands. It's much easier on you if you can get vehicle up on 4 stands. Some pipes need lots of clearance for removal and installation. If the system has been in the car for a long time, spray the clamped joints with a rust dissolving solutions such as WD-40® or Liquid Wrench®, and let it set according to the instructions on the can.

2. Remove the nuts from the U-bolts; don't be surprised if the U-bolts break while removing the nuts. Age and rust account for

this. Besides, you shouldn't reuse old U-bolts. When unbolting the headpipe from the exhaust manifold, make sure that the bolts are free before trying to remove them. If you snap a stud in the exhaust manifold, the stud will have to be removed with a bolt extractor, which often necessitates the removal of the manifold itself.

3. After the clamps are removed from the joints, first twist the parts at the joints to break loose rust and scale, then pull the components apart with a twisting motion. If the parts twist freely but won't pull apart, check the joint. The clamp may have been installed so tightly that it has caused a slight crushing of the joint. In this event, the best thing to do is secure a chisel designed for the purpose and, using the chisel and a hammer, peel back the female pipe end until the parts are freed.

4. Once the parts are freed, check the condition of the pipes which you had intended keeping. If their condition is at all in doubt, replace them too. You went to a lot of work to get one or more components out. You don't want to have to go through that again in the near future. If you are retaining a pipe, check the pipe end. If it was crushed by a clamp, it can be restored to its original diameter using a pipe expander, which can be rented at most good auto parts stores. Check, also, the condition of the exhaust system hangers. If ANY deterioration is noted, replace them. Oh, and one note about parts: use only parts designed for your vehicle. Don't use fits-all parts or flex pipes. The fits-all parts never fit and the flex pipes don't last very long.

5. When installing the new parts, coat the pipe ends with exhaust system lubricant. It makes fitting the parts much easier. It's also a good idea to assemble all the parts in position before clamping them. This will ensure a good fit, detect any problems and allow you to check all clearances between the parts and surrounding frame and floor members.

6. When you are satisfied with all fits and clearances, install the clamps. The headpipe-to-manifold nuts should be torqued to 20 ft. lbs. If the studs were rusty, wire-brush them clean and spray them with WD-40® or Liquid Wrench®. This will ensure a proper torque reading. Position the clamps on the slip points as illustrated. The slits in the female pipe ends should be under the U-bolts, not under the clamp end. Tighten the U-bolt nuts securely, without crushing the pipe. The pipe fit should be tight, so that you can't swivel the pipe by hand. Don't forget: always use new clamps. When the system is tight, recheck all clearances. Start the engine and check the joints for leaks. A leak can be felt by hand. MAKE CERTAIN THAT THE VEHICLE IS SECURE ON THE JACKSTANDS BEFORE GETTING UNDER IT WITH THE ENGINE RUNNING!! If any leaks are detected, tighten the clamp until the leak stops. If the pipe starts to deform before the leak stops, reposition the clamp and tighten it. If that still doesn't stop the leak, it may be that you don't have enough overlap on the pipe fit. Shut off the engine and try pushing the pipe together further. Be careful; the pipe gets hot quickly.

7. When everything is tight and secure, lower the vehicle and take it for a road test. Make sure there are no unusual sounds or vibration. Most new pipes are coated with a preservative, so the system will be pretty smelly for a day or two while the coating burns off.

Exhaust Pipe

REMOVAL AND INSTALLATION

1. Raise and support the vehicle.
2. Saturate the bolts and nuts with heat valve lubricant. Allow 5 minutes for penetration.
3. Remove the exhaust manifold-to-exhaust pipe nuts.
4. Remove the clamp bolt nuts holding the exhaust pipe to the support bracket.
5. Disconnect the exhaust pipe from the catalytic converter front flange and the exhaust manifold.

6. Remove the exhaust pipe.

7. Align and connect the exhaust pipe to the catalytic converter flange.

8. Connect the exhaust pipe to the exhaust manifold. Tighten the nuts to (33 N.m) 24 ft.lbs.

9. Install the clamp bolt and nuts to the support bracket. Tighten the nuts to (15 N.m) 125 in.lbs.

10. Install the exhaust pipe to the catalytic converter. Tighten the "U" bolt clamp nut to (41 N.m) 30 ft.lbs.

11. Lower the vehicle.

12. Start the engine and inspect for exhaust leaks and exhaust system contact with the body panels. Adjust the alignment, if needed.

Catalytic Converter

REMOVAL AND INSTALLATION

1. Raise and support the vehicle.

2. Saturate the bolts and nuts with heat valve lubricant. Allow 5 minutes for penetration.

3. Disconnect the downstream air injection tube.

4. Remove the clamp nuts holding the catalytic converter flanges to the exhaust pipe and the muffler.

5. To disconnect exhaust pipe and muffler from the catalytic converter flanges, you may have to loosen up other sections of the exhaust system.

6. Remove the catalytic converter.

7. Remove the heat shield.

8. Position the heat shield to the catalytic converter and install the clamps and nuts.

9. Align and connect the catalytic converter flanges to the exhaust pipe.

10. Install the muffler into the catalytic converter flange.

11. If other sections of the exhaust system where loosened in removal, refer to that section for the tightening procedures.

12. At the catalytic converter flange connections, install the clamp on the muffler and the exhaust pipe. Tighten the clamp nuts to (41 N.m) 30 ft.lbs.

13. Connect the downstream air injection tube. Tighten the nuts to (41 N.m) 30 ft.lbs.

14. Lower the vehicle.

15. Start the engine and inspect for exhaust leaks and exhaust system contact with the body panels. Adjust the alignment, if needed.

Muffler

REMOVAL AND INSTALLATION

1. Raise and support the vehicle.

2. Saturate the clamp nuts with heat valve lubricant. Allow 5 minutes for penetration.

3. Remove the muffler clamp nuts from the front and rear hanger.

4. Disconnect the muffler from the tail pipe. The tail pipe should be supported when the muffler is disconnected.

5. Remove the muffler from the extension pipe.

6. Install the muffler into the extension pipe. Install the clamp and tighten the nuts finger tight.

7. Install the tail pipe into the rear of the muffler.

8. Tighten the clamp nuts on the front and rear muffler hangers to (41 N.m) 30 ft.lbs.

9. Lower the vehicle.

10. Start the engine and inspect for exhaust leaks and exhaust system contact with the body panels. Adjust the alignment, if needed.

CONDITION	POSSIBLE CAUSE	CORRECTION
EXCESSIVE EXHAUST NOISE	(a) Leaks at pipe joints. (b) Burned or blown out muffler. (c) Burned or rusted out exhaust pipe. (d) Exhaust pipe leaking at manifold flange. (e) Exhaust manifold cracked or broken. (f) Leak between manifold and cylinder head. (g) Restriction in muffler or tail pipe.	(a) Tighten clamps at leaking joints. (b) Replace muffler assembly. (c) Replace exhaust pipe. (d) On all engines, tighten ball joint connection attaching bolt nuts. alternate tightening. (e) Replace manifold. (f) Tighten manifold to cylinder head stud nuts or bolts to specifications. (g) Remove restriction, if possible, or replace as necessary.
LEAKING EXHAUST GASES	(a) Leaks at pipe joints. (b) Damaged or improperly installed gaskets.	(a) Tighten U-bolts at leaking joints. (b) Replace gaskets as necessary.
ENGINE HARD TO WARM UP OR WILL NOT RETURN TO NORMAL IDLE	(a) Heat control valve frozen in the open position. (b) Blocked crossover passage in intake manifold.	(a) Free up manifold heat control valve using a suitable solvent. (b) Remove restriction or replace intake manifold.
HEAT CONTROL VALVE NOISY	(a) Thermostat broken. (b) Broken, weak or missing anti-rattle spring.	(a) Replace thermostat. (b) Replace spring.

Exhaust system diagnosis

Tail Pipe

REMOVAL AND INSTALLATION

1. Raise and support the vehicle.
2. Saturate the clamp nuts with heat valve lubricant. Allow 5 minutes for penetration.
3. Loosen the nuts on the muffler rear hanger.
4. Remove the tail pipe hanger bolt.
5. Remove the tail pipe.

6. Position the tail pipe into the muffler. Install the nuts onto the clamp bolt finger tight.
7. Install the tail pipe clamp hanger bolt. Tighten the bolt to (23 N.m) 200 in.lbs.
8. Tighten the muffler rear hanger nuts to (41 N.m) 30 ft.lbs.
9. Lower the vehicle.
10. Start the engine and inspect for exhaust leaks and exhaust system contact with the body panels. Adjust the alignment, if needed.

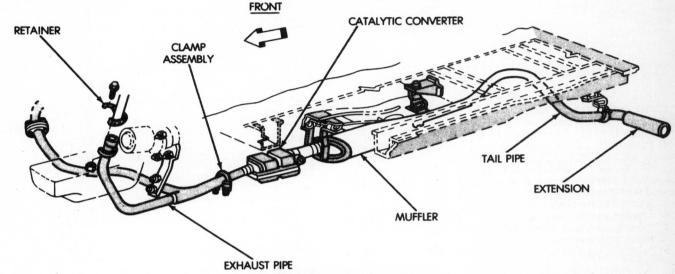

Exhaust system components

Ⓐ	11 N·m (100 in. lbs.)
Ⓑ	3 N·m (24 in. lbs.)
Ⓒ	23 N·m (200 in. lbs.)

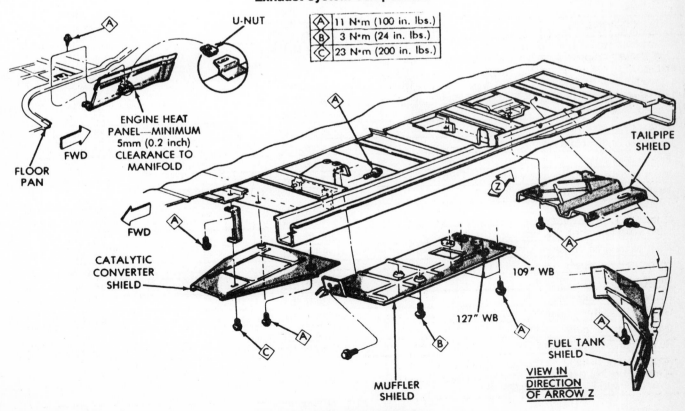

Exhaust system heat shields

Heat Shields
REMOVAL AND INSTALLATION

1. Raise and support the vehicle.
2. Remove the nuts or screws holding the heat shields to the floor pan, crossmember or bracket.

3. Slide the shield out around the exhaust system.
4. Position the heat shield to the floor pan, crossmember or bracket and install the nuts or screws.
5. Tighten the nuts, bolts or screws to the proper torque.
6. Lower the vehicle.

TORQUE SPECIFICATIONS

Component	English	Metric
Alternator		
Adusting bolt:	17 ft. lbs.	23 Nm
Mounting bolts:	30 ft. lbs.	41 Nm
Camshaft sprocket bolt:	35 ft. lbs.	48 Nm
Connecting rod nuts:	45 ft. lbs.	61 Nm
Crankshaft damper bolt:	135 ft. lbs.	183 Nm
Cylinder head bolts		
Step 1:	50 ft.lbs.	68 Nm
Step 2:	105 ft. lbs.	143 Nm
Downstream air injection tube nuts:	30 ft. lbs.	41 Nm
Engine mounts		
Front mount-to-frame:	50 ft. lbs.	68 Nm
Front mount-to-engine:	30 ft. lbs.	41 Nm
Rear mount-to-crossmember:	30 ft. lbs.	41 Nm
Rear mount-to-entension bracket:	75 ft. lbs.	102 Nm
Rear mount bracket-to-extension:	50 ft. lbs.	68 Nm
Exhaust System		
Catalytic converter-to-muffler nuts:	30 ft. lbs.	41 Nm
Exhaust manifold		
bolts:	20 ft. lbs.	27 Nm
nuts:	15 ft. lbs.	20 Nm
Exhaust pipes-to-manifold nuts:	24 ft. lbs.	33 Nm
Exhaust pipe clamp bolt and nuts:	125 inch lbs.	15 Nm
Exhaust pipe-to-catalytic converter nuts:	30 ft. lbs.	41 Nm
Muffler rear hanger nuts:	30 ft. lbs.	41 Nm
Tail pipe clamp hanger bolt:	200 inch lbs.	23 Nm
Fan-to-drive unit bolts:	17 ft. lbs.	23 Nm
Fan shroud screws:	95 inch lbs.	11 Nm
Flywheel or flex plate bolts:	55 ft. lbs.	75 Nm
Front cover bolts:	30 ft. lbs.	41 Nm
Front engine mount insulator nuts:	75 ft. lbs.	102 Nm
Intake manifold bolts		
Step 1:	25 ft.lbs.	34 Nm
Step 2:	40 ft. lbs.	54 Nm
Lifter yoke retainer bolts, V6:	200 inch lbs.	23 Nm
Main bearing cap nuts:	85 ft. lbs.	116 Nm
Oil cross-over pipe:	24 ft. lbs.	33 Nm
Oil pan bolts:	15 ft. lbs.	21 Nm
Oil pan drain plug:	20 ft. lbs.	27 Nm
Oil pump attaching bolts:	30 ft. lbs.	41 Nm
Oil pump cover bolts:	95 inch lbs.	11 Nm
Rear engine support bolts:	50 ft. lbs.	68 Nm
Rocker cover bolts		
models with a gasket:	80 inch lbs.	9 Nm
models with RTV silicone material		
nuts:	80 inch lbs.	9 Nm
studs:	115 inch lbs.	13 Nm

TORQUE SPECIFICATIONS

Component	English	Metric
Rocker shaft bolts:	200 inch lbs.	23 Nm
Starter mounting bolts:	50 ft. lbs.	68 Nm
Thermostat housing bolts:	200 inch lbs.	23 Nm
Throttle body-to-intake manifold bolts:	15 ft. lbs.	20 Nm
Torque converter housing-to-engine bolts		
⅜-16 bolts:	30 ft. lbs.	41 Nm
⁷⁄₁₆-14 bolts:	50 ft. lbs.	68 Nm
Transmission-to-bellhousing bolts		
⅜-16 bolts:	30 ft. lbs.	41 Nm
⁷⁄₁₆-14 bolts:	50 ft. lbs.	68 Nm
Water pump bolts:	30 ft. lbs.	41 Nm

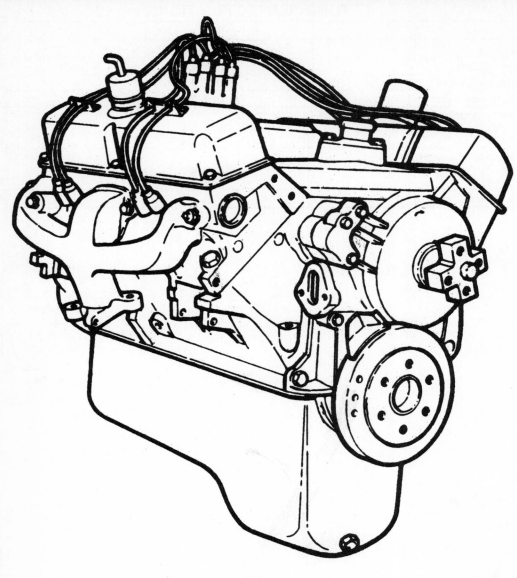

V8 engine

Emission Controls

4

QUICK REFERENCE INDEX

GENERAL INDEX

EMISSION CONTROLS

NOTE: Diagnostic procedures for components of Emission Controls are provided by the use of an on board diagnostic system. Refer to the Electronic Engine Control part of this section for fault code diagnostic procedures.

Crankcase Ventilation System

The crankcase emission control equipment consists of a positive crankcase ventilation (PCV) valve, a closed oil filler cap and the hoses that connect this equipment.

When the engine is running, a small portion of the gases which are formed in the combustion chamber leak by the piston rings and enter the crankcase. Since these gases are under pressure they tend to escape from the crankcase and enter into the atmosphere. If these gases are allowed to remain in the crankcase for any length of time, they would contaminate the engine oil and cause sludge to build up. If the gases are allowed to escape into the atmosphere, they would pollute the air, as they contain unburned hydrocarbons. The crankcase emission control equipment recycles these gases back into the engine combustion chamber, where they are burned.

Crankcase gases are recycled in the following manner. While the engine is running, clean filtered air is drawn into the crankcase through the intake air filter and then through a hose leading to the oil filler cap. As the air passes through the crankcase it picks up the combustion gases and carries them out of the crankcase, up through the PCV valve and into the intake manifold. After they enter the intake manifold they are drawn into the combustion chamber and are burned.

The most critical component of the system is the PCV valve. This vacuum-controlled valve regulates the amount of gases which are recycled into the combustion chamber. At low engine speeds the valve is partially closed, limiting the flow of gases into the intake manifold. As engine speed increases, the valve opens to admit greater quantities of the gases into the intake manifold. If the valve should become blocked or plugged, the gases will be prevented from escaping the crankcase by the normal route. Since these gases are under pressure, they will find their own way out of the crankcase. This alternate route is usually a weak oil seal or gasket in the engine. As the gas escapes by the gasket, it also creates an oil leak. Besides causing oil leaks, a clogged PCV valve also allows these gases to remain in the crankcase for an extended period of time, promoting the formation of sludge in the engine.

The above explanation and the troubleshooting procedure which follows applies to all of the gasoline engines installed in these vans, since all are equipped with PCV systems.

SERVICE

With the engine running, pull the PCV valve and hose from the valve rocker cover rubber grommet.

A hissing noise should be heard as air passes through the

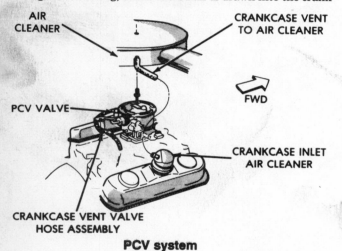

PCV system

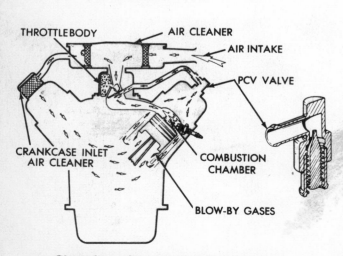

Closed crankcase ventilation system

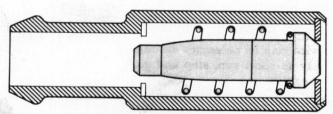

PCV valve position — engine off or engine backfire — no vapor flow

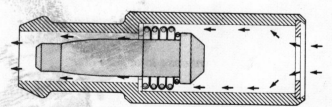

PCV valve position — high intake manifold — minimal vapor flow

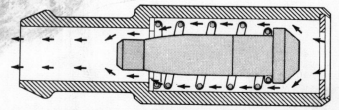

PCV valve position — moderate intake manifold vacuum — maximum vapor flow

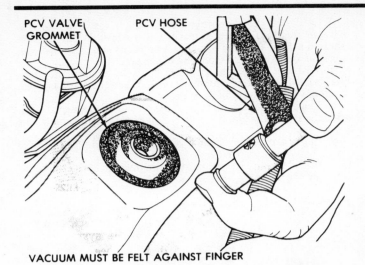

VACUUM MUST BE FELT AGAINST FINGER

Check vacuum at the PCV valve

valve and a strong vacuum should be felt when you place a finger over the valve inlet if the valve is working properly. While you have your finger over the PCV valve inlet, check for vacuum leaks in the hose and at the connections.

When the PCV valve is removed from the engine, a metallic licking noise should be heard when it is shaken. This indicates that the metal check ball inside the valve is still free and is not gummed up.

REMOVAL AND INSTALLATION

1. Pull the PCV valve and hose from the rubber grommet in the rocker cover.
2. Remove the PCV valve from the hose. Inspect the inside of the PCV valve. If it is dirty, clean it in a suitable, safe solvent. Install and test the cleaned valve for operation. Replace the valve with a new one if necessary.
To install, proceed as follows:
3. If the PCV valve hose was removed, connect it to the intake manifold.
4. Connect the PCV valve to its hose.
5. Install the PCV valve into the rubber grommet in the valve rocker cover.

Evaporative Emission Controls

OPERATION

Changes in atmospheric temperature cause fuel tanks to breathe, that is, the air within the tank expands and contracts with outside temperature changes. If an unsealed system was used, when the temperature rises, air would escape through the tank vent tube or the vent in the tank cap. The air which escapes contains gasoline vapors.

The Evaporative Emission Control System provides a sealed fuel system with the capability to store and condense fuel vapors. When the fuel evaporates in the fuel tank, the vapor passes through vent hoses or tubes to a carbon filled evaporative canister. When the engine is operating the vapors are drawn into the intake manifold.

The vapors are drawn into the engine at idle as well as at operating speeds. This system is called a Bi-level Purge System where there is a dual source of vacuum to remove fuel vapor from the canister. The source of vacuum at idle is a tee in the PCV system.

Evaporative Canister

A sealed, maintenance free evaporative canister is used. The canister is usually mounted under the hood on the right side of the vehicle behind the wheel well. The canister is filled with granules of an activated carbon mixture. Fuel vapors entering the canister are absorbed by the charcoal granules.

Fuel tank pressure vents fuel vapors into the canister. They are held in the canister until they can be drawn into the intake manifold. The canister purge solenoid allows the canister to be purged at a predetermined time and engine operating conditions.

Canister Purge Solenoid

Vacuum for the canister is controlled bt the Canister Purge Solenoid. The solenoid is operated by the engine controller. The controller regulates the solenoid by switching the ground circuit on and off based on engine operating conditions. When energized, the solenoid prevents vacuum from reaching the canister. When not energized the solenoid allows vacuum to flow through to the canister.

During warm up and for a specified time after hot starts, the engine controller energizes (grounds) the solenoid preventing vacuum from reaching the canister. When the engine temperature reaches the operating level of about 120°F (49°C), the engine controller removes the ground from the solenoid allowing vacuum to flow through the canister and purges vapors through the throttle body. During certain idle conditions, the purge solenoid may be grounded to control fuel mix calibrations.

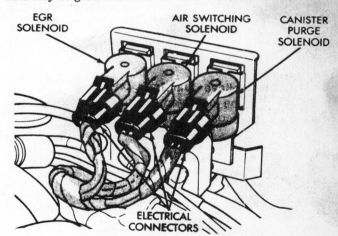

Vacuum solenoid identification

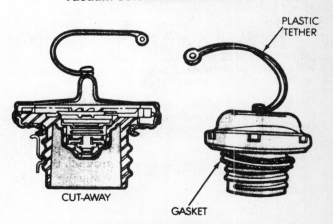

Pressure vacuum filler cap

Gas Tank Pressure-Vacuum Filler Cap

The fuel tank is sealed with a pressure-vacuum relief filler cap. The relief valves in the cap are a safety feature, preventing excessive pressure or vacuum in the fuel tank.

NOTE: To relieve fuel tank pressure, the filler cap must be removed before disconnecting any fuel system component.

Heated Inlet Air System

OPERATION

The heated air inlet system controls and maintains the temperature of the air entering the throttle body. By maintaining the inlet air temperature the air/fuel mixture can be calibrated much leaner to reduce hydrocarbon emissions. The system also improves engine warm up.

The system consists of: A diaphragm that operates the blend door in the air cleaner. A heated air temperature sensor. An air duct hose connected to the heat stove on the exhaust manifold.

The air cleaner blend door adjusts to outside air or heated air, or a position in between them. The adjustment of the throttle body inlet air temperature is performed by intake manifold vacuum. The heated air temperature sensor in the air cleaner senses air temperature. The sensor also controls the flow of vacuum to the blend door diaphragm. The amount of air allowed to enter from either outside or heated inlets is determined by the position of the blend door in the air cleaner snorkel.

SERVICE

An improperly operating heated air inlet system can affect driveability and the vehicle's exhaust emission.

To determine if the air inlet system is functioning properly:
1. Make sure that the duct hose (between the exhaust manifold stove and air cleaner, hose connector and all vacuum lines are in good condition and connected.
2. With a cold engine and the outside temperature less than 112°F (45°C), the heat control door in the air cleaner snorkel should be in the up, heat on position (closed, covering the outside intake of the snorkel).
3. With the engine running and warmed up — about 130°F (55°C), the air door should be opened allowing outside air to enter the air cleaner.
4. Remove the air cleaner from the engine. Allow the air cleaner to cool (about 112°F (45°C). Apply 20 in.Hg of vacuum (use a hand vacuum pump) to the temperature sensor. If the air door is not in the closed position covering the outside air inlet of the snorkel, the vacuum diaphragm should be suspect.
5. Apply 25 in.Hg of vacuum to the diaphragm. The pressure should not bleed down more than 10 in.Hg in 5 minutes. The door should not lift off the bottom of the snorkel when less than 2.5 in.Hg is applied to the diaphragm. The door should be in the full up position with no more than 4.0 in.Hg of vacuum is applied.
6. If the vacuum diaphragm performs correctly but the proper temperature is not maintained, replace the sensor. Repeat the tests after components have been replaced.

REMOVAL AND INSTALLATION

Vacuum Diaphragm

1. Remove the air cleaner from the vehicle.
2. Disconnect the vacuum hose from the blend door diaphragm.

3. Drill out the diaphragm mounting rivet. Tip the diaphragm slightly forward to disengage the lock. When the diaphragm is free, slide the complete assembly to one side to disengage the operating rod from the blend air door.
4. With the diaphragm removed from the air cleaner, check the blend air door for freedom of travel. When the door is raised, it should fall freely. Adjust as necessary.
5. Place the diaphragm in position on the air cleaner and engage the rod to the blend air door. Mount the diaphragm with a rivet.
6. Apply 4 in.Hg of vacuum to the diaphragm. The control door should operate freely.
7. Install the air cleaner, start the engine and check the blend air door operation.

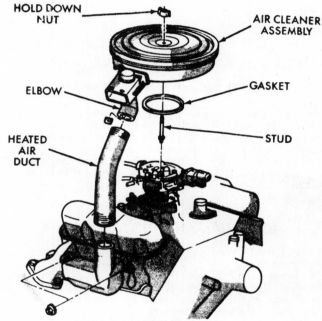

Heated air inlet system

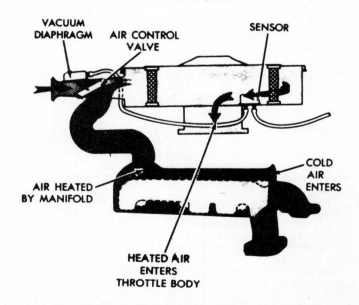

Heated air inlet system operation

Heated Air Temperature Sensor

1. Remove the air cleaner from the vehicle.
2. Remove and discard the sensor mounting clip. Remove the sensor.

3. Position the mounting gasket on the sensor. Install the sensor in position on the air cleaner.
4. Support the sensor and slide the new retainer into position. Make sure the sensor is secure and in the proper position with the mounting gasket to ensure a tight air seal.

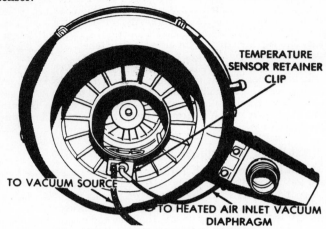

Heated air vacuum hose routing

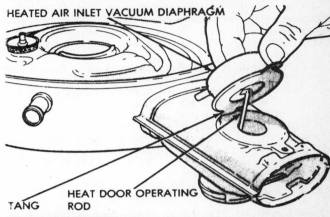

Removing/installing the vacuum diaphragm on the heated air intake

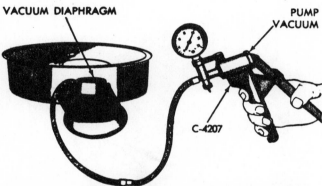

Testing the vacuum diaphragm on the heated air intake

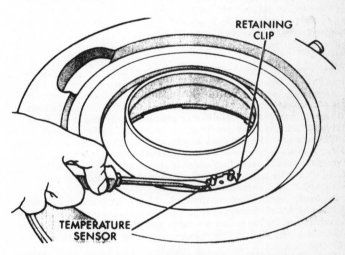

Removing the temperature sensor retaining clip — heated air system

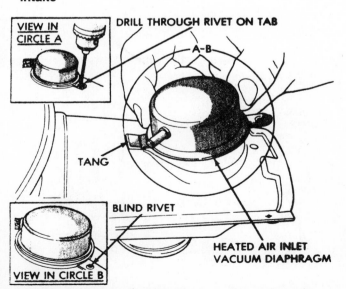

Removing the mounting rivet on the heated air diaphragm

Air temperature sensor installation — heated air inlet

Air Injection System

OPERATION

The air pump air injection system consists of a belt driven pump, an air control valve (either a diverter valve, relief or a switch/relief valve), rubber hoses, check valves to protect the hoses, and injection tubes and other components from hot exhaust gases.

The system adds a controlled amount of air to the exhaust gases aiding in the reduction of hydrocarbons and carbon monoxide in the exhaust gases. The system is designed so that the air injection will not interfere with the ability of the EGR system to control NOx emissions.

Air is injected at the exhaust ports for a short time during engine warm up. The air is then switched to a point downstream where it will assist the oxidation process in the catalyst but not interfere with exhaust gas recirculation. The switching is controlled by a vacuum solenoid.

Air Switching System

When the vehicle is started the air switching solenoid is energized and switches the air upstream. When the oxygen feedback system goes closed loop, the air will be switched to downstream until an open loop idle condition takes place. Open loop idle condition occurs when the vehicle speed is less than 3 mph (5 kmph), throttle switch is closed and about 5 seconds have elapsed when both conditions are met.

Air Injection Pump

An air pump is mounted on the front of the engine and is belt driven by crankshaft rotation. Intake air passes through a centrifugal fan at the front of the pump, where foreign materials are separated from the air by centrifugal force. Air is delivered to the air injection manifold and check valve tube assembly by a rubber hose through the diverter valve and switching valve.

The air injection system is not completely noiseless. Under normal conditions, noise rises in pitch as engine speed increases. To determine if excessive noise is the fault of the air injection, disconnect the drive belt and operate the engine. If noise now does not exist, proceed with diagnosis.

The only serviceable component of this pump is the centrifugal fan filter. Do not assume the pump is defective if it squeaks when turned by hand. Do not lubricate pump. Wipe all oil off of pump housing. Oil in the pump will cause rapid deterioration and failure.

If the engine or underhood compartment is to be cleaned with steam or high pressure detergent, the centrifugal fan filter should be masked off to prevent liquids from entering the pump.

Diverter Valve

The purpose of the diverter valve is to prevent backfire in the exhaust system during sudden deceleration. Sudden throttle closure at the beginning of the deceleration temporarily creates an air/fuel mixture too rich to burn. This mixture becomes burnable when it reaches the exhaust area and combines with injector air. The next firing of the engine will ignite this air/fuel mixture. The diverter valve senses the sudden increase in intake manifold vacuum causing the valve to open, allowing air from the air pump to pass through the valve and silencer to the atmosphere.

A pressure relief valve, incorporated in the same housing as the diverter valve, controls pressure within the system by diverting excessive pump output at higher engine speeds to the atmosphere through the silencer.

Switch/Relief Valve

The purpose of the switch/valve is twofold. First of all, the valve directs the air injection flow to either the exhaust port location or to the downstream injection point. In addition, the valve regulates, by controlling the output of the air pump at high speeds. When the pressure reaches a certain level, some of the output is vented to the atmosphere through the silencer.

Initially, air is injected at the upstream location as close to the exhaust valves as possible. As the engine temperature increases, exhaust gas recirculation will begin functioning. When this happens, the air injection point must be switched to the downstream location for the best possible operation.

The switch/relief valve is controlled by manifold vacuum and a vacuum solenoid. When the engine is cold, a manifold vacuum signal is sent to the switch/relief valve. A bleed orifice in the solenoid allows the vacuum signal to go to zero. Without a vacuum signal, the valve switches most of the air pump air to the downstream location.

Check Valve (Air Injection Tube)

A check valve is located in the injection tube assembles that lead to the exhaust manifolds on the eight cylinder engine and the cylinder head and exhaust pipe on the six cylinder engine. This valve has a one-way diaphragm which prevents hot exhaust gases from backing up into the hose and pump. This valve will protect the system in the event of pump belt failure, abnormally high exhaust system pressure, or air hose ruptures

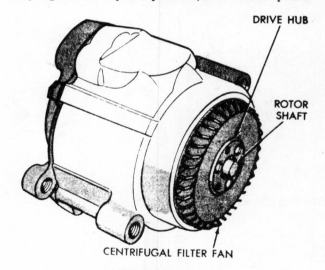

Air injection pump

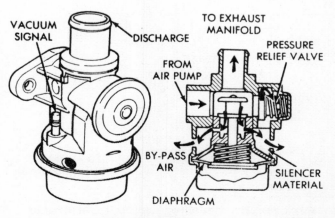

Diverter valve

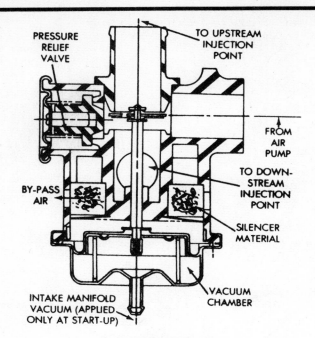

Switch/relief valve

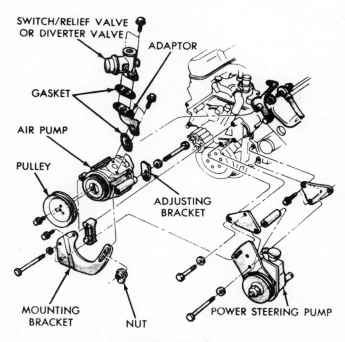

Air pump mounting

SERVICE

For satisfactory emission control and engine durability, it is important that the air pump be operating at all times (except when testing). For proper operation of the air pump, it is necessary that the air pump drive belt be in good condition and adjusted to the specified tension. Check the condition of the air pump belt and adjust the tension according to specifications.

Servicing of the air pump is limited to replacement of the cen-

trifugal fan filter or the entire pump. Do not disassemble pump for any reason. Do not clamp the pump in a vise or use a hammer or pry bar on the pump housing.

REMOVAL AND INSTALLATION

Air Injection Pump

1. Remove the air hose from the air pump outlet elbow, or control valve.
2. Loosen the air pump pivot and adjusting bolts and remove the belt.
3. Remove the pivot and adjusting bolts from the pump brackets and remove the pump, pulley and brackets as an assembly.
4. Remove the air pump pulley, brackets and diverter valve, switch relief valve or outlet elbow from the air pump.
5. Install the air pump pulley, diverter valve, switch/relief valve or outlet elbow and brackets on the air pump. Use a new gasket on the pump flange. Tighten the air pump pulley screws to 95 inch lbs. (11 N.m) and the outlet elbow screws to 95 inch lbs. (11 N.m) Tighten the bracket attachment bolts to 30 ft. lbs. (41 N.m).
6. Position the pump assembly on the engine and install the air hose to the diverter valve, switch/relief valve or pump outlet elbow. Install the pump pivot and adjusting bolts loosely.
7. Install the drive belt and adjust it to the proper tension. Do not, under any conditions, pry on the housing. Tighten the mounting bolts to 30 ft. lbs. (41 N.m).

Centrifugal Filter Fan

A damaged filter fan may be removed by inserting needle nose pliers between the plastic filter fins and breaking the fan from the hub. Care should be taken to prevent fragments from entering the air intake hole. Do not insert a screwdriver blade between the pump and filter. It is seldom possible to remove the fan without destroying it. Do not attempt to remove the metal drive hub.

Install the new filter fan by drawing it into position, using the pulley and bolts as tools. Draw the fan down evenly by alternately torquing the bolts, making certain that the outer edge of the fan slips into the housing. A slight amount of interference with the housing bore is normal. Do not attempt to install a fan by hammering or pressing it on. After a new fan is installed, it may squeal upon initial operation until its O.D. sealing lip has worn in. This may require 20–30 miles (32–48 km) of operation.

Diverter Valve

Servicing of the diverter valve is limited to replacement of the entire valve. Failure of the diverter valve will cause excessive noise. Either the relief valve or the diverter valve have failed if air escapes from the silencer at engine idle speed and the entire valve assembly should be replaced.

1. Remove the air and vacuum hoses from the diverter valve.
2. Remove the two screws securing the diverter valve to the mounting flange and remove the valve.
3. Remove any gasket material from the diverter valve and mounting surface.
4. Position a new gasket on the valve mounting surface.
5. Position the diverter valve on the mounting flange and secure it with the two screws. Tighten to 95 inch lbs. (11 N.m).
6. Install the air and vacuum hoses to the diverter valve.

Switch/Relief Valve

The switch/relief valve is not serviceable. If vacuum is applied to the valve and air injection is not upsteam (to exhaust manifold crossover tube) or if air injection is both upstream and downstream, the valve is faulty and must be replaced. The relief valve has failed if air escapes from the silencer at engine idle speed.

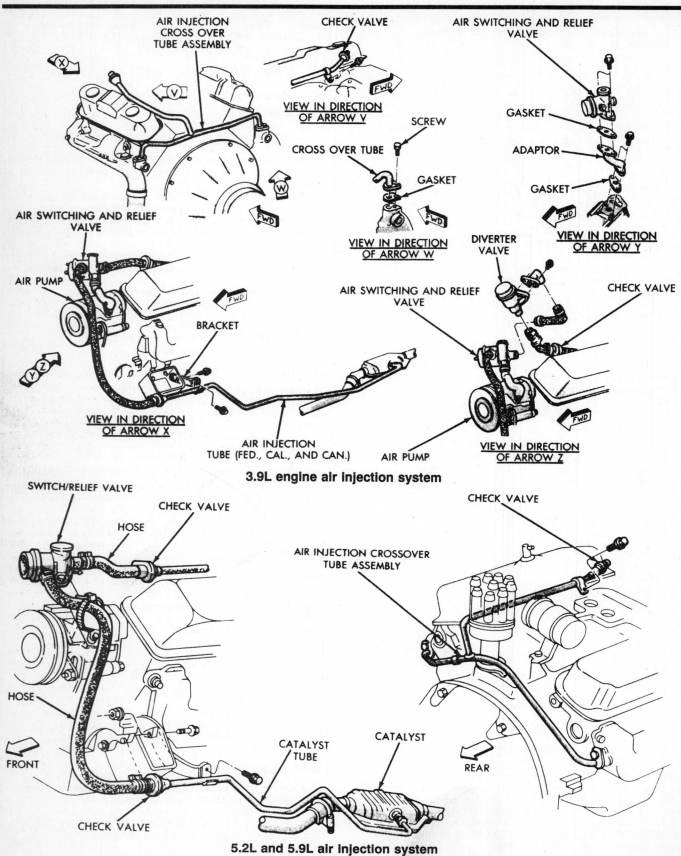

AIR INJECTION
CROSS OVER
TUBE ASSEMBLY

CHECK VALVE

AIR SWITCHING AND RELIEF
VALVE

X

V

FWD

**VIEW IN DIRECTION
OF ARROW V**

SCREW

GASKET

CROSS OVER TUBE

GASKET

FWD

**VIEW IN DIRECTION
OF ARROW W**

GASKET

ADAPTOR

GASKET

FWD

**VIEW IN DIRECTION
OF ARROW Y**

W

FWD

AIR SWITCHING AND RELIEF
VALVE

AIR PUMP

FWD

BRACKET

DIVERTER
VALVE

AIR SWITCHING AND RELIEF
VALVE

CHECK VALVE

Y Z

**VIEW IN DIRECTION
OF ARROW X**

AIR INJECTION
TUBE (FED., CAL., AND CAN.)

AIR PUMP

FWD

**VIEW IN DIRECTION
OF ARROW Z**

3.9L engine air injection system

SWITCH/RELIEF VALVE

CHECK VALVE

HOSE

CHECK VALVE

AIR INJECTION CROSSOVER
TUBE ASSEMBLY

HOSE

FRONT

REAR

CATALYST
TUBE

CATALYST

CHECK VALVE

5.2L and 5.9L air injection system

Condition	Possible Cause	Correction
EXCESSIVE BELT NOISE	(a) Loose belt.	(a) Tighten belt
	(b) Seized pump.	(b) Replace pump.
EXCESSIVE PUMP NOISE. CHIRPING	(a) Insufficient break-in.	(a) Recheck for noise after 1600 km. (1,000 miles) of operation.
EXCESSIVE PUMP NOISE CHIRPING, RUMBLING, OR KNOCKING	(a) Leak in hose.	(a) Locate source of leak using soap solution and correct.
	(b) Loose hose.	(b) Reassemble and replace or tighten hose clamp.
	(c) Hose touching other engine parts.	(c) Adjust hose position.
	(d) Switch/relief valve inoperative.	(d) Replace switch/relief valve.
	(e) Check valve inoperative.	(e) Replace check valve.
	(f) Pump mounting fasteners loose.	(f) Tighten mounting screws as specified.
	(g) Pump failure.	(g) Replace pump.
NO AIR SUPPLY (ACCELERATE ENGINE TO 1500 RPM AND OBSERVE AIR FLOW FROM HOSES. IF THE FLOW INCREASES AS THE RPM'S INCREASE, THE PUMP IS FUNCTIONING NORMALLY. IF NOT, CHECK POSSIBLE CAUSE.	(a) Loose drive belt.	(a) Tighten to specifications.
	(b) Leaks in supply hose.	(b) Locate leak and repair or replace as required.
	(c) Leak at fitting(s).	(c) Tighten or replace clamps.
	(d) Check valve inoperative.	(d) Replace check valve.
AIR SUPPLY UPSTREAM WITH NO VACUUM APPLIED TO SWITCH/RELIEF VALVE.	(a) Defective Air Switching Control Solenoid.	(a-1) With the engine running, connect voltmeter to the light blue with red tracer of the air switching solenoid connector and ground. Voltmeter should pulse between 0 and 14 volts. Solenoid is okay. If the voltage pulses between 0-2 volts replace solenoid dual assembly. Voltmeter not pulsating but reads within 1 volt of battery voltage, perform next test.
		(a-2) Turn the engine off. Disconnect the SBEC 60-way connector. Turn ignition switch to ON position. Connect a voltmeter to cavity 36 and ground. Voltmeter reads within 1 volt of battery voltage. Replace the computer. If voltmeter reads 0 volts (without vacuum actuated electrical switch), repair wire of cavity 36 for open circuit. If voltmeter reads 0 volts with vacuum actuated electrical switch, perform next test.
		(a-3) Turn ignition switch off. Disconnect the vacuum actuated electrical switch connector. Connect an ohmmeter between the terminals of the switch. Ohmmeter shows continuity, repair wire of SBEC connector cavity 36. Ohmmeter does not show continuity, replace switch.
		(a-4) Engine running connect voltmeter to green wire of the air switching solenoid connector and ground. Voltmeter reading within 1 volt of battery voltage, replace the computer. If voltmeter reads between 0 and 1 volt, repair wire in cavity 36 of SBEC 60-way connector for short to ground.
	(b) Switch/relief valve inoperative.	(b) Replace switch/relief valve.

Air pump diagnosis

1. Remove the air and vacuum hoses from the switch/relief valve.

2. Remove the two screws securing the switch/relief valve to the mounting flange and remove valve.

3. Remove any gasket material from the mounting surface and the switch/relief valve.

4. Position a new gasket on the mounting surface.

5. Position the switch/relief valve on the mounting surface and secure it with screws. Tighten to 95 inch lbs. (11 N.m).

6. Install the air and vacuum hoses to the switch/relief valve.

Check Valve (Injection Tube Assemblies)

The check valve is not repairable and if necessary, should be replaced with a new check valve. To determine if the valve has failed, remove the air hose from check valve inlet tube. If exhaust gas escapes from inlet tube, the valve has failed and must be replaced. If the exhaust manifold injection tube assembly joint is leaking, remove the injection tube assembly and replace the gasket material. If the tube nut is leaking torque it to 25–35 ft.lbs. (34–47 N.m).

NON THREADED TYPE CHECK VALVE

1. Release the clamp and disconnect the air hose from check valve inlet.

2. Remove the screws or tube nut securing the injection tube to the exhaust manifolds or exhaust pipe.

3. Remove the injection tube assembly from the engine.

4. Remove all gasket material from the exhaust manifold and injection tube flanges.

5. Install new gaskets on the exhaust manifold flanges and install a new injection tube assembly. Tighten the flange mounting and injection tube bracket screws to 100 inch lbs. (11.2 N.m). On tube nut joint assemblies, install the tube nut and tighten to 25–35 ft.lbs. (34–47 N.m).

6. Connect the air hoses to the check valve inlet and secure with clamps.

THREADED TYPE CHECK VALVE

1. Release the clamp and disconnect the air hose from the check valve inlet.

2. Unscrew the check valve from the injection tube assembly, using the hex at the threaded joint.

3. Install the check valve on tube assembly. Tighten using hex at the threaded joint, to 25 ft. lbs. (34 N.m).

4. Install the air hose to the check valve inlet and tighten the clamp.

NOTE: Removal of the lower tube assembly may require removal of the upper air pump and A/C compressor.

Exhaust Gas Recirculation

OPERATION

The EGR system reduces oxides of nitrogen (Nox) in engine exhaust and helps prevent spark knock. This is accomplished by allowing a predetermined amount of hot exhaust gas to recirculate and dilute the incoming fuel/air mixture. This dilution reduces peak flame temperature during combustion.

The EGR system is a backpressure type. A backpressure transducer measures the amount of exhaust backpressure on the exhaust side of the EGR valve and varies the strength of the vacuum signal applied to the EGR valve. The transducer uses this backpressure signal to provide the correct amount of Exhaust Gas Recirculation under all conditions. The 3.9L, 5.2L and 5.9L engines use manifold vacuum controlled by a vacuum solenoid.

EGR System On-Board Diagnostics (California Vehicles Only): All California vehicles with EGR systems have an On-Board Diagnostic System for the EGR system. The Diagnostic System uses a solenoid in the vacuum signal line to the EGR valve.

The Diagnostic System Check is activated only during selected engine/driving conditions to avoid mis-diagnosis, and checks the entire EGR system for failures. The engine controller monitors EGR system performance and registers a fault code if the system has failed or degraded, and the dash-mounted check engine light is turned on indicating immediate service is required.

If a malfunction is indicated by a check engine light and a fault code for EGR system, proper operation of the EGR system should be checked. If the EGR system is found to be functioning correctly, the on-board diagnostics system should then be checked.

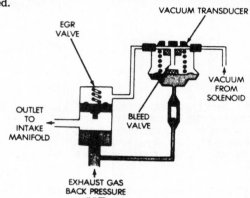

EGR system operation

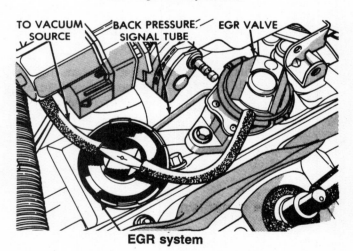

EGR system

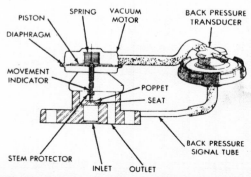

EGR and backpressure transducer assembly

CONDITION	POSSIBLE CAUSE	CORRECTION
EGR Valve Stem Does Not Move On System Test.	Cracked, leaking, disconnected or plugged hoses.	Verify correct hose connections. Replace hose harness if damaged or plugged hoses are found. Disconnect hose harness from EGR vacuum transducer and connect auxiliary vacuum supply. Raise and maintain engine RPM to 2000 RPM. Apply 10 in. of vacuum and check valve movement. If valve does not move replace EGR valve/transducer assembly. If valve opens approximately 3 mm (1/8 inch) apply vacuum to valve nipple and check for a leaking diaphragm. The valve should remain open 30 seconds or longer. If leakage occurs, replace EGR valve/transducer assembly. If valve does not leak check control system.
EGR Valve Stem Does Not Move on System Test. But, Operates Normally on External Vacuum Source.	Defective control system or plugged passages.	Remove throttle body. Inspect EGR vacuum source and associated passages in throttle body for obstructions. Clean passages as necessary. Use low air pressure to inspect for air flow through passages. Recheck for normal EGR operation.
Engine Will Not Idle. Dies Out on Return to Idle or Idle is Very Rough or Slow.	High EGR valve leakage in closed position.	If removal of vacuum hose from EGR valve does not correct rough idle, remove EGR valve/transducer assembly. Inspect poppet and ensure that it is seated. Replace valve/transducer if defective.

EGR system diagnosis

SERVICE

CAUTION

Apply the parking brake and block the wheels before performing idle check or adjustment, or any engine running tests or adjustment.

Exhaust Gas Recirculation (EGR) System

A failed or malfunctioning EGR system can cause engine spark knock, sags or hesitation, rough idle, and/or engine stalling. To assure proper operation of the EGR system all passages and moving parts must be free from plugging or sticking as a result of deposits. It is also important for the entire system to be free from leaks. Any gases or components found to be leaking should be replaced.

Proper operation of the EGR System can be checked with the following tests:

1. The engine should be warmed up and running at normal idle speed, with parking brake on and wheels blocked.
2. Allow the engine to idle in neutral with the throttle closed, then quickly accelerate the engine speed to approximately 2000 rpm, watching carefully the groove (movement indicator) on the EGR valve stem.
3. Movement of the valve stem during the period of acceleration should be visible by observing a change in relative location of the groove on the stem. This indicates that the control system is functioning correctly, and the EGR Gas Flow Test can then be performed.
4. If no movement of the valve stem is visible, the problem should be diagnosed using the test procedure described under "EGR DIAGNOSIS".

EGR Gas Flow Test

1. Connect a tachometer to the engine.
2. Remove the vacuum hose (or rubber elbow) from the EGR valve and connect a hand vacuum pump to the valve vacuum motor nipple.
3. Start the engine and slowly apply vacuum to EGR valve vacuum motor.
4. Engine rpm should drop as vacuum reaches 3 to 5 inches Hg and continue to drop as more vacuum is applied (engine may even stall). This means EGR gas is flowing through the system.
5. Successful completion of these tests indicates a fully functional EGR system.
6. If the engine speed does not drop, a defective EGR valve or plugged EGR passage is indicated. Remove the EGR valve and inspect. Intake manifold passages should be checked for deposits and cleaned if necessary.

REMOVAL AND INSTALLATION

EGR Valve

1. Disconnect the vacuum hose to EGR valve.
2. Remove the 2 attaching screws.
3. Remove the EGR valve. Remove the mounting gasket and discard. Clean the intake manifold surface and check for cracks.
4. Place a new EGR gasket on the intake manifold.
5. Install the EGR valve using the mounting screws and torque to 200 inch lbs. (23 N.m).
6. Connect the vacuum hose to EGR valve.

EGR Solenoid

1. Remove the vacuum hoses and electrical connectors from the solenoids.
2. Remove the fasteners and remove the solenoid pack.
3. Depress the tab on top of the solenoid to be replaced and slide the solenoid down out of the mounting bracket.

To install:

4. Place the new solenoid it position and slide in up into the bracket until the mounting tab is secure.
5. Mount the solenoid in position and secure it with the mounting fasteners.
6. Attach the vacuum lines and wiring connectors to the solenoids.

Catalytic Converters

The catalytic converter, mounted in the trucks exhaust system is a muffler-shaped device containing a ceramic honeycomb shaped material coated with alumina and impregnated with catalytically active precious metals such as platinum, palladium and rhodium.

The catalyst's job is to reduce air pollutants by oxidizing hydrocarbons (HC) and carbon monoxide (CO). Catalysts containing palladium and rhodium also oxidize nitrous oxides (NOx).

On some trucks, the catalyst is also fed by the secondary air system, via a small supply tube in the side of the catalyst.

No maintenance is possible on the converter, other than keeping the heat shield clear of flammable debris, such as leaves and twigs.

Other than external damage, the only significant damage possible to a converter is through the use of leaded gasoline, or by way of a too rich fuel/air mixture. Both of these problems will ruin the converter through contamination of the catalyst and will eventually plug the converter causing loss of power and engine performance. When this occurs, the catalyst must be replaced.

Emissions Maintenance Reminder Light

RESETTING

A special diagnostic tool such as a DRBII is required. The DRBII tool is connected to the vehicle's diagnostic connector. The tester is turned to the Emissions EMR tests, and the vehicle's ignition switch is turned to the RUN position. The tester is set to the EMR memory check, then to the Reset EMR light. These procedures reset the EMR timing in the computer and will turn off the reminder light. The ignition is then turned off and the tester disconnected from the vehicle.

Oxygen Sensor

OPERATION

The oxygen sensor is a device which produces an electrical voltage when exposed to the oxygen present in the exhaust gases. The sensor is mounted in the exhaust manifold. The oxygen sensor is electrically heated internally for faster switching when the engine is running. When there is a large amount of oxygen present (lean mixture), the sensor produces a low voltage. When

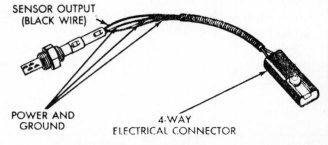

Heated oxygen sensor

there is a lesser amount present (rich mixture) it produces a higher voltage. By monitoring the oxygen content and converting it to electrical voltage, the sensor acts as a rich-lean switch. The voltage is transmitted to the engine controller. The controller signals the power module to trigger the fuel injector.

The oxygen sensor has a heating element that keeps the sensor at proper operating temperature during all operating modes. Maintaining correct sensor temperature at all times allows the system to enter into CLOSED LOOP operation sooner.

In CLOSED LOOP operation the engine controller monitors the sensor input (along with other inputs) and adjusts the injector pulse width accordingly. During OPEN LOOP operation the engine controller ignores the sensor input and adjusts the injector pulse to a preprogrammed value based on other inputs.

On 3.9L engines, the oxygen sensor is located in the exhaust Y-pipe. On 5.2L and 5.9L engines, it is located above the outlet of the left exhaust manifold.

REMOVAL AND INSTALLATION

—————— CAUTION ——————

The exhaust manifold/pipe get extremely hot during engine operation and if touched can cause severe burns. If servicing the oxygen sensor when the engine is hot, avoid contact with the exhaust manifold or pipe.

NOTE: A special tool C-4907, or equivalent is required to remove the oxygen sensor. Do not pull on the oxygen sensor wires.

1. Disconnect the wiring harness from the oxygen sensor.
2. Remove the sensor using the appropriate tool.

NOTE: The oxygen sensor threads are coated with an anti-seize compound. The compound must be removed from the mounting boss threads, either in the exhaust manifold or Y-pipe. An 18mm x 1.5 x 6E tap is required.

3. Clean the threads of the mount to remove any old anti-seize compound.
4. If the old sensor is to be reused, apply anti-seize compound to it's threads. New sensors come with the compound already applied.
5. Install and tighten the sensor to 20 ft. lbs. (27 N.m). Connect the wiring harness.

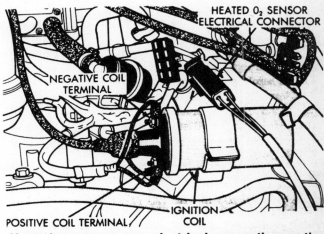

Heated oxygen sensor electrical connection on the 3.9L engine

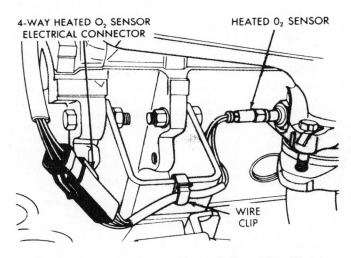

Heated oxygen sensor location on the 5.2L engine

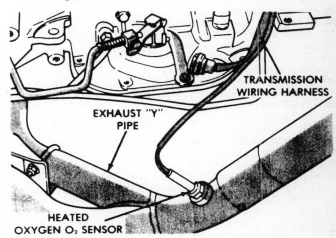

Heated oxygen sensor location on the 3.9L engine

ELECTRONIC ENGINE CONTROLS

Components

The engine controller tests many of its own input and output circuits. If a fault is found in a major system, this information is stored in the memory.

Engine Controller

The engine controller is a digital computer containing a microprocessor. It receives input signals from various switches and sensors. Based on these inputs, the engine controller regulates various engine and vehicle operations through different system components that are referred to as "Engine Controller Outputs".

The engine controller adjusts pulse width (air-fuel ratio) based on inputs indicating:
- exhaust gas oxygen content
- coolant temperature
- manifold absolute pressure
- battery voltage
- engine speed
- throttle position
- throttle body temperature (5.2L and 5.9L engines only)
- battery voltage
- park/neutral switch (neutral safety switch — automatic transmissions only)
- speed control

Manifold absolute pressure and engine speed are the primary inputs that determine fuel injector pulse width. However, the throttle position sensor and oxygen sensor (when in closed loop operation) inputs also are important inputs that determine injector pulse width. These sensors and switches are considered "Engine Controller Inputs".

The engine controller adjusts ignition timing based upon inputs it receives from sensors that react to:
- engine RPM (the distributor pick-up)
- manifold absolute pressure
- coolant temperature
- throttle position
- throttle body temperature (5.2L and 5.9L engines only)
- park/neutral switch (neutral safety switch — automatic transmission only)
- vehicle speed
- idle contact switch.

These sensors are considered "Engine Controller Inputs". The engine controller adjusts idle speed based on inputs it receives from sensors that react to: throttle position, vehicle speed, park neutral switch, coolant temperature and from inputs it receives from the air conditioning switch, brake switch, and the idle contact (closed throttle) switch.

Based on inputs that it receives, the engine controller also controls ignition coil dwell, and evaporative canister purge operation. The engine controller also adjusts the alternator charge rate through control of the alternator field and provides speed control operation.

Engine Controller Inputs
- air conditioning switch and low pressure switch
- ASD sense (Z2 sense)
- battery voltage
- brake switch
- coolant temperature sensor
- cruise control (on/off, accel/resume, coast/set)
- engine speed (distributor pick-up)
- idle contact (closed throttle) switch
- throttle body temperature sensor (5.2L and 5.9L engines only)
- manifold absolute pressure (MAP) sensor

- park/neutral switch (neutral safety switch)
- overdrive override switch
- oxygen sensor
- throttle position sensor
- vehicle distance sensor

Engine Controller Outputs
- air conditioning clutch relay
- air switching solenoid
- alternator
- auto shutdown (ASD) relay
- canister purge solenoid
- check engine lamp
- diagnostic connector
- EGR solenoid
- EMR lamp
- fuel injectors
- idle speed control (ISC) actuator
- ignition coil
- overdrive and lock-up torque convertor solenoids
- shift indicator lamp
- tachometer

The automatic shut down (ASD) relay is mounted externally, but is switched on and off by the engine controller. When the relay is energized it provides power for the electric fuel pump, fuel injector, ignition coil and exhaust gas oxygen sensor heater element. The engine controller monitors the distributor pick-up signal to determine engine speed and compute injector synchronization. If the engine controller does not receive a distributor signal when the ignition witch is in the RUN position, it will de-energize the ASD relay.

The engine controller contains a power supply which converts battery voltage to a regulated 8.0 volts to power the distributor pick-up. The engine controller also provides a five (5) volt supply for the Manifold Absolute Pressure (MAP) sensor and Throttle Position Sensor (TPS).

A/C Switch — Engine Controller Input

When the A/C switch is in the ON position and the low pressure cut out switch is closed, the engine controller receives an input indicating that the air conditioning had been selected. After receiving this input the engine controller activates the A/C compressor clutch by grounding the A/C clutch relay, and maintains idle speed to a scheduled RPM through control of the Idle Speed Control Actuator.

Battery Voltage — Engine Controller Input

The engine controller monitors the battery voltage input to determine fuel injector pulse width and alternator field control. If battery voltage is low the engine controller will increase injector pulse width (period of time that the injector is energized). This is done to compensate for the reduced flow through the injector. The reduced flow is a result of the increased amount of time required to fully open the injector due to lower battery voltage.

Brake Switch — Engine Controller Input

When the brake switch is activated, the engine controller receives an input indicating that the brakes are being applied. The torque convertor is unlocked in response to this input.

Coolant Temperature Sensor — Engine Controller Input

The coolant temperature sensor is installed next to the thermostat housing in the intake manifold water jacket passage and provides an input voltage to the engine controller. As coolant temperature varies the coolant temperature sensors resistance changes resulting in a different input voltage to the engine controller.

When the engine is cold, the engine controller will demand slightly richer air-fuel mixtures and higher idle speeds until normal operating temperatures are reached.

Engine Speed (Distributor Pick-Up) — Engine Controller Input

The engine speed input is supplied to the engine controller by the distributor pick-up. The distributor pick-up is a Hall Effect device.

A shutter (sometimes referred to as an interrupter) is attached to the distributor shaft. The shutter contains one blade per engine cylinder (6 blades on 3.9L engines and 8 blades on 5.2L and 5.9L engines). A switch plate is mounted to the distributor housing above the shutter. The switch plate contains the distributor pick-up (a Hall Effect device and magnet) through which the shutter blades rotate. As the shutter blades pass through the pick-up, they interrupt the magnetic field. The Hall effect device in the pick-up senses the change in the magnetic field and switches on and off (creating pulses), generating the input signal to the engine controller. the engine controller calculates engine speed through the number of pulses generated. The engine speed input is also used to control coil dwell.

Idle Contact Switch (Closed Throttle Switch) — Engine Controller Input

The idle contact switch is integral with the idle speed actuator (ISC) actuator and provides an input signal to the engine controller. This input enables the engine controller to increase or decrease the throttle stop angle (by extending or retracting the ISC actuator) in response to engine operating conditions. The idle contact switch is sometimes referred to as the closed throttle switch.

Manifold Absolute Pressure (MAP) Sensor — Engine Controller Input

The MAP sensor reacts to absolute pressure in the intake manifold and provides an input voltage to the engine controller. As engine load changes manifold pressure varies, causing the MAP sensors resistance to change. The change in MAP sensor resistance results in a different input voltage level supplies the engine controller. The input voltage level supplies the engine controller with information relating to ambient barometric pressure during engine start-up (cranking) and to engine load while the engine is running. The engine controller uses this input along with inputs from other sensors to adjust air-fuel mixture.

The MAP sensor is mounted on the throttle body and is connected to the throttle body with a vacuum hose and to the engine controller electrically.

Oxygen Sensor — Engine Controller Input

The oxygen sensor is located in the exhaust down pipe on 3.9L engines and on 5.2L and 5.9L engines above the left exhaust manifold outlet. The oxygen sensor provides an input voltage to the engine controller relating the oxygen content of the exhaust gas. The engine controller uses this information to fine tune the air-fuel ratio by adjusting injector pulse width.

The oxygen sensor produces voltages from 0 to 1 volt, depending upon the oxygen content of the exhaust gas in the exhaust manifold. When a large amount of oxygen is present (caused by a lean air-fuel mixture), the sensor produces a low voltage. When there is a lesser amount present (rich air-fuel mixture) it produces a higher voltage. By monitoring the oxygen content and converting it to electrical voltage, the sensor acts as a rich-lean switch.

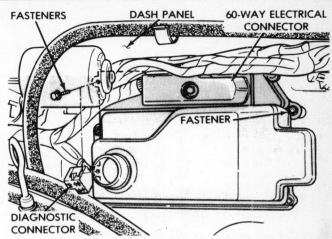

Single board engine controller SBEC — 1990-91

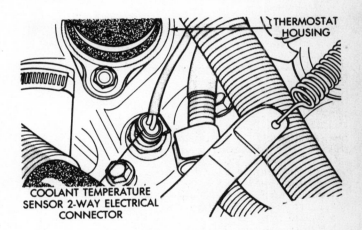

Coolant temperature sensor

The oxygen sensor is equipped with a heating element that keeps the sensor at proper operating temperature during all operating modes. Maintaining correct sensor temperature at all times allows the system to enter into closed loop operation sooner.

In "Closed Loop" operation the Engine controller monitors the oxygen sensor input (along with other inputs) and adjusts the injector pulse width accordingly. During "Open Loop" operation the Engine controller ignores the oxygen sensor input and adjusts injector pulse width to a preprogrammed value (based on other sensor inputs).

Park/Neutral Switch — Engine Controller Input

The park/neutral switch is located on the automatic transmission housing and provides an input to the engine controller that indicates the automatic transmission is in Park, Neutral or a drive gear selection. This input is used to determine idle speed (varying with gear selection), fuel injector pulse width, and ignition timing advance. The park neutral switch is sometimes referred to as the neutral safety switch.

Throttle Body Temperature Sensor — Engine Controller Input (5.2L and 5.9L Engines)

Vehicles equipped with a 5.2L or 5.9L engine have a throttle body temperature sensor. The sensor monitors throttle body temperature which is the same as fuel temperature. It is mount-

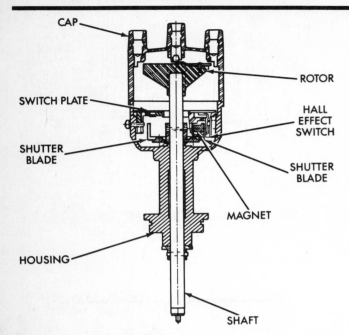

Distributor pick-up

CAP
ROTOR
SWITCH PLATE
HALL EFFECT SWITCH
SHUTTER BLADE
SHUTTER BLADE
MAGNET
HOUSING
SHAFT

ed in the throttle body. This sensor provides information on fuel temperature which allows the Engine controller to enrichen the air fuel mixture for a hot restart condition.

Throttle Position Sensor (TPS) — Engine Controller Input

The Throttle Position Sensor (TPS) is mounted on the throttle body below the MAP sensor and connected to the throttle blade shaft. THe TPS is a variable resistor that provides the engine controller with an input signal (voltage) that represents throttle blade position. As the position of the throttle blade changes, the resistance of the TPS changes.

The engine controller supplies approximately 5 volts to the TPS. The TPS output voltage (input signal to the engine controller) represents the throttle blade position. The engine controller receives an input signal voltage from the TPS varying in an approximate range of from 1 volt at minimum throttle opening (idle) to 4 volts at wide open throttle. Along with inputs from other sensors, the engine controller uses the TPS input to determine current engine operating conditions and adjust fuel injector pulse width and ignition timing.

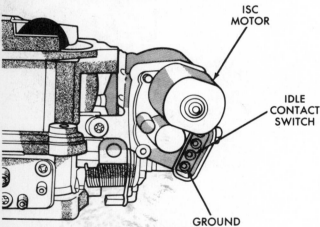

ISC MOTOR
IDLE CONTACT SWITCH
GROUND

Idle contact switch

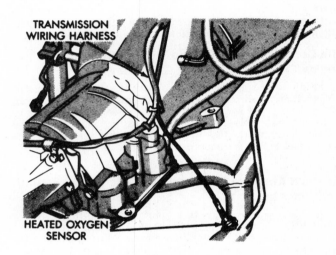

TRANSMISSION WIRING HARNESS
HEATED OXYGEN SENSOR

Heated oxygen sensor location — 3.9L engine

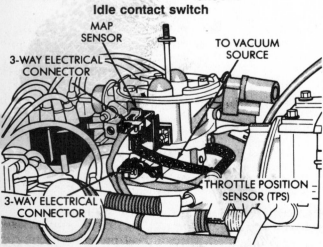

MAP SENSOR
TO VACUUM SOURCE
3-WAY ELECTRICAL CONNECTOR
3-WAY ELECTRICAL CONNECTOR
THROTTLE POSITION SENSOR (TPS)

Manifold absolute pressure (MAP) sensor location

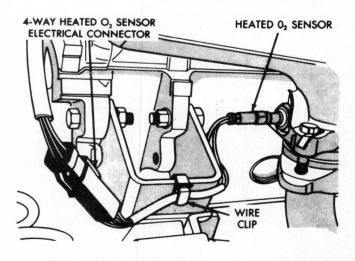

4-WAY HEATED O₂ SENSOR ELECTRICAL CONNECTOR
HEATED O₂ SENSOR
WIRE CLIP

Heated oxygen sensor location — 5.2L and 5.9L engines

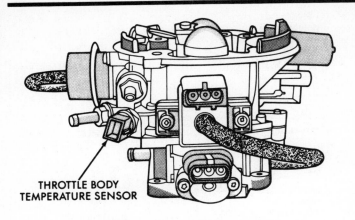

Throttle body temperature sensor

THROTTLE BODY
TEMPERATURE SENSOR

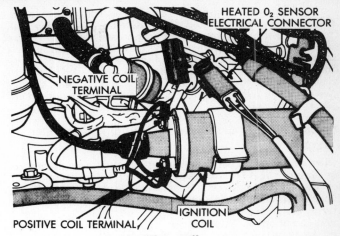

HEATED O₂ SENSOR
ELECTRICAL CONNECTOR

NEGATIVE COIL
TERMINAL

POSITIVE COIL TERMINAL

IGNITION
COIL

Ignition coil

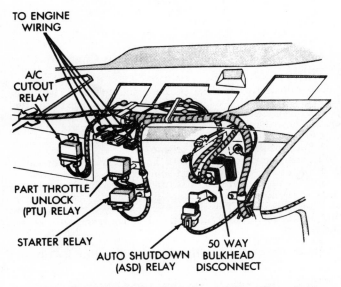

TO ENGINE
WIRING

A/C
CUTOUT
RELAY

PART THROTTLE
UNLOCK
(PTU) RELAY

STARTER RELAY

AUTO SHUTDOWN
(ASD) RELAY

50 WAY
BULKHEAD
DISCONNECT

A/C clutch, ASD, starter and part throttle unlock relays

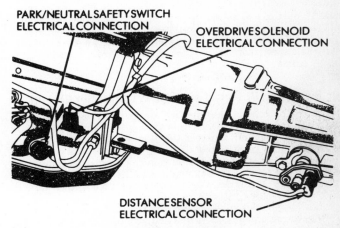

PARK/NEUTRAL SAFETY SWITCH
ELECTRICAL CONNECTION

OVERDRIVE SOLENOID
ELECTRICAL CONNECTION

DISTANCE SENSOR
ELECTRICAL CONNECTION

Overdrive solenoid

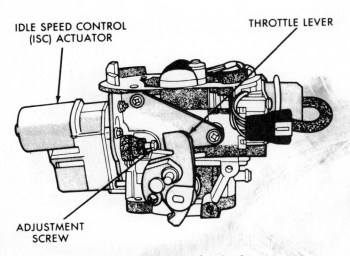

IDLE SPEED CONTROL
(ISC) ACTUATOR

THROTTLE LEVER

ADJUSTMENT
SCREW

Idle speed control actuator

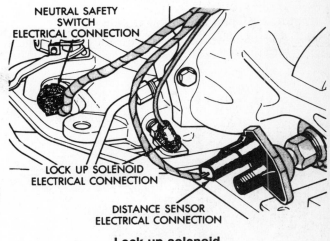

NEUTRAL SAFETY
SWITCH
ELECTRICAL CONNECTION

LOCK UP SOLENOID
ELECTRICAL CONNECTION

DISTANCE SENSOR
ELECTRICAL CONNECTION

Lock-up solenoid

Vehicle Distance (Speed) Sensor — Engine Controller Input

The distance sensor is located in the extension housing of the transmission. The sensor input is used by the engine controller to determine vehicle speed and distance traveled.

The distance sensor generates 8 pulses per sensor revolution. These signals are interpreted along with a closed throttle signal from the throttle position sensor by the engine controller. These inputs are used to determine if a closed throttle deceleration or a normal idle (vehicle stopped) condition exists. Under deceleration conditions, the engine controller adjusts the ISC motor to maintain a desired MAP value. Under idle conditions, the engine controller adjusts the ISC motor to maintain a desired engine speed.

Air Conditioning Clutch Relay — Engine Controller Output

The engine controller operates the air conditioning compressor through the A/C clutch relay. By switching the ground path for relay on and off, the engine controller is able to cycle the air conditioning compressor clutch based on changes in engine operating conditions.

Air conditioning compressor clutch engagement is controlled by the engine controller through the air conditioning clutch relay. The relay is energized when the A/C switch is closed (A/C has been selected) and the blower motor switch is in the on position. If, during A/C operation, the engine controller senses low idle speeds or a wide open throttle condition, it will de-energize the relay. This prevents air conditioning clutch engagement until the idle speed increases or the wide open throttle condition no longer exists.

Air Switching Solenoid — Engine Controller Output

The engine controller controls the upstream or downstream discharge of air from the air pump through the air switching solenoid. The solenoid controls the flow of vacuum to the air switching/relief valve of the air pump system.

When the air switching system is in the downstream mode of operation the solenoid is not supplied to the air switching/relief valve. During downstream operation air pump output is directed to the catalytic convertor.

When the air switching system is in the upstream mode of operation, the solenoid is energized and vacuum is supplied to the air switching/relief valve. During upstream operation air pump output is directed to the exhaust manifolds.

Alternator — Engine Controller Output

The engine controller regulates the charging system voltage within a range of 12.9 to 15.0 volts.

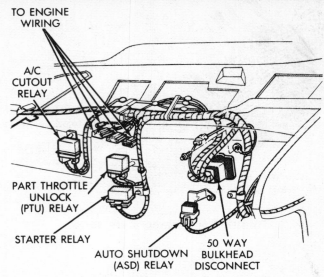

Engine compartment relay identification

Auto Shutdown (ASD) Relay — Engine Controller Output

The Auto Shutdown Relay (ASD) is mounted on the left inner fender well. The ASD relay supplies battery voltage to the fuel pump, fuel injector, ignition coil, and oxygen sensor heating element. The ground circuit for the ASD relay is controlled by the engine controller. The engine controller operates the relay by switching the ground circuit on and off.

The engine controller monitors the distributor pick-up signal to determine engine speed and compute injector synchronization. If the engine controller does not receive a distributor signal when the ignition switch is in the "Run" position it will de-energize the ASD relay. This disconnects battery voltage from the fuel pump, fuel injectors, ignition coil, and oxygen sensor heater element.

Canister Purge Solenoid — Engine Controller Output

Vacuum for the Evaporative Canister is controlled by the Canister Purge Solenoid. The Solenoid is regulated by the engine controller based on engine operating conditions. The engine controller operates the solenoid by switching the ground

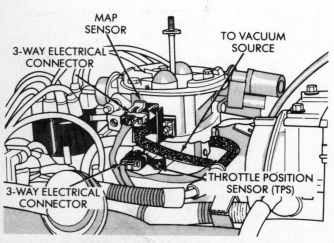

Manifold absolute pressure (MAP) sensor location

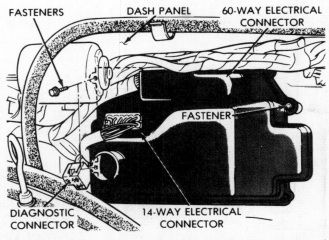

Single module engine controller (1989) SMEC electrical connectors

circuit on and off. When energized, the solenoid prevents vacuum from reaching the evaporative canister. When not energized the solenoid allows vacuum to flow through to canister.

During warm-up and for a specified time period after hot starts the engine controller grounds the purge solenoid causing it to energize, preventing vacuum from reaching the charcoal canister valve. The engine controller removes the ground to the solenoid when the engine reaches a specified temperature and the time delay interval has occurred. The solenoid is de-energized. Vacuum flows to the canister purge valve to purge fuel vapors through the throttle body.

The Purge Solenoid will also be energized during certain idle conditions, to update the fuel delivery calibration.

Check Engine Lamp — Engine Controller Output

The Check Engine Lamp illuminates at the bottom of the instrument panel each time the ignition key is turned on and stays on for three seconds as a bulb test.

If the engine controller receives an incorrect signal or no signal from certain sensors or emission related systems (California vehicles only) the lamp is turned on. This is a warning that the engine controller has gone into a limp-in mode in an attempt to keep the system operating. It signals an immediate need for service.

The lamp can also be used to display fault codes. Cycle the ignition switch on, off, on, off, on within five seconds and any fault codes stored in the engine controller memory will be displayed in a series of flashes representing digits.

EGR Solenoid — Engine Controller Output

Vacuum for the exhaust gas recirculation (EGR) valve function is switched on and off by the EGR solenoid. The solenoid is controlled by the engine controller. The solenoid is located on the engine right valve cover.

When the solenoid is energized by the engine controller, it prevents vacuum from reaching the EGR valve transducer and EGR valve. The solenoid is energized during engine warm-up, closed throttle (idle), wide open throttle and rapid acceleration/deceleration. If the solenoid wire connector is disconnected, the EGR valve function will be operational at all times.

On California vehicles, there is an On-Board Diagnostics test that is performed by the engine controller. The test will check the EGR system for failures. The engine controller monitors EGR system performance and registers a fault code if the system has failed or is degraded and the dash-mounted check engine light is turned on indicating immediate service is required.

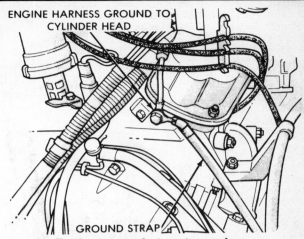
Engine ground strap to engine

EMR Lamp — Engine Controller Output

The emission maintenance reminder (EMR) lamp informs the vehicle owner that scheduled maintenance is required for certain emission system components. The engine controller determines vehicle mileage by monitoring the vehicle distance sensor. The EMR lamp illuminates at 96,600 km, 132,000 km, 192,000 km (60,000, 82,500, and 120,000 miles).

Fuel Injectors — Engine Controller Output

The Fuel Injector is an electric solenoid driven by the engine controller. Based on sensor inputs the engine controller determines injector pulse width (how long the injector is energized) and when the fuel injector should operate. When electrical current is supplied to the injector, a spring loaded pintle is lifted from its seat. This allows fuel to flow through past the pintle and orifice. This action causes the fuel to from a 30° cone shaped spray pattern before entering the air stream in the throttle body.

Fuel is supplied to the injector constantly at regulated 14.5 psi, the unused fuel is returned to the fuel tank.

Fuel Pump — Engine Controller Output

The engine controller operates the fuel pump through the auto shutdown (ASD) relay. When the relay is energized by the engine controller, battery voltage is supplied to the fuel pump.

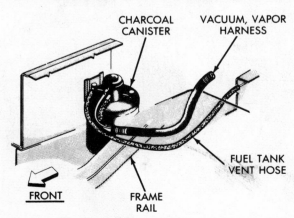
Vapor canister

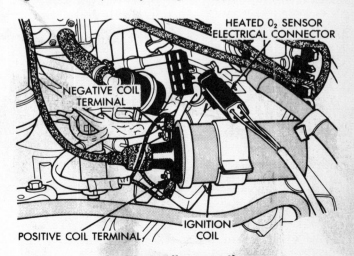
Ignition coil connections

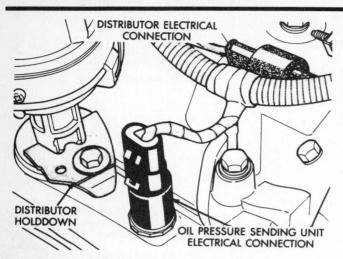

Distributor and oil pressure switch electrical connections

Idle Speed Control (ISC) Actuator — Engine Controller Output

The Idle Speed Control Actuator (ISC) motor is mounted to the throttle body and controlled by the engine controller. The throttle lever rests against an adjustment screw at the end of the actuator (plunger). The actuator extends or retracts to control engine idle speed and to set throttle stop angle during deceleration. Based on inputs from the various engine control system sensors and switches the engine controller supplies current and a ground path to the ISC motor to adjust the actuator position for the particular operating conditions.

Do not attempt to correct a high idle speed condition by changing the adjustment screw position. When the engine is shut off, the engine controller extends the ISC actuator to its maximum position to preset a "fast idle" for the next start-up. Turning the adjustment screw inward will not change the idle speed of a warm engine, but can cause cold start problems due to restricted air flow. The ISC actuator also contains the idle contact switch.

Ignition Coil — Engine Controller Output

The engine controller controls ignition coil firing through the auto shutdown (ASD) relay. When the relay is energized by the engine controller, battery voltage is supplied to the ignition coil positive terminal. The engine controller will de-energize the ASD relay if it does not receive input from the distributor pick-up.

Overdrive Solenoid — Engine Controller Outputs

On vehicles equipped with overdrive, the engine controller regulates the 3–4 overdrive upshift and downshift through the Overdrive solenoid.

Lock-up Torque Convertor Solenoids — Engine Controller Outputs

On vehicles equipped with either an A-999 or A-500 transmission the engine controller regulates torque converter lock-up through the Lock-up solenoid.

MODES OF OPERATION

As input signals to the engine controller change, the engine controller adjusts its response to the output devices. For example, the engine controller must calculate a different injector pulse width and ignition timing for idle than it does for wide open throttle (WOT). There are several different modes of operation that determine how the engine controller responds to the various input signals.

Modes of operation are of two different types, Open Loop and Closed Loop. During Open Loop modes the engine controller receives input signals and responds only according to preset engine controller programming. Input from the oxygen sensor is not monitored during Open Loop modes.

During Closed Loop modes the engine controller does monitor the oxygen sensor input. This input indicates to the engine controller whether or not the calculated injector pulse width results in the ideal air-fuel ration of 14.7 parts air to 1 part fuel. By monitoring the exhaust oxygen content through the oxygen sensor, the engine controller can "fine tune" the injector pulse width to achieve optimum fuel economy combined with low emission engine performance.

The dual point fuel injection system, has the following modes of operation:

- Ignition switch ON
- Engine start-up
- Engine warm-up
- Cruise
- Acceleration
- Deceleration
- Wide Open Throttle
- Ignition switch OFF

The ignition switch on, engine start-up (crank), engine warm-up, acceleration, deceleration and wide open throttle modes are Open Loop modes. The cruise mode, with the engine at operating temperature is a Closed Loop mode.

Ignition Switch On Mode

This is an Open Loop mode. When the dual point fuel injection System is activated by the ignition switch, the following actions occur:

- Engine controller determines atmospheric air pressure from the MAP sensor input to determine basic fuel strategy.
- The engine controller monitors the coolant temperature sensor input. The engine controller modifies fuel strategy based on this input.

When the key is in the ON position, the auto shutdown (ASD) relay is energized for 0.5 seconds. If the engine controller does not receive an input from the distributor pick-up within three seconds it will de-energize the ASD relay.

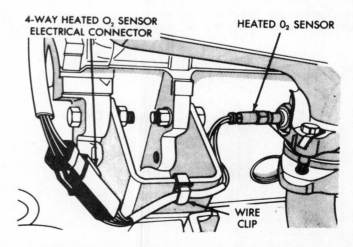

Heated oxygen sensor electrical connection — V8 engines

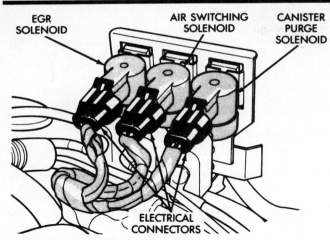

EGR, air switching and canister purge solenoids electrical connectors

Engine Start — Up Mode

This is an Open Loop mode. The following actions occur when the starter motor is engaged.

If the engine controller receives a distributor signal it will energize the auto shut down (ASD) relay to supply battery voltage to the fuel pump, fuel injector, ignition coil, and oxygen sensor heating element. If the engine controller does not receive a distributor input, the ASD relay will not be energized.

When the engine is operating and idling within $\pm$ 64 RPM of the target RPM, the engine controller compares the current MAP value with the atmospheric pressure value it received during the Key-On mode. If a minimum difference between the two is not detected, a pneumatic fault code is set into memory.

Once the ASD relay has been energized the engine controller:

• Supplies a ground path to the injectors and the injectors. 3.9L engines are pulsed six times per engine revolution instead of the normal three pulses per revolution. 5.2L and 5.9L engines are pulsed 8 times instead of the normal 4 times.

• Determines injector pulse width based on coolant temperature, barometric pressure (MAP sensor), and the number of engine revolutions since cranking was initiated.

• Monitors coolant temperature sensor, distributor pick-up, MAP sensor, throttle body temperature sensor (5.2L and 5.9L engines) and throttle position sensor. The engine controller determines correct ignition timing based on these inputs.

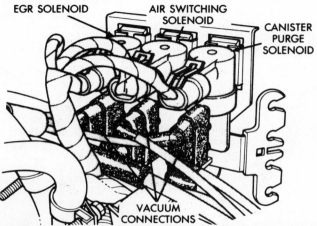

EGR, air switching and canister purge solenoids vacuum connections

Engine Warm — Up Mode

This is a Open Loop mode. The following inputs are received by the engine controller:

• coolant temperature
• idle contact switch (closed throttle switch)
• manifold absolute pressure
• engine speed (distributor pick-up)
• throttle body temperature sensor (5.2L and 5.9L engines only)
• throttle position
• park/neutral switch (neutral safety switch — automatic transmission)
• A/C switch
• battery voltage
• vehicle distance (speed) sensor

The engine controller provides a ground path for the injectors to precisely control injector pulse width (by switching the ground on and off). On 3.9L engines the injectors are fired three times per engine revolution. The engine controller adjusts engine idle speed, throttle stop angle, and ignition timing.

For vehicles equipped with a manual transmission, the upshift indicator lamp is controlled by the engine controller according to coolant temperature, engine speed and load.

Cruise Mode

When the engine is at operating temperature this is a Closed Loop mode. During cruising speed the following inputs are received by the engine controller:

• coolant temperature
• manifold absolute pressure
• engine speed
• throttle position
• exhaust gas oxygen content
• park/neutral switch (neutral safety switch) (automatic transmission)
• A/C control positions
• brake switch
• vehicle distance (speed) sensor
• speed control
• brake switch
• overdrive override switch

The engine controller provides a ground path for the injectors to precisely control injectors pulse width. On 3.9L engines the injectors are fired three times per engine revolution. On 5.2L and 5.9L engines the injectors are fired 4 times per engine revolution. The engine controller adjusts engine idle speed, throttle stop angle and ignition timing. The engine controller adjusts the

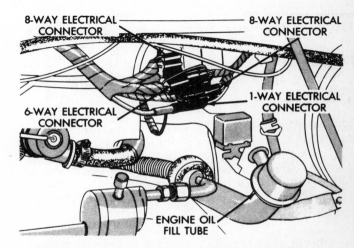

Engine harness to main harness connection

air/fuel mixture ratio according to the oxygen content in the exhaust gas.

On vehicles equipped with a manual transmission, the upshift indicator lamp is controlled by the engine controller according to engine speed and load.

Acceleration Mode

This is an OPEN LOOP mode when MAP and engine RPM conditions are exceeded. The engine controller recognizes an abrupt increase in throttle position or MAP pressure as a demand for increased engine output and vehicle acceleration. The engine controller increases injector pulse width in response to increased fuel demand.

Deceleration Mode

When the engine is at operating temperature this is a CLOSED LOOP mode. However, if the MAP sensor input is below the minimum programmed level, the system will enter OPEN LOOP. The system will return to closed loop when the MAP sensor input rises above the minimum level. During deceleration the following inputs are received by the engine controller:
- coolant temperature
- manifold absolute pressure
- idle contact switch
- engine speed
- throttle position
- exhaust gas oxygen content
- park/neutral switch
- AC control positions

A closed throttle input along with an abrupt decrease in manifold pressure indicates to the controller that the vehicle is in a hard deceleration. The engine controller may reduce injector firing to once per engine revolution to lean the air/fuel mixture as sensed through the oxygen sensor during hard deceleration.

The engine controller grounds the EGR and evaporative purge solenoids stopping EGR and canister purge functions. The engine controller may cycle air switching solenoid for short periods of time in response to the MAP sensor sending a high vacuum signal.

Wide Open Throttle Mode

This is a OPEN LOOP mode. During wide open throttle operation, the following inputs are received by the engine controller:
- coolant temperature
- manifold absolute pressure
- engine speed
- throttle position

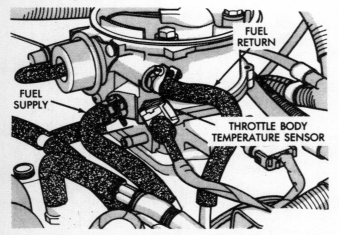

Throttle body temperature sensor location — V8 engines

- AC switch
- overdrive override switch

When the engine controller senses wide open throttle condition through the TPS it will provide a ground for the EGR solenoid and evaporative canister purge solenoid preventing EGR and canister purge functions.

The exhaust gas oxygen content input is not accepted by the engine controller and it will adjust injector pulse width to supply a predetermined amount of additional fuel.

Ignition Switch Off Mode

This is an OPEN LOOP mode. When the ignition switch is turned to the OFF position, the engine controller ceases to provide a ground for the ASD relay and extends the ISC actuator in anticipation of the next startup. When the ASD relay is not energized, battery voltage is shut off from the fuel pump, fuel injector, ignition coil and oxygen sensor heating element. All fuel injection stops.

ON BOARD DIAGNOSTICS

A visual inspection for loose, disconnected, or incorrectly routed wires and hose should be made to save unnecessary test and diagnostic time.

The engine controller has been programmed to monitor many different circuits. If a problem is sensed in a monitored circuit often enough to indicate an actual problem, a fault is stored in the engine controller memory for eventual display. If the problem is repaired or ceases to exist, the engine controller cancels the fault code after 51 engine starts.

Certain criteria must be met for a fault to be entered into the engine controller memory. The criteria may be a specific range of engine RPM, engine temperature, and/or input voltage to the engine controller.

It is possible that a fault for a monitored circuit may not be entered into the memory even though a malfunction has occurred. This may happen because one of the fault criteria has not been met.

There are several operating conditions that the engine controller does not monitor and set faults for.

MONITORED CIRCUITS

The engine controller can detect certain fault conditions:
- Open or Shorted Circuit: The engine controller can determine if the sensor output (input to the controller) is within the proper range and if the circuit is open or shorted
- Output Device Current Flow: The engine controller senses whether the output devices are hooked up. If there is a problem with the circuit, the controller senses whether the circuit is open, shorted to ground, or shorted high.
- Oxygen Sensor: The engine controller can determine if the oxygen sensor is switching between rich and lean once the system has entered closed loop.

NON-MONITORED CIRCUITS

The engine controller does not monitor the following circuits, systems and conditions that could have malfunctions that result in driveability problems. Fault codes may not be displayed for these conditions. However, problems with these systems may cause fault codes to be displayed for other systems. Foe example, a fuel pressure problem will not register a fault directly, but could cause a rich or lean condition. This would cause an oxygen sensor fault to be stored in the engine controller.
- Fuel Pressure: Fuel pressure is controlled by the vacuum assisted fuel pressure regulator. The engine controller cannot detect a clogged fuel pump inlet filter, clogged inline fuel filter, or a pinched fuel supply or return line. However, these could result in a rich or lean condition causing an oxygen sensor fault to be stored in the engine controller.
- Engine Timing: The engine controller cannot detect an incorrectly indexed timing chain, camshaft sprocket and crank-

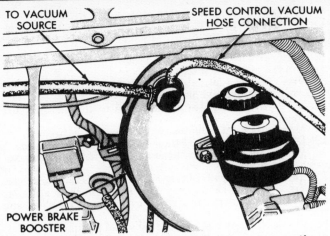

Power brake and speed control vacuum connections

shaft sprocket. The engine controller also cannot detect an incorrectly indexed distributor. However, these could result in a rich or lean condition causing an oxygen sensor fault to be stored in the engine controller.
- Cylinder Compression: The engine controller cannot detect uneven, low, or high engine cylinder compression.
- Exhaust System: The engine controller cannot detect a plugged, restricted or leaking exhaust system.
- Fuel Injector Malfunctions: The engine controller cannot determine if the fuel injector is clogged, the pintle is sticking or the wrong injector is installed. However, these could result in a rich or lean condition causing and oxygen sensor fault to be stored in the engine controller.
- Excessive Oil Consumption: Although the engine controller monitors the exhaust system oxygen content through the oxygen sensor when the system is in closed loop, it cannot determine excessive oil consumption.
- Throttle Body Air Flow: The engine controller cannot detect a clogged or restricted air cleaner inlet of filter element.
- Evaporative System: The engine controller will not detect a restricted, plugged or loaded evaporative purge canister.
- Vacuum Assist: Leaks or restrictions in the vacuum circuits of vacuum assisted engine control system devices are not monitored by the engine controller. However, these could result in a MAP sensor fault being stored in the engine controller.
- Engine Controller System Ground: The engine controller cannot determine a poor system ground. However a fault code may be generated as a result of this condition.

- Engine Controller Connector Engagement: The engine controller cannot determine spread or damaged connector pins. However, a fault code may be generated as a result of this condition.

High and Low Limits

The engine controller compares input signal voltage from each input device with established high and low limits that are programmed into it for that device. If the input voltage is not within specifications and other fault criteria are met, a fault will be stored in the memory. Other fault code criteria might include engine RPM limits or input voltages from other sensors or switches. The other inputs might have to be sensed by the controller when it senses a high or low input voltage from the control system device in question.

NOTE: A fault indicates that the engine controller has recognized an abnormal signal in the system. Fault codes indicate the result of the failure but never identify the failed component directly.

VISUAL INSPECTION

A visual inspection for loose, disconnected, or incorrectly routed wires and hoses should be made before attempting to diagnose or service the fuel injection system. A visual check will help find these conditions. It also saves unnecessary test and diagnostic time. A thorough visual inspection include the following:
1. Verify that the 14-way and/or 60-way connector is fully inserted into the socket of the engine controller.
2. Verify that the hoses and electrical connections are securely attached to the vapor canister and purge solenoid.
3. Verify that the alternator electrical wiring and the drive belt are correctly installed.
4. Verify that the engine ground strap is attached at the engine dash panel.
5. Check the ignition coil electrical connections and verify that the 3-way connector is attached to the distributor.
6. Verify that the connector is attached to the heated oxygen sensor.
7. Verify that the vacuum and electrical connections at the EGR solenoid are secure.
8. Verify that the vacuum and electrical connections to the MAP sensor are secure.
9. Verify that the throttle body wiring connection to the main harness is secure.

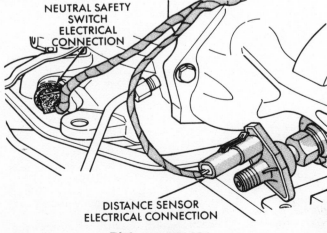

Distance sensor

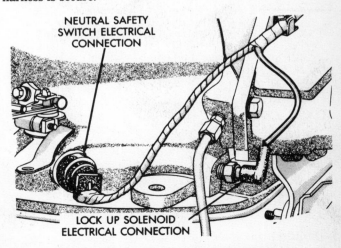

Transmission electrical connections

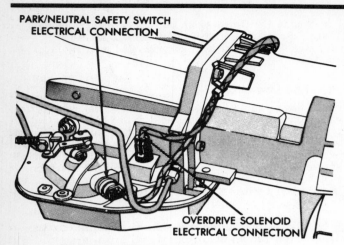

PARK/NEUTRAL SAFETY SWITCH
ELECTRICAL CONNECTION

OVERDRIVE SOLENOID
ELECTRICAL CONNECTION

Transmission electrical connections

10. Check the transmission neutral safety switch, lockup solenoid and distance sensor connections.

11. Verify that the connector is secure at the AIS motor.

12. Verify that the connectors to the throttle position sensor and throttle body temperature sensor are secure.

13. Verify that the fuel injector connector is secure.

14. Verify that the hose from the PCV valve is securely attached to the intake manifold vacuum port. Make sure the orifice plug is in the valve cover nipple.

15. Verify that all throttle body vacuum connections are secure and the hoses are intact.

16. Verify that the heated air door vacuum connection at the air cleaner is secure and not leaking.

17. Verify that the hoses are attached to the backpressure transducer.

18. Make sure all distributor connections are secure.

19. Check the connector at the engine temperature sensor.

20. Verify that the engine harness to main harness connector is tight and secure.

21. Verify that the power brake vacuum connections are tight.

22. Verify that all relays are firmly seated in their electrical connectors. Make sure the battery connections are clean and tight.

23. Check all hoses and wiring at the fuel pump.

Fault Codes

Stored fault codes can be displayed on the Check Engine Lamp. To display fault code flashes, cycle the ignition key switch to ON-OFF-ON-OFF-ON within five seconds. Any fault codes stored in the engine controller memory will be displayed in a series of flashes representing digits.

CHECK ENGINE LAMP FAULT CODE	DRB II DISPLAY	DESCRIPTION OF FAULT CONDITION
11	IGN REFERENCE SIGNAL	No Distributor reference signal detected during engine cranking.
12	No. of Key-ons since last fault or since faults were erased.	Direct battery input to controller disconnected within the last 50-100 ignition key-ons.
13† **	MAP PNEUMATIC SIGNAL	No variation in MAP sensor signal is detected.
	or MAP PNEUMATIC CHANGE	No difference is recognized between the engine MAP reading and the stored barometric pressure reading.
14† **	MAP VOLTAGE TOO LOW	MAP sensor input below minimum acceptable voltage.
15 **	or MAP VOLTAGE TOO HIGH	MAP sensor input above maximum acceptable voltage.
15 **	VEHICLE SPEED SIGNAL	No distance sensor signal detected during road load conditions.
16† **	BATTERY INPUT SENSE	Battery voltage sense input not detected during engine running.
17	LOW ENGINE TEMP	Engine coolant temperature remains below normal operating temperatures during vehicle travel (Thermostat).
21**	OXYGEN SENSOR SIGNAL	Neither rich or lean condition is detected from the oxygen sensor input.
22† **	COOLANT VOLTAGE LOW	Coolant temperature sensor input below the minimum acceptable voltage.
	or COOLANT VOLTAGE HIGH	Coolant temperature sensor input above the maximum acceptable voltage.
23	T/B TEMP VOLTAGE LOW	Throttle body temperature sensor input below the minimum acceptable voltage. (5.2L and 5.9L only)
	or T/B TEMP VOLTAGE HI	Throttle body temperature sensor input above the maximum acceptable voltage. (5.2L and 5.9L only)
24† **	TPS VOLTAGE LOW	Throttle position sensor input below the minimum acceptable voltage.
	or TPS VOLTAGE HIGH	Throttle position sensor input above the maximum acceptable voltage.
25**	ISC MOTOR CIRCUITS	A shorted condition detected in one or more of the ISC control circuits.
26	INJ 1 PEAK CURRENT	High resistance condition detected in the INJ 1 injector output circuit.
	or INJ 2 PEAK CURRENT	High resistance condition detected in the INJ 2 injector output circuit.
27	INJ 1 CONTROL CKT	INJ 1 injector output driver stage does not respond properly to the control signal.
	or INJ 2 CONTROL CKT	INJ 2 injector output driver stage does not respond properly to the control signal.
31**	PURGE SOLENOID CKT	An open or shorted condition detected in the purge solenoid circuit.
32**	EGR SOLENOID CIRCUIT	An open or shorted condition detected in the EGR solenoid circuit. (California emissions only)
	or EGR SYSTEM FAILURE	Required change in Fuel/Air ratio not detected during diagnostic test. (California emissions only)
33	A/C CLUTCH RELAY CKT	An open or shorted condition detected in the A/C clutch relay circuit.
34	S/C SERVO SOLENOIDS	An open or shorted condition detected in the speed control vacuum or vent solenoid circuits.
35	IDLE SWITCH SHORTED	Idle contact switch input circuit shorted to ground.
	or IDLE SWITCH OPENED	Idle contact switch input circuit opened.

1989 Fault Codes

CHECK ENGINE LAMP FAULT CODE	DRB II DISPLAY	DESCRIPTION OF FAULT CONDITION
36	AIR SWITCH SOLENOID	An open or shorted condition detected in the air switching solenoid circuit.
37	PTU SOLENOID CIRCUIT	An open or shorted condition detected in the torque converter part throttle unlock solenoid circuit. (Automatic transmission only)
41	CHARGING SYSTEM CKT	Output driver stage for alternator field does not respond properly to the voltage regulator control signal.
42	ASD RELAY CIRCUIT	An open or shorted condition detected in the auto shutdown relay circuit.
	or Z1 VOLTAGE SENSE	No Z1 voltage sensed when the auto shutdown relay is energized
43	IGNITION CONTROL CKT	Output driver stage for ignition coil does not respond properly to the dwell control signal.
44	FJ2 VOLTAGE SENSE	No FJ2 voltage present at the logic board during controller operation.
45	OVERDRIVE SOLENOID	An open or shorted condition detected in the overdrive solenoid circuit. (Automatic transmission only)
46**	BATTERY VOLTAGE HIGH	Battery voltage sense input above target charging voltage during engine operation.
47	BATTERY VOLTAGE LOW	Battery voltage sense input below target charging voltage during engine operation.
51**	AIR FUEL AT LIMIT	Oxygen sensor signal input indicates lean fuel/air ratio condition during engine operation.
52**	AIR FUEL AT LIMIT	Oxygen sensor signal input indicates rich fuel/air ratio condition during engine operation.
	or EXCESSIVE LEANING	Adaptive fuel value leaned excessively due to a sustained rich condition.
53	INTERNAL SELF-TEST	Internal engine controller fault condition detected,
55		Completion of fault code display on the CHECK ENGINE lamp.
62	EMR MILEAGE ACCUM	Unsuccessful attempt to update EMR mileage in the controller EEPROM.
63	EEPROM WRITE DENIED	Unsuccessful attempt to write to an EEPROM location by the controller.
	FAULT CODE ERROR	An unrecognized fault ID received by DRB II.

† Check Engine Lamp On
**Check Engine Lamp On (California Only)

1989 Fault Codes

CAV	WIRE COLOR	DESCRIPTION
1	DG/RD*	MAP SENSOR
2		
3	TN/WT*	COOLANT SENSOR
4	BK/LB*	SENSOR RETURN
5	BK/WT*	SIGNAL GROUND
6		
7	WT/LG*	SPEED CONTROL RESUME SWITCH
8	YL/RD*	SPEED CONTROL ON/OFF SWITCH
9	BR/RD*	SPEED CONTROL SET SWITCH
10	DG/BK*	Z1 INPUT (VOLTAGE SENSE)
11		
12	DB/WT*	FJ2
13	VT/WT*	5 VOLT SUPPLY
14	DG/OR*	ALTERNATOR FIELD CONTROL
15	LB/RD*	POWER GROUND
16	LB/RD*	POWER GROUND
17	BR/WT*	ISC (IDLE/SPEED CONTROL ACTUATOR)
18		
19	GY/RD*	ISC (IDLE SPEED CONTROL ACTUATOR)
20		
21	BK/RD*	THROTTLE BODY TEMPERATURE SENSOR
22	OR/DB*	THROTTLE POSITION SENSOR
23	BK/DG*	OXYGEN SENSOR
24		
25		
26	VT	CLOSED THROTTLE SWITCH
27		
28		
29	WT/PK*	BRAKE SWITCH
30	BR/YL*	PARK/NEUTRAL SWITCH
31	LG	SCI RECEIVE
32	GY/WT*	INJECTOR CONTROL 2
33	VT/YL*	INJECTOR CONTROL 1
34	YL	DWELL CONTROL
35		
36		

CAV	WIRE COLOR	DESCRIPTION
37	GY/PK*	EMISSION MAINTENANCE REMINDER LAMP
38	OR/WT*	OVERDRIVE TRANSMISSION LOCK-OUT SWITCH
39	BK/OR*	AIR SWITCHING SOLENOID
40	GY/YL*	EGR SOLENOID
41	RD	DIRECT BATTERY
42		
43		
44		
45	BR	A/C SWITCH SENSE
46		
47	GY/BK*	DISTRIBUTOR REFERENCE PICKUP
48	WT/OR*	VEHICLE DISTANCE SENSOR SIGNAL
49		
50		
51	PK	SCI TRANSMIT
52	OR	9 VOLT INPUT
53	TN/RD*	SPEED CONTROL VACUUM SOLENOID
54	PK/BK*	PURGE SOLENOID
55	OR/BK*	LOCK-UP TORQUE CONVERTER OR SHIFT IND LAMP — MANUAL TRANS
56	DB/OR*	A/C CLUTCH RELAY
57		
58	DB/YL*	AUTO SHUT DOWN RELAY
59	BK/PK*	CHECK ENGINE LAMP
60	LG/RD*	SPEED CONTROL VENT SOLENOID

WIRE COLOR CODES					
BK	BLACK	LB	LIGHT BLUE	VT	VIOLET
BR	BROWN	LG	LIGHT GREEN	WT	WHITE
DB	DARK BLUE	OR	ORANGE	YL	YELLOW
DG	DARK GREEN	PK	PINK	*	WITH TRACER
GY	GRAY	RD	RED		
		TN	TAN		

CONNECTOR TERMINAL SIDE SHOWN

1989 – 60 way single module engine controller (SMEC) wiring connector

Pin	Circuit	Color	Function
1	N-6	OR	9 Volt Output
2	K-5	BK/WT	Signal Ground
3	K-14	DB/WT	FJ2 Output
4	J-2	DB	12-Volt
5	Y-1	GY/WT	Injector Control 2
6	J-9-A	BK	Power Ground
7	J-9-B	BK	Power Ground
8	K-16	VT/YL	Injector Control Signal
9	Y-11	WT	Injector Driver #1
10	Y-12	TN	Injector Driver 2
11	R-31	DG/OR	Voltage Regulator Signal
12	J-5	BK/YL	Ignition Coil Driver
13	K-15	YL	Anti Dwell Signal
14	R-3	DG	Regulator Control

TERMINAL SIDE

1989 — 14 way (SMEC) connector

4-Way Heated Oxygen Sensor Connector

Pin	Circuit	Color	Function
1	N11	18 BK/DG*	Oxygen switching output

2 Coolant Temperature Sensor Connector

Pin	Circuit	Color	Function
1	N5	18 BK/LB*	Sensor ground input
2	K10	20 TN	Sensor signal output

Check Engine Lamp I.P. Printed Circuit Board

Pin	Circuit	Color	Function
1	J2	14 DB	Ignition switch input
2	K3	20 BK/PK*	Ground input

2-Way Distance Sensor Connector

Pin	Circuit	Color	Function
1	N5	18 BK/LB*	Sensor ground input
2	G7	20 WT/OR*	Sensor signal output

3-Way MAP Sensor Connector

Pin	Circuit	Color	Function
1	K8	20 VT	5 volts input
2	K4	20 DG/R*	Sensor signal output
3	N5	18 BK/LB*	Sensor ground input

2-Way Canister Purge Solenoid Canister

Pin	Circuit	Color	Function
1	J2	18 DB	Ignition switch input
2	K1	20 PK/BK*	Solenoid ground input

3-Way Distributor Connector

Pin	Circuit	Color	Function
1	N5	18 BK/LB*	Ground input
2	N6	18 OR	8 volts input
3	N7	18 GY BK*	Signal output

2-Way EGR Solenoid Connector

Pin	Circuit	Color	Function
1	J2	18 DB	Ignition switch input
2	S6	20 GY/YL*	Solenoid ground input

2-Way Air Switching Solenoid Connector

Pin	Circuit	Color	Function
1	J2	14 DB	Ignition switch input
2	N23	20 BK/OR*	Solenoid ground input

3-Way Throttle Position Sensor (TPS) Connector

Pin	Circuit	Color	Function
1	N5	18 BK/LB*	Sensor ground input
2	K7	18 OR/DB*	Sensor signal output
3	K8	20 VT/WT*	5 volt input

4-Way ISC Actuator Connector

Pin	Color	Circuit	Function
1	N1	18 GY/RD*	ISC Actuator
2	N2	18 BR/WT*	ISC Actuator
3	N31	18 VT	ISC Actuator
4	K5	18 BK*	ISC Actuator

3-Way and 1-Way A/C Clutch Relay Connecto

Pin	Circuit	Color	Function
1	N13	20 DB/OR*	Ground input
2	C2	14 BK	A/C Damped Pressure Switch
3	J2	14 DB	Ignition switch input
4	C26	18 BR	A/C clutch ground

4-Way Auto Shutdown (ASD) Relay Connector

Pin	Circuit	Color	Function
1	Z1	14 DG/BK*	Ignition and Fuel System Feed
2	K19	20 DB	ASD Signal
3	J1	14 RD	Battery Feed
4	K14	18 DB	Ignition Switch Input

2-Way Throttle Body Temperature Sensor Connector

Pin	Circuit	Color	Function
1	N5	18 BK/LB*	Sensor ground input
2	K13	18 BK/RD*	Sensor signal output

*Denotes tracer

1989 Pin connector identification

CAV	WIRE COLOR	DESCRIPTION
1	DG/RD*	MAP SENSOR
2	TN/WT*	COOLANT SENSOR
3	RD/WT*	DIRECT BATTERY
4	BK/LB*	SENSOR RETURN
5	BK/WT*	SIGNAL GROUND
6	VT/WT*	5-VOLT OUTPUT (MAP AND TPS)
7	OR	8-VOLT OUTPUT (DISTRIBUTOR PICK-UP)
8	DG/BK*	Z-1 INPUT
9	DB	J2
10		
11	LB/RD*	POWER GROUND
12	LB/RD*	POWER GROUND
13		
14		
15	TN	INJECTOR DRIVER #2
16	WT/DB*	INJECTOR DRIVER #1
17		
18		
19	BK/GY*	IGNITION COIL DRIVER
20	DG	ALTERNATOR FIELD CONTROL
21	BK/RD*	THROTTLE BODY TEMPERATURE SENSOR (5.2L AND 5.9L)
22	OR/LB*	THROTTLE POSITION SENSOR (TPS)
23		
24	GY/BK*	IGNITION (DISTRIBUTOR) REFERENCE PICK-UP
25	PK	SCI TRANSMIT
26		
27	BR	A/C SWITCH SENSE
28	VT	IDLE CONTACT SWITCH
29	WT/PK*	BRAKE SWITCH
30	BR/YL*	PARK NEUTRAL SWITCH
31		
32	BK/PK*	CHECK ENGINE LAMP
33	TN/RD*	SPEED CONTROL VACUUM SOLENOID
34	DB/OR*	A/C CLUTCH RELAY
35	GY/YL*	EGR SOLENOID
36	BK/OR*	AIR SWITCHING SOLENOID

CAV	WIRE COLOR	DESCRIPTION
37		
38		
39		
40	BR/WT*	ISC MOTOR CLOSED
41	BK/DG*	OXYGEN SENSOR SIGNAL
42		
43	GY/LB*	TACHOMETER SIGNAL OUTPUT (VEHICLES WITH TACHOMETER)
44		
45	LG	SCI RECEIVE
46		
47	WT/OR*	DISTANCE SENSOR SIGNAL
48	BR/RD*	SPEED CONTROL SET SWITCH
49	YL/RD*	SPEED CONTROL ON/OFF SWITCH
50	WT/LG*	SPEED CONTROL RESUME SWITCH
51	DB/YL*	AUTO SHUTDOWN (ASD) RELAY
52	PK/BK*	PURGE SOLENOID
53	LG/RD*	SPEED CONTROL VENT SOLENOID
54	OR/BK*	SHIFT INDICATOR LIGHT (MANUAL TRANSMISSION ONLY)
54	OR/BK*	PART THROTTLE UNLOCK SOLENOID (AUTO TRANSMISSION)
55	OR/WT*	OVERDRIVE SOLENOID (AUTO TRANSMISSION ONLY)
56	GY/PK*	EMISSION MAINTENANCE REMINDER (EMR) LIGHT
57		
58		
59		
60	GY/RD*	ISC MOTOR OPEN

WIRE COLOR CODES				
BK	BLACK	LB	LIGHT BLUE	VT VIOLET
BR	BROWN	LG	LIGHT GREEN	WT WHITE
DB	DARK BLUE	OR	ORANGE	YL YELLOW
DG	DARK GREEN	PK	PINK	* WITH TRACER
GY	GRAY	RD	RED	
TN	TAN			

CONNECTOR TERMINAL SIDE SHOWN

1990 – 60 way single board engine controller (SBEC) identification

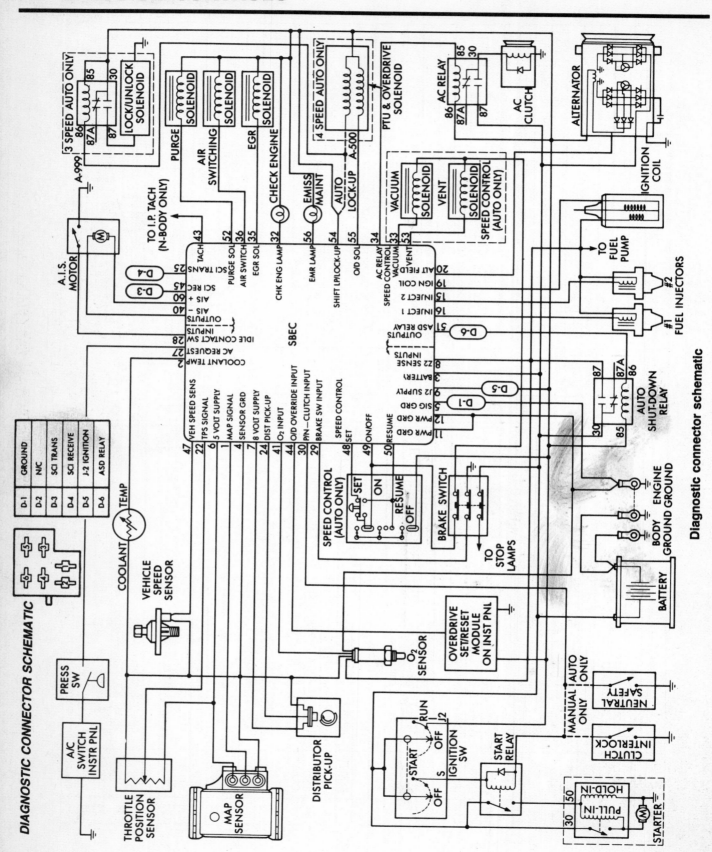

Diagnostic connector schematic

Diagnostic connector schematic

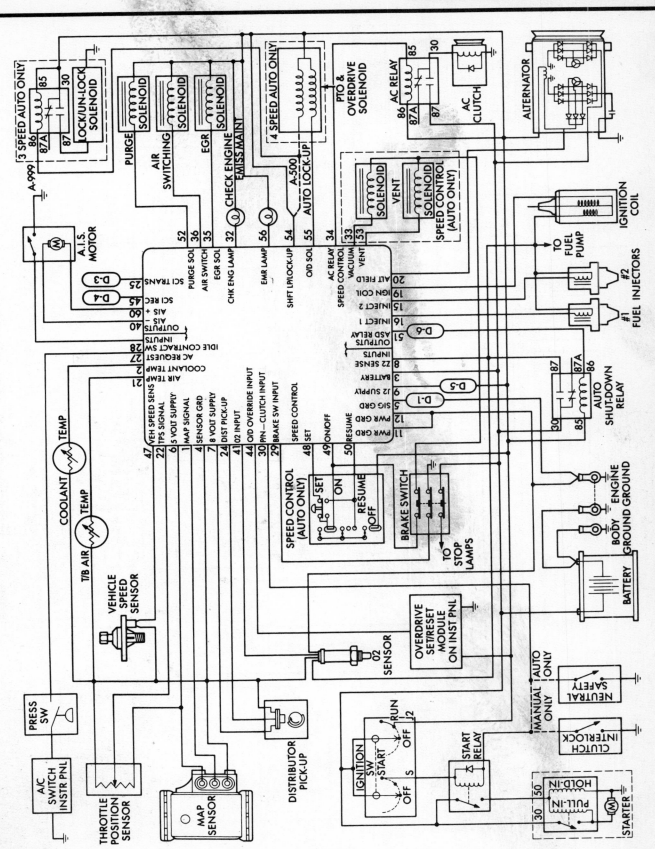

1990 — SMEC controller schematic — 5.2L engine

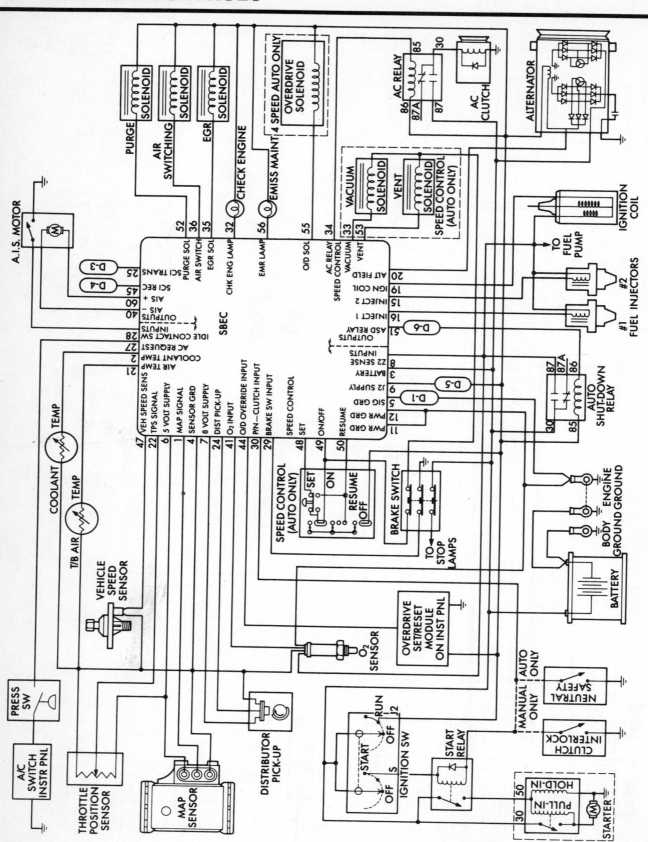

1990 — SMEC controller schematic — 5.9L normal duty engine

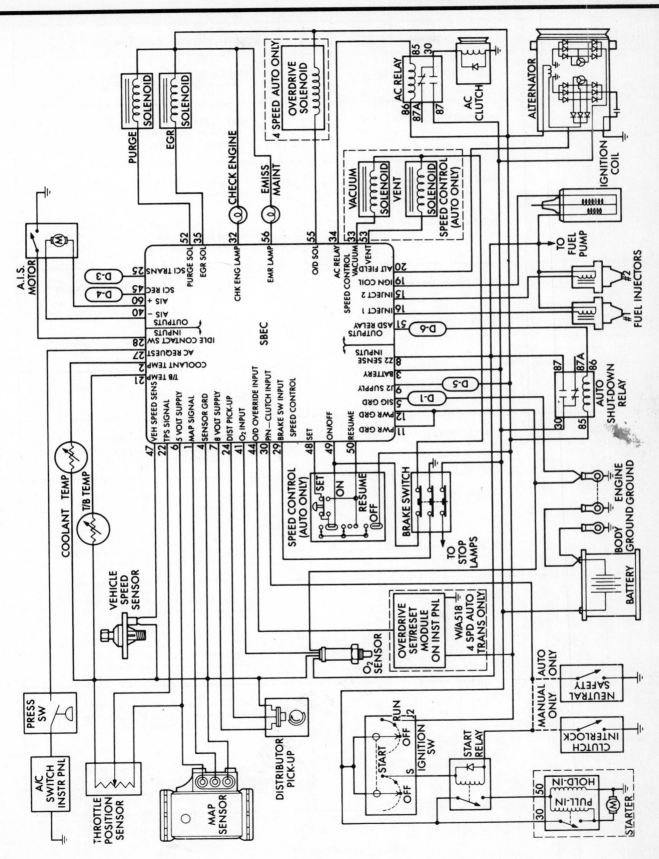

1990 — SMEC controller schematic — 5.9L heavy duty engine

Fault Code	DRBII Display	Description of Fault Condition
11	Ign Reference Signal	No distributor reference signal detected during engine cranking.
12	N/A (See Key-On Info)	Direct battery input to controller disconnected within the last 50 ignition key on cycles.
13	MAP Pneumatic Signal or MAP Voltage Too Low	No variation in MAP sensor signal is detected. No difference is recognized between the engine MAP reading and the stored barometric pressure reading.
14	MAP Voltage Too Low or MAP Voltage Too High	MAP sensor input below minimum acceptable voltage. MAP sensor input above maximum acceptable voltage.
15	Vehicle Speed Signal	No distance sensor signal detected during road load conditions.
17	Low Engine Temp	Engine coolant temperature remains below normal operating temperatures during vehicle travel (thermostat).
21	Oxygen Sensor Signal or O_2 Sensor Shorted High	Neither rich or lean condition is detected from the oxygen sensor input. Oxygen sensor input voltage maintained above normal operating range.
22	Coolant Voltage Low or Coolant Voltage High	Coolant temperature sensor input below the minimum acceptable voltage. Coolant temperature sensor input above the maximum acceptable voltage.
23	T/B Temp Voltage Low or T/B Temp Voltage High	Throttle body temperature sensor input below the minimum acceptable voltage (5.2L and 5.9L engines). Throttle body temperature sensor input above the maximum acceptable voltage (5.2L and 5.9L Engines).

1990 Fault Codes

FAULT CODE	DRBII DISPLAY	DESCRIPTION OF FAULT CONDITION
24	TPS Voltage Low or TPS Voltage High	Throttle position sensor input below the minimum acceptable voltage. Throttle position sensor input above the maximum acceptable voltage.
25	AIS Motor Circuits	A shorted condition detected in one or more of the AIS control circuits.
27	INJ 1 Control Ckt	Injector output driver #1 or #2 does not respond properly to the control signal.
31	Purge Solenoid Ckt	An open or shorted condition detected in the purge solenoid circuit.
32	EGR Solenoid Circuit or EGR System Failure	An open or shorted condition detected in the EGR solenoid circuit. Required change in air-fuel ratio not detected during diagnostic test (California emission packages only).
33	A/C Clutch Relay Ckt	An open or shorted condition detected in the A/C clutch relay circuit.
34	S/C Servo Solenoids	An open or shorted condition detected in the speed control vacuum or vent solenoid circuits.
35	Idle Switch Shorted or Idle Switch Opened	Idle contact switch input circuit shorted to ground. Idle contact switch input circuit opened.
36	Air Switch Solenoid	An open or shorted condition detected in the air switching solenoid circuit.
37	PTU Solenoid Circuit	An open or shorted condition detected in the torque convertor part throttle unlock circuit (Engine packages with A-999 or A-500 automatic transmissions only).

1990 Fault Codes

FAULT CODE	DRBII DISPLAY	DESCRIPTION OF FAULT CONDITION
41	Alternator Field Ckt	An open or shorted condition detected in the alternator control circuit.
42	ASD Relay Circuit or Z1 Voltage Sense	An open or shorted condition detected in the auto shutdown relay circuit. No Z1 voltage sensed when the auto shutdown relay circuit.
45	Overdrive Solenoid	An open or shorted condition detected in the overdrive solenoid circuit (engine packages with A-500 or A-518 automatic transmissions only).
46	Battery Voltage High	Battery voltage sensor input above target charging voltage during engine operation.
47	Charging Output Low	Battery voltage sense input below target charging voltage during engine operation and no significant change in voltage detected during active test of alternator output.
51	Lean F/A Condition	Oxygen sensor signal input indicates lean fuel/air ratio condition during engine operation.
52	Rich F/A Condition or Excessive Leaning	Oxygen sensor signal input indicates rich fuel/air ratio condition during engine operation. Adaptive fuel valve leaned excessively due to a sustained rich condition.
53	Internal Self Test	Internal engine controller fault condition detected.
62	EMR Miles Not Stored	Unsuccessful attempt to update EMR mileage in the controller EEPROM.
63	EEPROM Write Denied	Unsuccessful attempt to write to an EEPROM location by the controller.
55	N/A	Completion of fault code display on the CHECK ENGINE lamp.

1990 Fault Codes

CAV	WIRE COLOR	DESCRIPTION
1	DG/RD*	MAP SENSOR
2	TN	COOLANT SENSOR
3	RD	DIRECT BATTERY
4	BK/LB*	SENSOR RETURN
5	BK/WT*	SIGNAL GROUND
6	VT/WT*	5-VOLT OUTPUT (MAP AND TPS)
7	OR	8-VOLT OUTPUT (DISTRIBUTOR PICK-UP)
8	DG/BK*	AUTO SHUTDOWN (ASD) SENSE
9	DB	IGNITION
10		
11	LB/RD*	POWER GROUND
12	LB/RD*	POWER GROUND
13		
14		
15	TN	INJECTOR DRIVER #2
16	WT/DB*	INJECTOR DRIVER #1
17		
18		
19	BK/GY*	IGNITION COIL DRIVER
20	DG	ALTERNATOR FIELD CONTROL
21	BK/RD*	THROTTLE BODY TEMPERATURE SENSOR (5.2L AND 5.9L)
22	OR/LB*	THROTTLE POSITION SENSOR (TPS)
23		
24	GY	IGNITION (DISTRIBUTOR) REFERENCE PICK-UP
25	PK	SCI TRANSMIT
26		
27	BR	A/C SWITCH SENSE
28	VT	IDLE CONTACT SWITCH (CLOSED THROTTLE SWITCH)
29	WT/PK*	BRAKE SWITCH
30	BR/YL*	PARK NEUTRAL SWITCH
31		
32	BK/PK*	CHECK ENGINE LAMP
33	TN/RD*	SPEED CONTROL VACUUM SOLENOID
34	DB/OR*	A/C CLUTCH RELAY
35	GY/YL*	EGR SOLENOID
36	BK/OR*	AIR SWITCHING SOLENOID

CAV	WIRE COLOR	DESCRIPTION
37		
38		
39		
40	BR	ISC MOTOR CLOSED
41	BK/DG*	OXYGEN SENSOR SIGNAL
42		
43	GY/LB*	TACHOMETER SIGNAL OUTPUT (VEHICLES WITH TACHOMETER)
44	OR/WT*	OVERDRIVE LOCKOUT CONNECTOR
45	LG	SCI RECEIVE
46		
47	WT/OR*	DISTANCE SENSOR SIGNAL
48	BR/RD*	SPEED CONTROL SET SWITCH
49	YL/RD*	SPEED CONTROL ON/OFF SWITCH
50	WT/LG*	SPEED CONTROL RESUME SWITCH
51	DB/YL*	AUTO SHUTDOWN (ASD) RELAY
52	PK	PURGE SOLENOID
53	LG/RD*	SPEED CONTROL VENT SOLENOID
54	OR/BK*	SHIFT INDICATOR LIGHT (MANUAL TRANSMISSION ONLY)
54	OR/BK*	PART THROTTLE UNLOCK SOLENOID (AUTO TRANSMISSION)
55	OR/LG*	OVERDRIVE SOLENOID (AUTO TRANSMISSION ONLY)
56	GY/PK*	EMISSION MAINTENANCE REMINDER (EMR) LIGHT
57		
58		
59		
60	GY/RD*	ISC MOTOR OPEN

WIRE COLOR CODES

BK	BLACK	LB	LIGHT BLUE	VT	VIOLET
BR	BROWN	LG	LIGHT GREEN	WT	WHITE
DB	DARK BLUE	OR	ORANGE	YL	YELLOW
DG	DARK GREEN	PK	PINK	*	WITH TRACER
GY	GRAY	RD	RED		

CONNECTOR TERMINAL SIDE SHOWN

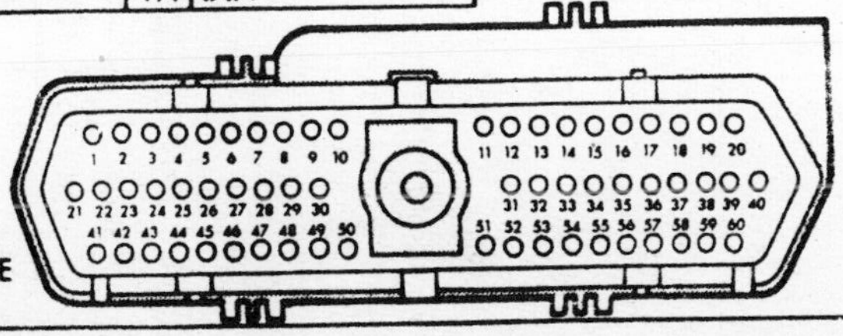

1991 — 60 way single board engine controller (SBEC) identification

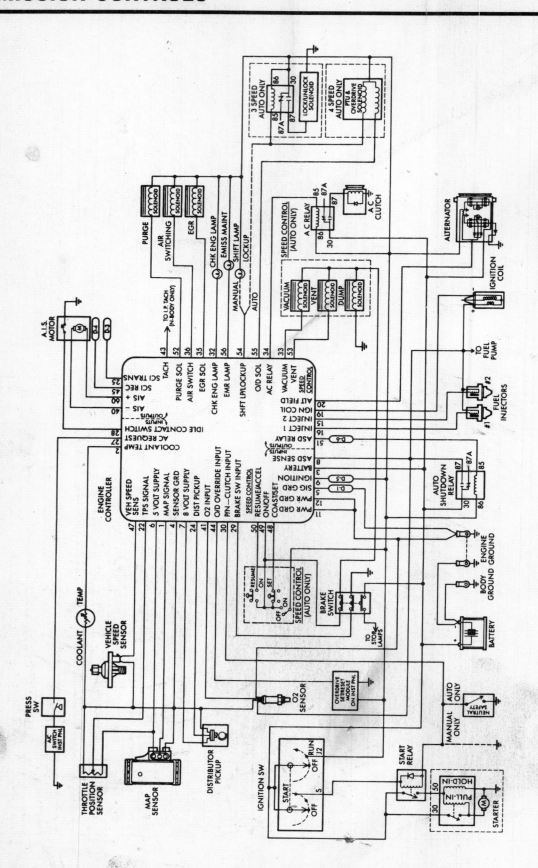

1991 – SMEC controller schematic – 3.9L engine

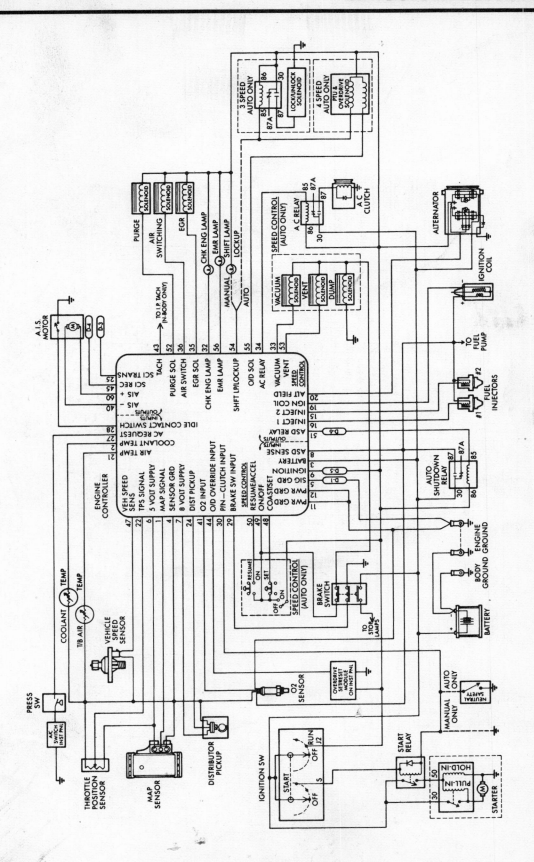

1991 – SMEC controller schematic – 5.2L engine

1991 — SMEC controller schematic — 5.9L normal duty engine

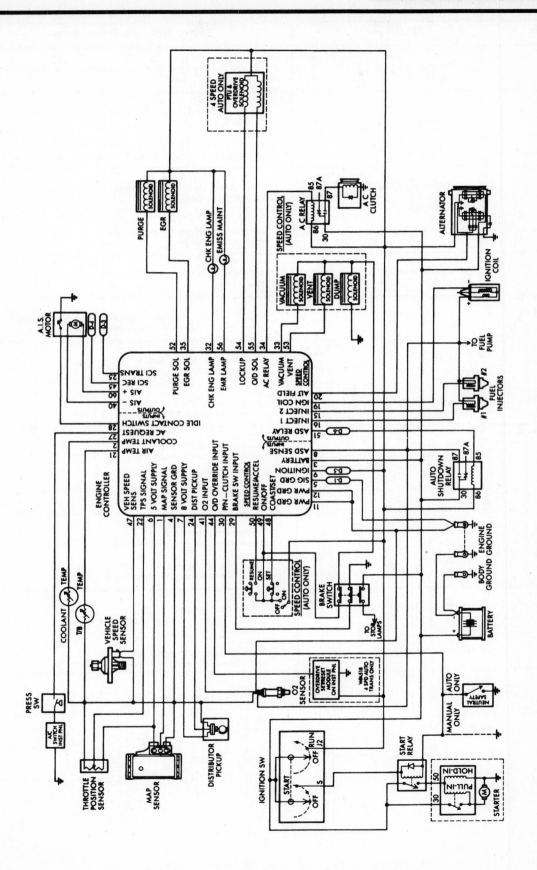

1991 — SMEC controller schematic — 5.9L heavy duty engine

Fault Code	DRBII Display	Description of Fault Condition
11	No Reference Signal During Cranking	No distributor reference signal detected during engine cranking.
13 + **	Slow Change in Idle MAP Signal or No Change in MAP From Start to Run	No variation in MAP sensor signal is detected. No difference is recognized between the engine MAP reading and the barometric pressure reading at start-up.
14 + **	MAP Voltage Too Low or MAP Voltage Too High	MAP sensor input below minimum acceptable voltage. MAP sensor input above maximum acceptable voltage.
15 **	No Vehicle Speed Signal	No distance sensor signal detected during road load conditions.
17	Engine is Cold Too Long	Engine coolant temperature remains below normal operating temperatures during vehicle travel (thermostat).
21 **	O_2 Signal Stays at Center or O_2 Signal Shorted to Voltage	Neither rich or lean condition is detected from the oxygen sensor input. Oxygen sensor input voltage maintained above normal operating range.
22 + **	Coolant Sensor Voltage Too Low or Coolant Sensor Voltage Too High	Coolant temperature sensor input below the minimum acceptable voltage. Coolant temperature sensor input above the maximum acceptable voltage.
23 (5.2L or 5.9L only)	Throttle Body Temp Voltage High or Throttle Body Temp Voltage Low	Throttle body temperature input above maximum acceptable voltage. Throttle body temperature input below minimum acceptable voltage.
24 + **	TPS Voltage High or TPS Voltage Low	Throttle position sensor (TPS) input above the maximum acceptable voltage. Throttle position sensor (TPS) input below the minimum acceptable voltage.

1991 Fault Codes

Fault Code	DRBII Display	Description of Fault Condition
25 * *	Automatic Idle Speed (ISC Actuator) Motor Circuits	A shorted condition detected in one or more of the automatic idle speed actuator circuits.
27 + * *	Inj #1 Control Circuit	Injector output driver does not respond properly to the control signal.
31 * *	Purge Solenoid Circuit.	An open or shorted condition detected in the purge solenoid circuit.
32 * *	EGR Solenoid Circuit or EGR System Failure	An open or shorted condition detected in the EGR solenoid circuit (except 5.9L Heavy Duty). Required change in air-fuel ratio not detected during diagnostic test (California emissions packages only)
33	A/C Clutch Relay Circuit	An open or shorted condition detected in the A/C clutch relay circuit.
34	Speed Control Solenoid Circuits	An open or shorted condition detected in the speed control vacuum or vent solenoid circuits.
35	Idle Switch Shorted Low or Idle Switch Open Circuit	Idle contact switch input circuit shorted to ground. Idle contact switch input circuit opened.
36 * *	Air Switching Solenoid Circuit (3.9L and 5.2L)	An open or shorted condition detected in the air switching solenoid circuit.
37	Torque Converter Lockup Solenoid Circuit	An open or shorted condition detected in the torque converter lockup solenoid circuit (vehicles with automatic transmissions only).
41 + * *	Alternator Field Not Switching Properly	Alternator field not switching properly.
42	Auto Shutdown Relay Control Circuit or No ASD Relay Voltage Sense at Controller	An open or short condition detected in the auto shutdown relay circuit. No ASD voltage sensed at engine controller.

1991 Fault Codes

VACUUM DIAGRAMS

Each vehicle is equipped with a VECI (Vehicle Emission Control Information) label. The label is located in the engine compartment. Information on the label covers:
- engine family and displacement
- evaporative family
- emission control system
- certification application

- engine timing specifications
- idle speed if adjustable
- spark plug gap
- engine vacuum schematic

NOTE: If the specifications or vacuum schematic differ from those shown in this section, use the specifications shown on the label.

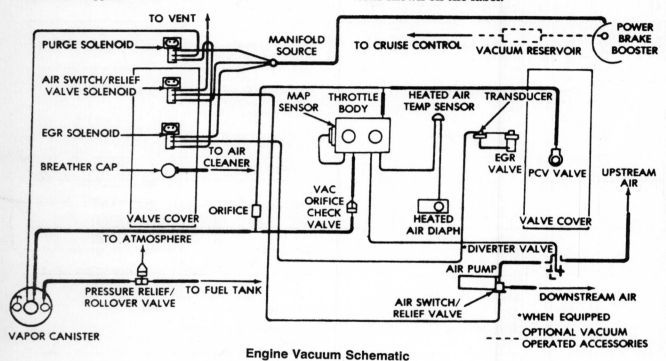

Engine Vacuum Schematic

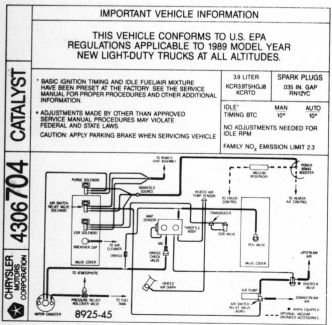

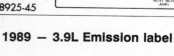

1989 — 3.9L Emission label

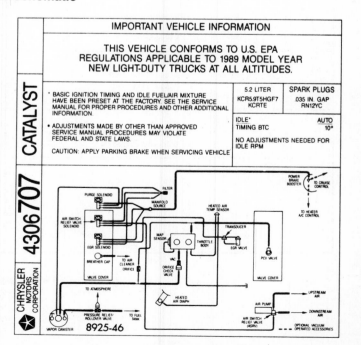

1989 — 5.2L Emission label

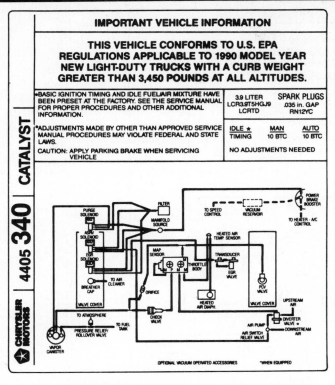

IMPORTANT VEHICLE INFORMATION

THIS VEHICLE CONFORMS TO U.S. EPA REGULATIONS APPLICABLE TO 1990 MODEL YEAR NEW LIGHT-DUTY TRUCKS WITH A CURB WEIGHT GREATER THAN 3,450 POUNDS AT ALL ALTITUDES.

•BASIC IGNITION TIMING AND IDLE FUEL/AIR MIXTURE HAVE BEEN PRESET AT THE FACTORY. SEE THE SERVICE MANUAL FOR PROPER PROCEDURES AND OTHER ADDITIONAL INFORMATION.	3.9 LITER LCR3.9T5HGJ9 LCRTD	SPARK PLUGS .035 in. GAP RN12YC
•ADJUSTMENTS MADE BY OTHER THAN APPROVED SERVICE MANUAL PROCEDURES MAY VIOLATE FEDERAL AND STATE LAWS.	IDLE ★ TIMING	MAN 10 BTC / AUTO 10 BTC
CAUTION: APPLY PARKING BRAKE WHEN SERVICING VEHICLE	NO ADJUSTMENTS NEEDED	

4405 340 CATALYST

CHRYSLER MOTORS

OPTIONAL VACUUM OPERATED ACCESSORIES   *WHEN EQUIPPED

1990 — Federal spec. emission label

IMPORTANT VEHICLE INFORMATION

AIR, EGR, HO2S, TWC + OC, TBI

THIS VEHICLE CONFORMS TO U.S. EPA AND STATE OF CALIFORNIA REGULATIONS APPLICABLE TO 1990 MODEL YEAR NEW LIGHT-DUTY TRUCKS AND MEDIUM-DUTY VEHICLES PROVIDED THAT THIS VEHICLE IS ONLY INTRODUCED INTO COMMERCE FOR SALE IN THE STATE OF CALIFORNIA.

•BASIC IGNITION TIMING AND IDLE FUEL/AIR MIXTURE HAVE BEEN PRESET AT THE FACTORY. SEE THE SERVICE MANUAL FOR PROPER PROCEDURES AND OTHER ADDITIONAL INFORMATION.	5.9 LITER LCR5.9T5HGD6 LCRTE	SPARK PLUGS .035 in. GAP RN12YC
•ADJUSTMENTS MADE BY OTHER THAN APPROVED SERVICE MANUAL PROCEDURES MAY VIOLATE FEDERAL AND STATE LAWS.	IDLE ★ TIMING	AUTO 10 BTC
CAUTION: APPLY PARKING BRAKE WHEN SERVICING VEHICLE	NO ADJUSTMENTS NEEDED RHC/CO/NOx STDS. .50/9.0/1.0	

4405 353 CATALYST

CHRYSLER MOTORS

59T5HGDD

OPTIONAL VACUUM OPERATED ACCESSORIES

1990 — California spec. emission label

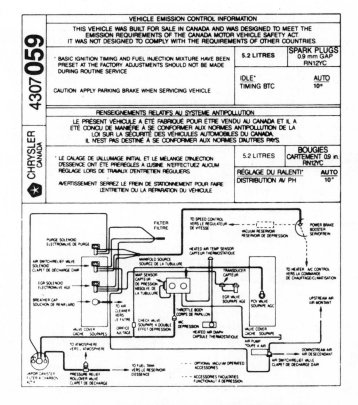

VEHICLE EMISSION CONTROL INFORMATION

THIS VEHICLE WAS BUILT FOR SALE IN CANADA AND WAS DESIGNED TO MEET THE EMISSION REQUIREMENTS OF THE CANADA MOTOR VEHICLE SAFETY ACT. IT WAS NOT DESIGNED TO COMPLY WITH THE REQUIREMENTS OF OTHER COUNTRIES.

· BASIC IGNITION TIMING AND FUEL INJECTION MIXTURE HAVE BEEN PRESET AT THE FACTORY. ADJUSTMENTS SHOULD NOT BE MADE DURING ROUTINE SERVICE	5.2 LITRES	SPARK PLUGS 0.9 mm GAP RN12YC
CAUTION APPLY PARKING BRAKE WHEN SERVICING VEHICLE	IDLE* TIMING BTC	AUTO 10°

4307 059

RENSEIGNEMENTS RELATIFS AU SYSTEME ANTIPOLLUTION

LE PRÉSENT VÉHICULE A ÉTÉ FABRIQUÉ POUR ÊTRE VENDU AU CANADA ET IL A ÉTÉ CONÇU DE MANIÈRE À SE CONFORMER AUX NORMES ANTIPOLLUTION DE LA LOI SUR LA SÉCURITÉ DES VÉHICULES AUTOMOBILES DU CANADA. IL N'EST PAS DESTINÉ À SE CONFORMER AUX NORMES D'AUTRES PAYS.

· LE CALAGE DE L'ALLUMAGE INITIAL ET LE MELANGE D'INJECTION D'ESSENCE ONT ÉTÉ PRÉRÉGLÉS À L'USINE. N'EFFECTUEZ AUCUN RÉGLAGE LORS DE TRAVAUX D'ENTRETIEN RÉGULIERS.	5.2 LITRES	BOUGIES CARTEMENT 0.9 in. RN12YC
AVERTISSEMENT SERREZ LE FREIN DE STATIONNEMENT POUR FAIRE L'ENTRETIEN OU LA RÉPARATION DU VÉHICULE.	RÉGLAGE DU RALENTI* DISTRIBUTION AV PH	AUTO 10°

CHRYSLER CANADA

1990 — Canada spec. emission label

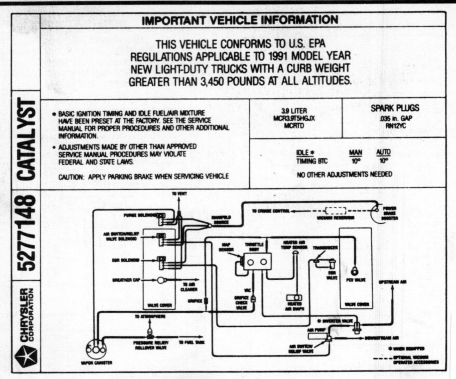

1991 — Federal spec. emission label

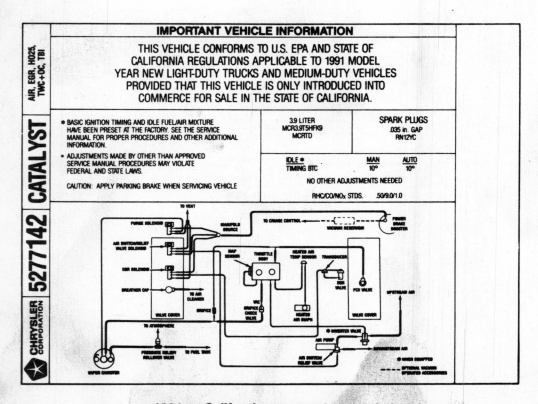

1991 — California spec. emission label

5

Fuel System

QUICK REFERENCE INDEX

GENERAL INDEX

THROTTLE BODY FUEL INJECTION SYSTEM

CAUTION

The fuel injection system is under a constant pressure of approximately 14.5 psi. Before servicing any part of the fuel injection system, the system pressure must be released. Use a clean shop towel to catch any fuel spray and take precautions to avoid the risk of fire.

Description

A TBI (Throttle Body Injection) system is used on all engines. The V6 and V8 use dual point injection (two injectors mounted on the throttle body.

The TBI system is controlled by a pre-programmed digital computer known as the Single Module Engine Controller (SMEC) on 1989 models, or a Single Board Engine Controller (SBEC) on 1990-91 models. The computer controls ignition timing, air/fuel ratio, emission control devices, charging system and idle speed. The computer constantly varies timing, fuel delivery and idle speed to meet changing engine operating conditions.

Various sensors provide the input necessary for the logic module to correctly regulate the fuel flow at the fuel injector. These include the manifold absolute pressure, throttle position, oxygen sensor, coolant temperature, charge temperature, vehicle speed (distance) sensors and throttle body temperature. In addition to the sensors, various switches also provide important information. These include the neutral-safety, heated backlite, air conditioning, air conditioning clutch switches, and an electronic idle switch.

All inputs to the logic module (SMEC/SBEC) are converted into signals which are used to calculate and adjust the fuel flow at the injector or ignition timing or both. The computer accomplishes this by sending signals to the power module.

The SMEC/SBEC tests many of its own input and output circuits. If a fault is found in a major system this information is stored in the logic module or SMEC. Information on this fault can be displayed to a technician by means of the instrument panel power loss (check engine) lamp or by connecting a diagnostic read out test monitor and reading a numbered display code which directly relates to a specific fault. The electronic diagnostics are described in detail in Section 4.

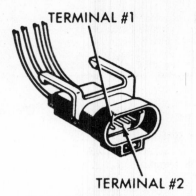

Fuel injector harness connector

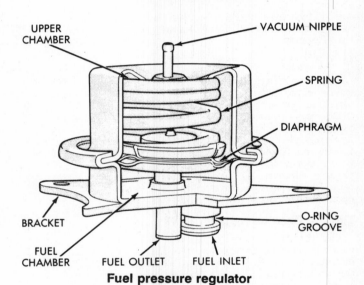

Fuel pressure regulator

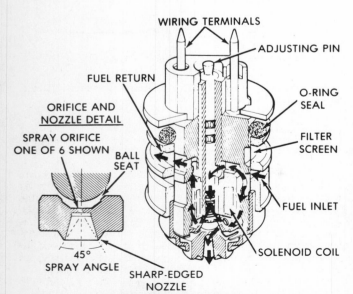

Fuel injector components

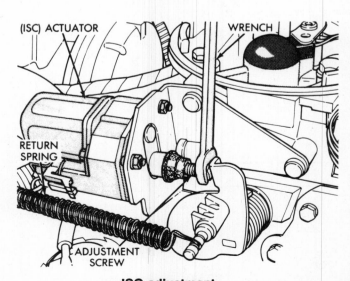

ISC adjustment

COMPONENTS

Throttle Body

The throttle body assembly replaces a conventional carburetor and is mounted on top of the intake manifold. The throttle body houses the fuel injector(s), pressure regulator, throttle position sensor, automatic idle speed motor and throttle body temperature sensor. Air flow through the throttle body is controlled by a manually operated throttle blade located in the base of the throttle body. The throttle body itself provides the chamber for metering atomizing and distributing fuel throughout the air entering the engine.

Fuel Injector

The fuel injector is an electric solenoid driven by the power module, but controlled by the SMEC/SBEC. The SMEC, based on ambient, mechanical, and sensor input, determines when and how long the power module should operate the injector. When an electric current is supplied to the injector, a spring loaded ball is lifted from its seat. This allows fuel to flow through six spray orifices and deflects off the sharp edge of the injector nozzle. This action causes the fuel to form a 45° cone shaped spray pattern before entering the air stream in the throttle body.

Fuel Pressure Regulator

The pressure regulator is a mechanical device located downstream of the fuel injector on the throttle body. Its function is to maintain a constant 14.5 psi across the fuel injector tip. The regulator uses a spring loaded rubber diaphragm to uncover a fuel return port. When the fuel pump becomes operational, fuel flows past the injector into the regulator, and is restricted from flowing any further by the blocked return port. When fuel pressure reaches the predetermined setting, it pushes on the diaphragm, compressing the spring, and uncovers the fuel return port. The diaphragm and spring will constantly move from an open to closed position to keep the fuel pressure constant.

Throttle Position Sensor (TPS)

The throttle position sensor (TPS) is an electric resistor which is activated by the movement of the throttle shaft. It is mounted on the throttle body and senses the angle of the throttle blade opening. The voltage that the sensor produces increases or decreases according to the throttle blade opening. This voltage is transmitted to the controller, where it is used along with data from other sensors to adjust the air/fuel ratio to

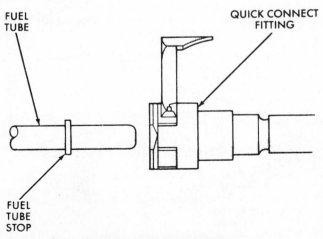

Quick connect fuel line fitting

FUEL TUBE
QUICK CONNECT FITTING
FUEL TUBE STOP

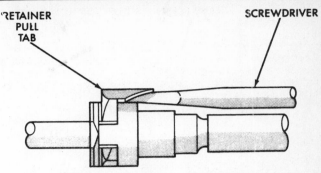

Disassembly with a screwdriver

RETAINER PULL TAB
SCREWDRIVER

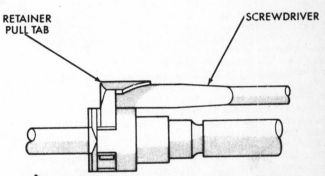

Releasing the quick connect fitting retainer

RETAINER PULL TAB
SCREWDRIVER

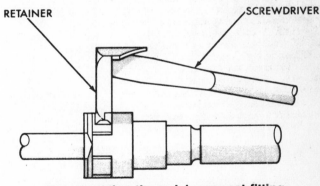

Disconnecting the quick connect fitting

RETAINER
SCREWDRIVER

varying conditions and during acceleration, deceleration, idle, and wide open throttle operations.

Automatic Idle Speed (AIS) Motor

The automatic idle speed motor (AIS) is operated by the computer. Data from the throttle position sensor, speed sensor, coolant temperature sensor, and various switch operations, (electric backlite, air conditioning, safety/neutral, brake) are used by the module to adjust engine idle to an optimum during all idle conditions. The AIS adjusts the air portion of the air/fuel mixture through an air bypass on the back of the throttle body. Basic (no load) idle is determined by the minimum air flow through the throttle body. The AIS opens or closes off the air bypass as an increase or decrease is needed due to engine loads or ambient conditions. The module senses an air/fuel change and increases or decreases fuel proportionally to change engine idle. Deceleration die out is also prevented by increasing engine idle when the throttle is closed quickly after a driving (speed) condition.

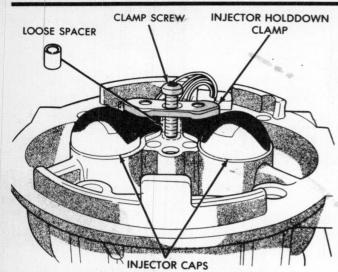

Removing the fuel injector cap hold down

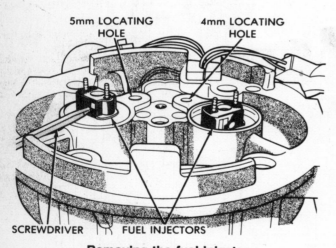

Removing the fuel injector

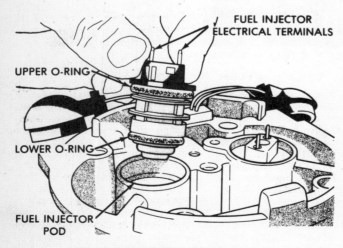

Servicing the fuel injector

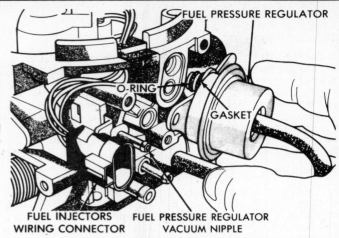

Servicing the fuel pressure regulator

Throttle Body Temperature Sensor

The throttle body temperature sensor is a device that monitors throttle body temperature which is the same as fuel temperature. It is mounted in the throttle body. This sensor provides information on fuel temperature which allows the SMEC to provide the correct air fuel mixture for a hot restart condition.

Fuel Pump

The fuel pump used in this system is a positive displacement, roller vane immersible pump with a permanent magnet electric motor. The fuel is drawn in through a filter sock and pushed through the electric motor to the outlet. The pump contains two check valves. One valve is used to relieve internal fuel pump pressure and regulate maximum pump output. The other check valve, located near the pump outlet, restricts fuel movement in either direction when the pump is not operational. Voltage to operate the pump is supplied through the auto shutdown relay (ASD).

Service Precautions

When working around any part of the fuel system, take precautionary steps to prevent possible fire and/or explosion:

a. Disconnect the negative battery terminal, except when testing with battery voltage is required.

b. Whenever possible, use a flashlight instead of a drop light to inspect fuel system components or connections.

c. Keep all open flames and smoking material out of the area and make sure there is adequate ventilation to remove fuel vapors.

d. Use a clean shop cloth to catch fuel when opening a fuel system. Dispose of gasoline-soaked rags properly.

e. Relieve the fuel system pressure before any service procedures are attempted that require disconnecting a fuel line.

f. Use eye protection.

g. Always keep a dry chemical (class B) fire extinguisher near the area.

QUICK-CONNECT FUEL FITTINGS

The hoses used on fuel injected vehicles are of special construction to prevent contamination of the fuel system. Quick-connect fittings are used at several fuel system connections. These fittings are equipped with a retainer tab that must be released before disconnecting the quick-connect. If the retainer on the removable pull tab is not bent down, the connector will be damaged.

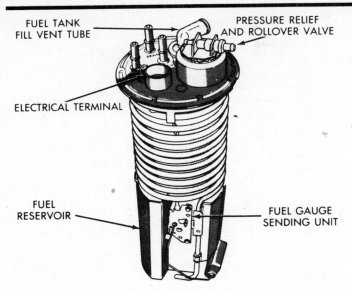

Fuel pump module

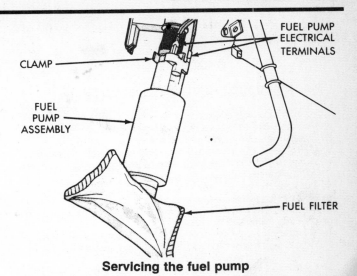

Servicing the fuel pump

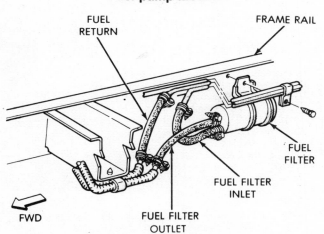

Fuel filter location

Fuel System Pressure Release

1. Loosen the gas cap to release tank pressure.
2. Remove the wiring harness connector from the injector.
3. Ground one terminal of the injector.
4. Connect a jumper wire to the second terminal and touch the battery positive post for no longer than ten seconds. This releases system pressure.
5. Remove the jumper wire and continue fuel system service.

Fuel System Pressure Test

CAUTION

Fuel system pressure must be released as previously described each time a fuel hose is to be disconnected. Take precautions to avoid the risk of fire.

1. Remove the fuel intake hose from the throttle body and connect fuel system pressure testers C-3292, and C-4749, C-4799B or the equivalent, between the fuel filter hose and the throttle body.
2. Start the engine and read the gauge. Pressure should be 14.5 psi.

NOTE: Diagnostic tester C-4805, or equivalent can be used. With the ignition in RUN, depress the ATM button. This activates the fuel pump and pressurizes the system

3. If the fuel pressure is below specifications:
   a. Install the tester between the fuel filter hose and the fuel line.
   b. Start the engine. If the pressure is now correct, replace the fuel filter. If no change is observed, gently squeeze the return hose. If the pressure increases, replace the pressure regulator. If no change is observed, the problem is either a plugged pump filter sock or a defective fuel pump.
4. If the pressure is above specifications:
   a. Remove the fuel return hose from the throttle body. Connect a substitute hose and place the other end of the hose in a clean container.
   b. Start the engine. If the pressure is now correct, check for a restricted fuel return line. If no change is observed, replace the fuel regulator.

Fuel Filter

REMOVAL AND INSTALLATION

NOTE: Do not use conventional fuel filters, hoses or clamps when servicing this fuel system. They are not compatible with the injection system and could fail, causing personal injury or damage to the vehicle. Use only hoses and clamps specifically designed for fuel injection.

1. Relieve the fuel pressure.
2. Disconnect the negative battery cable.
3. The filter is located on the frame rail toward the rear of the vehicle. Raise the vehicle and support safely. Remove the filter retaining screw and remove the filter assembly from the mounting plate.
4. Loosen the outlet hose clamp on the filter and inlet hose clamp on the rear fuel tube.
5. Wrap a shop towel around the hoses to absorb fuel. Remove the hoses from the filter and fuel tube and discard the clamps and the filter.
To install:
6. Install the inlet hose on the fuel tube and tighten the new clamp to 10 inch lbs.
7. Install the outlet hose on the filter outlet fitting and tighten the new clamp to 10 inch lbs.
8. Position the filter assembly on the mounting plate and tighten the mounting screw to 75 inch lbs. (8 Nm).

9. Connect the negative battery cable, start the engine and check for leaks.

Throttle Body

REMOVAL AND INSTALLATION

1. Remove the air cleaner.
2. Perform the fuel system pressure release.
3. Disconnect the negative battery cable.
4. Disconnect the vacuum hoses and electrical connectors.
5. Remove the throttle cable and, if so equipped, speed control and kickdown cables.
6. Remove the return spring.
7. Remove the fuel intake and return hoses.
8. Remove the throttle body mounting screws and lift the throttle body from the engine.
9. When installing the throttle body, use a new gasket. Install the throttle body and torque the mounting screws to 175 inch lbs.
10. Install the fuel intake and return hoses using new original equipment type clamps.
11. Install the return spring.
12. Install the throttle cable and, if so equipped, install the kickdown and speed control cables.
13. Install the wiring connectors and vacuum hoses.
14. Install the air cleaner.
15. Reconnect the negative battery cable.

Fuel Fittings

REMOVAL AND INSTALLATION

1. Remove the air cleaner assembly.
2. Perform the fuel system pressure release.
3. Disconnect the negative battery cable.
4. Loosen the fuel intake and return hose clamps. Wrap a shop towel around each hose, twist and pull off each hose.
5. Remove each fitting and note the inlet diameter. Remove the copper washers.
To install:
6. Replace the copper washers with new washers.
7. Install the fuel fittings in the proper ports and torque to 175 inch lbs.
8. Using new original equipment type hose clamps, install the fuel return and supply hoses.
9. Reconnect the negative battery cable.
10. Test for leaks using ATM tester C-4805 or equivalent. With the ignition in the **RUN** position depress the ATM button. This will activate the pump and pressurize the system. Check for leaks.
11. Reinstall the air cleaner assembly.

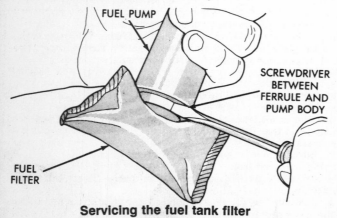

FUEL PUMP

SCREWDRIVER BETWEEN FERRULE AND PUMP BODY

FUEL FILTER

Servicing the fuel tank filter

Fuel Pressure Regulator

REMOVAL AND INSTALLATION

1. Remove the air cleaner assembly.
2. Perform the fuel system pressure release.
3. Disconnect the negative battery cable.
4. Remove the three screws attaching the pressure regulator to the throttle body. Place a shop towel around the inlet chamber to contain any fuel remaining in the system.
5. Pull the pressure regulator from the throttle body.
6. Carefully remove the O-ring from the pressure regulator and remove the gasket.
7. To install, place a new gasket on the pressure regulator and carefully install a new O-ring.
8. Position the pressure regulator on the throttle body press it into place.
9. Install the three screws and torque them to 40 inch lbs.
10. Connect the negative battery cable.
11. Test for leaks using ATM tester C-4805 or equivalent. With the ignition in the **RUN** position depress ATM button. This will activate the pump and pressurize the system. Check for leaks.
12. Reinstall the air cleaner assembly.

Fuel Injectors

REMOVAL AND INSTALLATION

1. Remove the air cleaner assembly.
2. Perform the fuel system pressure release.
3. Disconnect the negative battery cable.
4. Remove the fuel pressure regulator.
5. Remove the Torx® screw holding down the injector cap.
6. With two small screwdrivers, lift the cap off the injector using the slots provided.
7. Using a small screwdriver placed in the hole in the front of the electrical connector, gently pry the injector from pod.
8. Make sure the injector lower O-ring has been removed from the pod.
To install:
9. Place a new lower O-ring on the injector and a new O-ring on the injector cap. The injector will have the upper O-ring already installed.
10. Put the injector cap on injector. (Injector and cap are keyed). The cap should sit on the injector without interference.

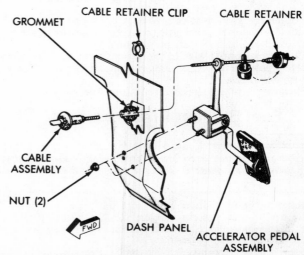

CABLE RETAINER CLIP

GROMMET

CABLE RETAINER

CABLE ASSEMBLY

NUT (2)

FWD

DASH PANEL

ACCELERATOR PEDAL ASSEMBLY

Accelerator pedal removal/installation

Apply a light coating of castor oil or petroleum jelly on the O-rings. Place the assembly in the pod.

11. Rotate the cap and injector to line up the attachment hole.
12. Push down on the cap until it contacts the injector pod.
13. Install the Torx® screw and torque it to 35-45 inch lbs.
14. Install the fuel pressure regulator.
15. Connect the negative battery cable.
16. Test for leaks using ATM tester C-4805 or equivalent. With the ignition in the **RUN** position depress the ATM button. This will activate the pump and pressurize the system. Check for leaks.
17. Reinstall the air cleaner assembly.

Throttle Position Sensor
REMOVAL AND INSTALLATION

1. Disconnect the negative battery cable.
2. Remove the air cleaner.
3. Disconnect the three way connector at the throttle position sensor.
4. Remove the two screws mounting the throttle position sensor to the throttle body.
5. Lift the throttle position sensor off the throttle shaft.
To install:
6. Install the throttle position sensor on the throttle body. Position the connector toward the rear of vehicle.
7. Connect the three way connector at the throttle position sensor.
8. Install the air cleaner.
9. Connect the negative battery cable.
10. Check operation with sensor read test #5.

Throttle Body Temperature Sensor

REMOVAL AND INSTALLATION

1. Remove the air cleaner.
2. Disconnect the throttle cables from the throttle body linkage.
3. Remove the two screws from the throttle cable bracket and lay the bracket aside.

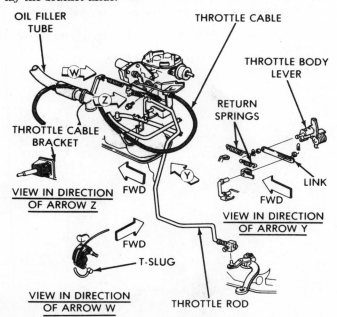

Throttle cable attachment

4. Disconnect the wiring connector.
5. Unscrew the sensor.
To install:
6. Apply heat transfer compound to the tip portion of the new sensor.
7. Install the sensor and torque it to 100 inch lbs.
8. Connect the wiring connector.
9. Install the throttle cable bracket with two screws.
10. Connect the throttle cables to the throttle body linkage and install the clips.
11. Install the air cleaner.

Automatic Idle Speed (AIS) Motor Assembly

REMOVAL AND INSTALLATION

1. Remove the air cleaner.
2. Disconnect the negative battery cable.
3. Disconnect the four pin connector on the AIS.
4. Remove the temperature sensor from the throttle body housing.
5. Remove the two Torx® head screws.
6. Remove the AIS from the throttle body housing, making sure that the O-ring is with the AIS.
To install:
7. Be sure that the pintle is in the retracted position. If the pintle measures more than 1 in. (25mm), it must be retracted by using the appropriate ATM test code.

NOTE: The battery must be connected for this operation.

8. Install new a O-ring on the AIS.
9. Install the AIS into the housing making sure the O-ring is in place.
10. Install the two Torx® head screws.
11. Connect the four pin connector to the AIS.
12. Install the temperature sending unit into the throttle body housing.
13. Connect the negative battery cable.

Fuel Pump

The TBI system uses an electric fuel pump mounted in the fuel tank.

REMOVAL AND INSTALLATION

─────── CAUTION ───────
Perform the fuel pressure release procedure described above!

1. Remove the fuel tank from the truck.
2. Remove the locking ring and lift out the fuel pump module.
3. Remove the sending unit attaching screws from the mounting bracket located on the drain tube.
4. Disconnect the wires from the sending unit and remove the sending unit.
5. Remove the drain tube from the mounting lug at the bottom of the reservoir.
6. Remove the lower-most coil of the drain tube from the mounting lugs on top of the reservoir. Be careful to avoid unsnapping the return line check valve cover from the bottom of the reservoir.
7. Release the pump mounting bracket from the reservoir. Press the bracket with both thumbs toward the center of the reservoir.
8. Remove the pump mounting bracket and rubber collar from the hose. Cut the hose clamp on the supply line and discard

the clamp. Remove the pump/filter assembly. Pry the filter from the pump.

To install:

9. Press a new filter onto the pump.
10. Using a new clamp, attach the supply hose.
11. Position the pump mounting bracket and rubber collar on the supply hose between the bulge in the hose and the pump.
12. Position the pump in the reservoir so that the filter aligns with the cavity in the reservoir.
13. Snap the pump bracket into the reservoir.
14. Position the coil tube on the reservoir so that the drain

tube aligns with the mounting lugs on the reservoir.
15. Snap the lower-most coil into the mounting lugs on top of the reservoir.
16. Snap the drain tube into the lugs on the bottom of the reservoir.
17. Connect the wires to the new sending unit.
18. Align the index tab on the level unit with the index hole in the mounting bracket.
19. Install the level unit screws.
20. Install the assembly in the tank.

FUEL TANK

REMOVAL AND INSTALLATION

1. Disconnect the battery ground cable.
2. Remove the fuel tank filler cap.
3. Raise and safely support the vehicle. Pump all fuel from the tank into an approved holding tank, and raise the vehicle.
4. Disconnect the fuel line and wire lead to the gauge unit. Remove the ground strap.
5. Remove the vent hose shield and the hose clamps from the hoses running to the vapor vent tube.
6. Remove the filler tube hose clamps and disconnect the hose from the tank.
7. Place a transmission jack under the center of the tank and apply sufficient pressure to support the tank.
8. Disconnect the two J-bolts and remove the retaining straps at the rear of the tank. Lower the tank from the vehicle. Feed the two vent tube hoses and filler tube vent hose through the grommets in the frame as the tank is being lowered. Remove the tank gauge unit.

To install:

9. Inspect the fuel filter, and if it is clogged or damaged, replace it.
10. Insert a new gasket in the recess of the fuel gauge and slide the gauge into the tank. Align the positioning tangs on the gauge with those on the tank. Install the lock ring and tighten securely.
11. Position the tank on a transmission jack and hoist it into place, feeding the vent hoses through the grommets on the way up.
12. Connect the J-bolts and retaining straps, and tighten to 40 inch lbs. Remove the jack.
13. Connect the filler tube and all vent hoses.
14. Connect the fuel supply line, ground strap, and gauge unit wire lead.
15. Refill the tank and inspect it for leaks. Connect the battery ground cable.
16. Reconnect the battery cable ground cable.

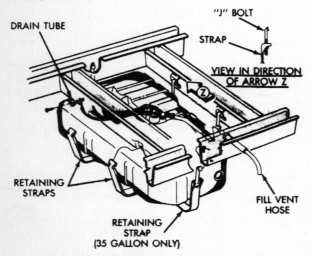

Fuel tank installation

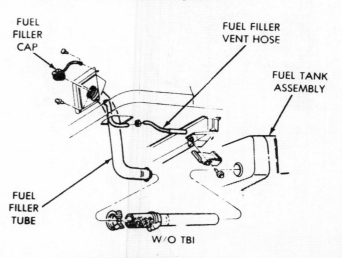

Fuel tank filler tubes

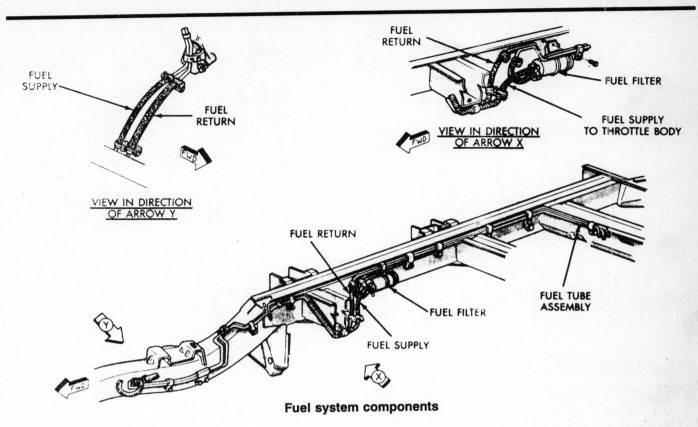

Fuel system components

TORQUE SPECIFICATIONS

Component	English	Metric
Fuel filter:	75 inch lbs.	8 Nm
Throttle body mounting bolts:	175 inch lbs.	20 Nm
Throttle body fuel fittings:	175 inch lbs.	20 Nm
Throttle body temperature sensor:	110 inch lbs.	12 Nm
Fuel pressure regulator:	40 inch lbs.	5 Nm
Fuel injector clamp:	35 inch lbs.	4 Nm
Throttle position sensor:	27 inch lbs.	3 Nm
Throttle body vacuum manifold:	35 inch lbs.	4 Nm
Oxygen sensor:	20 ft. lbs.	27 Nm

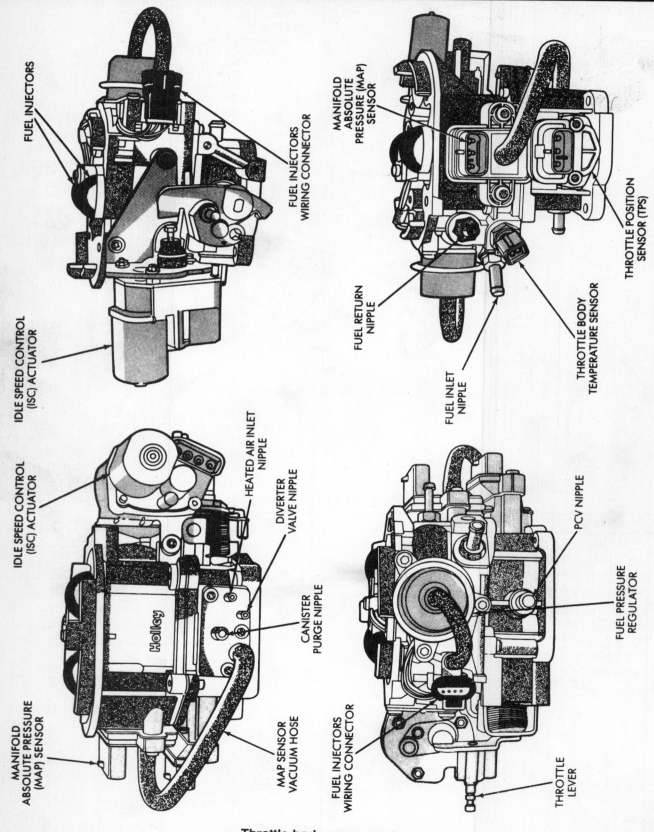

FUEL INJECTORS

FUEL INJECTORS WIRING CONNECTOR

IDLE SPEED CONTROL (ISC) ACTUATOR

MANIFOLD ABSOLUTE PRESSURE (MAP) SENSOR

THROTTLE POSITION SENSOR (TPS)

FUEL RETURN NIPPLE

FUEL INLET NIPPLE

THROTTLE BODY TEMPERATURE SENSOR

IDLE SPEED CONTROL (ISC) ACTUATOR

HEATED AIR INLET NIPPLE

DIVERTER VALVE NIPPLE

CANISTER PURGE NIPPLE

MANIFOLD ABSOLUTE PRESSURE (MAP) SENSOR

MAP SENSOR VACUUM HOSE

FUEL INJECTORS WIRING CONNECTOR

PCV NIPPLE

FUEL PRESSURE REGULATOR

THROTTLE LEVER

Holley

Throttle body components

6 Chassis Electrical

QUICK REFERENCE INDEX

GENERAL INDEX

UNDERSTANDING AND TROUBLESHOOTING ELECTRICAL SYSTEMS

At the rate which both import and domestic manufacturers are incorporating electronic control systems into their production lines, it won't be long before every new vehicle is equipped with one or more on-board computer. These electronic components (with no moving parts) should theoretically last the life of the vehicle, provided nothing external happens to damage the circuits or memory chips.

While it is true that electronic components should never wear out, in the real world malfunctions do occur. It is also true that any computer-based system is extremely sensitive to electrical voltages and cannot tolerate careless or haphazard testing or service procedures. An inexperienced individual can literally do major damage looking for a minor problem by using the wrong kind of test equipment or connecting test leads or connectors with the ignition switch ON. When selecting test equipment, make sure the manufacturers instructions state that the tester is compatible with whatever type of electronic control system is being serviced. Read all instructions carefully and double check all test points before installing probes or making any test connections.

The following section outlines basic diagnosis techniques for dealing with computerized automotive control systems. Along with a general explanation of the various types of test equipment available to aid in servicing modern electronic automotive systems, basic repair techniques for wiring harnesses and connectors is given. Read the basic information before attempting any repairs or testing on any computerized system, to provide the background of information necessary to avoid the most common and obvious mistakes that can cost both time and money. Although the replacement and testing procedures are simple in themselves, the systems are not, and unless one has a thorough understanding of all components and their function within a particular computerized control system, the logical test sequence these systems demand cannot be followed. Minor malfunctions can make a big difference, so it is important to know how each component affects the operation of the overall electronic system to find the ultimate cause of a problem without replacing good components unnecessarily. It is not enough to use the correct test equipment; the test equipment must be used correctly.

Safety Precautions

------- **CAUTION** -------

Whenever working on or around any computer based microprocessor control system, always observe these general precautions to prevent the possibility of personal injury or damage to electronic components.

- Never install or remove battery cables with the key ON or the engine running. Jumper cables should be connected with the key OFF to avoid power surges that can damage electronic control units. Engines equipped with computer controlled systems should avoid both giving and getting jump starts due to the possibility of serious damage to components from arcing in the engine compartment when connections are made with the ignition ON.
- Always remove the battery cables before charging the battery. Never use a high output charger on an installed battery or attempt to use any type of "hot shot" (24 volt) starting aid.
- Exercise care when inserting test probes into connectors to insure good connections without damaging the connector or spreading the pins. Always probe connectors from the rear (wire) side, NOT the pin side, to avoid accidental shorting of terminals during test procedures.
- Never remove or attach wiring harness connectors with the ignition switch ON, especially to an electronic control unit.

- Do not drop any components during service procedures and never apply 12 volts directly to any component (like a solenoid or relay) unless instructed specifically to do so. Some component electrical windings are designed to safely handle only 4 or 5 volts and can be destroyed in seconds if 12 volts are applied directly to the connector.
- Remove the electronic control unit if the vehicle is to be placed in an environment where temperatures exceed approximately 176°F (80°C), such as a paint spray booth or when arc or gas welding near the control unit location in the car.

ORGANIZED TROUBLESHOOTING

When diagnosing a specific problem, organized troubleshooting is a must. The complexity of a modern automobile demands that you approach any problem in a logical, organized manner. There are certain troubleshooting techniques that are standard:

1. Establish when the problem occurs. Does the problem appear only under certain conditions? Were there any noises, odors, or other unusual symptoms?

2. Isolate the problem area. To do this, make some simple tests and observations; then eliminate the systems that are working properly. Check for obvious problems such as broken wires, dirty connections or split or disconnected vacuum hoses. Always check the obvious before assuming something complicated is the cause.

3. Test for problems systematically to determine the cause once the problem area is isolated. Are all the components functioning properly? Is there power going to electrical switches and motors? Is there vacuum at vacuum switches and/or actuators? Is there a mechanical problem such as bent linkage or loose mounting screws? Doing careful, systematic checks will often turn up most causes on the first inspection without wasting time checking components that have little or no relationship to the problem.

4. Test all repairs after the work is done to make sure that the problem is fixed. Some causes can be traced to more than one component, so a careful verification of repair work is important to pick up additional malfunctions that may cause a problem to reappear or a different problem to arise. A blown fuse, for example, is a simple problem that may require more than another fuse to repair. If you don't look for a problem that caused a fuse to blow, for example, a shorted wire may go undetected.

Experience has shown that most problems tend to be the result of a fairly simple and obvious cause, such as loose or corroded connectors or air leaks in the intake system; making careful inspection of components during testing essential to quick and accurate troubleshooting. Special, hand held computerized testers designed specifically for diagnosing these systems are available from a variety of aftermarket sources, as well as from the vehicle manufacturer, but care should be taken that any test equipment being used is designed to diagnose that particular computer controlled system accurately without damaging the control unit (ECU) or components being tested.

NOTE: Pinpointing the exact cause of trouble in an electrical system can sometimes only be accomplished by the use of special test equipment. The following describes commonly used test equipment and explains how to put it to best use in diagnosis. In addition to the information covered below, the manufacturer's instructions booklet provided with the tester should be read and clearly understood before attempting any test procedures.

TEST EQUIPMENT

Jumper Wires

Jumper wires are simple, yet extremely valuable, pieces of test equipment. Jumper wires are merely wires that are used to bypass sections of a circuit. The simplest type of jumper wire is merely a length of multistrand wire with an alligator clip at each end. Jumper wires are usually fabricated from lengths of standard automotive wire and whatever type of connector (alligator clip, spade connector or pin connector) that is required for the particular vehicle being tested. The well equipped tool box will have several different styles of jumper wires in several different lengths. Some jumper wires are made with three or more terminals coming from a common splice for special purpose testing. In cramped, hard-to-reach areas it is advisable to have insulated boots over the jumper wire terminals in order to prevent accidental grounding, sparks, and possible fire, especially when testing fuel system components.

Jumper wires are used primarily to locate open electrical circuits, on either the ground (–) side of the circuit or on the hot (+) side. If an electrical component fails to operate, connect the jumper wire between the component and a good ground. If the component operates only with the jumper installed, the ground circuit is open. If the ground circuit is good, but the component does not operate, the circuit between the power feed and component is open. You can sometimes connect the jumper wire directly from the battery to the hot terminal of the component, but first make sure the component uses 12 volts in operation. Some electrical components, such as fuel injectors, are designed to operate on about 4 volts and running 12 volts directly to the injector terminals can burn out the wiring. By inserting an inline fuseholder between a set of test leads, a fused jumper wire can be used for bypassing open circuits. Use a 5 amp fuse to provide protection against voltage spikes. When in doubt, use a voltmeter to check the voltage input to the component and measure how much voltage is being applied normally. By moving the jumper wire successively back from the lamp toward the power source, you can isolate the area of the circuit where the open is located. When the component stops functioning, or the power is cut off, the open is in the segment of wire between the jumper and the point previously tested.

CAUTION

Never use jumpers made from wire that is of lighter gauge than used in the circuit under test. If the jumper wire is of too small gauge, it may overheat and possibly melt. Never use jumpers to bypass high resistance loads (such as motors) in a circuit. Bypassing resistances, in effect, creates a short circuit which may, in turn, cause damage and fire. Never use a jumper for anything other than temporary bypassing of components in a circuit.

---

12 Volt Test Light

The 12 volt test light is used to check circuits and components while electrical current is flowing through them. It is used for voltage and ground tests. Twelve volt test lights come in different styles but all have three main parts; a ground clip, a probe, and a light. The most commonly used 12 volt test lights have pick-type probes. To use a 12 volt test light, connect the ground clip to a good ground and probe wherever necessary with the pick. The pick should be sharp so that it can penetrate wire insulation to make contact with the wire, without making a large hole in the insulation. The wrap-around light is handy in hard to reach areas or where it is difficult to support a wire to push a probe pick into it. To use the wrap around light, hook the wire to probed with the hook and pull the trigger. A small pick will be forced through the wire insulation into the wire core.

CAUTION

Do not use a test light to probe electronic ignition spark plug or coil wires. Never use a pick-type test light to probe wiring on computer controlled systems unless specifically instructed to do so. Any wire insulation that is pierced by the test light probe should be taped and sealed with silicone after testing.

Like the jumper wire, the 12 volt test light is used to isolate opens in circuits. But, whereas the jumper wire is used to bypass the open to operate the load, the 12 volt test light is used to locate the presence of voltage in a circuit. If the test light glows, you know that there is power up to that point; if the 12 volt test light does not glow when its probe is inserted into the wire or connector, you know that there is an open circuit (no power). Move the test light in successive steps back toward the power source until the light in the handle does glow. When it does glow, the open is between the probe and point previously probed.

NOTE: The test light does not detect that 12 volts (or any particular amount of voltage) is present; it only detects that some voltage is present. It is advisable before using the test light to touch its terminals across the battery posts to make sure the light is operating properly.

Self-Powered Test Light

The self-powered test light usually contains a 1.5 volt penlight battery. One type of self-powered test light is similar in design to the 12 volt test light. This type has both the battery and the light in the handle and pick-type probe tip. The second type has the light toward the open tip, so that the light illuminates the contact point. The self-powered test light is dual purpose piece of test equipment. It can be used to test for either open or short circuits when power is isolated from the circuit (continuity test). A powered test light should not be used on any computer controlled system or component unless specifically instructed to do so. Many engine sensors can be destroyed by even this small amount of voltage applied directly to the terminals.

Open Circuit Testing

To use the self-powered test light to check for open circuits, first isolate the circuit from the vehicle's 12 volt power source by disconnecting the battery or wiring harness connector. Connect the test light ground clip to a good ground and probe sections of the circuit sequentially with the test light. (start from either end of the circuit). If the light is out, the open is between the probe and the circuit ground. If the light is on, the open is between the probe and end of the circuit toward the power source.

Short Circuit Testing

By isolating the circuit both from power and from ground, and using a self-powered test light, you can check for shorts to ground in the circuit. Isolate the circuit from power and ground. Connect the test light ground clip to a good ground and probe any easy-to-reach test point in the circuit. If the light comes on, there is a short somewhere in the circuit. To isolate the short, probe a test point at either end of the isolated circuit (the light should be on). Leave the test light probe connected and open connectors, switches, remove parts, etc., sequentially, until the light goes out. When the light goes out, the short is between the last circuit component opened and the previous circuit opened.

NOTE: The 1.5 volt battery in the test light does not provide much current. A weak battery may not provide enough power to illuminate the test light even when a complete circuit is made (especially if there are high resistances in the circuit). Always make sure that the test battery is strong. To check the battery, briefly touch the ground clip to the probe; if the light glows brightly the

battery is strong enough for testing. **Never use a self-powered test light to perform checks for opens or shorts when power is applied to the electrical system under test. The 12 volt vehicle power will quickly burn out the 1.5 volt light bulb in the test light.**

Voltmeter

A voltmeter is used to measure voltage at any point in a circuit, or to measure the voltage drop across any part of a circuit. It can also be used to check continuity in a wire or circuit by indicating current flow from one end to the other. Voltmeters usually have various scales on the meter dial and a selector switch to allow the selection of different voltages. The voltmeter has a positive and a negative lead. To avoid damage to the meter, always connect the negative lead to the negative (–) side of circuit (to ground or nearest the ground side of the circuit) and connect the positive lead to the positive (+) side of the circuit (to the power source or the nearest power source). Note that the negative voltmeter lead will always be black and that the positive voltmeter will always be some color other than black (usually red). Depending on how the voltmeter is connected into the circuit, it has several uses.

A voltmeter can be connected either in parallel or in series with a circuit and it has a very high resistance to current flow. When connected in parallel, only a small amount of current will flow through the voltmeter current path; the rest will flow through the normal circuit current path and the circuit will work normally. When the voltmeter is connected in series with a circuit, only a small amount of current can flow through the circuit. The circuit will not work properly, but the voltmeter reading will show if the circuit is complete or not.

Available Voltage Measurement

Set the voltmeter selector switch to the 20V position and connect the meter negative lead to the negative post of the battery. Connect the positive meter lead to the positive post of the battery and turn the ignition switch ON to provide a load. Read the voltage on the meter or digital display. A well charged battery should register over 12 volts. If the meter reads below 11.5 volts, the battery power may be insufficient to operate the electrical system properly. This test determines voltage available from the battery and should be the first step in any electrical trouble diagnosis procedure. Many electrical problems, especially on computer controlled systems, can be caused by a low state of charge in the battery. Excessive corrosion at the battery cable terminals can cause a poor contact that will prevent proper charging and full battery current flow.

Normal battery voltage is 12 volts when fully charged. When the battery is supplying current to one or more circuits it is said to be "under load". When everything is off the electrical system is under a "no-load" condition. A fully charged battery may show about 12.5 volts at no load; will drop to 12 volts under medium load; and will drop even lower under heavy load. If the battery is partially discharged the voltage decrease under heavy load may be excessive, even though the battery shows 12 volts or more at no load. When allowed to discharge further, the battery's available voltage under load will decrease more severely. For this reason, it is important that the battery be fully charged during all testing procedures to avoid errors in diagnosis and incorrect test results.

Voltage Drop

When current flows through a resistance, the voltage beyond the resistance is reduced (the larger the current, the greater the reduction in voltage). When no current is flowing, there is no voltage drop because there is no current flow. All points in the circuit which are connected to the power source are at the same voltage as the power source. The total voltage drop always equals the total source voltage. In a long circuit with many connectors, a series of small, unwanted voltage drops due to corrosion at the connectors can add up to a total loss of voltage which impairs the operation of the normal loads in the circuit.

INDIRECT COMPUTATION OF VOLTAGE DROPS

1. Set the voltmeter selector switch to the 20 volt position.
2. Connect the meter negative lead to a good ground.
3. Probe all resistances in the circuit with the positive meter lead.
4. Operate the circuit in all modes and observe the voltage readings.

DIRECT MEASUREMENT OF VOLTAGE DROPS

1. Set the voltmeter switch to the 20 volt position.
2. Connect the voltmeter negative lead to the ground side of the resistance load to be measured.
3. Connect the positive lead to the positive side of the resistance or load to be measured.
4. Read the voltage drop directly on the 20 volt scale.

Too high a voltage indicates too high a resistance. If, for example, a blower motor runs too slowly, you can determine if there is too high a resistance in the resistor pack. By taking voltage drop readings in all parts of the circuit, you can isolate the problem. Too low a voltage drop indicates too low a resistance. If, for example, a blower motor runs too fast in the MED and/or LOW position, the problem can be isolated in the resistor pack by taking voltage drop readings in all parts of the circuit to locate a possibly shorted resistor. The maximum allowable voltage drop under load is critical, especially if there is more than one high resistance problem in a circuit because all voltage drops are cumulative. A small drop is normal due to the resistance of the conductors.

HIGH RESISTANCE TESTING

1. Set the voltmeter selector switch to the 4 volt position.
2. Connect the voltmeter positive lead to the positive post of the battery.
3. Turn on the headlights and heater blower to provide a load.
4. Probe various points in the circuit with the negative voltmeter lead.
5. Read the voltage drop on the 4 volt scale. Some average maximum allowable voltage drops are:

FUSE PANEL—7 volts
IGNITION SWITCH—5 volts
HEADLIGHT SWITCH—7 volts
IGNITION COIL (+)—5 volts
ANY OTHER LOAD—1.3 volts

NOTE: Voltage drops are all measured while a load is operating; without current flow, there will be no voltage drop.

Ohmmeter

The ohmmeter is designed to read resistance (ohms) in a circuit or component. Although there are several different styles of ohmmeters, all will usually have a selector switch which permits the measurement of different ranges of resistance (usually the selector switch allows the multiplication of the meter reading by 10, 100, 1000, and 10,000). A calibration knob allows the meter to be set at zero for accurate measurement. Since all ohmmeters are powered by an internal battery (usually 9 volts), the ohmmeter can be used as a self-powered test light. When the ohmmeter is connected, current from the ohmmeter flows through the circuit or component being tested. Since the ohmmeter's internal resistance and voltage are known values, the amount of current flow through the meter depends on the resistance of the circuit or component being tested.

The ohmmeter can be used to perform continuity test for opens or shorts (either by observation of the meter needle or as a self-powered test light), and to read actual resistance in a circuit. It should be noted that the ohmmeter is used to check the resistance of a component or wire while there is no voltage applied to the circuit. Current flow from an outside voltage source (such as the vehicle battery) can damage the ohmmeter, so the circuit or component should be isolated from the vehicle electrical system before any testing is done. Since the ohmmeter uses its own voltage source, either lead can be connected to any test point.

NOTE: When checking diodes or other solid state components, the ohmmeter leads can only be connected one way in order to measure current flow in a single direction. Make sure the positive (+) and negative (–) terminal connections are as described in the test procedures to verify the one-way diode operation.

In using the meter for making continuity checks, do not be concerned with the actual resistance readings. Zero resistance, or any resistance readings, indicate continuity in the circuit. Infinite resistance indicates an open in the circuit. A high resistance reading where there should be none indicates a problem in the circuit. Checks for short circuits are made in the same manner as checks for open circuits except that the circuit must be isolated from both power and normal ground. Infinite resistance indicates no continuity to ground, while zero resistance indicates a dead short to ground.

RESISTANCE MEASUREMENT

The batteries in an ohmmeter will weaken with age and temperature, so the ohmmeter must be calibrated or "zeroed" before taking measurements. To zero the meter, place the selector switch in its lowest range and touch the two ohmmeter leads together. Turn the calibration knob until the meter needle is exactly on zero.

NOTE: All analog (needle) type ohmmeters must be zeroed before use, but some digital ohmmeter models are automatically calibrated when the switch is turned on. Self-calibrating digital ohmmeters do not have an adjusting knob, but its a good idea to check for a zero readout before use by touching the leads together. All computer controlled systems require the use of a digital ohmmeter with at least 10 meagohms impedance for testing. Before any test procedures are attempted, make sure the ohmmeter used is compatible with the electrical system or damage to the on-board computer could result.

To measure resistance, first isolate the circuit from the vehicle power source by disconnecting the battery cables or the harness connector. Make sure the key is OFF when disconnecting any components or the battery. Where necessary, also isolate at least one side of the circuit to be checked to avoid reading parallel resistances. Parallel circuit resistances will always give a lower reading than the actual resistance of either of the branches. When measuring the resistance of parallel circuits, the total resistance will always be lower than the smallest resistance in the circuit. Connect the meter leads to both sides of the circuit (wire or component) and read the actual measured ohms on the meter scale. Make sure the selector switch is set to the proper ohm scale for the circuit being tested to avoid misreading the ohmmeter test value.

----- CAUTION -----

Never use an ohmmeter with power applied to the circuit. Like the self-powered test light, the ohmmeter is designed to operate on its own power supply. The normal 12 volt automotive electrical system current could damage the meter.

Ammeters

An ammeter measures the amount of current flowing through a circuit in units called amperes or amps. Amperes are units of electron flow which indicate how fast the electrons are flowing through the circuit. Since Ohms Law dictates that current flow in a circuit is equal to the circuit voltage divided by the total circuit resistance, increasing voltage also increases the current level (amps). Likewise, any decrease in resistance will increase the amount of amps in a circuit. At normal operating voltage, most circuits have a characteristic amount of amperes, called "current draw" which can be measured using an ammeter. By referring to a specified current draw rating, measuring the amperes, and comparing the two values, one can determine what is happening within the circuit to aid in diagnosis. An open circuit, for example, will not allow any current to flow so the ammeter reading will be zero. More current flows through a heavily loaded circuit or when the charging system is operating.

An ammeter is always connected in series with the circuit being tested. All of the current that normally flows through the circuit must also flow through the ammeter; if there is any other path for the current to follow, the ammeter reading will not be accurate. The ammeter itself has very little resistance to current flow and therefore will not affect the circuit, but it will measure current draw only when the circuit is closed and electricity is flowing. Excessive current draw can blow fuses and drain the battery, while a reduced current draw can cause motors to run slowly, lights to dim and other components to not operate properly. The ammeter can help diagnose these conditions by locating the cause of the high or low reading.

Multimeters

Different combinations of test meters can be built into a single unit designed for specific tests. Some of the more common combination test devices are known as Volt/Amp testers, Tach/Dwell meters, or Digital Multimeters. The Volt/Amp tester is used for charging system, starting system or battery tests and consists of a voltmeter, an ammeter and a variable resistance carbon pile. The voltmeter will usually have at least two ranges for use with 6, 12 and 24 volt systems. The ammeter also has more than one range for testing various levels of battery loads and starter current draw and the carbon pile can be adjusted to offer different amounts of resistance. The Volt/Amp tester has heavy leads to carry large amounts of current and many later models have an inductive ammeter pickup that clamps around the wire to simplify test connections. On some models, the ammeter also has a zero-center scale to allow testing of charging and starting systems without switching leads or polarity. A digital multimeter is a voltmeter, ammeter and ohmmeter combined in an instrument which gives a digital readout. These are often used when testing solid state circuits because of their high input impedance (usually 10 megohms or more).

The tach/dwell meter combines a tachometer and a dwell (cam angle) meter and is a specialized kind of voltmeter. The tachometer scale is marked to show engine speed in rpm and the dwell scale is marked to show degrees of distributor shaft rotation. In most electronic ignition systems, dwell is determined by the control unit, but the dwell meter can also be used to check the duty cycle (operation) of some electronic engine control systems. Some tach/dwell meters are powered by an internal battery, while others take their power from the car battery in use. The battery powered testers usually require calibration much like an ohmmeter before testing.

Special Test Equipment

A variety of diagnostic tools are available to help troubleshoot and repair computerized engine control systems. The most sophisticated of these devices are the console type engine analyzers that usually occupy a garage service bay, but there are several types of aftermarket electronic testers available that will allow quick circuit tests of the engine control system by plugging directly into a special connector located in the engine compartment or under the dashboard. Several tool and equipment manufacturers offer simple, hand held testers that measure various circuit voltage levels on command to check all system components for proper operation. Although these testers usually cost about $300–500, consider that the average computer control unit (or ECM) can cost just as much and the money saved by not replacing perfectly good sensors or components in an attempt to correct a problem could justify the purchase price of a special diagnostic tester the first time it's used.

These computerized testers can allow quick and easy test measurements while the engine is operating or while the car is being driven. In addition, the on-board computer memory can be read to access any stored trouble codes; in effect allowing the computer to tell you where it hurts and aid trouble diagnosis by pinpointing exactly which circuit or component is malfunctioning. In the same manner, repairs can be tested to make sure the problem has been corrected. The biggest advantage these special testers have is their relatively easy hookups that minimize or eliminate the chances of making the wrong connections and getting false voltage readings or damaging the computer accidentally.

NOTE: It should be remembered that these testers check voltage levels in circuits; they don't detect mechanical problems or failed components if the circuit voltage falls within the preprogrammed limits stored in the tester PROM unit. Also, most of the hand held testers are designed to work only on one or two systems made by a specific manufacturer.

A variety of aftermarket testers are available to help diagnose different computerized control systems. Owatonna Tool Company (OTC), for example, markets a device called the OTC Monitor which plugs directly into the assembly line diagnostic link (ALDL). The OTC tester makes diagnosis a simple matter of pressing the correct buttons and, by changing the internal PROM or inserting a different diagnosis cartridge, it will work on any model from full size to subcompact, over a wide range of years. An adapter is supplied with the tester to allow connection to all types of ALDL links, regardless of the number of pin terminals used. By inserting an updated PROM into the OTC tester, it can be easily updated to diagnose any new modifications of computerized control systems.

Wiring Harnesses

The average automobile contains about ½ mile of wiring, with hundreds of individual connections. To protect the many wires from damage and to keep them from becoming a confusing tangle, they are organized into bundles, enclosed in plastic or taped together and called wire harnesses. Different wiring harnesses serve different parts of the vehicle. Individual wires are color coded to help trace them through a harness where sections are hidden from view.

A loose or corroded connection or a replacement wire that is too small for the circuit will add extra resistance and an additional voltage drop to the circuit. A ten percent voltage drop can result in slow or erratic motor operation, for example, even though the circuit is complete. Automotive wiring or circuit conductors can be in any one of three forms:

1. Single strand wire
2. Multistrand wire
3. Printed circuitry

Single strand wire has a solid metal core and is usually used inside such components as alternators, motors, relays and other devices. Multistrand wire has a core made of many small strands of wire twisted together into a single conductor. Most of the wiring in an automotive electrical system is made up of multistrand wire, either as a single conductor or grouped together in a harness. All wiring is color coded on the insulator, either as a solid color or as a colored wire with an identification stripe. A printed circuit is a thin film of copper or other conductor that is printed on an insulator backing. Occasionally, a printed circuit is sandwiched between two sheets of plastic for more

protection and flexibility. A complete printed circuit, consisting of conductors, insulating material and connectors for lamps or other components is called a printed circuit board. Printed circuitry is used in place of individual wires or harnesses in places where space is limited, such as behind instrument panels.

Wire Gauge

Since computer controlled automotive electrical systems are very sensitive to changes in resistance, the selection of properly sized wires is critical when systems are repaired. The wire gauge number is an expression of the cross section area of the conductor. The most common system for expressing wire size is the American Wire Gauge (AWG) system.

Wire cross section area is measured in circular mils. A mil is 1/1000 in. (0.001 in.) (0.0254mm); a circular mil is the area of a circle one mil in diameter. For example, a conductor ¼ in. (6mm) in diameter is 0.250 in. or 250 mils. The circular mil cross section area of the wire is 250 squared (250^2) or 62,500 circular mils. Imported car models usually use metric wire gauge designations, which is simply the cross section area of the conductor in square millimeters (mm6).

Gauge numbers are assigned to conductors of various cross section areas. As gauge number increases, area decreases and the conductor becomes smaller. A 5 gauge conductor is smaller than a 1 gauge conductor and a 10 gauge is smaller than a 5 gauge. As the cross section area of a conductor decreases, resistance increases and so does the gauge number. A conductor with a higher gauge number will carry less current than a conductor with a lower gauge number.

NOTE: Gauge wire size refers to the size of the conductor, not the size of the complete wire. It is possible to have two wires of the same gauge with different diameters because one may have thicker insulation than the other.

12 volt automotive electrical systems generally use 10, 12, 14, 16 and 18 gauge wire. Main power distribution circuits and larger accessories usually use 10 and 12 gauge wire. Battery cables are usually 4 or 6 gauge, although 1 and 2 gauge wires are occasionally used. Wire length must also be considered when making repairs to a circuit. As conductor length increases, so does resistance. An 18 gauge wire, for example, can carry a 10 amp load for 10 feet without excessive voltage drop; however if a 15 foot wire is required for the same 10 amp load, it must be a 16 gauge wire.

An electrical schematic shows the electrical current paths when a circuit is operating properly. It is essential to understand how a circuit works before trying to figure out why it doesn't. Schematics break the entire electrical system down into individual circuits and show only one particular circuit. In a schematic, no attempt is made to represent wiring and components as they physically appear on the vehicle; switches and other components are shown as simply as possible. Face views of harness connectors show the cavity or terminal locations in all multi-pin connectors to help locate test points.

If you need to backprobe a connector while it is on the component, the order of the terminals must be mentally reversed. The wire color code can help in this situation, as well as a keyway, lock tab or other reference mark.

NOTE: Wiring diagrams are not included in this book. As trucks have become more complex and available with longer option lists, wiring diagrams have grown in size and complexity. It has become almost impossible to provide a readable reproduction of a wiring diagram in a book this size. Information on ordering wiring diagrams from the vehicle manufacturer can be found in the owner's manual.

WIRING REPAIR

Soldering is a quick, efficient method of joining metals permanently. Everyone who has the occasion to make wiring repairs should know how to solder. Electrical connections that are soldered are far less likely to come apart and will conduct electricity much better than connections that are only "pig-tailed" together. The most popular (and preferred) method of soldering is with an electrical soldering gun. Soldering irons are available in many sizes and wattage ratings. Irons with higher wattage ratings deliver higher temperatures and recover lost heat faster. A small soldering iron rated for no more than 50 watts is recommended, especially on electrical systems where excess heat can damage the components being soldered.

There are three ingredients necessary for successful soldering; proper flux, good solder and sufficient heat. A soldering flux is necessary to clean the metal of tarnish, prepare it for soldering and to enable the solder to spread into tiny crevices. When soldering, always use a resin flux or resin core solder which is non-corrosive and will not attract moisture once the job is finished. Other types of flux (acid core) will leave a residue that will attract moisture and cause the wires to corrode. Tin is a unique metal with a low melting point. In a molten state, it dissolves and alloys easily with many metals. Solder is made by mixing tin with lead. The most common proportions are 40/60, 50/50 and 60/40, with the percentage of tin listed first. Low priced solders usually contain less tin, making them very difficult for a beginner to use because more heat is required to melt the solder. A common solder is 40/60 which is well suited for all-around general use, but 60/40 melts easier, has more tin for a better joint and is preferred for electrical work.

Soldering Techniques

Successful soldering requires that the metals to be joined be heated to a temperature that will melt the solder—usually 360–460°F (182–238°C). Contrary to popular belief, the purpose of the soldering iron is not to melt the solder itself, but to heat the parts being soldered to a temperature high enough to melt the solder when it is touched to the work. Melting flux-cored solder on the soldering iron will usually destroy the effectiveness of the flux.

NOTE: Soldering tips are made of copper for good heat conductivity, but must be "tinned" regularly for quick transference of heat to the project and to prevent the solder from sticking to the iron. To "tin" the iron, simply heat it and touch the flux-cored solder to the tip; the solder will flow over the hot tip. Wipe the excess off with a clean rag, but be careful as the iron will be hot.

After some use, the tip may become pitted. If so, simply dress the tip smooth with a smooth file and "tin" the tip again. An old saying holds that "metals well cleaned are half soldered." Flux-cored solder will remove oxides but rust, bits of insulation and oil or grease must be removed with a wire brush or emery cloth. For maximum strength in soldered parts, the joint must start off clean and tight. Weak joints will result in gaps too wide for the solder to bridge.

If a separate soldering flux is used, it should be brushed or swabbed on only those areas that are to be soldered. Most solders contain a core of flux and separate fluxing is unnecessary. Hold the work to be soldered firmly. It is best to solder on a wooden board, because a metal vise will only rob the piece to be soldered of heat and make it difficult to melt the solder. Hold the soldering tip with the broadest face against the work to be soldered. Apply solder under the tip close to the work, using enough solder to give a heavy film between the iron and the piece being soldered, while moving slowly and making sure the solder melts properly. Keep the work level or the solder will run to the lowest part and favor the thicker parts, because these require more heat to melt the solder. If the soldering tip overheats

(the solder coating on the face of the tip burns up), it should be retinned. Once the soldering is completed, let the soldered joint stand until cool. Tape and seal all soldered wire splices after the repair has cooled.

Wire Harness and Connectors

The on-board computer (ECM) wire harness electrically connects the control unit to the various solenoids, switches and sensors used by the control system. Most connectors in the engine compartment or otherwise exposed to the elements are protected against moisture and dirt which could create oxidation and deposits on the terminals. This protection is important because of the very low voltage and current levels used by the computer and sensors. All connectors have a lock which secures the male and female terminals together, with a secondary lock holding the seal and terminal into the connector. Both terminal locks must be released when disconnecting ECM connectors.

These special connectors are weather-proof and all repairs require the use of a special terminal and the tool required to service it. This tool is used to remove the pin and sleeve terminals. If removal is attempted with an ordinary pick, there is a good chance that the terminal will be bent or deformed. Unlike standard blade type terminals, these terminals cannot be straightened once they are bent. Make certain that the connectors are properly seated and all of the sealing rings in place when connecting leads. On some models, a hinge-type flap provides a backup or secondary locking feature for the terminals. Most secondary locks are used to improve the connector reliability by retaining the terminals if the small terminal lock tangs are not positioned properly.

Molded-on connectors require complete replacement of the connection. This means splicing a new connector assembly into the harness. All splices in on-board computer systems should be soldered to insure proper contact. Use care when probing the connections or replacing terminals in them as it is possible to short between opposite terminals. If this happens to the wrong terminal pair, it is possible to damage certain components. Always use jumper wires between connectors for circuit checking and never probe through weatherproof seals.

Open circuits are often difficult to locate by sight because corrosion or terminal misalignment are hidden by the connectors. Merely wiggling a connector on a sensor or in the wiring harness may correct the open circuit condition. This should always be considered when an open circuit or a failed sensor is indicated. Intermittent problems may also be caused by oxidized or loose connections. When using a circuit tester for diagnosis, always probe connections from the wire side. Be careful not to damage sealed connectors with test probes.

All wiring harnesses should be replaced with identical parts, using the same gauge wire and connectors. When signal wires are spliced into a harness, use wire with high temperature insulation only. With the low voltage and current levels found in the system, it is important that the best possible connection at all wire splices be made by soldering the splices together. It is seldom necessary to replace a complete harness. If replacement is necessary, pay close attention to insure proper harness routing. Secure the harness with suitable plastic wire clamps to prevent vibrations from causing the harness to wear in spots or contact any hot components.

NOTE: Weatherproof connectors cannot be replaced with standard connectors. Instructions are provided with replacement connector and terminal packages. Some wire harnesses have mounting indicators (usually pieces of colored tape) to mark where the harness is to be secured.

In making wiring repairs, it's important that you always replace damaged wires with wires that are the same gauge as the wire being replaced. The heavier the wire, the smaller the gauge number. Wires are color-coded to aid in identification and whenever possible the same color coded wire should be used for replacement. A wire stripping and crimping tool is necessary to install solderless terminal connectors. Test all crimps by pulling on the wires; it should not be possible to pull the wires out of a good crimp.

Wires which are open, exposed or otherwise damaged are repaired by simple splicing. Where possible, if the wiring harness is accessible and the damaged place in the wire can be located, it is best to open the harness and check for all possible damage. In an inaccessible harness, the wire must be bypassed with a new insert, usually taped to the outside of the old harness.

When replacing fusible links, be sure to use fusible link wire, NOT ordinary automotive wire. Make sure the fusible segment is of the same gauge and construction as the one being replaced and double the stripped end when crimping the terminal connector for a good contact. The melted (open) fusible link segment of the wiring harness should be cut off as close to the harness as possible, then a new segment spliced in as described. In the case of a damaged fusible link that feeds two harness wires, the harness connections should be replaced with two fusible link wires so that each circuit will have its own separate protection.

NOTE: Most of the problems caused in the wiring harness are due to bad ground connections. Always check all vehicle ground connections for corrosion or looseness before performing any power feed checks to eliminate the chance of a bad ground affecting the circuit.

Repairing Hard Shell Connectors

Unlike molded connectors, the terminal contacts in hard shell connectors can be replaced. Weatherproof hard-shell connectors with the leads molded into the shell have non-replaceable terminal ends. Replacement usually involves the use of a special terminal removal tool that depress the locking tangs (barbs) on the connector terminal and allow the connector to be removed from the rear of the shell. The connector shell should be replaced if it shows any evidence of burning, melting, cracks, or breaks. Replace individual terminals that are burnt, corroded, distorted or loose.

NOTE: The insulation crimp must be tight to prevent the insulation from sliding back on the wire when the wire is pulled. The insulation must be visibly compressed under the crimp tabs, and the ends of the crimp should be turned in for a firm grip on the insulation.

The wire crimp must be made with all wire strands inside the crimp. The terminal must be fully compressed on the wire strands with the ends of the crimp tabs turned in to make a firm grip on the wire. Check all connections with an ohmmeter to insure a good contact. There should be no measurable resistance between the wire and the terminal when connected.

Mechanical Test Equipment

Vacuum Gauge

Most gauges are graduated in inches of mercury (in.Hg), although a device called a manometer reads vacuum in inches of water (in. H_2O). The normal vacuum reading usually varies between 18 and 22 in.Hg at sea level. To test engine vacuum, the vacuum gauge must be connected to a source of manifold vacuum. Many engines have a plug in the intake manifold which can be removed and replaced with an adapter fitting. Connect the vacuum gauge to the fitting with a suitable rubber hose or, if no manifold plug is available, connect the vacuum gauge to any device using manifold vacuum, such as EGR valves, etc. The vacuum gauge can be used to determine if enough vacuum is reaching a component to allow its actuation.

Hand Vacuum Pump

Small, hand-held vacuum pumps come in a variety of designs. Most have a built-in vacuum gauge and allow the component to be tested without removing it from the vehicle. Operate the pump lever or plunger to apply the correct amount of vacuum required for the test specified in the diagnosis routines. The level of vacuum in inches of Mercury (in.Hg) is indicated on the pump gauge. For some testing, an additional vacuum gauge may be necessary.

Intake manifold vacuum is used to operate various systems and devices on late model vehicles. To correctly diagnose and solve problems in vacuum control systems, a vacuum source is necessary for testing. In some cases, vacuum can be taken from the intake manifold when the engine is running, but vacuum is normally provided by a hand vacuum pump. These hand vacuum pumps have a built-in vacuum gauge that allow testing while the device is still attached to the component. For some tests, an additional vacuum gauge may be necessary.

HEATING AND AIR CONDITIONING

Heater or Heater A/C Blower Motor

REMOVAL AND INSTALLATION

Without Air Conditioning

1. Disconnect the negative battery ground cable.
2. Disconnect the blower motor wiring at the wiring connectors.

3. Remove the backing plate-to-heater housing screws and remove the blower motor assembly..
4. Remove the spring clip that retains the blower motor fan wheel to the blower motor shaft.
5. Remove the air vent hose from the blower motor housing.
6. Remove the blower motor backing plate by unfastening and removing the mounting nuts.
7. Inspect all blower motor mounting seals. Replace any seals that are damaged.

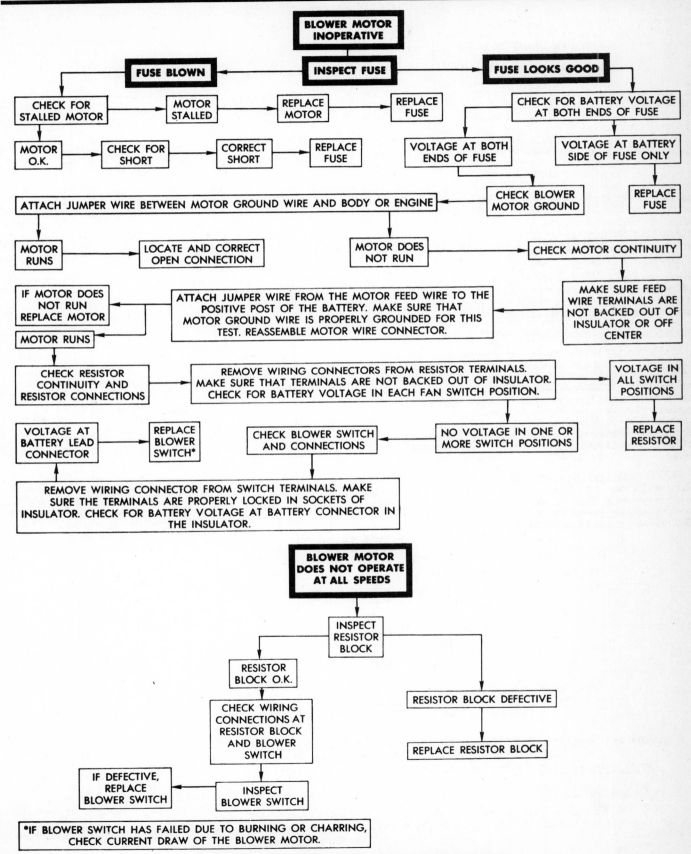

Heater blower motor and control system diagnosis

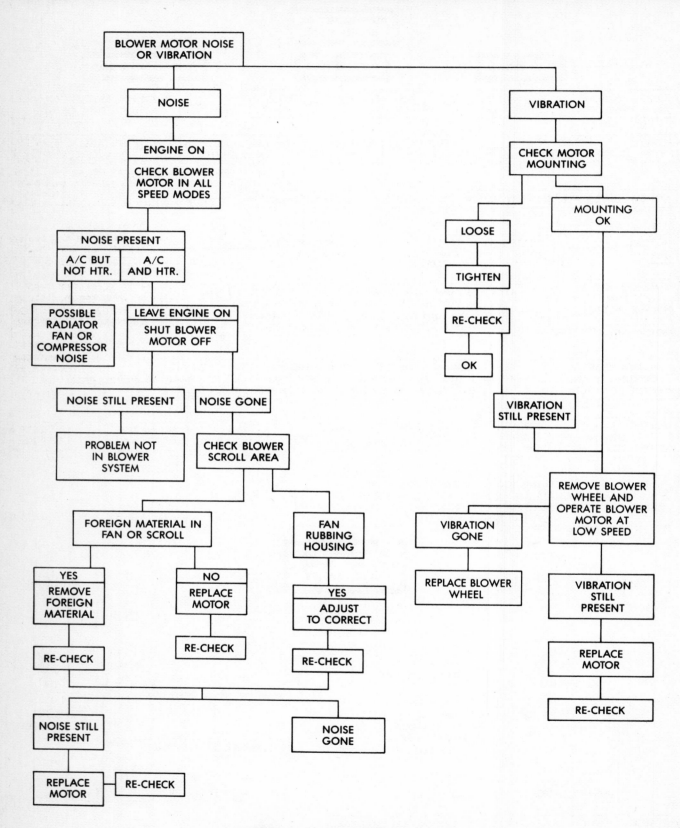

Blower motor noise/vibration diagnosis

8. Install the blower motor backing plate, vent hose and fan wheel.

9. Place the blower motor assembly in position and secure it with the mounting screws.

10. Connect the wiring harness connectors.

11. Connect the battery cable and check the blower motor operation.

With Air Conditioning

1. Disconnect the negative battery cable. Remove the top half of the shroud by removing the 4 screws: 2 from the radiator support and the 2 holding the shroud halves together.

NOTE: On 6-cylinder engines, the upper right screw is hidden behind the discharge line muffler.

2. Move the top half of the shroud out of the way. Remove the blower motor cooler tube from the blower.

3. Disconnect the blower motor wiring connector, which is a two wire connector located near the blower motor.

4. Remove the blower motor mounting plate screws or nuts and washers.

5. While holding the suction and discharge lines inboard and upward, pull the blower motor from the housing.

6. Inspect and replace any damaged sealing material.

7. Place the blower motor in position and secure it with the mounting screws or nuts and washers.

8. Install the blower motor vent cooling tube and connect the wiring harness.

9. Install the top shroud. Connect the battery cable and check the blower motor performance.

Heater Core

REMOVAL AND INSTALLATION

Without Air Conditioning

1. Disconnect the negative battery ground cable.
2. Drain the cooling system.

CAUTION

When draining the coolant, keep in mind that cats and dogs are attracted by the ethylene glycol antifreeze, and are quite likely to drink any that is left in an uncovered container or in puddles on the ground. This will prove fatal in sufficient quantity. Always drain the coolant into a sealable container. Coolant should be reused unless it is contaminated or several years old.

---

3. Disconnect the heater hoses at the core tubes.

4. Disconnect the temperature control cable at the heater core cover and air door crank.

5. Disconnect the wiring connector at the blower motor.

6.. Remove the heater case mounting nuts and bolts from the side cowl and firewall. Remove the heater case from the van.

7. Remove the back plate from the heater assembly. Remove the heater core cover.

8. Remove the heater core retaining screws and lift the core from the case.

9. Replace any damaged sealing material.

10. Lower the core into the case and install the core retaining screws.

11. Install the heater core cover and backing plate.

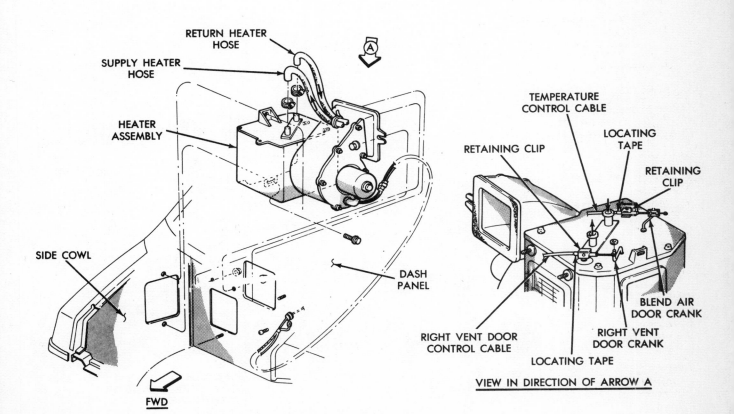

Heater assembly

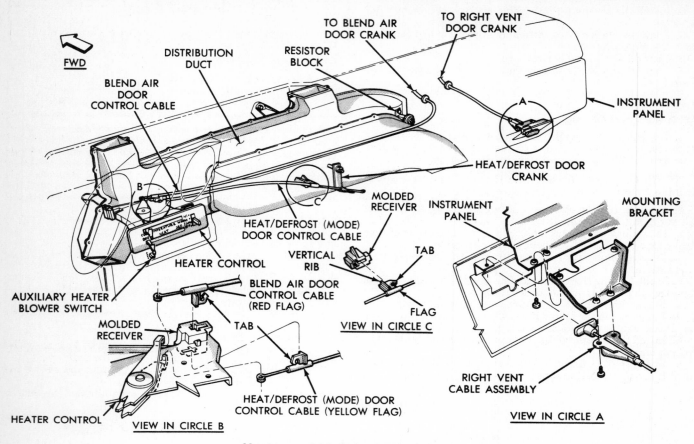

Heater control and cable routing

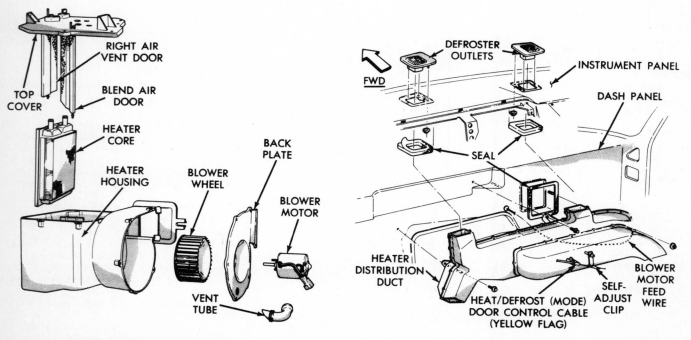

Heater disassembly/assembly

Heat distribution duct system

12. Install the heater case in the van and install the heater case mounting nuts and bolts.

13. Connect the wiring at the blower resistor.

15. Connect the temperature control cable at the heater core cover and air door crank.

16. Connect the heater hoses at the core tubes.

17. Fill the cooling system.

18. Connect the battery ground cable. Check the heating system operation.

With Air Conditioning

1. Discharge the system. See Section 1.
2. Disconnect the negative battery ground cable.
3. Drain the cooling system. Disconnect the freeze control connector from the wiring harness at the H-expansion valve (years equipped).

CAUTION

When draining the coolant, keep in mind that cats and dogs are attracted by the ethylene glycol antifreeze, and are quite likely to drink any that is left in an uncovered container or in puddles on the ground. This will prove fatal in sufficient quantity. Always drain the coolant into a sealable container. Coolant should be reused unless it is contaminated or several years old.

---

4. Cover the alternator with a plastic bag.
5. Disconnect the heater hoses at the core tubes.
6. Using a back-up wrench on the fittings, disconnect the refrigerant lines at the H-valve. Cap all openings at once!
7. Remove the mounting screws from the filter-drier bracket and swing the piping out of the way, towards the center of the van. Cap all openings at once!
8. Remove the temperature control cable from the case cover.
9. Remove the glove box.
10. Remove the spot cooler bezel and appearance shield.
11. Working through the glove box, remove the evaporator housing-to-dash panel attaching screws and nuts.
12. Remove the mounting screws from the flange connection at the blower housing. Separate the evaporator housing from the blower motor housing and carefully remove it from the van.
13. Remove the cover from the housing.
14. Remove the screw from the strap on the heater core tubes and pull the core from the housing.

To install:

15. Put the core in the housing. Install the screw in the strap on the heater core tubes.
16. Install the cover on the housing. Place the housing into position in the van.
17. Join the evaporator housing to the blower motor housing. Install the mounting screws in the flange connection at the blower housing.
18. Working through the glove box, install the evaporator housing-to-dash panel attaching screws and nuts.
19. Install the spot cooler bezel and appearance shield.
20. Install the glove box.
21. Install the temperature control cable on the case cover.
22. Install the mounting screws from the filter-drier bracket.
23. Connect the refrigerant lines at the H-valve.
24. Connect the heater hoses at the core tubes.
25. Uncover the alternator.
26. Fill the cooling system.
27. Connect the freeze control wiring harness. Connect the battery ground.
28. Evacuate, charge and leak test the system. See Section 1.

Heater Water Control Valve

REMOVAL AND INSTALLATION

1. Drain the cooling system so that the level (of the system) is below the control valve.

CAUTION

When draining the coolant, keep in mind that cats and dogs are attracted by the ethylene glycol antifreeze, and are quite likely to drink any that is left in an uncovered container or in puddles on the ground. This will prove fatal in sufficient quantity. Always drain the coolant into a sealable container. Coolant should be reused unless it is contaminated or several years old.

---

2. Loosen the heater hose clamps and remove the hoses.
3. Disconnect the vacuum line to the valve vacuum control.
4. Remove the mounting screws and remove the valve.
5. On models with an auxiliary heater; the water control valve is "teed" from the main control valve. The removal and installation procedure is the same except for more hose clamps to loosen and an extra vacuum line to remove and connect.
6. Place the control in position and secure it. Connect the heater hoses (replace any hoses that show wear) and secure the hose clamps. Connect the vacuum line(s).
7. Fill the cooling system and check the control valve operation.

Control Head

REMOVAL AND INSTALLATION

1. Disconnect the battery ground cable.
2. Remove the control unit bezel.
3. Remove the control unit screws.
4. Pull the control unit out just far enough to disconnect the wiring, cables and hoses and then remove the unit.
5. Place the control unit in position. Connect the wiring, cables and hoses.
6. Install the mounting screws and bezel. Connect the battery cable.

Evaporator Blower Motor

Refer to the Heater or Heater A/C Blower Motor procedure which appears at the beginning of this Section.

Evaporator Core

REMOVAL AND INSTALLATION

1. Discharge the system. See Section 1.
2. Disconnect the battery ground.
3. Drain the cooling system. Disconnect the freeze control wiring connector at the H-valve.

CAUTION

When draining the coolant, keep in mind that cats and dogs are attracted by the ethylene glycol antifreeze, and are quite likely to drink any that is left in an uncovered container or in puddles on the ground. This will prove fatal in sufficient quantity. Always drain the coolant into a sealable container. Coolant should be reused unless it is contaminated or several years old.

---

4. Cover the alternator with a plastic bag.
5. Disconnect the heater hoses at the core tubes.
6. Using a back-up wrench on the fittings, disconnect the refrigerant lines at the H-valve. Cap all openings at once!
7. Remove the mounting screws from the filter-drier bracket and swing the piping out of the way, towards the center of the van. Cap all openings at once!
8. Remove the temperature control cable from the case cover.
9. Remove the glove box.
10. Remove the spot cooler bezel and appearance shield.
11. Working through the glove box, remove the evaporator housing-to-dash panel attaching screws and nuts.

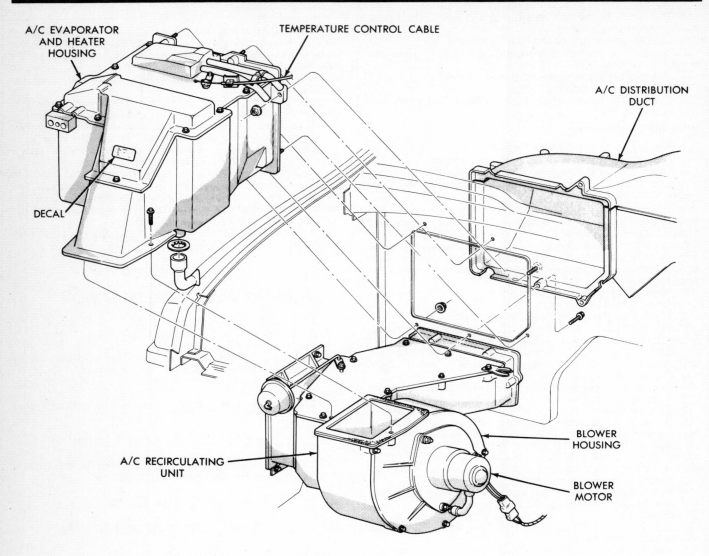

A/C EVAPORATOR AND HEATER HOUSING

TEMPERATURE CONTROL CABLE

A/C DISTRIBUTION DUCT

DECAL

A/C RECIRCULATING UNIT

BLOWER HOUSING

BLOWER MOTOR

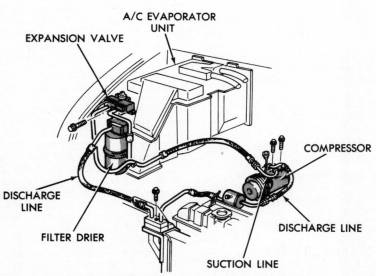

A/C EVAPORATOR UNIT

EXPANSION VALVE

COMPRESSOR

DISCHARGE LINE

FILTER DRIER

DISCHARGE LINE

SUCTION LINE

Air conditioning and heater assembly

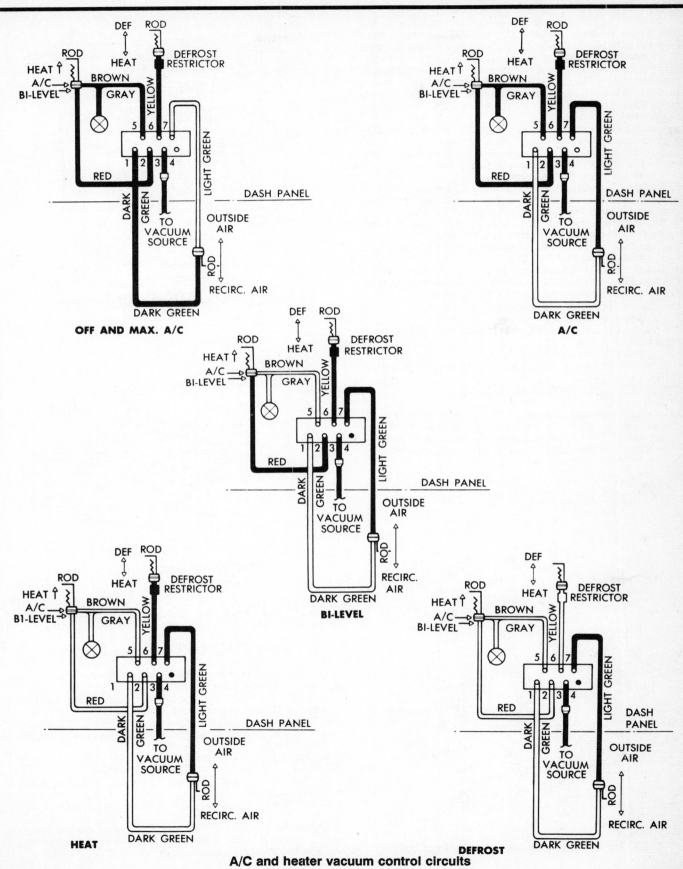

A/C and heater vacuum control circuits

12. Remove the mounting screws from the flange connection at the blower housing. Separate the evaporator housing from the blower motor housing and carefully remove it from the van.

13. Remove the cover from the housing.

14. Remove the mounting screw from the strap on the heater core tubes and pull the core from the housing.

15. Remove the attaching screw from under the plumbing attachment plate and pull the evaporator core from the housing.

To install:

15. Put the evaporator core in the housing. Install the attaching screw under the plumbing attachment plate.

16. Put the heater core in the housing. Install the mounting screw in the strap on the heater core tubes.

17. Install the cover on the housing.

18. Position the assembly in the van. Join the evaporator housing to the blower motor housing. Install the mounting screws in the flange connection at the blower housing.

19. Working through the glove box, install the evaporator housing-to-dash panel attaching screws and nuts.

20. Install the spot cooler bezel and appearance shield.

21. Install the glove box.

22. Install the temperature control cable on the case cover.

23. Install the mounting screws from the filter-drier bracket.

24. Connect the refrigerant lines at the H-valve.

25. Connect the heater hoses at the core tubes.

26. Uncover the alternator.

27. Connect the freeze control wiring connector. Fill the cooling system.

28. Connect the battery ground.

29. Evacuate, charge and leak test the system. See Section 1.

Expansion Valve

An H-type expansion valve is used to meter refrigerant onto the evaporator in accordance with cooling requirements. A low-pressure cut-off switch is mounted on the expansion valve assembly. The cut-off switch is wired in series with the compres-

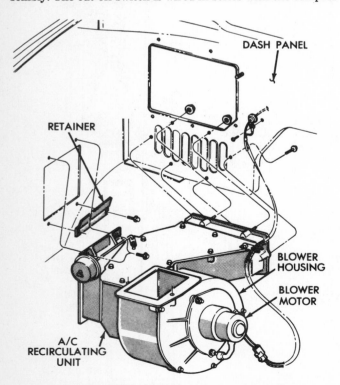

A/C blower housing removal/installation

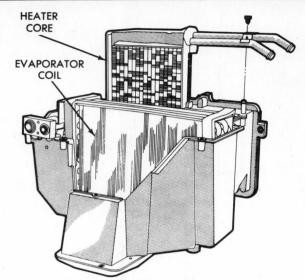

A/C heater core removal/installation

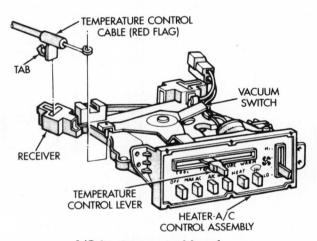

A/C heater control head

sor clutch. The switch cuts off the power supply to the clutch when refrigerant pressure drops off below a given point.

REMOVAL AND INSTALLATION

1. Discharge the refrigerant system. Refer to Section 1 for the proper safe procedure.

2. Disconnect the negative battery cable. Disconnect the wires from the low pressure cut-off switch.

3. Remove the center refrigerant line assembly mounting plate screw. Carefully pull the mount and lines toward the front of the van, away from the H-valve.

4. Remove the valve mounting screws, while holding the valve so that it will not fall. Remove the valve.

5. Clean the mounting surface on the line mounting assembly and evaporator.

6. Install new mounting gaskets, at the evaporator and line mount.

7. Remove the sealing cap from the evaporator side of the new expansion valve. Install the H-valve to the evaporator. Tighten the mounting screws to 30 inch lbs. (11.3 N.m).

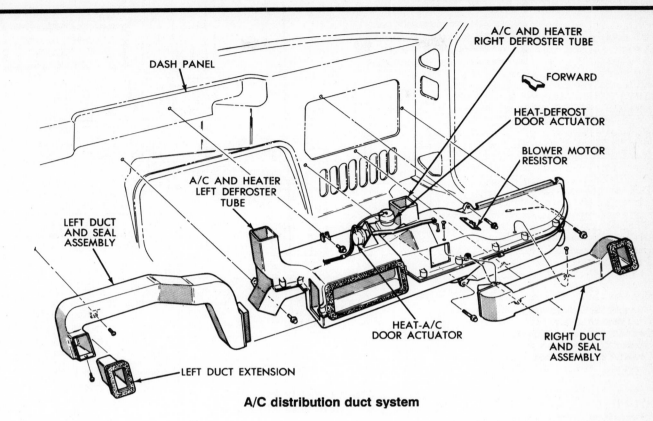

A/C distribution duct system

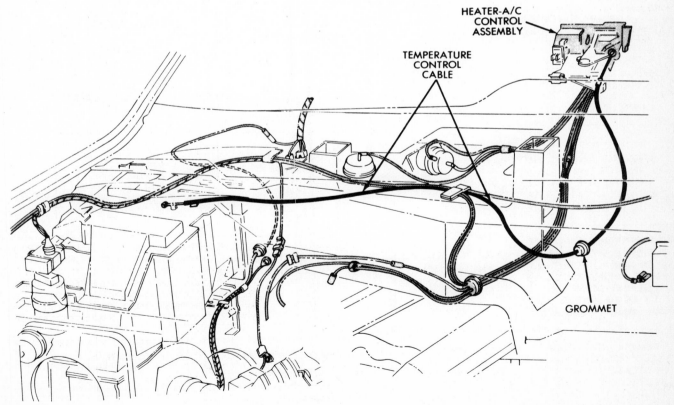

Temperature control cable routing

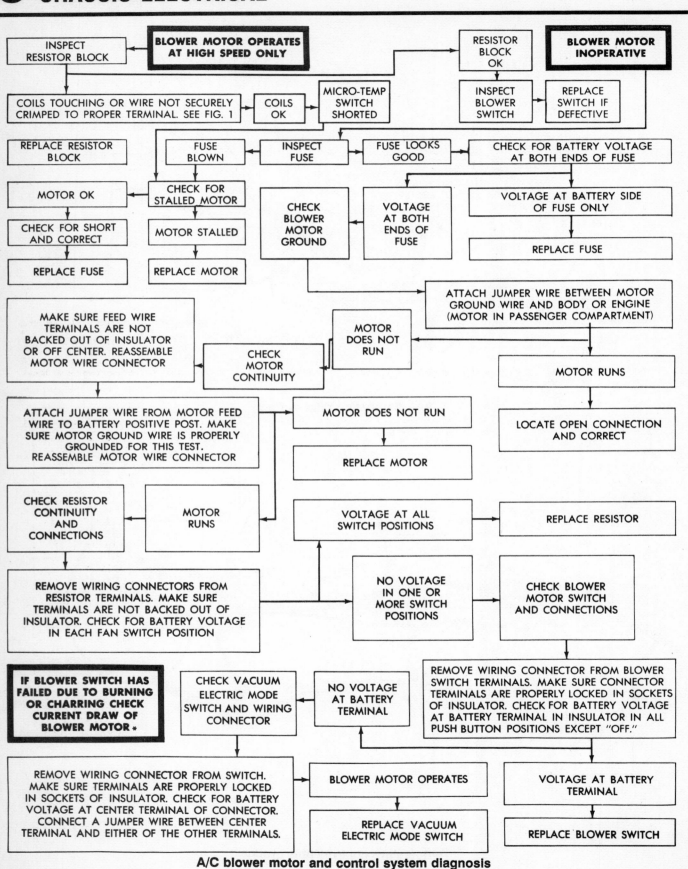

A/C blower motor and control system diagnosis

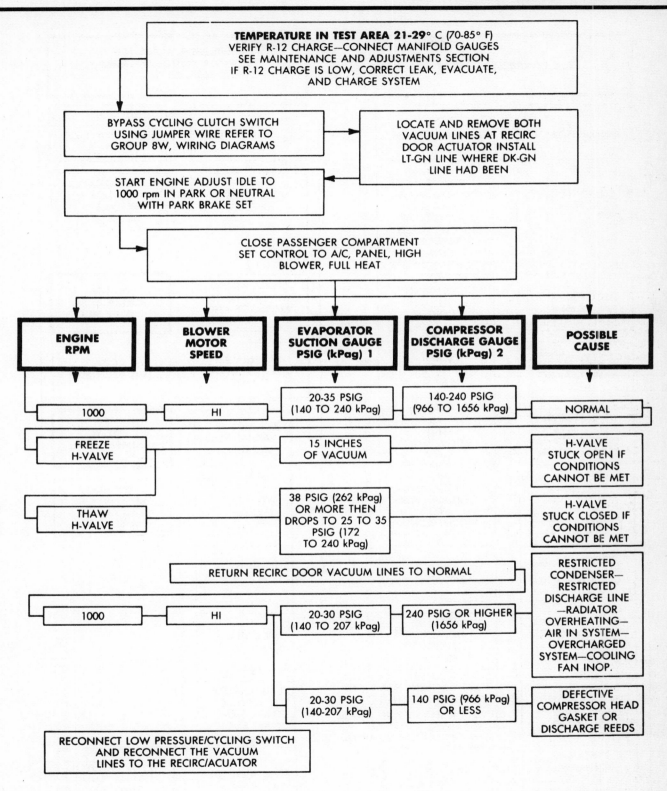

A/C refrigerant system diagnosis

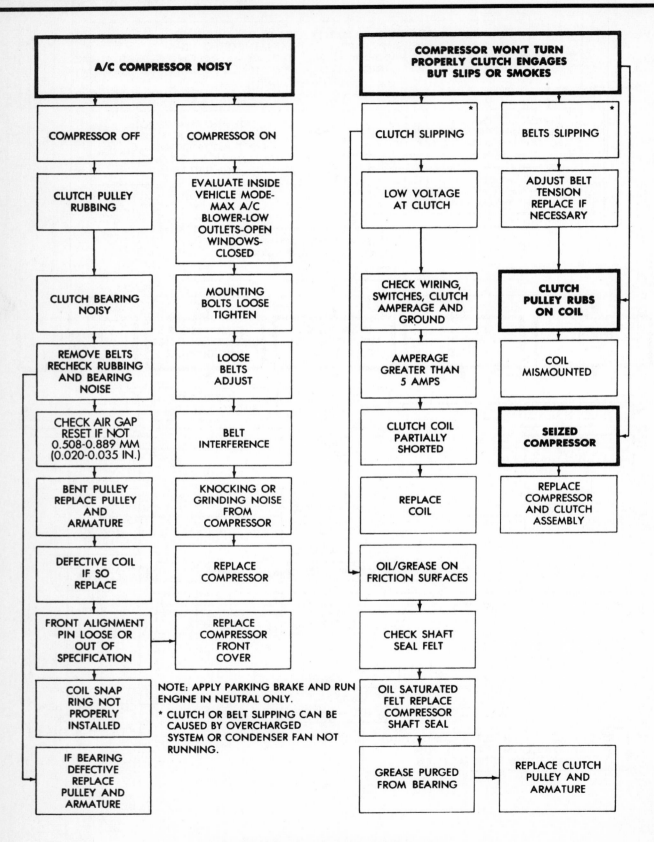

A/C compressor and clutch diagnosis

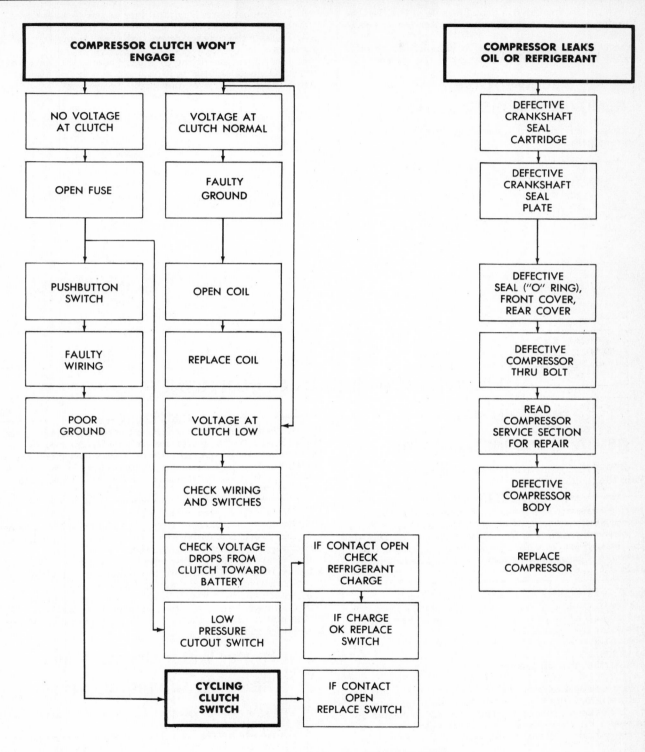

COMPRESSOR CLUTCH WON'T ENGAGE

NO VOLTAGE AT CLUTCH

OPEN FUSE

PUSHBUTTON SWITCH

FAULTY WIRING

POOR GROUND

VOLTAGE AT CLUTCH NORMAL

FAULTY GROUND

OPEN COIL

REPLACE COIL

VOLTAGE AT CLUTCH LOW

CHECK WIRING AND SWITCHES

CHECK VOLTAGE DROPS FROM CLUTCH TOWARD BATTERY

LOW PRESSURE CUTOUT SWITCH

IF CONTACT OPEN CHECK REFRIGERANT CHARGE

IF CHARGE OK REPLACE SWITCH

CYCLING CLUTCH SWITCH

IF CONTACT OPEN REPLACE SWITCH

COMPRESSOR LEAKS OIL OR REFRIGERANT

DEFECTIVE CRANKSHAFT SEAL CARTRIDGE

DEFECTIVE CRANKSHAFT SEAL PLATE

DEFECTIVE SEAL ("O" RING), FRONT COVER, REAR COVER

DEFECTIVE COMPRESSOR THRU BOLT

READ COMPRESSOR SERVICE SECTION FOR REPAIR

DEFECTIVE COMPRESSOR BODY

REPLACE COMPRESSOR

NOTE: APPLY PARKING BRAKE AND RUN ENGINE IN NEUTRAL ONLY.

A/C compressor and clutch diagnosis

8. Remove the sealing cap from the line side of the H-valve. Mount the line assembly to the H-valve. Tighten the screws to 200 inch lbs. (23 N.m).

9. Connect the low pressure cut-off switch and the negative battery cable. Evacuate, charge and leak test the system.

Receiver/Dryer

REMOVAL AND INSTALLATION

1. Discharge the refrigerant system. See Section 1 for proper and safe procedures.

2. Disconnect the refrigerant lines at the receiver/dryer.

3. Remove the mounting bolts and the receiver/dryer.

4. Position the receiver/dryer. Install and tighten the mounting bolts.

5. Inspect the line fitting O-rings (if equipped), replace as required. Connect and tighten the lines.

6. Evacuate, charge and leak test the system.

AUXILIARY HEATER/AIR CONDITIONER

Blower Motor

REMOVAL AND INSTALLATION

1. Disconnect the negative battery ground cable.

2. Drain the cooling system.

CAUTION

When draining the coolant, keep in mind that cats and dogs are attracted by the ethylene glycol antifreeze, and are quite likely to drink any that is left in an uncovered container or in puddles on the ground. This will prove fatal in sufficient quantity. Always drain the coolant into a sealable container. Coolant should be reused unless it is contaminated or several years old.

3. Disconnect the hoses at the core (under vehicle).

4. Remove the nuts that mount the heater to the floor pan.

5. Disconnect the blower wiring at the blower and remove the heater assembly from inside the van.

6. Remove the cover screws and lift off the cover. The core is attached to the cover.

7. Remove the attaching screws and lift out the blower motor.

8. Replace any sealing material that is damaged. Place the blower motor into position and secure the mounting bolts.

9. Attach the heater cover and heater core to the heater. Install the heater into the van and connect the electrical wiring and the heater hoses. Fill the cooling system. Connect the battery cable.

Heater Core

REMOVAL AND INSTALLATION

1. Disconnect the negative battery ground cable.

2. Drain the cooling system.

CAUTION

When draining the coolant, keep in mind that cats and dogs are attracted by the ethylene glycol antifreeze, and are quite likely to drink any that is left in an uncovered container or in puddles on the ground. This will prove fatal in sufficient quantity. Always drain the coolant into a sealable container. Coolant should be reused unless it is contaminated or several years old.

3. Disconnect the hoses at the core.

4. Remove the nuts that mount the heater to the floor pan.

5. Disconnect the blower wiring at the blower and remove the heater assembly from inside the van.

6. Remove the cover screws and lift off the cover. The core is attached to the cover. Remove the heater core from the cover.

7. Replace any sealing material that is damaged.

8. Attach the heater core to the cover. Install the cover assembly. Install the heater into the van and connect the electrical wiring and the heater hoses. Fill the cooling system. Connect the battery cable.

Heater Water Control Valve

REMOVAL AND INSTALLATION

1. Drain the cooling system so that the level (of the system) is below the control valve.

2. Loosen the heater hose clamps and remove the hoses.

3. Disconnect the vacuum line to the valve vacuum control.

4. Remove the mounting screws and remove the valve.

5. On models with an auxiliary heater; the water control valve is "teed" from the main control valve. The removal and installation procedure is the same except for more hose clamps to loosen and an extra vacuum line to remove and connect.

6. Place the control in position and secure it. Connect the heater hoses (replace any hoses that show wear) and secure the hose clamps. Connect the vacuum line(s).

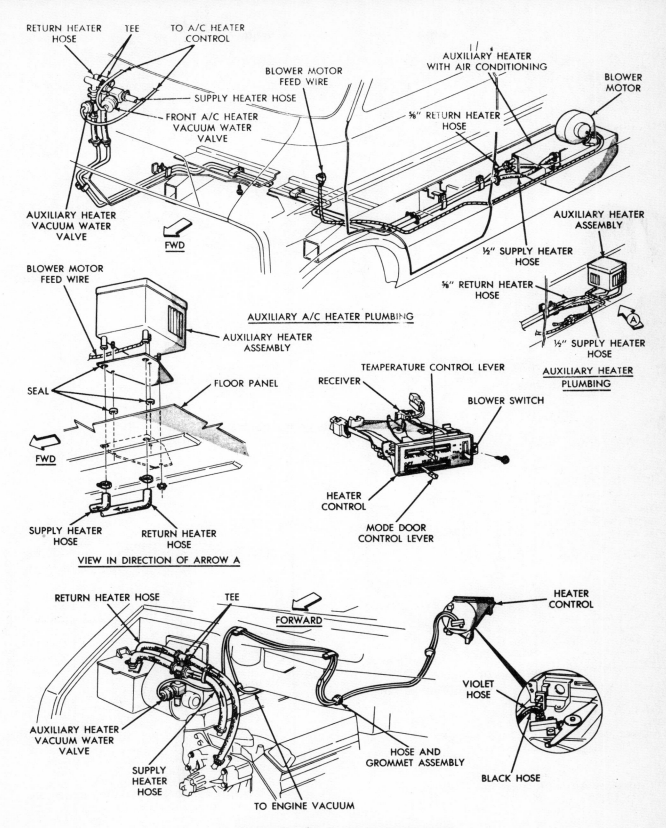

RETURN HEATER HOSE
TEE
TO A/C HEATER CONTROL
SUPPLY HEATER HOSE
FRONT A/C HEATER VACUUM WATER VALVE
BLOWER MOTOR FEED WIRE
AUXILIARY HEATER WITH AIR CONDITIONING
⅜" RETURN HEATER HOSE
BLOWER MOTOR
AUXILIARY HEATER VACUUM WATER VALVE
FWD
AUXILIARY HEATER ASSEMBLY
½" SUPPLY HEATER HOSE
⅜" RETURN HEATER HOSE
½" SUPPLY HEATER HOSE
AUXILIARY A/C HEATER PLUMBING
AUXILIARY HEATER PLUMBING

BLOWER MOTOR FEED WIRE
AUXILIARY HEATER ASSEMBLY
SEAL
FLOOR PANEL
FWD
SUPPLY HEATER HOSE
RETURN HEATER HOSE
VIEW IN DIRECTION OF ARROW A

TEMPERATURE CONTROL LEVER
RECEIVER
BLOWER SWITCH
HEATER CONTROL
MODE DOOR CONTROL LEVER

RETURN HEATER HOSE
TEE
FORWARD
HEATER CONTROL
AUXILIARY HEATER VACUUM WATER VALVE
VIOLET HOSE
SUPPLY HEATER HOSE
HOSE AND GROMMET ASSEMBLY
BLACK HOSE
TO ENGINE VACUUM

Auxiliary heater pluming and controls

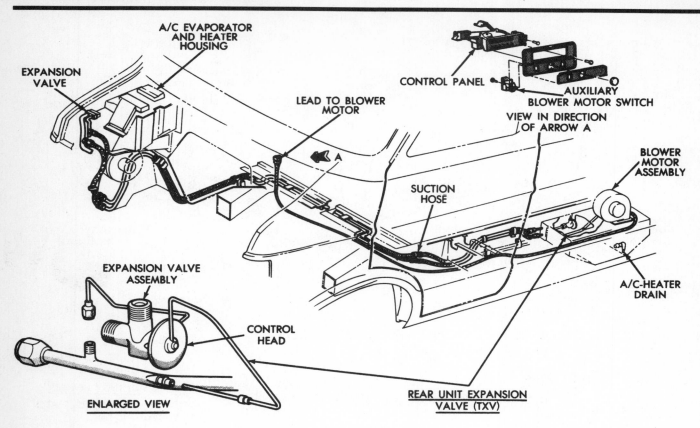

Auxiliary A/C plumbing and controls

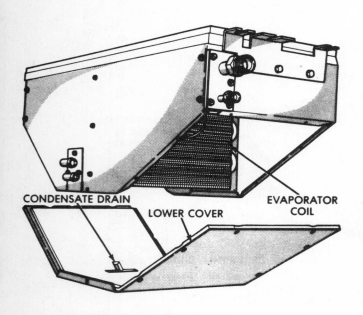

Auxiliary A/C with lower cover removed

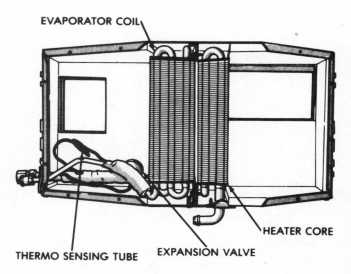

Auxiliary A/C evaporator housing assembly

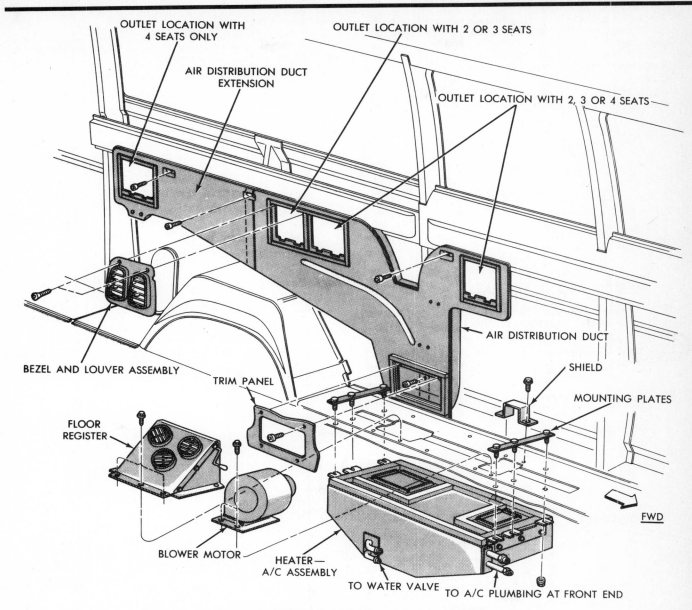

OUTLET LOCATION WITH 4 SEATS ONLY

AIR DISTRIBUTION DUCT EXTENSION

OUTLET LOCATION WITH 2 OR 3 SEATS

OUTLET LOCATION WITH 2, 3 OR 4 SEATS

AIR DISTRIBUTION DUCT

SHIELD

MOUNTING PLATES

BEZEL AND LOUVER ASSEMBLY

TRIM PANEL

FLOOR REGISTER

BLOWER MOTOR

HEATER— A/C ASSEMBLY

TO WATER VALVE

TO A/C PLUMBING AT FRONT END

FWD

Auxiliary A/C assembly and duct system

7. Fill the cooling system and check the control valve operation.

Control Head

REMOVAL AND INSTALLATION

1. Disconnect the battery ground cable.
2. Remove the control unit bezel.
3. Remove the control unit screws.
4. Pull the control unit out just far enough to disconnect the wiring, cables and hoses and then remove the unit.
5. Place the control unit in position. Connect the wiring, cables and hoses.
6. Install the mounting screws and bezel. Connect the battery cable.

Evaporator Blower Motor

REMOVAL AND INSTALLATION

1. Disconnect the battery ground cable.
2. Disconnect the wiring at the blower.
3. Remove the blower motor-to-floor screws and lift out the blower motor.
4. Install and secure the blower motor. Connect the electrical wiring and the battery cable.

Evaporator Core

REMOVAL AND INSTALLATION

1. Discharge the refrigerant system. See Section 1.

2. From under the van, using a back-up wrench on the fittings, disconnect the refrigerant lines from the core tubes. Cover all openings at once!

3. Remove the auxiliary unit lower cover.

4. Remove the seal and cover plate.

5. Remove the core.

6. Install and secure the core. Secure the cover plate and lower cover. Connect the refrigerant lines. Evacuate, charge and leak test the system. See Section 1.

RADIO

REMOVAL AND INSTALLATION

NOTE: Do not operate the radio with the speaker leads disconnected. Damage to the transistors may occur.

1. Disconnect the negative battery cable. Remove the instrument cluster hood and bezel. Pull the bezel off of the retaining clips.

2. Remove the radio mounting screws.

3. Remove the radio ground strap screw.

4. Pull the radio from the instrument panel just far enough to disconnect the wiring and antenna lead.

5. Connect the wiring and antenna lead to the radio. Connect the radio ground strap. Place the radio into position and secure it with the mounting screws. Install the bezel and instrument cluster hood. Connect the battery cable.

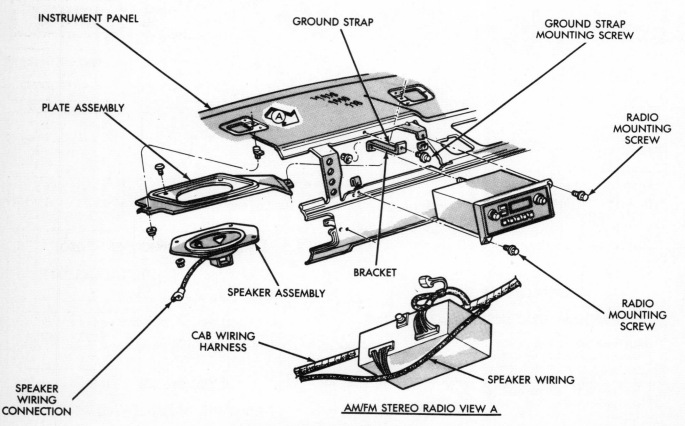

Radio mounting and connectors

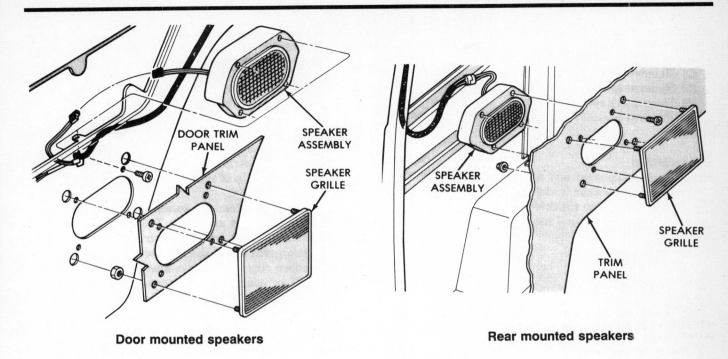

Door mounted speakers

Rear mounted speakers

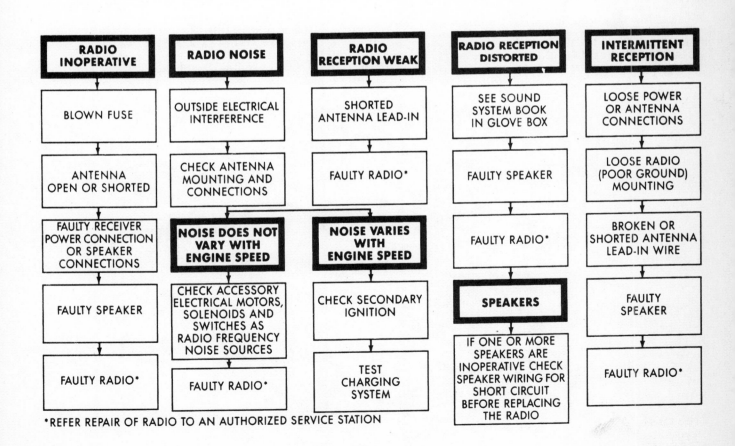

RADIO INOPERATIVE	RADIO NOISE	RADIO RECEPTION WEAK	RADIO RECEPTION DISTORTED	INTERMITTENT RECEPTION
BLOWN FUSE	OUTSIDE ELECTRICAL INTERFERENCE	SHORTED ANTENNA LEAD-IN	SEE SOUND SYSTEM BOOK IN GLOVE BOX	LOOSE POWER OR ANTENNA CONNECTIONS
ANTENNA OPEN OR SHORTED	CHECK ANTENNA MOUNTING AND CONNECTIONS	FAULTY RADIO*	FAULTY SPEAKER	LOOSE RADIO (POOR GROUND) MOUNTING
FAULTY RECEIVER POWER CONNECTION OR SPEAKER CONNECTIONS	**NOISE DOES NOT VARY WITH ENGINE SPEED** / **NOISE VARIES WITH ENGINE SPEED**		FAULTY RADIO*	BROKEN OR SHORTED ANTENNA LEAD-IN WIRE
FAULTY SPEAKER	CHECK ACCESSORY ELECTRICAL MOTORS, SOLENOIDS AND SWITCHES AS RADIO FREQUENCY NOISE SOURCES / CHECK SECONDARY IGNITION		**SPEAKERS**	FAULTY SPEAKER
FAULTY RADIO*	FAULTY RADIO* / TEST CHARGING SYSTEM		IF ONE OR MORE SPEAKERS ARE INOPERATIVE CHECK SPEAKER WIRING FOR SHORT CIRCUIT BEFORE REPLACING THE RADIO	FAULTY RADIO*

*REFER REPAIR OF RADIO TO AN AUTHORIZED SERVICE STATION

Radio diagnosis

WINDSHIELD WIPERS

Blade and Arm

REMOVAL AND INSTALLATION

To remove the arm from the pivot, lift the arm to permit the latch to be pulled out to the holding position and remove the arm from the pivot with a rocking motion.

When installing the arms, the at-rest position of the blades should be determined before pushing the arm onto the pivot. The driver's side blade should be 47mm (1.84 in.) above the windshield weatherstripping at the heel of the blade; the passenger's side blade should be 57mm (2.24 in.) above the weatherstripping.

Windshield Wiper Motor

REMOVAL AND INSTALLATION

1. Raise the hood. Disconnect the negative battery ground cable.
2. Unplug the wiring at the motor.
3. Remove the wiper motor mounting bolts.
4. Lower the motor to gain access to the crank arm-to-drive link bushing.
5. Disconnect the crank arm from the drive link by prying the bushing off.
6. Remove the motor. Remove the nut securing the crank arm to the motor. Remove the crank arm.
7. Install the crank arm. Tighten the retaining nut to 95 inch lbs. (11 N.m). Connect the drive link to the crank arm. Position the wiper motor and install and secure the mounting bolts. Tighten the mounting bolts to 65 inch lbs. (7 N.m). Connect the wiring harness to the wiper motor. Connect the battery cable.

Removing the windshield wiper arm

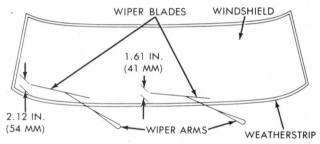

ADJUST WIPER ARM SO TIP OF BLADE IS ABOVE THE WEATHERSTRIP IN PARK POSITION AS SHOWN ± .94

Adjusting the wiper arms

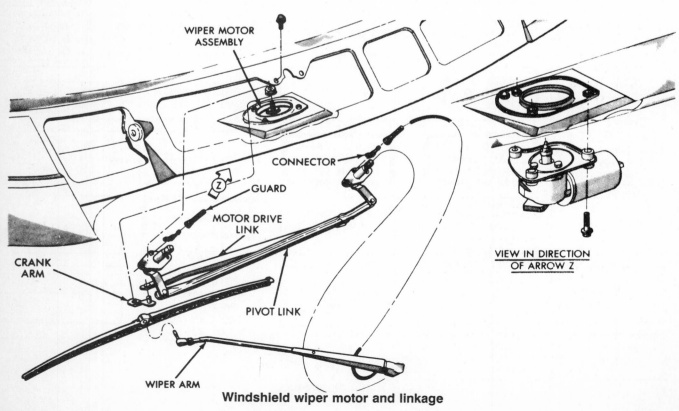

Windshield wiper motor and linkage

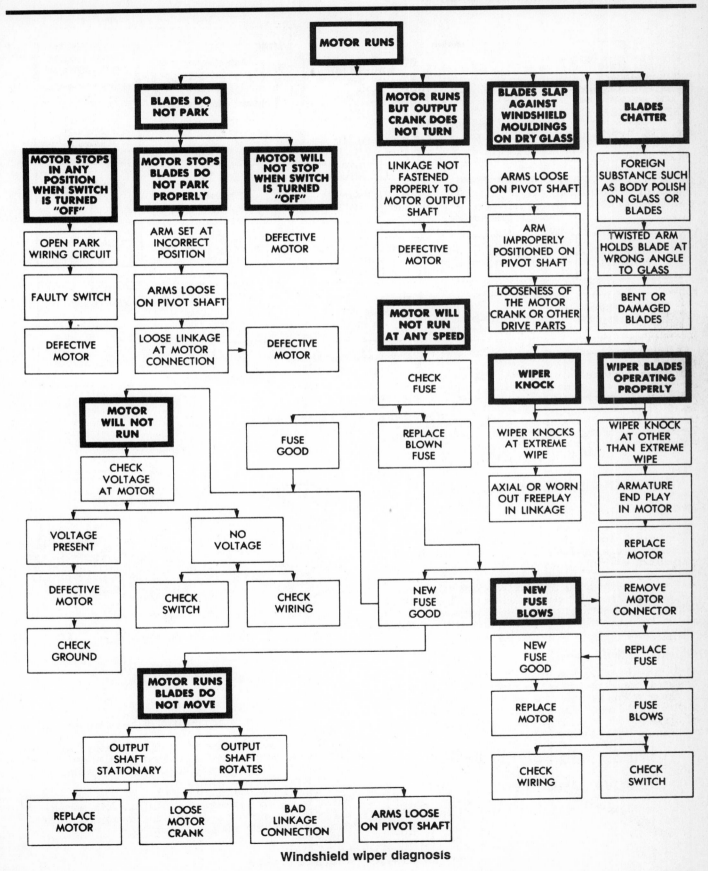

Windshield wiper diagnosis

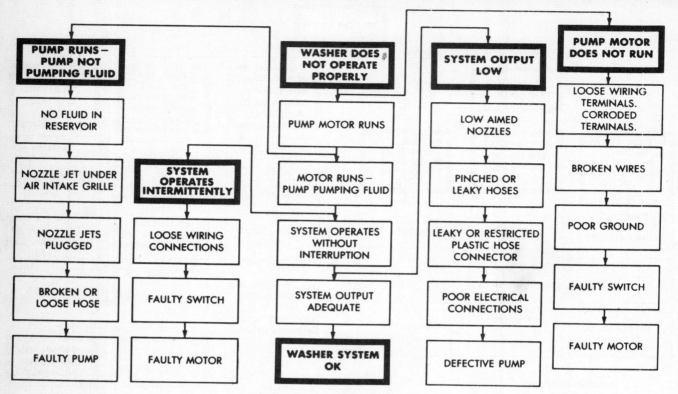

PUMP RUNS— PUMP NOT PUMPING FLUID		WASHER DOES NOT OPERATE PROPERLY	SYSTEM OUTPUT LOW	PUMP MOTOR DOES NOT RUN
NO FLUID IN RESERVOIR		PUMP MOTOR RUNS	LOW AIMED NOZZLES	LOOSE WIRING TERMINALS. CORRODED TERMINALS.
NOZZLE JET UNDER AIR INTAKE GRILLE	SYSTEM OPERATES INTERMITTENTLY	MOTOR RUNS— PUMP PUMPING FLUID	PINCHED OR LEAKY HOSES	BROKEN WIRES
NOZZLE JETS PLUGGED	LOOSE WIRING CONNECTIONS	SYSTEM OPERATES WITHOUT INTERRUPTION	LEAKY OR RESTRICTED PLASTIC HOSE CONNECTOR	POOR GROUND
BROKEN OR LOOSE HOSE	FAULTY SWITCH	SYSTEM OUTPUT ADEQUATE	POOR ELECTRICAL CONNECTIONS	FAULTY SWITCH
FAULTY PUMP	FAULTY MOTOR	WASHER SYSTEM OK	DEFECTIVE PUMP	FAULTY MOTOR

Windshield washer diagnosis

Wiper Linkage

REMOVAL AND INSTALLATION

Drive Link

1. Remove the wiper arms and washer hoses.
2. Remove the cowl grille.
3. Reach through the access hole, remove the drive link from the crank arm and connecting pivot links by prying the retainer bushings apart.
4. Remove the drive link through the access hole.
5. Place the drive link in position on the crank arm and connecting pivot link pins. Snap them together carefully so the bushings are not damaged (use pliers if necessary). Install the grille cover, wiper arms and washer hoses.

Connecting Pivot Link

1. Remove the cowl grille.
2. Reach through the access hole, remove the connecting link from the drive link by prying the retainer bushings apart.
3. Remove the connecting link through the access hole.
4. Place the link on the drive and snap it into position over the bushings. Install the cowl grille.

Windshield Washer Pump

REMOVAL AND INSTALLATION

1. Remove the screws mounting the washer reservoir to the mounting bracket.

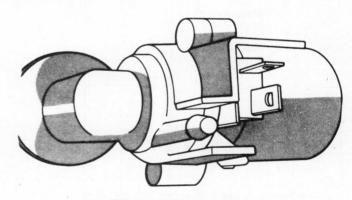

Windshield washer pump

2. Hold the reservoir and disconnect the wiring leads to the washer pump.
3. Remove the pump mounting nut and mounting washer from the reservoir but reaching through the filler cap opening.
4. Remove the pump from the bottom of the washer reservoir. Discard the rubber grommet.
5. Install a new mounting grommet. Install the washer pump and secure it with the washer and nut. Connect the wiring leads and secure the reservoir to its mounting bracket. Fill the reservoir. Check for leaks and pump operation.

INSTRUMENTS AND SWITCHES

Instrument Cluster

REMOVAL AND INSTALLATION

1. Disconnect the negative battery ground cable.
2. Remove the screws securing the instrument panel hood and bezel. Pull the bezel off of the upper retaining clips.
3. On model equipped, disconnect the message center electrical connector. Disconnect the gear shift pointer cable from the steering column. Remove the cluster mounting screws.
4. Pull the cluster out just far enough to disconnect the speedometer cable.
5. Unplug the wiring connectors at the back of the cluster. Remove the cluster.
6. Positon the cluster at the panel opening. Connect the wiring harness connectors and speedometer.
7. Place the cluster into position and secure it with the mounting screws. Connect the gearshift pointer cable to arm on the steering column. Connect the message center wiring harness. Install the cluster panel hood and bezel. Connect the battery cable.

Oil Pressure or Volt/Amp Gauge

REMOVAL AND INSTALLATION

1. Disconnect the negative battery ground cable.
2. Remove the screws retaining the instrument cluster hood and bezel. Pull the assembly off over the upper retaining clips.
3. Disconnect the message center wiring harness (if equipped).
4. Remove the small screws that fasten the small lens and shroud to the instrument cluster. (Cluster removal may be necessary on some models). Remove the gauge mask.
5. Remove the gauge mounting screws or nuts and remove the gauge.
6. Install and secure the gauge in its mounted position. Install the gauge mask, small lens and shroud assembly (install the cluster assembly if removed). Connect the message center wiring harness. Install the cluster hood and bezel and connect the battery cable.

Temperature or Fuel Gauge

REMOVAL AND INSTALLATION

1. Disconnect the negative battery ground cable.
2. Remove the screws retaining the instrument cluster hood and bezel. Pull the assembly off over the upper retaining clips.
3. Disconnect the message center wiring harness (if equipped).
4. Remove the screws that fasten the large lens and shroud to the instrument cluster. (Cluster removal may be necessary on some models). Remove the gauge mask.
5. Remove the gauge mounting screws or nuts and remove the gauge.
6. Install and secure the gauge in its mounted position. Install the gauge mask, small lens and shroud assembly (install the cluster assembly if removed). Connect the message center wiring harness. Install the cluster hood and bezel and connect the battery cable.

Printed Circuit Board

REMOVAL AND INSTALLATION

1. Remove the instrument cluster assembly.
2. On 1989 models, remove the screws attaching the lens and shroud assemblies to the instrument cluster, and remove the assemblies. Remove the voltage limiter and lamp socket assemblies.Remove all of the gauges except the speedometer. Remove the printed circuit mounting screws and the circuit board.
3. On 1990-91 models, remove the lamp sockets. Remove the circuit board mounting screws and the circuit board.
4. On 1989 moels, install the circuit board and secure it with the mounting screws. Install the gauges, lamp socket assemblies and voltage limiter. Install the lens and shrouds.
5. On 1990-91 models, install the circuit board and tighten the mounting screws. Install the lamp socket assemblies.
6. Install the instrument cluster.

Engine Cover

REMOVAL AND INSTALLATION

The in van engine cover may be removed by positioning the front seats to the full rear position, if the vehicle is equipped with swivel seats, the passenger's seat may be swiveled rearward for improved clearance. Remove the ashtray, disengage the forward engine cover latches and remove all the latching screws that hold the cover to the floor.

After finishing the required servicing, place the cover in position and secure it with the floor mounting screws and foward latches. Put the ashtray in position and return the front seats to their normal positions.

Windshield Wiper Switch

REMOVAL AND INSTALLATION

The wiper switch is incorporated into the turn signal switch stalk. For switch removal and installation, see Section 8.

Headlight Switch

REMOVAL AND INSTALLATION

1. Disconnect the negative battery ground cable.
2. Remove the left side lower steering column cover. Remove the screws that mount the hood release handle and lower the handle. (The proceeding two operations may not be necessary on 1989 models). Working under the instrument panel, depress the locking buttom on the switch and pull the knob and stem from the switch.
3. Remove the instrument cluster hood and bezel.
4. Remove the switch bezel mounting screws.
5. Remove the switch mounting nut, remove the switch and disconnect the wiring.
6. Place the light switch into position after connect the wiring harness. Secure the switch with the mounting nut.
7. Install the switch bezel and secure it with the mounting screws. Install the instrument cluster hood and bezel. Push the headlamp switch stem into the switch until it is locked into position. Install the hood release hand and mounting screws, and the lower steering column cover (models equipped). Connect the battery cable.

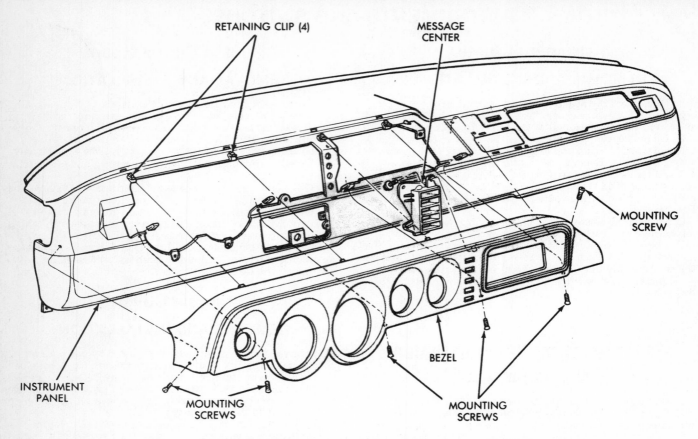

Instrument cluster hood, message center and bezel

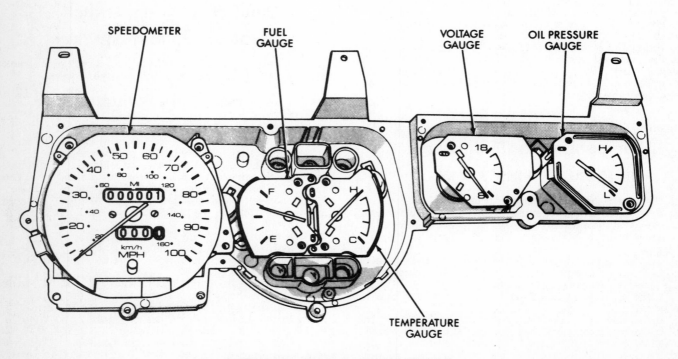

Instrument cluster with mask removed

MOUNTING SCREWS (3)

MOUNTING SCREWS (6)

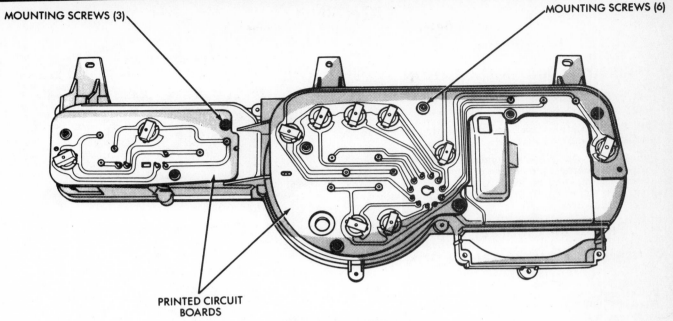

PRINTED CIRCUIT BOARDS

Printed circuit board

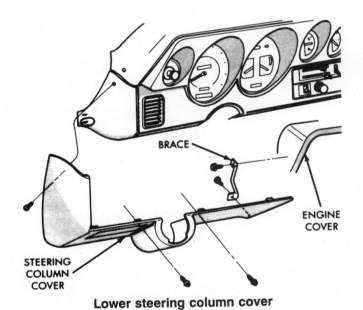

BRACE

ENGINE COVER

STEERING COLUMN COVER

Lower steering column cover

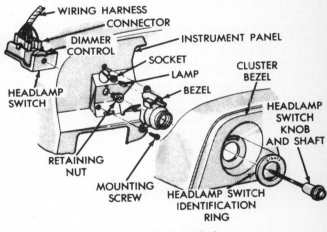

WIRING HARNESS

CONNECTOR

DIMMER CONTROL

INSTRUMENT PANEL

SOCKET

LAMP

BEZEL

CLUSTER BEZEL

HEADLAMP SWITCH

HEADLAMP SWITCH KNOB AND SHAFT

RETAINING NUT

MOUNTING SCREW

HEADLAMP SWITCH IDENTIFICATION RING

Headlamp switch

Speedometer Cable

REMOVAL AND INSTALLATION

1990-91 models use an electronic speedometer. A wiring harness extends from the speedometer to a distance sensor at the transmission, and an adapter and pinion in the transmission. A signal is sent from the distance sensor to the speedometer controlling the speedometer "needle" reading.

1989 models may use the cable driven system. Replacement is done by:
1. Disconnect the negative battery ground cable.

2. Disconnect the fusible link located under the hood.
3. If equipped with a trip odometer, remove the reset knob. Remove the plastic pins fastening the lens to the instrument cluster assembly and remove the lens and mask.
4. Remove the speedometer assembly mounting screws. Depress the tab on the plastic ferrule and disconnect the speedometer from the speedometer cable.
5. Remove the cable core. Install the new cable core into the housing make sure it seats fully.
6. Place the speedometer in position after connecting the cable. Secure the assembly with the mounting screws. Install the mask, lens and trip odometer knob. Connect the fusible link and battery cable.

LIGHTING

Headlights

REMOVAL AND INSTALLATION

1. Remove the headlight bezel.
2. Remove the headlight mounting ring screws.

NOTE: Don't mistake the headlight aiming screws for the headlight mounting screws!

3. Pull out the headlight slowly and disconnect the wiring.
4. Connect the harness plug to the prongs at the rear of the headlamp. Hold the lamp in position, make sure the locating lugs are in their slots, install and secure the retaining ring. Install the headlamp bezel.

Signal and Marker Lights

REMOVAL AND INSTALLATION

1. Remove the lens screws.
2. Pull off the lens, turn the bulb and pull it out of its socket.
3. Place the bulb into the socket and turn it to secure the mounting lugs. Install and secure the lens.

Front Turn Signal and Parking Lights

REMOVAL AND INSTALLATION

1. Remove the lens screws.
2. Pull off the lens, turn the bulb and pull it out of its socket.
3. Place the bulb into the socket and turn it to secure the mounting lugs. Install and secure the lens.

Side Marker Lights

REMOVAL AND INSTALLATION

1. Remove the lens screws.
2. Pull off the lens, turn the bulb and pull it out of its socket.
3. Place the bulb into the socket and turn it to secure the mounting lugs. Install and secure the lens.
4. Installation is the reverse of removal.

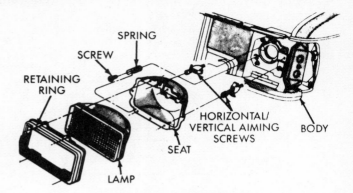

Sealed beam replacement

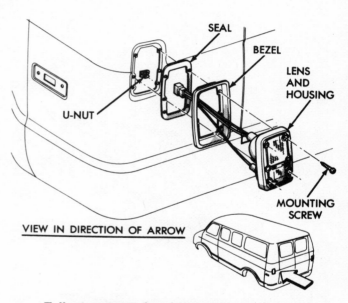

Tail, stop, turn signal and back-up lamp

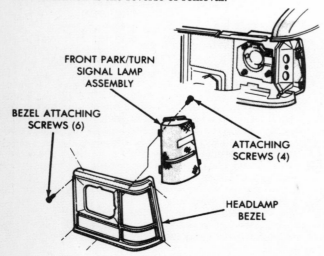

Headlamp bezel, park and turn signal lamp

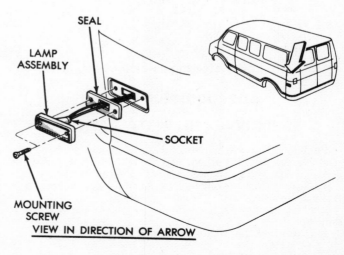

Rear side marker lamp

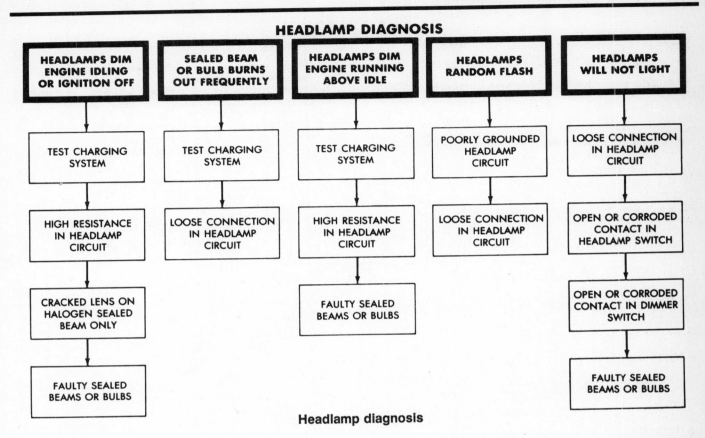

HEADLAMP DIAGNOSIS

Headlamp diagnosis

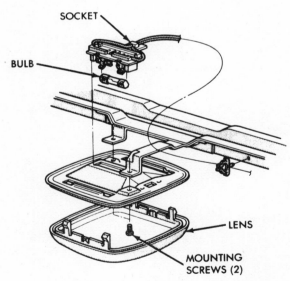

Dome/cargo lamp

Rear Turn Signal, Brake and Parking Lights

REMOVAL AND INSTALLATION

1. Remove the lens screws.
2. Pull off the lens, turn the bulb and pull it out of its socket.
3. Place the bulb into the socket and turn it to secure the mounting lugs. Install and secure the lens.

Dome/Reading Light

REMOVAL AND INSTALLATION

Gently squeeze the front and rear of the dome light lens together and pull down to remove it. Pull the bulb down and out of the socket. Push the new bulb into the socket and install the dome cover.

Remove the screws that mount the lamp housing. Remove the housing, and rotate the buld out of the mounting socket. Install a new bulb and secure the housing.

TRAILER WIRING

Wiring

Wiring the van for towing is fairly easy. There are a number of good wiring kits available and these should be used, rather than trying to design your own. All trailers will need brake lights and turn signals as well as tail lights and side marker lights. Most states require extra marker lights for overly wide trailers. Also, most states have recently required back-up lights for trailers, and most trailer manufacturers have been building trailers with back-up lights for several years.

Additionally, some Class I, most Class II and just about all Class III trailers will have electric brakes.

Add to this number an accessories wire, to operate trailer internal equipment or to charge the trailer's battery, and you can have as many as seven wires in the harness.

Determine the equipment on your trailer and buy the wiring

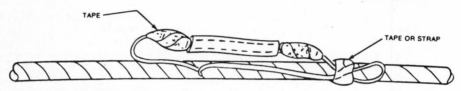

TYPICAL REPAIR USING THE SPECIAL #17 GA. (9.00" LONG-YELLOW) FUSE LINK REQUIRED FOR THE AIR/COND. CIRCUITS (2) #687E and #261A LOCATED IN THE ENGINE COMPARTMENT

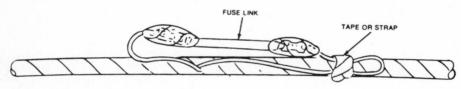

TYPICAL REPAIR FOR ANY IN-LINE FUSE LINK USING THE SPECIFIED GAUGE FUSE LINK FOR THE SPECIFIC CIRCUIT

TYPICAL REPAIR USING THE EYELET TERMINAL FUSE LINK OF THE SPECIFIED GAUGE FOR ATTACHMENT TO A CIRCUIT WIRE END

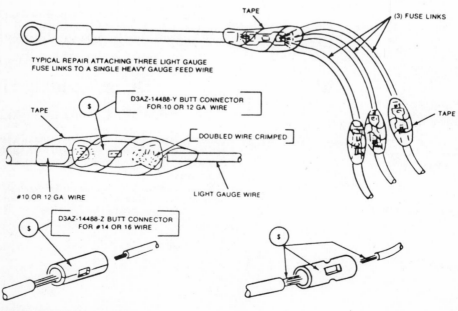

FUSIBLE LINK REPAIR PROCEDURE

General fusible link repair

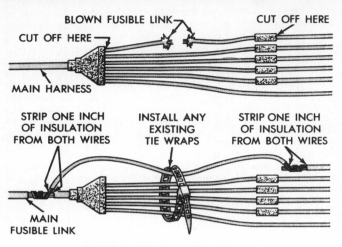

Fusible link at harness repair

Wire and Gauge	Color Code	Color
12 Ga.	BK	Black
14 Ga.	RD	Red
16 Ga.	DB	Dark Blue
18 Ga.	GY	Gray
20 Ga.	OR	Orange
22 Ga.	WT	White

Fusible link chart

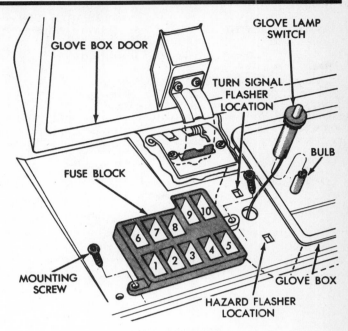

Fuse box location

kit necessary. The kit will contain all the wires needed, plus a plug adapter set which included the female plug, mounted on the bumper or hitch, and the male plug, wired into, or plugged into the trailer harness.

When installing the kit, follow the manufacturer's instructions. The color coding of the wires is standard throughout the industry.

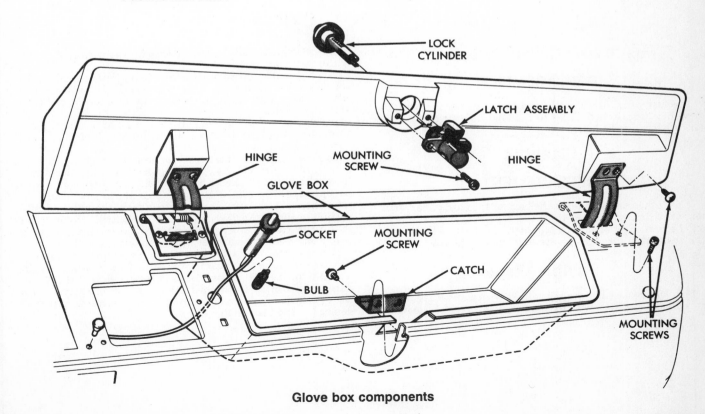

Glove box components

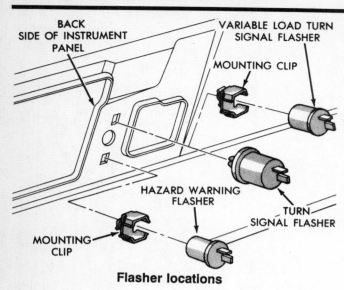

Flasher locations

One point to note, some domestic vehicles, and most imported vehicles, have separate turn signals. On most domestic vehicles, the brake lights and rear turn signals operate with the same bulb. For those vehicles with separate turn signals, you can purchase an isolation unit so that the brake lights won't blink whenever the turn signals are operated, or, you can go to your local electronics supply house and buy four diodes to wire in series with the brake and turn signal bulbs. Diodes will isolate the brake and turn signals. The choice is yours. The isolation units are simple and quick to install, but far more expensive than the diodes. The diodes, however, require more work to install properly, since they require the cutting of each bulb's wire and soldering in place of the diode.

One final point, the best kits are those with a spring loaded cover on the vehicle mounted socket. This cover prevents dirt and moisture from corroding the terminals. Never let the vehicle socket hang loosely. Always mount it securely to the bumper or hitch.

CIRCUIT PROTECTION

Fuses and Circuit Breakers

REPLACEMENT

The fuse panel box is to the right of the glove box with the glove box door opened. Refer to the following wiring diagrams for fuse location and size.

Flashers

REPLACEMENT

The flashers are located at the back of the dash panel. To replace the flasher(s), open the glove box door. Depress the top and bottom edges of the box together, lift upwards and remove the box. Reach through the opening and change the flasher(s).

Fusible Links

DO NOT replace blown fusible links with standard wire. Use only fusible type wire with hypalon insulation or damage to the

electrical system could occur. Make sure the correct gauge of wiring is used.

When a fusible link blows it is very important to find out the cause. Do not just replace the link to correct the problem. The fusible links are placed in the system for protection against dead shorts to ground.

In some instances the link may be blown and it will not show through the insulation. Check the entire length of the fusible wire when the link is suspect of failure.

To repair a blown fusible; disconnect the negative battery ground cable. Cut off any remaining portion of the blown link flush with the multiple connection insulator. Take care not to cut any of the other links. Remove about one inch of insulation from the main harness about one inch from the multiple connection insulator. Remove one inch of insulation from one end of the new fusible link and wrap it around the main harness wire that was stripped. Heat splice the wires together with a high temperature soldering gun. Apply resin type solder until it flows freely. Allow the soldering to cool and wrap the splice with a minimum of three layers of electrical tape. Repeat the procedure with the other end of the fusible link and wire harness. Tie the new fusible link to the existing ones to prevent chafing or damage to the insulation.

AMPS	FUSE	COLOR CODE
3	VT	VIOLET
4	PK	PINK
5	TN	TAN
10	RD	RED
20	YL	YELLOW
25	NAT	NATURAL
30	LG	LIGHT GREEN

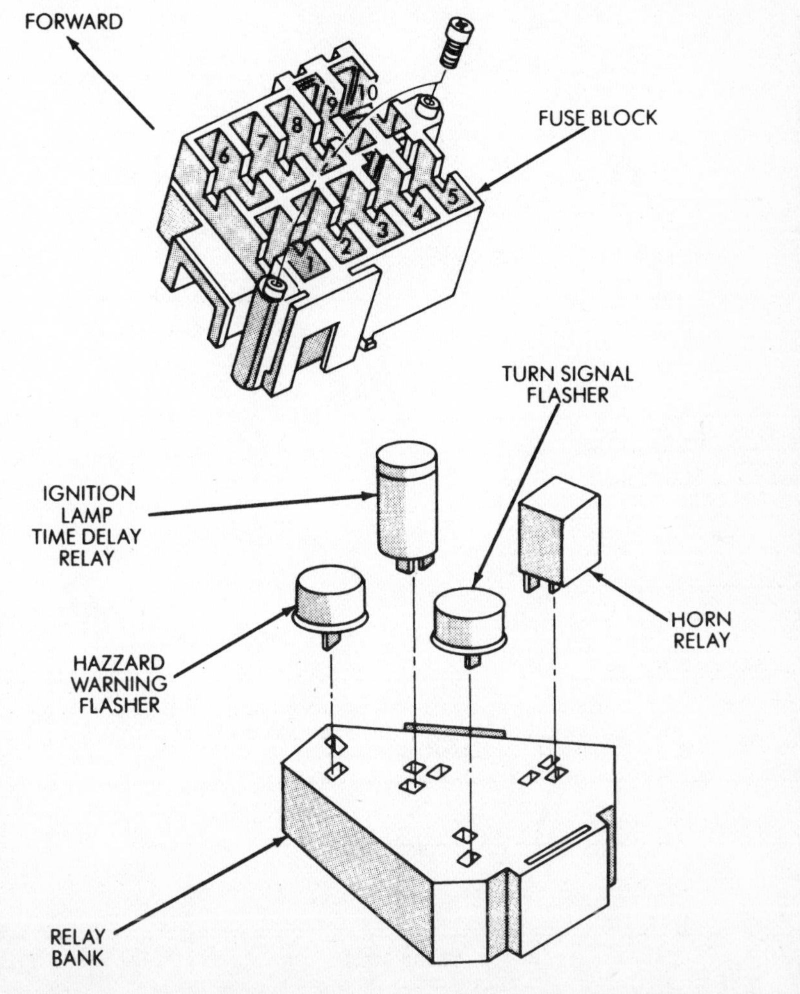

CAUTION: WHEN REPLACING A BLOWN FUSE, IT IS IMPORTANT TO REPLACE IT WITH A FUSE HAVING THE CORRECT AMPERAGE RATING. THE USE OF A FUSE WITH A RATING OTHER THAN INDICATED MAY RESULT IN A DANGEROUS ELECTRICAL OVERLOAD. IF A PROPERLY RATED FUSE CONTINUES TO BLOW, IT INDICATES A PROBLEM THAT SHOULD BE CORRECTED.

FUSE BLOCK

FORWARD

HORN RELAY

TURN SIGNAL FLASHER

IGNITION LAMP TIME DELAY RELAY

HAZZARD WARNING FLASHER

RELAY BANK

CAVITY	FUSE/COLOR	ITEMS FUSED	MODE
1	3 AMP	ILLUMINATION LAMPS, RADIO, A/C & HEATER, CIGAR LIGHTER, AUXILIARY A/C HEATER, REAR DEFOGGER & CLOCK DISPLAY DIMMING	IGNITION ACC & RUN
2	10 AMP	RADIO AND CLOCK	
3	20 AMP	WINDSHIELD WIPER SWITCH	
4	25 AMP	BACK-UP LAMPS, TURN SIGNALS, VAN CONVERSION RELAY, RWAL MODULE	
5	5 AMP	GAUGES (AMPS, FUEL, TEMPERATURE & OIL) OIL LAMP, EMR LAMP, BRAKE WARNING LAMP, REAR DEFOGGER SWITCH & RELAY, LOW FLUID LAMP, SEAT BELT WARNING LAMP, SPEEDOMETER, RWAL LAMP, O/D MODULE	IGNITION RUN AND START
6	20 AMP	ILLUMINATION LAMPS, FUSE #1, PARK, TAIL, SIDE MARKER & LICENSE LAMP, CLOCK DISPLAY INTENSITY, HORNS	
7	25 AMP	CIGAR LIGHTER, AUXILIARY A/C & HEATER, KEY-IN & HEADLAMP ON BUZZER, STOP LAMPS RWAL MODULE	BATTERY FEED
8	20 AMP	RADIO & CLOCK MEMORY, DOME LAMP, COURTESY READING LAMP, GLOVE BOX LAMP, IGNITION TIME DELAY RELAY AND LAMPS, POWER MIRRORS UNDERHOOD LAMP	
9	30 AMP	AIR CONDITIONING BLOWER MOTOR AND HEATER BLOWER MOTOR, REAR HEAT/AC BLOWER MOTOR	IGNITION ACC & RUN
10	20 AMP	HAZARD FLASHERS	

1990 Fuse block and relay bank

CAVITY	FUSE/COLOR	ITEMS FUSED
1	NOT USED	
2	20 AMP YELLOW	RADIO ELECTRONICS, GLOVE BOX; TIME DELAY; DOOR COURTESY; DOME; READING; AND UNDER-HOOD LAMPS
3	20 AMP YELLOW	HAZARD FLASHER
4		NOT USED
5		NOT USED
6	10 AMP RED	BACK-UP LAMPS, HEATED REAR WINDOW SWITCH
7	30 AMP LT. GREEN	A/C & HEATER BLOWER MOTOR, AUXILIARY A/C HEATER RELAY
8	30 AMP C/BRKR	POWER WINDOWS
9	15 AMP LT. BLUE	CIGAR LIGHTER
10	20 AMP YELLOW	WINDSHIELD WIPER SWITCH
11	10 AMP RED	RADIO AND CLOCK
12	20 AMP YELLOW	TURN SIGNAL FLASHER, VAN CONVERSION RELAY, REAR WHEEL ANTI-LOCK MODULE
13	20 AMP YELLOW	HORNS
14	20 AMP YELLOW	HEADLAMP SWITCH
15	20 AMP YELLOW	STOP LAMPS, KEY-IN & HEADLAMP ON BUZZER, REAR WHEEL ANTI-LOCK, AUXILIARY AC & HEATER
16	30 AMP C/BRKR	POWER DOOR LOCK
17	5 AMP TAN	GAUGES, BRAKE LAMP, SEAT BELT BUZZER AND LAMP, EMR LAMP, LOW WASHER FLUID LAMP, LOW OIL LAMP, CHECK ENGINE LAMP, ANTI-LOCK LAMP, O/D CONTROL MODULE

CAVITY	FUSE/COLOR	ITEMS FUSED
18	2 AMP GRAY	SPEED CONTROL
19		NOT USED
20	3 AMP VIOLET	CLUSTER LAMPS, RADIO LAMPS, A/C HEATER CONTROL LAMP, CIGAR LIGHTER LAMP, HEATED REAR WINDOW SWITCH LAMP, REAR A/C HEATER SWITCH LAMP

AMPS	FUSE	COLOR CODE
2	GY	GREY
3	VT	VIOLET
5	TN	TAN
10	RD	RED
15	LB	LIGHT BLUE
20	YL	YELLOW
30	LG	LIGHT GREEN

1991 Fuse block and relay bank

AMPS	FUSE	COLOR CODE
3	VT	VIOLET
4	PK	PINK
5	TN	TAN
10	RD	RED
20	YL	YELLOW
25	NAT	NATURAL
30	LG	LIGHT GREEN

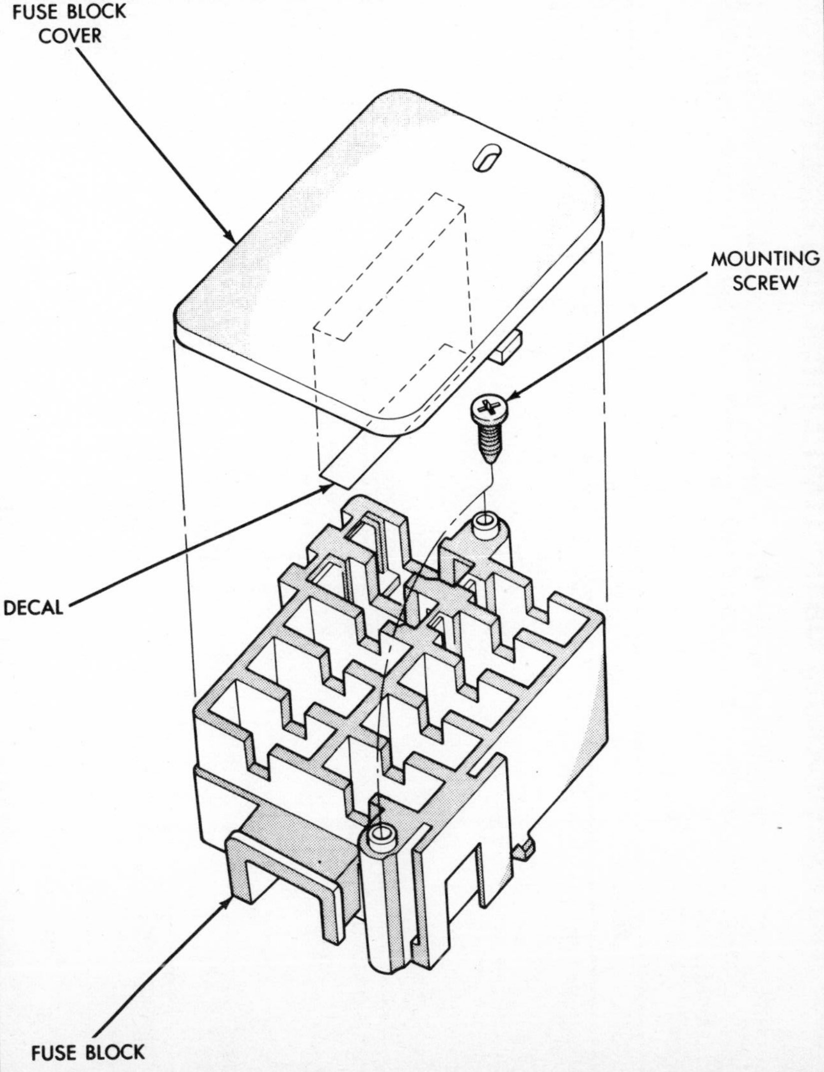

CAUTION: WHEN REPLACING A BLOWN FUSE, IT IS IMPORTANT TO REPLACE IT WITH A FUSE HAVING THE CORRECT AMPERAGE RATING. THE USE OF A FUSE WITH A RATING OTHER THAN INDICATED MAY RESULT IN A DANGEROUS ELECTRICAL OVERLOAD. IF A PROPERLY RATED FUSE CONTINUES TO BLOW, IT INDICATES A PROBLEM THAT SHOULD BE CORRECTED.

MOUNTING SCREW

FUSE BLOCK COVER

DECAL

FUSE BLOCK

CAVITY	FUSE/COLOR	ITEMS FUSED	MODE
1	3 AMP	ILLUMINATION LAMPS, RADIO, A/C & HEATER, CIGAR LIGHTER, AUXILIARY A/C HEATER, REAR DEFOGGER & CLOCK DISPLAY DIMMING	
2	10 AMP	RADIO AND CLOCK	IGNITION ACC & RUN
3	20 AMP	WINDSHIELD WIPER SWITCH	
4	20 AMP	BACK-UP LAMPS, TURN SIGNALS, & A/C CLUTCH	
5	5 AMP	GAUGES (AMPS, FUEL, TEMPERATURE & OIL) OIL LAMP, EMR LAMP, BRAKE WARNING LAMP, REAR DEFOGGER SWITCH & RELAY, WINDOW LIFT RELAY, SEAT BELT WARNING LAMP & BUZZER, SPEED CONTROL, EMR FEED & OVERDRIVE MODULE	IGNITION RUN AND START
6	20 AMP	ILLUMINATION LAMPS, FUSE #1, PARK, TAIL, SIDE MARKER & LICENSE LAMP, CLOCK DISPLAY INTENSITY HORNS	
7	25 AMP	CIGAR LIGHTER, AUXILIARY A/C & HEATER, KEY-IN & HEADLAMP ON BUZZER, STOP LAMPS	BATTERY FEED
8	20 AMP	RADIO & CLOCK MEMORY, DOME LAMP, COURTESY READING LAMP, GLOVE BOX LAMP, IGNITION TIME DELAY RELAY AND LAMPS	
9	30 AMP	AIR CONDITIONING BLOWER MOTOR AND HEATER BLOWER MOTOR	IGNITION ACC & RUN
10	20 AMP	HAZARD FLASHERS	

1989 Fuse block

Troubleshooting Basic Turn Signal and Flasher Problems

Most problems in the turn signals or flasher system can be reduced to defective flashers or bulbs, which are easily replaced. Occasionally, problems in the turn signals are traced to the switch in the steering column, which will require professional service.

F = Front   R = Rear   ● = Lights off   o = Lights on

Problem		Solution
Turn signals light, but do not flash		• Replace the flasher
No turn signals light on either side		• Check the fuse. Replace if defective. • Check the flasher by substitution • Check for open circuit, short circuit or poor ground
Both turn signals on one side don't work		• Check for bad bulbs • Check for bad ground in both housings
One turn signal light on one side doesn't work		• Check and/or replace bulb • Check for corrosion in socket. Clean contacts. • Check for poor ground at socket
Turn signal flashes too fast or too slow		• Check any bulb on the side flashing too fast. A heavy-duty bulb is probably installed in place of a regular bulb. • Check the bulb flashing too slow. A standard bulb was probably installed in place of a heavy-duty bulb. • Check for loose connections or corrosion at the bulb socket
Indicator lights don't work in either direction		• Check if the turn signals are working • Check the dash indicator lights • Check the flasher by substitution

Troubleshooting Basic Turn Signal and Flasher Problems

Most problems in the turn signals or flasher system can be reduced to defective flashers or bulbs, which are easily replaced. Occasionally, problems in the turn signals are traced to the switch in the steering column, which will require professional service.

F = Front   R = Rear   ● = Lights off   o = Lights on

Problem		Solution
One indicator light doesn't light	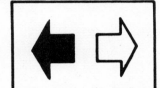	• On systems with 1 dash indicator: See if the lights work on the same side. Often the filaments have been reversed in systems combining stoplights with taillights and turn signals. Check the flasher by substitution • On systems with 2 indicators: Check the bulbs on the same side Check the indicator light bulb Check the flasher by substitution

Troubleshooting the Heater

Problem	Cause	Solution
Blower motor will not turn at any speed	• Blown fuse • Loose connection • Defective ground • Faulty switch • Faulty motor • Faulty resistor	• Replace fuse • Inspect and tighten • Clean and tighten • Replace switch • Replace motor • Replace resistor
Blower motor turns at one speed only	• Faulty switch • Faulty resistor	• Replace switch • Replace resistor
Blower motor turns but does not circulate air	• Intake blocked • Fan not secured to the motor shaft	• Clean intake • Tighten security
Heater will not heat	• Coolant does not reach proper temperature • Heater core blocked internally • Heater core air-bound • Blend-air door not in proper position	• Check and replace thermostat if necessary • Flush or replace core if necessary • Purge air from core • Adjust cable
Heater will not defrost	• Control cable adjustment incorrect • Defroster hose damaged	• Adjust control cable • Replace defroster hose

Troubleshooting Basic Dash Gauge Problems

Problem	Cause	Solution
Coolant Temperature Gauge		
Gauge reads erratically or not at all	• Loose or dirty connections • Defective sending unit • Defective gauge	• Clean/tighten connections • Bi-metal gauge: remove the wire from the sending unit. Ground the wire for an instant. If the gauge registers, replace the sending unit. • Magnetic gauge: disconnect the wire at the sending unit. With ignition ON gauge should register COLD. Ground the wire; gauge should register HOT.
Ammeter Gauge—Turn Headlights ON (do not start engine). Note reaction		
Ammeter shows charge Ammeter shows discharge Ammeter does not move	• Connections reversed on gauge • Ammeter is OK • Loose connections or faulty wiring • Defective gauge	• Reinstall connections • Nothing • Check/correct wiring • Replace gauge
Oil Pressure Gauge		
Gauge does not register or is inaccurate	• On mechanical gauge, Bourdon tube may be bent or kinked • Low oil pressure • Defective gauge • Defective wiring • Defective sending unit	• Check tube for kinks or bends preventing oil from reaching the gauge • Remove sending unit. Idle the engine briefly. If no oil flows from sending unit hole, problem is in engine. • Remove the wire from the sending unit and ground it for an instant with the ignition ON. A good gauge will go to the top of the scale. • Check the wiring to the gauge. If it's OK and the gauge doesn't register when grounded, replace the gauge. • If the wiring is OK and the gauge functions when grounded, replace the sending unit
All Gauges		
All gauges do not operate All gauges read low or erratically All gauges pegged	• Blown fuse • Defective instrument regulator • Defective or dirty instrument voltage regulator • Loss of ground between instrument voltage regulator and car • Defective instrument regulator	• Replace fuse • Replace instrument voltage regulator • Clean contacts or replace • Check ground • Replace regulator

Troubleshooting Basic Dash Gauge Problems

Problem	Cause	Solution
Warning Lights		
Light(s) do not come on when ignition is ON, but engine is not started	• Defective bulb • Defective wire • Defective sending unit	• Replace bulb • Check wire from light to sending unit • Disconnect the wire from the sending unit and ground it. Replace the sending unit if the light comes on with the ignition ON.
Light comes on with engine running	• Problem in individual system • Defective sending unit	• Check system • Check sending unit (see above)

Troubleshooting Basic Windshield Wiper Problems

Problem	Cause	Solution
Electric Wipers		
Wipers do not operate— Wiper motor heats up or hums	• Internal motor defect • Bent or damaged linkage • Arms improperly installed on linking pivots	• Replace motor • Repair or replace linkage • Position linkage in park and reinstall wiper arms
Electric Wipers		
Wipers do not operate— No current to motor	• Fuse or circuit breaker blown • Loose, open or broken wiring • Defective switch • Defective or corroded terminals • No ground circuit for motor or switch	• Replace fuse or circuit breaker • Repair wiring and connections • Replace switch • Replace or clean terminals • Repair ground circuits
Wipers do not operate— Motor runs	• Linkage disconnected or broken	• Connect wiper linkage or replace broken linkage
Vacuum Wipers		
Wipers do not operate	• Control switch or cable inoperative • Loss of engine vacuum to wiper motor (broken hoses, low engine vacuum, defective vacuum/fuel pump) • Linkage broken or disconnected • Defective wiper motor	• Repair or replace switch or cable • Check vacuum lines, engine vacuum and fuel pump • Repair linkage • Replace wiper motor
Wipers stop on engine acceleration	• Leaking vacuum hoses • Dry windshield • Oversize wiper blades • Defective vacuum/fuel pump	• Repair or replace hoses • Wet windshield with washers • Replace with proper size wiper blades • Replace pump

Troubleshooting Basic Lighting Problems

Problem	Cause	Solution
Lights		
One or more lights don't work, but others do	· Defective bulb(s) · Blown fuse(s) · Dirty fuse clips or light sockets · Poor ground circuit	· Replace bulb(s) · Replace fuse(s) · Clean connections · Run ground wire from light socket housing to car frame
Lights burn out quickly	· Incorrect voltage regulator setting or defective regulator · Poor battery/alternator connections	· Replace voltage regulator · Check battery/alternator connections
Lights go dim	· Low/discharged battery · Alternator not charging · Corroded sockets or connections · Low voltage output	· Check battery · Check drive belt tension; repair or replace alternator · Clean bulb and socket contacts and connections · Replace voltage regulator
Lights flicker	· Loose connection · Poor ground · Circuit breaker operating (short circuit)	· Tighten all connections · Run ground wire from light housing to car frame · Check connections and look for bare wires
Lights "flare"—Some flare is normal on acceleration—if excessive, see "Lights Burn Out Quickly"	· High voltage setting	· Replace voltage regulator
Lights glare—approaching drivers are blinded	· Lights adjusted too high · Rear springs or shocks sagging · Rear tires soft	· Have headlights aimed · Check rear springs/shocks · Check/correct rear tire pressure
Turn Signals		
Turn signals don't work in either direction	· Blown fuse · Defective flasher · Loose connection	· Replace fuse · Replace flasher · Check/tighten all connections
Right (or left) turn signal only won't work	· Bulb burned out · Right (or left) indicator bulb burned out · Short circuit	· Replace bulb · Check/replace indicator bulb · Check/repair wiring
Flasher rate too slow or too fast	· Incorrect wattage bulb · Incorrect flasher	· Flasher bulb · Replace flasher (use a variable load flasher if you pull a trailer)
Indicator lights do not flash (burn steadily)	· Burned out bulb · Defective flasher	· Replace bulb · Replace flasher
Indicator lights do not light at all	· Burned out indicator bulb · Defective flasher	· Replace indicator bulb · Replace flasher

WIRING DIAGRAMS

CAV	WIRE COLOR	DESCRIPTION	CAV	WIRE COLOR	DESCRIPTION
1	RD/WT*	BATTERY FEED TO HEADLAMP SWITCH	37	DB/OR*	A/C ELECTRONIC CLUTCH CYCLING SWITCH
2	RD/PK*	BATTERY FEED (VAN CONVERSION)	38	BK/PK	CHECK ENGINE LAMP
3			39		
4			40		
5	VT/WT*	HEADLAMP LOW BEAM	41	WT/PK*	BRAKE SENSOR
6	RD/OR*	HEADLAMP HIGH BEAM	42	GY/PK*	EMISSION MAINTENANCE REMINDER LAMP
7	DB/RD*	SPEED CONTROL	43	RD	AMPMETER
8	BR/RD*	SPEED CONTROL	44	BK/GY*	AMPMETER
9	WT/RD*	SPEED CONTROL	45	BR/WT*	WINDSHIELD WIPER LOW SPEED
10	YL/RD*	SPEED CONTROL	46	RD/YL*	WINDSHIELD WIPER HIGH SPEED
11			47	DB	IGNITION RUN & START (PIN 1)
12	OR/WT*	OVERDRIVE SWITCH	48	BK/RD	REAR WINDOW DEFOGGER
13	OR/DB*	OVERDRIVE SOLENOID	49	RD	IGNITION SWITCH (PIN B1)
14			50	PK/BK*	IGNITION SWITCH (PIN B3)
15					
16					
17	VT/YL*	TEMPERATURE SENDING UNIT OR SWITCH			
18	GY	OIL PRESSURE LAMP SWITCH			
19	GY/YL*	OIL PRESSURE SENDING UNIT			
20					
21	LG	LEFT FRONT TURN SIGNAL LAMP			
22	TN	RIGHT FRONT TURN SIGNAL LAMP			
23	PK/WT*	HAZARD FLASHER			
24					
25					
26	BK/YL*	PARKING LAMPS			
27	WT	BACK-UP LAMP FEED			
28	VT/BK*	BACK-UP LAMP GROUND			
29	BR	WINDSHIELD WIPER WASHER MOTOR			
30	DG/BK	ELECTRIC FUEL PUMP FEED			
31	YL	IGNITION START (PIN 5)			
32	DG/RD*	HORNS			
33	BK/RD	UNDERHOOD LAMP			
34	GY/BK*	BRAKE WARNING LAMP			
35	DG	WINDSHIELD WIPER SWITCH			
36	DB	WINDSHIELD WIPER SWITCH			

WIRE COLOR CODES	
BK	BLACK
BR	BROWN
DB	DARK BLUE
DG	DARK GREEN
GY	GRAY
LB	LIGHT BLUE
LG	LIGHT GREEN
OR	ORANGE
PK	PINK
RD	RED
TN	TAN
VT	VIOLET
WT	WHITE
YL	YELLOW
*	WITH TRACER

50 WAY
BULKHEAD DISCONNECT

1989 50-Way bulkhead disconnect

CAV	WIRE COLOR	DESCRIPTION	CAV	WIRE COLOR	DESCRIPTION
1	DG/RD*	MAP SENSOR	37	GY/PK*	EMISSION MAINTENANCE REMINDER LAMP
2			38	OR/WT*	OVERDRIVE TRANSMISSION LOCK-OUT SWITCH
3	TN/WT*	COOLANT SENSOR	39	BK/OR*	AIR SWITCHING SOLENOID
4	BK/LB*	SENSOR RETURN	40	GY/YL*	EGR SOLENOID
5	BK/WT*	SIGNAL GROUND	41	RD	DIRECT BATTERY
6			42		
7	WT/LG*	SPEED CONTROL RESUME SWITCH	43		
8	YL/RD*	SPEED CONTROL ON/OFF SWITCH	44		
9	BR/RD*	SPEED CONTROL SET SWITCH	45	BR	A/C SWITCH SENSE
10	DG/BK*	Z1 INPUT (VOLTAGE SENSE)	46		
11			47	GY/BK*	DISTRIBUTOR REFERENCE PICKUP
12	DB/WT*	FJ2	48	WT/OR*	VEHICLE DISTANCE SENSOR SIGNAL
13	VT/WT*	5 VOLT SUPPLY	49		
14	DG/OR*	ALTERNATOR FIELD CONTROL	50		
15	LB/RD*	POWER GROUND	51	PK	SCI TRANSMIT
16	LB/RD*	POWER GROUND	52	OR	9 VOLT INPUT
17	BR/WT*	ISC (IDLE SPEED CONTROL ACTUATOR)	53	TN/RD*	SPEED CONTROL VACUUM SOLENOID
18			54	PK/BK*	PURGE SOLENOID
19	GY/RD*	ISC (IDLE SPEED CONTROL ACTUATOR)	55	OR/BK*	LOCK-UP TORQUE CONVERTER OR SHIFT IND LAMP — MANUAL TRANS
20			56	DB/OR*	A/C CLUTCH RELAY
21	BK/RD*	THROTTLE BODY TEMPERATURE SENSOR	57		
22	OR/DB*	THROTTLE POSITION SENSOR	58	DB/YL*	AUTO SHUT DOWN RELAY
23	BK/DG*	OXYGEN SENSOR	59	BK/PK*	CHECK ENGINE LAMP
24			60	LG/RD*	SPEED CONTROL VENT SOLENOID
25					
26	VT	CLOSED THROTTLE SWITCH			
27					
28					
29	WT/PK*	BRAKE SWITCH			
30	BR/YL*	PARK/NEUTRAL SWITCH			
31	LG	SCI RECEIVE			
32	GY/WT*	INJECTOR CONTROL 2			
33	VT/YL*	INJECTOR CONTROL 1			
34	YL	DWELL CONTROL			
35					
36					

WIRE COLOR CODES		LB	LIGHT BLUE	VT	VIOLET
BK	BLACK	LG	LIGHT GREEN	WT	WHITE
BR	BROWN	OR	ORANGE	YL	YELLOW
DB	DARK BLUE	PK	PINK	*	WITH TRACER
DG	DARK GREEN	RD	RED		
GY	GRAY	TN	TAN		

CONNECTOR
TERMINAL SIDE
SHOWN

1989 60-Way single module engine connector 3.9L and 5.2L engine

CAV	WIRE COLOR	DESCRIPTION	CAV	WIRE COLOR	DESCRIPTION
1			20	BK/LB*	GROUND FROM LEFT FRONT DOOR JAMB SWITCH
2	BK BK	HORN & KEY-IN LAMP GROUND (STANDARD COLUMN)	21		
3	BK/RD*	GROUND FROM HORN SWITCH	22		
4	LG	LEFT FRONT TURN SIGNAL	23		
5	TN*	RIGHT FRONT TURN SIGNAL	24	YL/RD*	KEY-IN LAMP FEED (W/OPTIONS)
6	PK/BK*	FEED FROM HAZARD FLASHER	25	BK	KEY-IN LAMP GROUND WIRE (TILT COLUMN)
7	RD/BK*	FEED FROM TURN SIGNAL FLASHER			
8	DG/YL*	LEFT REAR TURN SIGNAL			
9	BR/YL*	RIGHT REAR TURN SIGNAL			
10	WT/TN*	FEED FROM BRAKE SWITCH			
11					
12	BR*	WINDSHIELD WIPER LOW SPEED			
13	RD/YL*	WINDSHIELD WIPER HIGH SPEED			
14	DG	WINDSHIELD WIPER MOTOR PARK RETURN & INTERMITTENT WIPE SIGNAL			
15	DB DB	WINDSHIELD WIPER SWITCH FEED & MOTOR PARK FEED			
16	BR	WINDSHIELD WASHER MOTOR			
17	BK BK	INTERMITTENT WIPE SWITCH GROUND			
18	BK/LG*	INTERMITTENT WIPE MODULE GROUND			
19	LB	KEY-IN BUZZER			

WIRE COLOR CODES			
BK	BLACK	OR	ORANGE
BR	BROWN	PK	PINK
DB	DARK BLUE	RD	RED
DG	DARK GREEN	TN	TAN
GY	GRAY	VT	VIOLET
LB	LIGHT BLUE	WT	WHITE
LG	LIGHT GREEN	YL	YELLOW
		*	WITH TRACER

25 WAY CONNECTOR

WIRE END OF CONNECTOR SHOWN

25-Way instrument panel to body connector

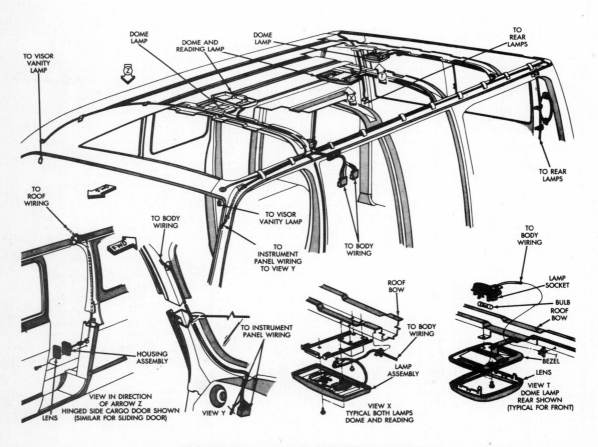

Body wiring and component identification

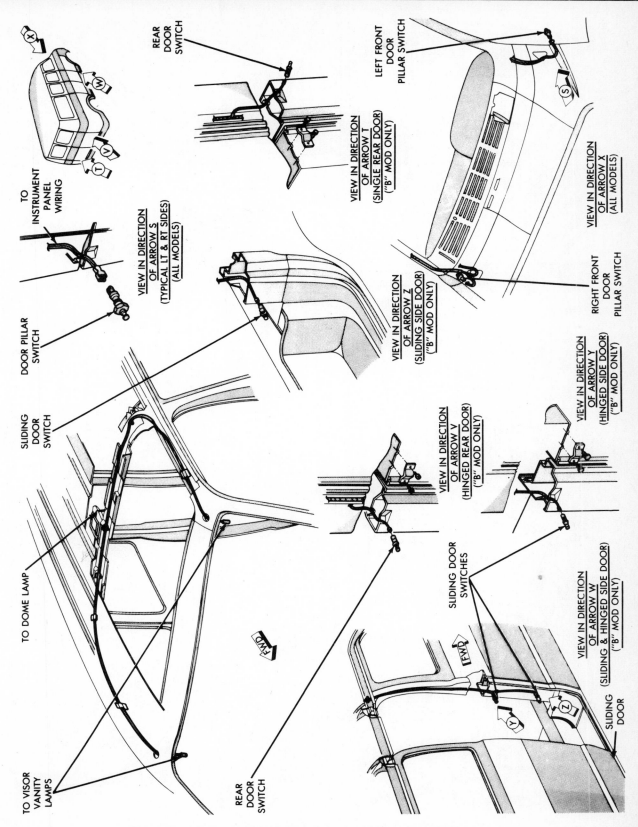

REAR DOOR SWITCH

LEFT FRONT DOOR PILLAR SWITCH

VIEW IN DIRECTION OF ARROW T (SINGLE REAR DOOR) ("B" MOD ONLY)

VIEW IN DIRECTION OF ARROW X (ALL MODELS)

TO INSTRUMENT PANEL WIRING

VIEW IN DIRECTION OF ARROW S (TYPICAL LT & RT SIDES) (ALL MODELS)

RIGHT FRONT DOOR PILLAR SWITCH

DOOR PILLAR SWITCH

VIEW IN DIRECTION OF ARROW Z (SLIDING SIDE DOOR) ("B" MOD ONLY)

VIEW IN DIRECTION OF ARROW Y (HINGED SIDE DOOR) ("B" MOD ONLY)

SLIDING DOOR SWITCH

VIEW IN DIRECTION OF ARROW V (HINGED REAR DOOR) ("B" MOD ONLY)

SLIDING DOOR SWITCHES

TO DOME LAMP

FWD

VIEW IN DIRECTION OF ARROW W (SLIDING & HINGED SIDE DOOR) ("B" MOD ONLY)

FWD

SLIDING DOOR

TO VISOR VANITY LAMPS

REAR DOOR SWITCH

Body courtesy lamp wiring and component identification

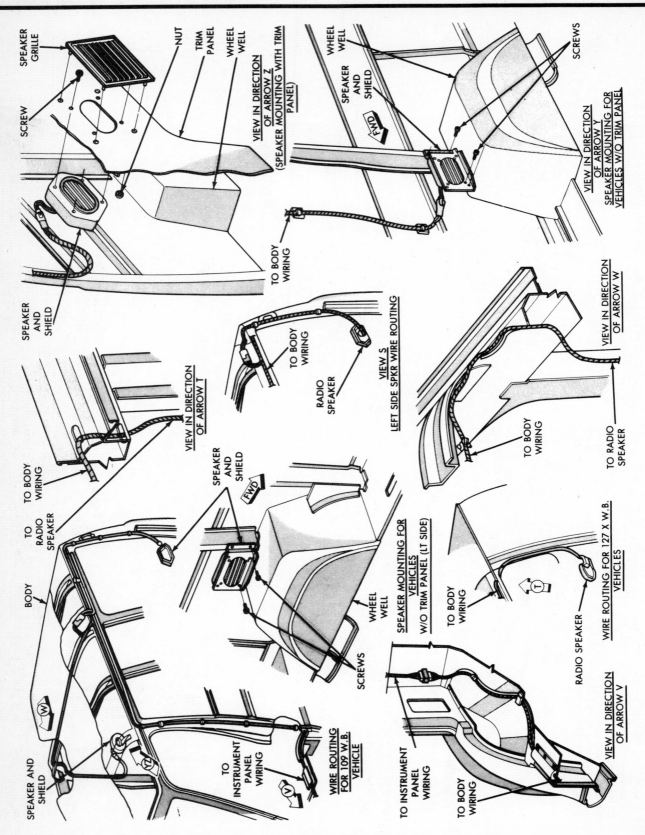

Body radio speaker wiring and component identification

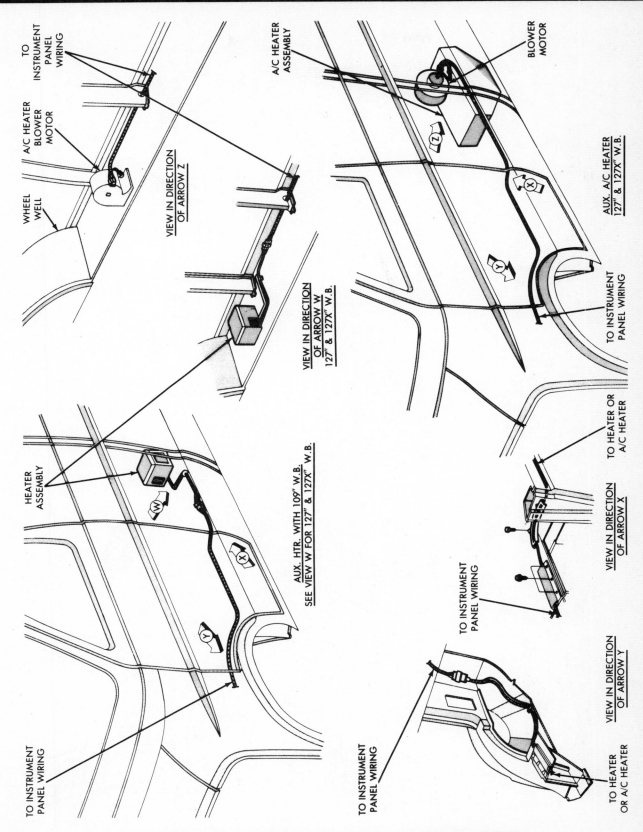

TO INSTRUMENT PANEL WIRING

A/C HEATER BLOWER MOTOR

WHEEL WELL

VIEW IN DIRECTION OF ARROW Z

A/C HEATER ASSEMBLY

BLOWER MOTOR

AUX. A/C HEATER 127″ & 127X″ W.B.

TO INSTRUMENT PANEL WIRING

VIEW IN DIRECTION OF ARROW W 127″ & 127X″ W.B.

HEATER ASSEMBLY

AUX. HTR. WITH 109″ W.B. SEE VIEW W FOR 127″ & 127X″ W.B.

TO INSTRUMENT PANEL WIRING

TO HEATER OR A/C HEATER

VIEW IN DIRECTION OF ARROW X

TO INSTRUMENT PANEL WIRING

TO INSTRUMENT PANEL WIRING

VIEW IN DIRECTION OF ARROW Y

TO HEATER OR A/C HEATER

Body rear heater and A/C heater wiring and component identification

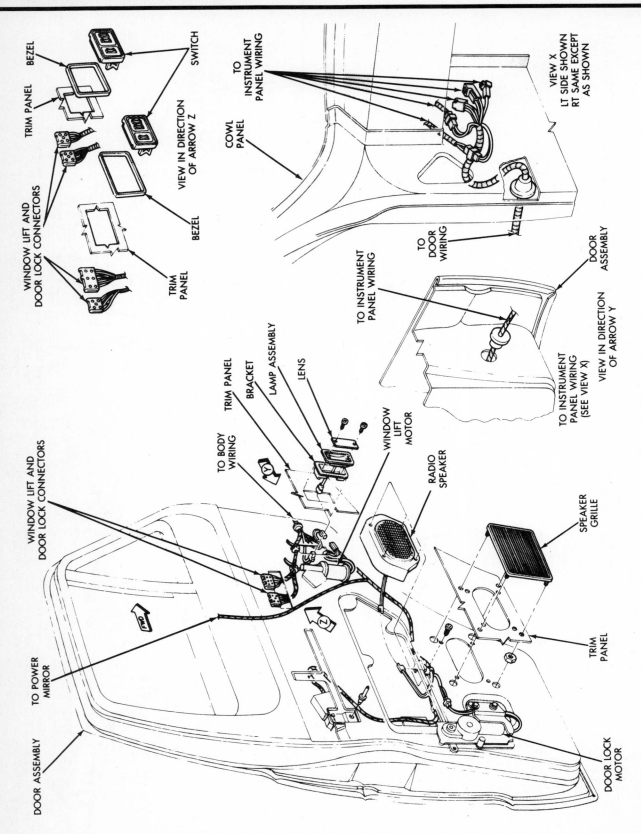

Body front door wiring and component identification

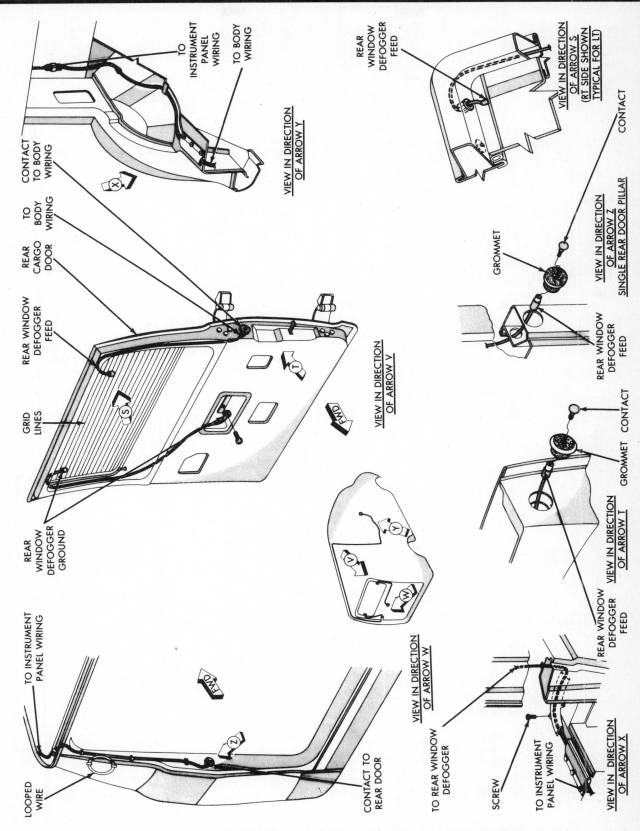

TO INSTRUMENT PANEL WIRING

TO BODY WIRING

CONTACT TO BODY WIRING

VIEW IN DIRECTION OF ARROW Y

TO BODY WIRING

REAR CARGO DOOR

REAR WINDOW DEFOGGER FEED

GRID LINES

REAR WINDOW DEFOGGER GROUND

VIEW IN DIRECTION OF ARROW V

FWD

TO INSTRUMENT PANEL WIRING

LOOPED WIRE

FWD

CONTACT TO REAR DOOR

VIEW IN DIRECTION OF ARROW W

TO REAR WINDOW DEFOGGER

REAR WINDOW DEFOGGER FEED

VIEW IN DIRECTION OF ARROW S (RT SIDE SHOWN TYPICAL FOR LT)

CONTACT

VIEW IN DIRECTION OF ARROW Z SINGLE REAR DOOR PILLAR

GROMMET

REAR WINDOW DEFOGGER FEED

GROMMET CONTACT

VIEW IN DIRECTION OF ARROW T

REAR WINDOW DEFOGGER FEED

SCREW

TO INSTRUMENT PANEL WIRING

VIEW IN DIRECTION OF ARROW X

Body rear window defogger wiring and component identification

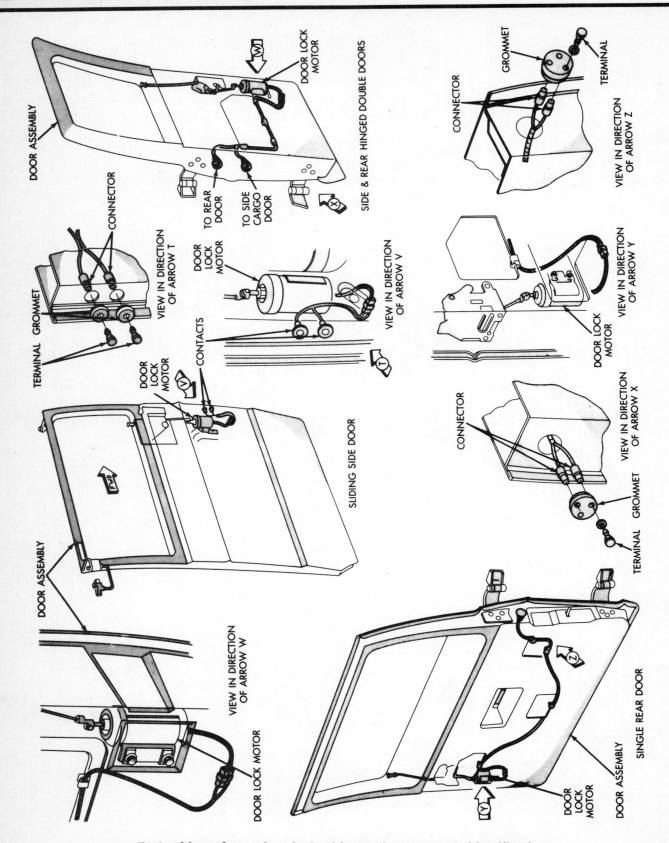

Body side and rear door lock wiring and component identification

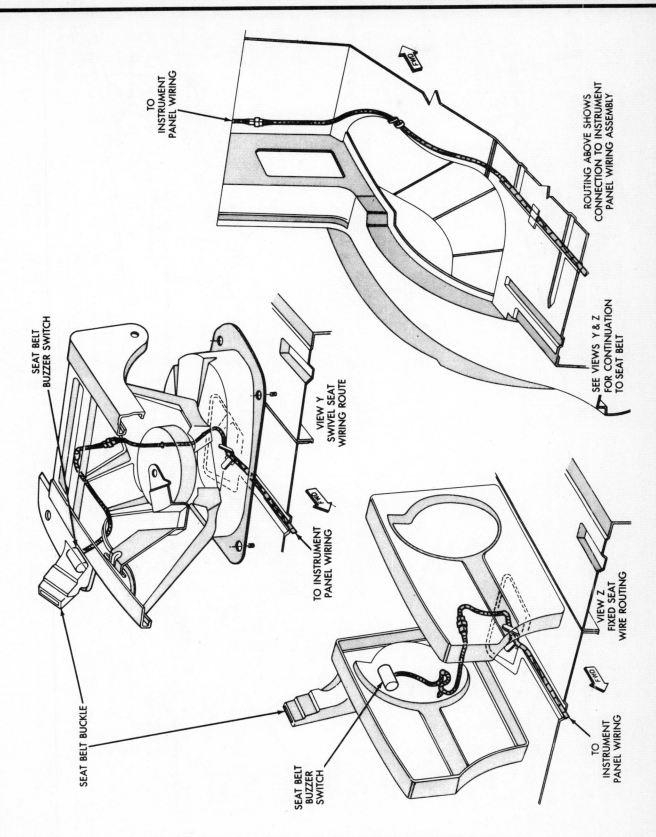

TO INSTRUMENT PANEL WIRING

ROUTING ABOVE SHOWS CONNECTION TO INSTRUMENT PANEL WIRING ASSEMBLY

SEE VIEWS Y & Z FOR CONTINUATION TO SEAT BELT

SEAT BELT BUZZER SWITCH

VIEW Y SWIVEL SEAT WIRING ROUTE

TO INSTRUMENT PANEL WIRING

SEAT BELT BUCKLE

SEAT BELT BUZZER SWITCH

VIEW Z FIXED SEAT WIRE ROUTING

TO INSTRUMENT PANEL WIRING

Body seat belt buzzer wiring and component identification

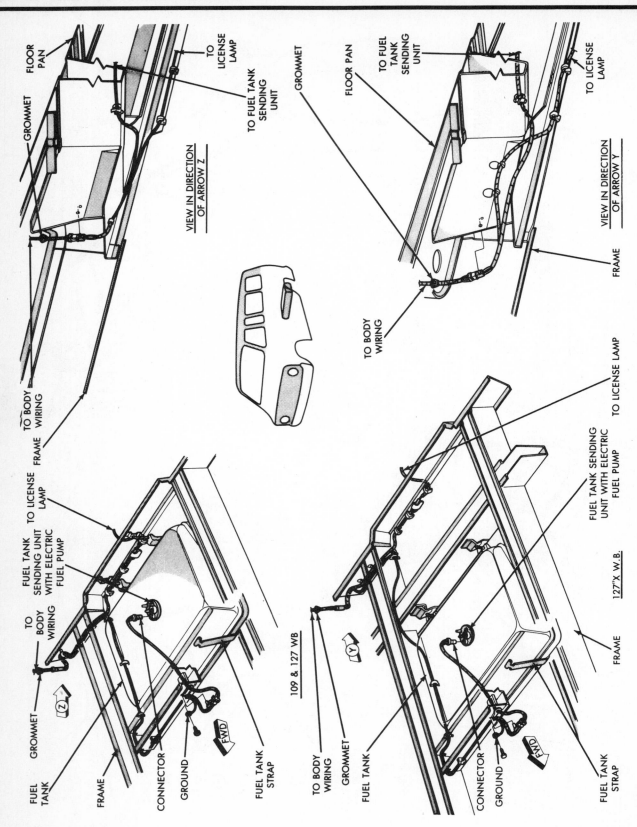

Body frame wiring and component identification

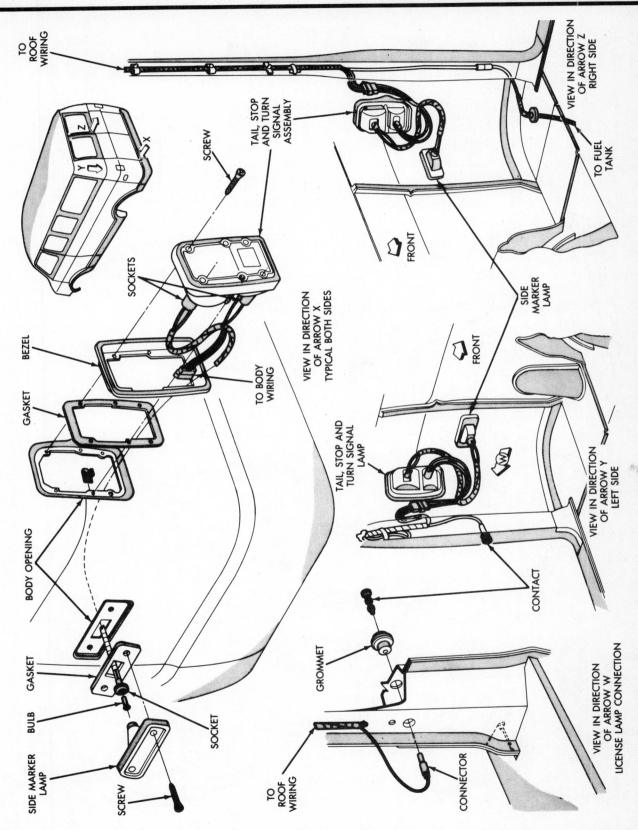

Body tail, stop, turn signal and side marker wiring and component identification

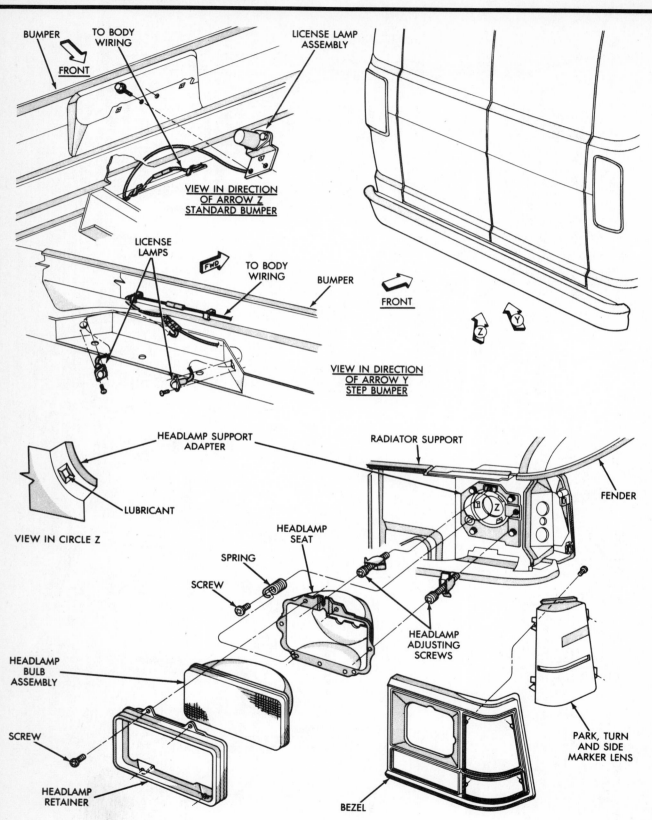

Body headlamp assembly components and license plate wiring

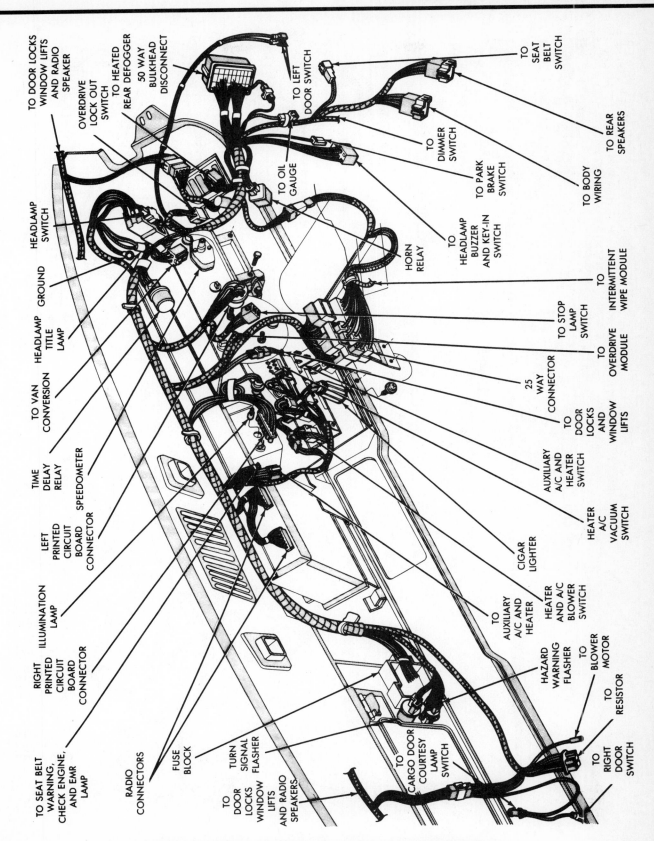

1989 Instrument panel wiring and component identification

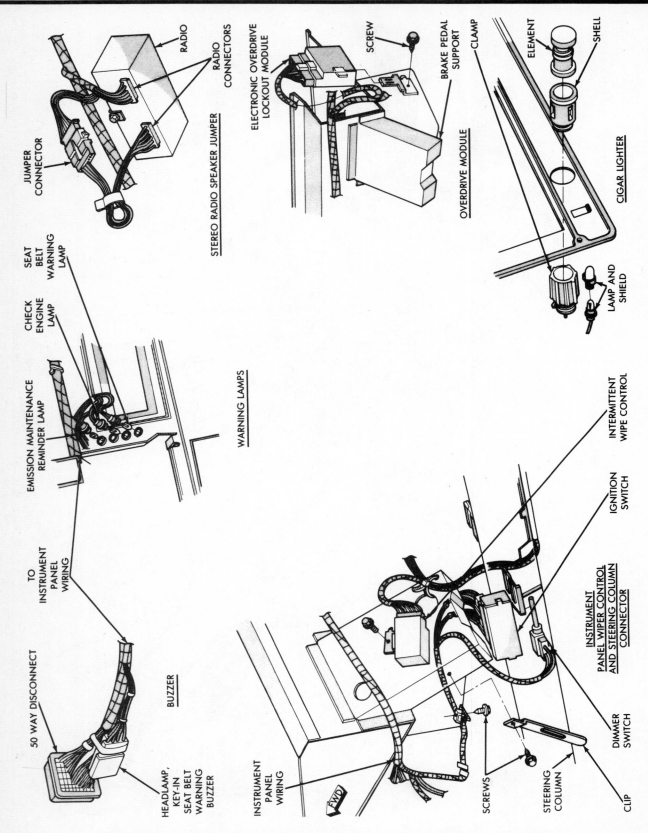

RADIO

RADIO CONNECTORS

JUMPER CONNECTOR

STEREO RADIO SPEAKER JUMPER

ELECTRONIC OVERDRIVE LOCKOUT MODULE

SCREW

BRAKE PEDAL SUPPORT

OVERDRIVE MODULE

CLAMP

ELEMENT

SHELL

CIGAR LIGHTER

LAMP AND SHIELD

SEAT BELT WARNING LAMP

CHECK ENGINE LAMP

EMISSION MAINTENANCE REMINDER LAMP

WARNING LAMPS

TO INSTRUMENT PANEL WIRING

BUZZER

50 WAY DISCONNECT

HEADLAMP, KEY-IN SEAT BELT WARNING BUZZER

INSTRUMENT PANEL WIRING

SCREWS

STEERING COLUMN

CLIP

DIMMER SWITCH

INSTRUMENT PANEL WIPER CONTROL AND STEERING COLUMN CONNECTOR

IGNITION SWITCH

INTERMITTENT WIPE CONTROL

1989 Instrument panel accessory wiring and component identification

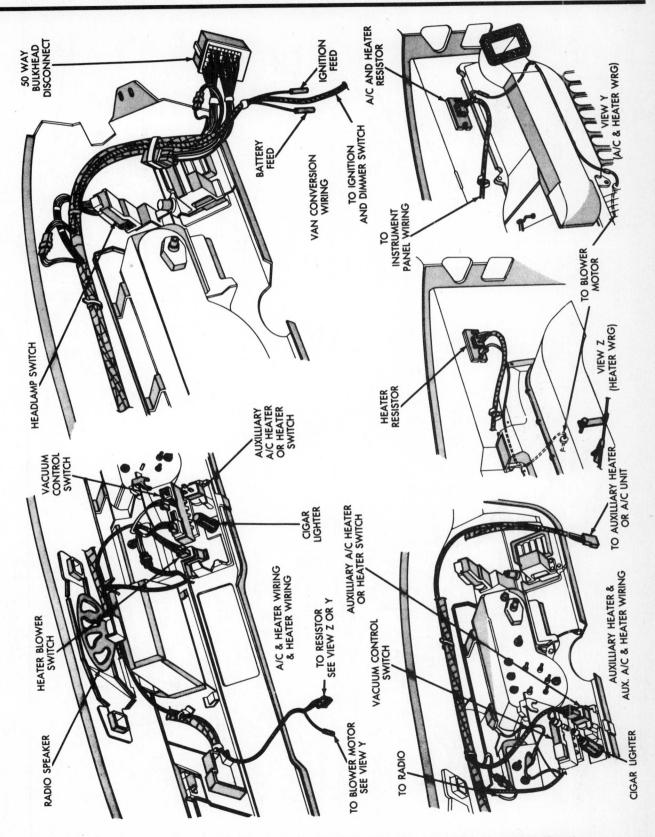

Instrument panel heater and A/C and van conversion wiring and component identification

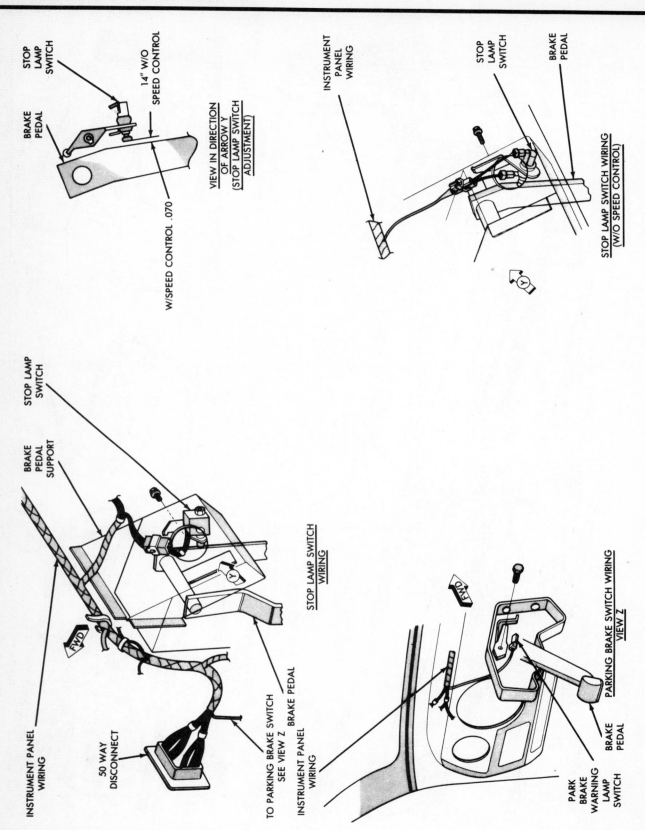

STOP LAMP SWITCH

BRAKE PEDAL

14" W/O SPEED CONTROL

VIEW IN DIRECTION OF ARROW Y (STOP LAMP SWITCH ADJUSTMENT)

W/SPEED CONTROL .070

INSTRUMENT PANEL WIRING

STOP LAMP SWITCH

BRAKE PEDAL

STOP LAMP SWITCH WIRING (W/O SPEED CONTROL)

Y

STOP LAMP SWITCH

BRAKE PEDAL SUPPORT

INSTRUMENT PANEL WIRING

50 WAY DISCONNECT

TO PARKING BRAKE SWITCH SEE VIEW Z

BRAKE PEDAL

INSTRUMENT PANEL WIRING

STOP LAMP SWITCH WIRING

FWD

FWD

PARKING BRAKE SWITCH WIRING VIEW Z

BRAKE PEDAL

PARK BRAKE WARNING LAMP SWITCH

1989 Instrument panel speed control stop lamp and park brake switch wiring and component identification

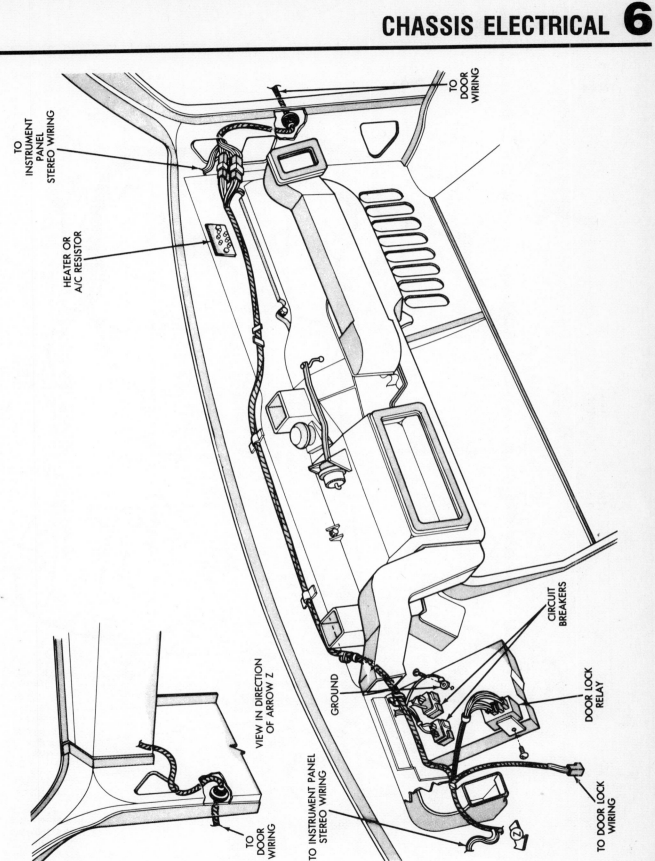

TO DOOR WIRING

TO INSTRUMENT PANEL STEREO WIRING

HEATER OR A/C RESISTOR

VIEW IN DIRECTION OF ARROW Z

TO DOOR WIRING

TO INSTRUMENT PANEL STEREO WIRING

GROUND

CIRCUIT BREAKERS

DOOR LOCK RELAY

TO DOOR LOCK WIRING

1989 Instrument panel window lift and door lock wiring and component identification

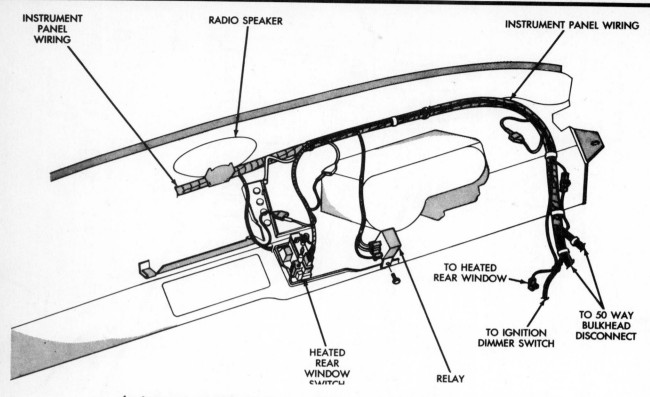

INSTRUMENT PANEL WIRING

RADIO SPEAKER

INSTRUMENT PANEL WIRING

TO HEATED REAR WINDOW

TO IGNITION DIMMER SWITCH

TO 50 WAY BULKHEAD DISCONNECT

HEATED REAR WINDOW SWITCH

RELAY

Instrument panel heated rear window and component identification

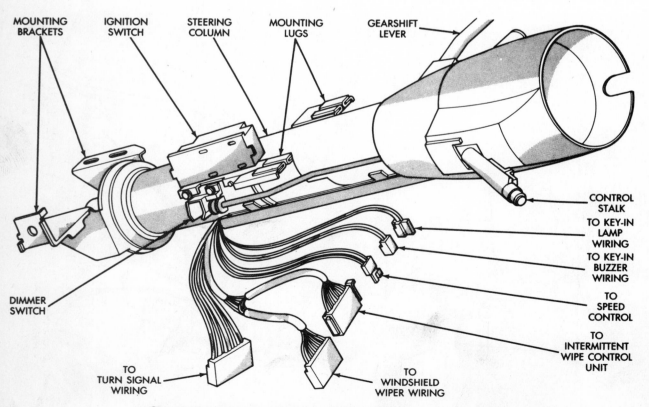

MOUNTING BRACKETS

IGNITION SWITCH

STEERING COLUMN

MOUNTING LUGS

GEARSHIFT LEVER

CONTROL STALK

TO KEY-IN LAMP WIRING

TO KEY-IN BUZZER WIRING

TO SPEED CONTROL

TO INTERMITTENT WIPE CONTROL UNIT

DIMMER SWITCH

TO TURN SIGNAL WIRING

TO WINDSHIELD WIPER WIRING

Steering column wiring and component identification

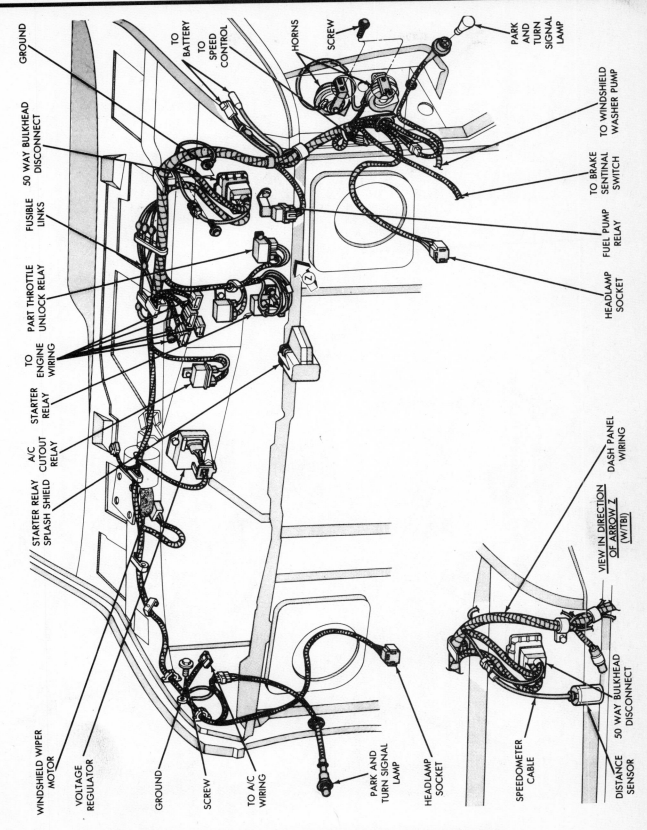

1989 Engine compartment wiring and component identification

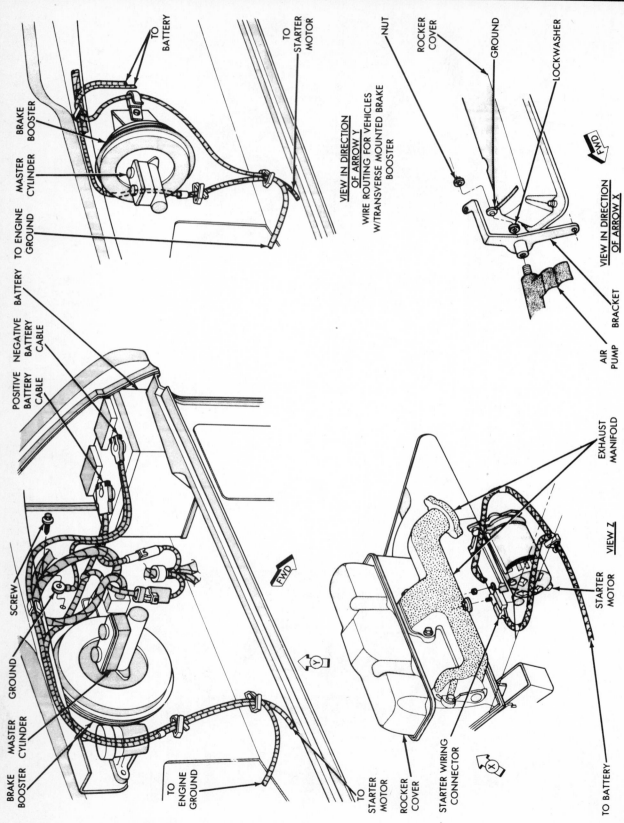

Engine compartment starter wiring and component identification

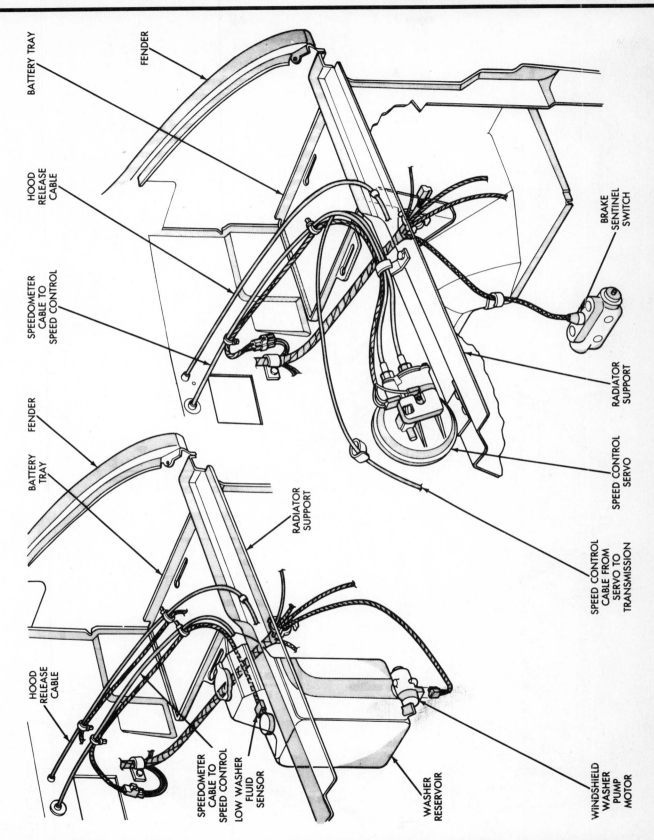

BATTERY TRAY

FENDER

HOOD RELEASE CABLE

SPEEDOMETER CABLE TO SPEED CONTROL

FENDER

BATTERY TRAY

RADIATOR SUPPORT

HOOD RELEASE CABLE

SPEEDOMETER CABLE TO SPEED CONTROL

LOW WASHER FLUID SENSOR

WASHER RESERVOIR

WINDSHIELD WASHER PUMP MOTOR

SPEED CONTROL CABLE FROM SERVO TO TRANSMISSION

SPEED CONTROL SERVO

RADIATOR SUPPORT

BRAKE SENTINEL SWITCH

1989 Engine compartment speed control and windshield washer wiring and component identification

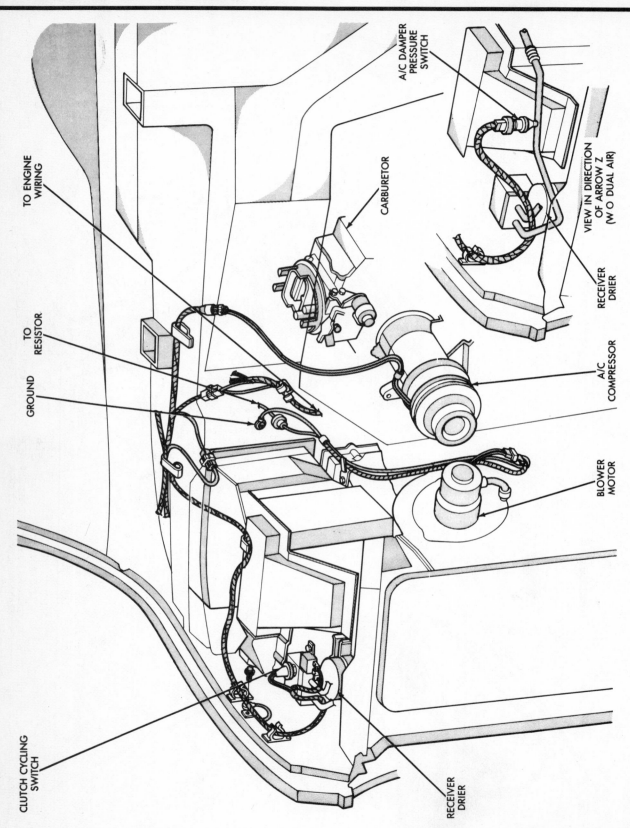

TO ENGINE WIRING

TO RESISTOR

GROUND

CARBURETOR

A/C DAMPER PRESSURE SWITCH

VIEW IN DIRECTION OF ARROW Z (W O DUAL AIR)

RECEIVER DRIER

A/C COMPRESSOR

BLOWER MOTOR

CLUTCH CYCLING SWITCH

RECEIVER DRIER

1989 Engine compartment A/C and heater wiring and component identification

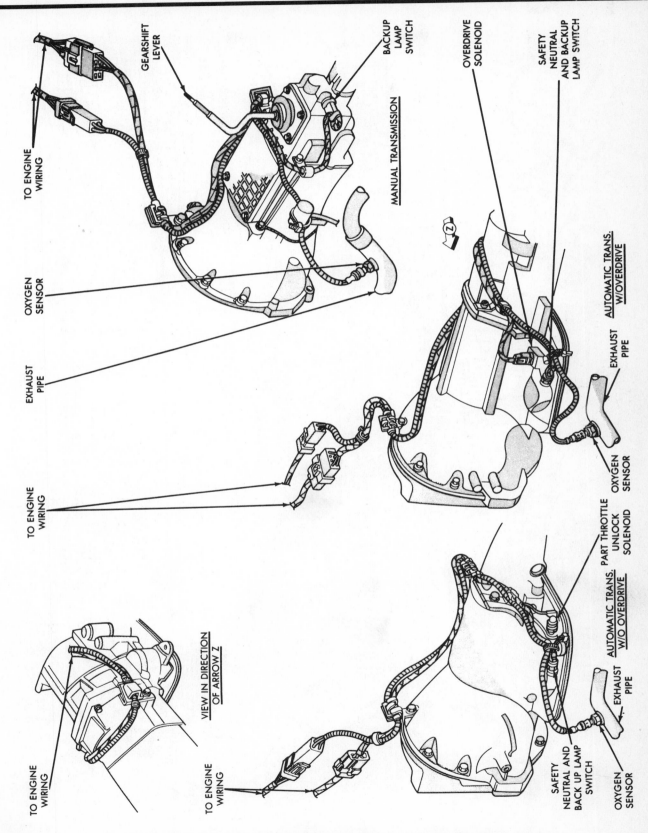

TO ENGINE WIRING

GEARSHIFT LEVER

BACKUP LAMP SWITCH

OVERDRIVE SOLENOID

SAFETY NEUTRAL AND BACKUP LAMP SWITCH

MANUAL TRANSMISSION

OXYGEN SENSOR

EXHAUST PIPE

AUTOMATIC TRANS. W/OVERDRIVE

EXHAUST PIPE

OXYGEN SENSOR

TO ENGINE WIRING

PART THROTTLE UNLOCK SOLENOID

AUTOMATIC TRANS. W/O OVERDRIVE

VIEW IN DIRECTION OF ARROW Z

TO ENGINE WIRING

TO ENGINE WIRING

EXHAUST PIPE

SAFETY NEUTRAL AND BACK UP LAMP SWITCH

OXYGEN SENSOR

Engine compartment transmission and component identification

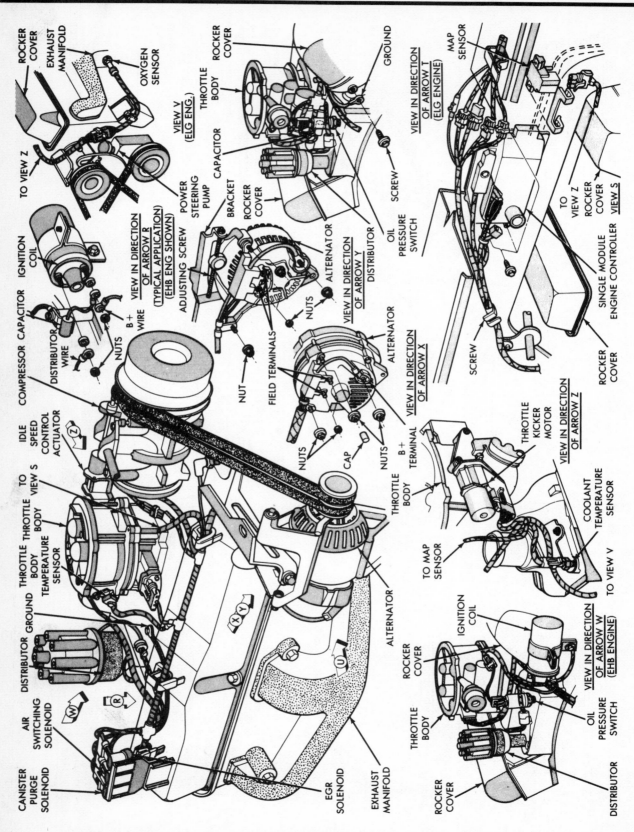

1989 Engine wiring and component identification 3.9L and 5.2L EFI engine

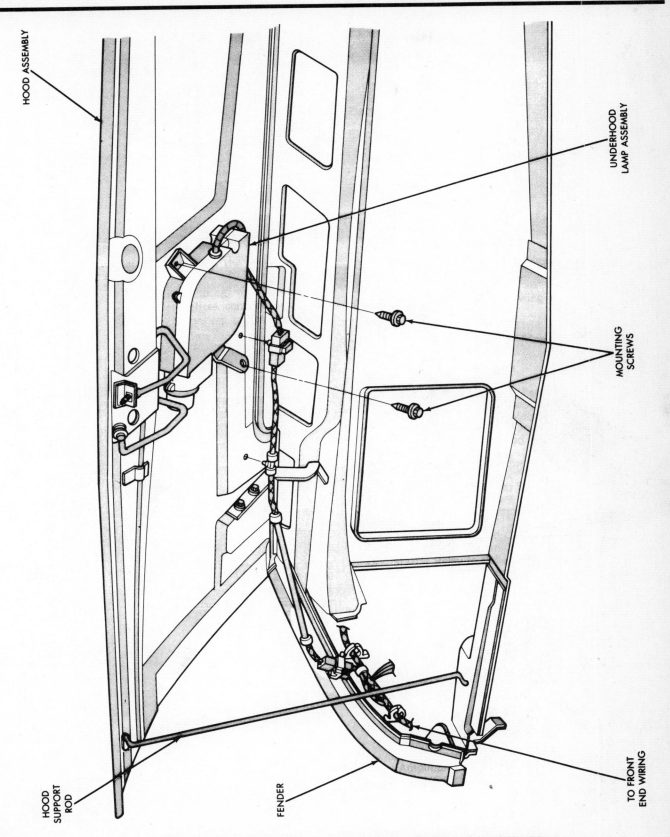

HOOD ASSEMBLY

UNDERHOOD LAMP ASSEMBLY

MOUNTING SCREWS

HOOD SUPPORT ROD

FENDER

TO FRONT END WIRING

Engine compartment underhood lamp wiring and component identification

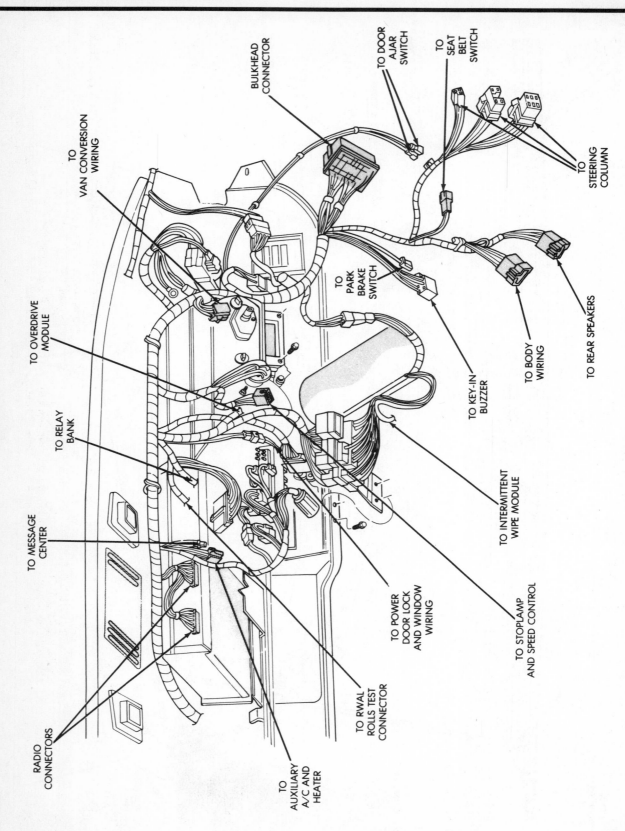

1990 Instrument panel wiring

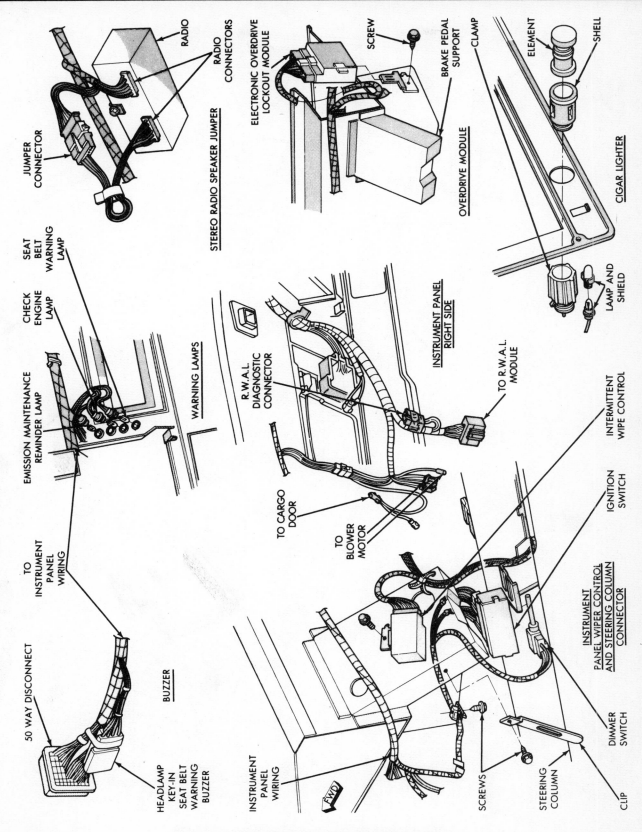

RADIO

RADIO CONNECTORS

JUMPER CONNECTOR

STEREO RADIO SPEAKER JUMPER

SCREW

ELECTRONIC OVERDRIVE LOCKOUT MODULE

BRAKE PEDAL SUPPORT

CLAMP

OVERDRIVE MODULE

ELEMENT

SHELL

CIGAR LIGHTER

LAMP AND SHIELD

SEAT BELT WARNING LAMP

CHECK ENGINE LAMP

EMISSION MAINTENANCE REMINDER LAMP

WARNING LAMPS

INSTRUMENT PANEL RIGHT SIDE

R.W.A.L. DIAGNOSTIC CONNECTOR

TO R.W.A.L. MODULE

TO CARGO DOOR

TO BLOWER MOTOR

INTERMITTENT WIPE CONTROL

IGNITION SWITCH

TO INSTRUMENT PANEL WIRING

BUZZER

50 WAY DISCONNECT

HEADLAMP KEY-IN SEAT BELT WARNING BUZZER

INSTRUMENT PANEL WIRING

INSTRUMENT PANEL WIPER CONTROL AND STEERING COLUMN CONNECTOR

FWD

SCREWS

STEERING COLUMN

DIMMER SWITCH

CLIP

1990 Instrument panel accessory wiring

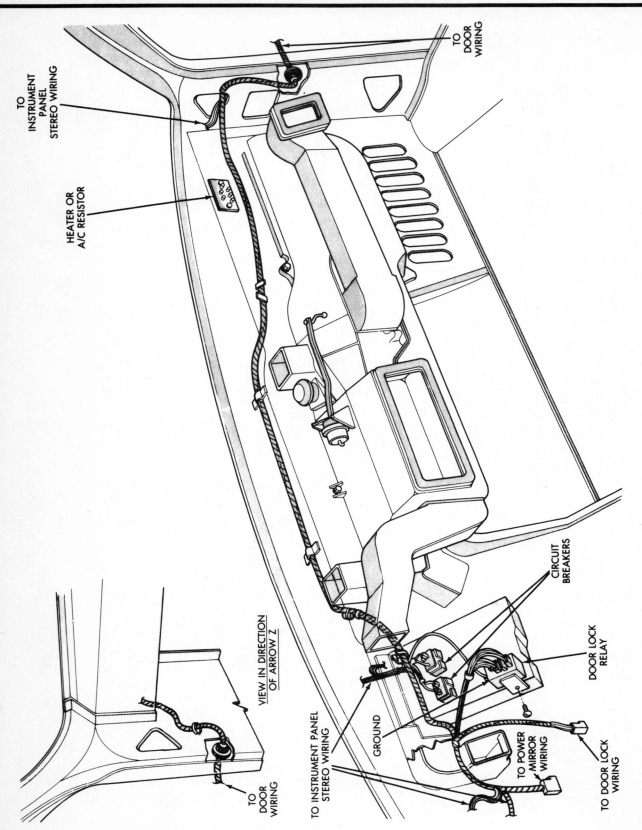

TO DOOR WIRING

TO INSTRUMENT PANEL STEREO WIRING

HEATER OR A/C RESISTOR

CIRCUIT BREAKERS

DOOR LOCK RELAY

VIEW IN DIRECTION OF ARROW Z

GROUND

TO POWER MIRROR WIRING

TO DOOR LOCK WIRING

TO DOOR WIRING

TO INSTRUMENT PANEL STEREO WIRING

1990 Instrument panel power windows and door lock wiring

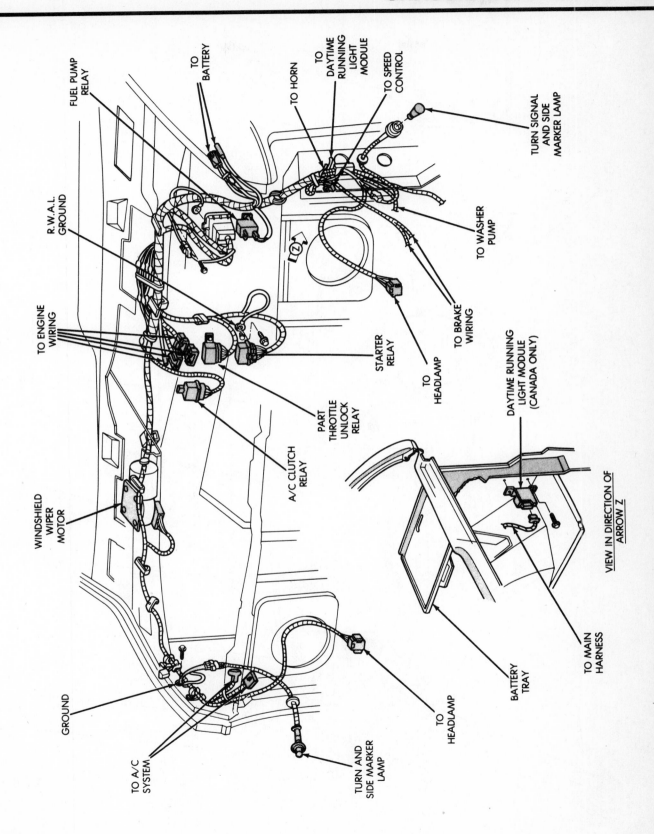

FUEL PUMP RELAY

TO BATTERY

TO HORN

TO DAYTIME RUNNING LIGHT MODULE

TO SPEED CONTROL

TURN SIGNAL AND SIDE MARKER LAMP

R.W.A.L. GROUND

TO WASHER PUMP

TO BRAKE WIRING

TO ENGINE WIRING

STARTER RELAY

TO HEADLAMP

PART THROTTLE UNLOCK RELAY

A/C CLUTCH RELAY

DAYTIME RUNNING LIGHT MODULE (CANADA ONLY)

VIEW IN DIRECTION OF ARROW Z

WINDSHIELD WIPER MOTOR

BATTERY TRAY

TO MAIN HARNESS

GROUND

TO A/C SYSTEM

TURN AND SIDE MARKER LAMP

TO HEADLAMP

1990 Engine compartment wiring

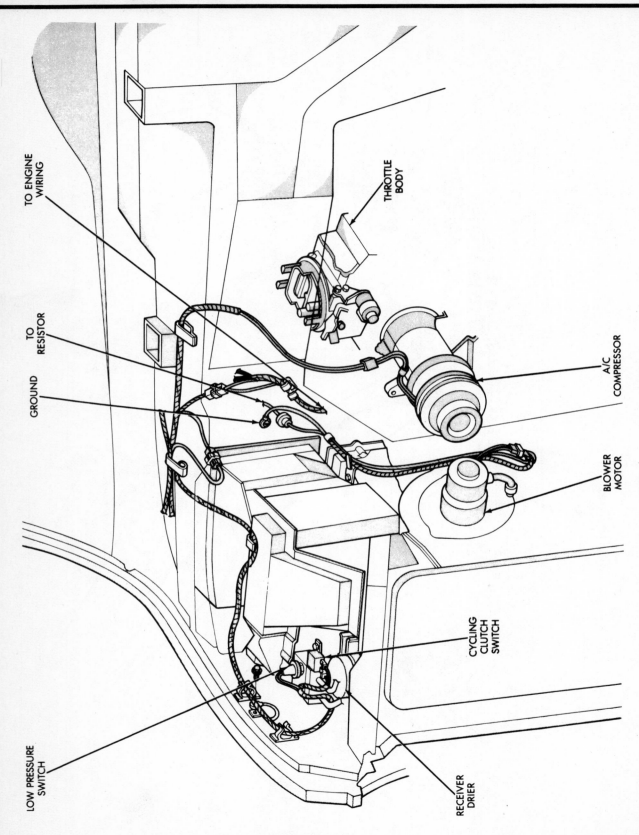

TO ENGINE WIRING

THROTTLE BODY

TO RESISTOR

GROUND

A/C COMPRESSOR

BLOWER MOTOR

CYCLING CLUTCH SWITCH

LOW PRESSURE SWITCH

RECEIVER DRIER

1990 A/C and heater wiring

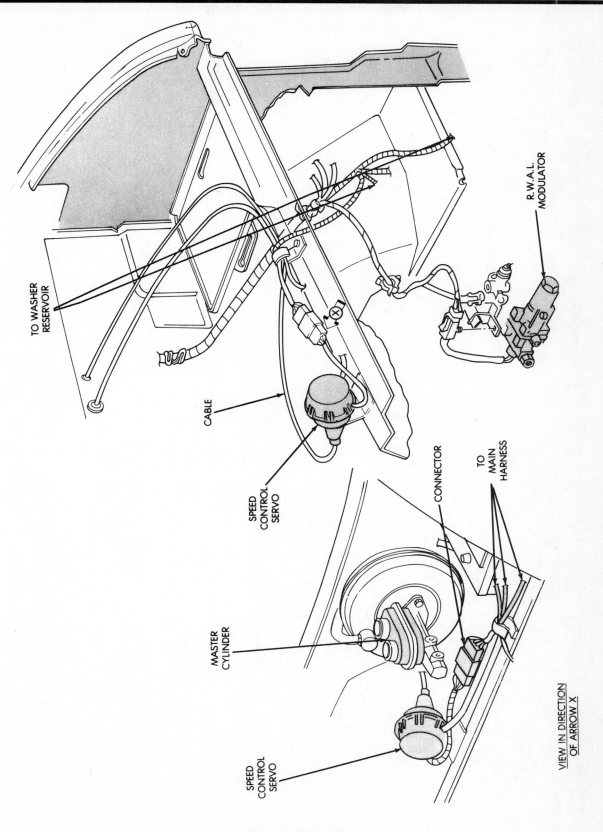

1990 Speed control wiring

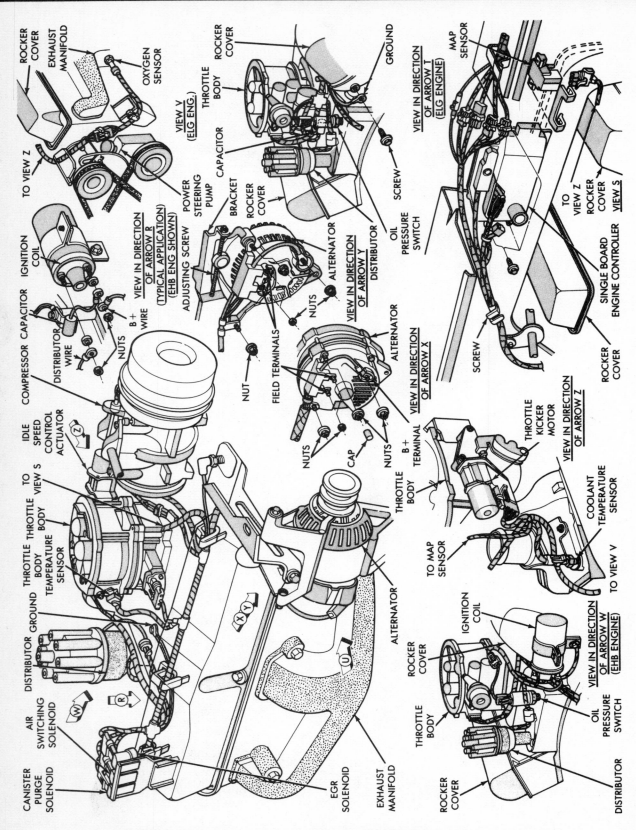

1990 Engine wiring and component identification 3.9L and 5.2L engines

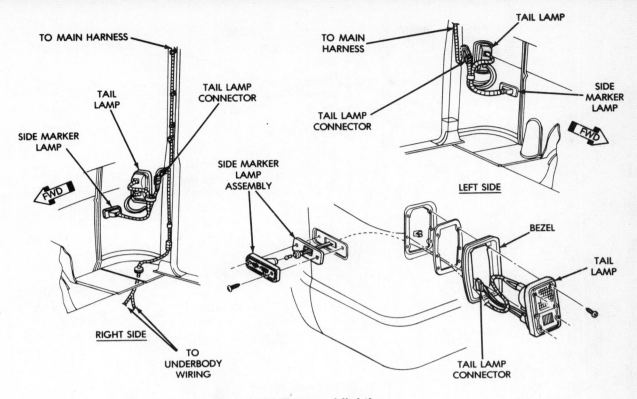

1991 Rear end lighting

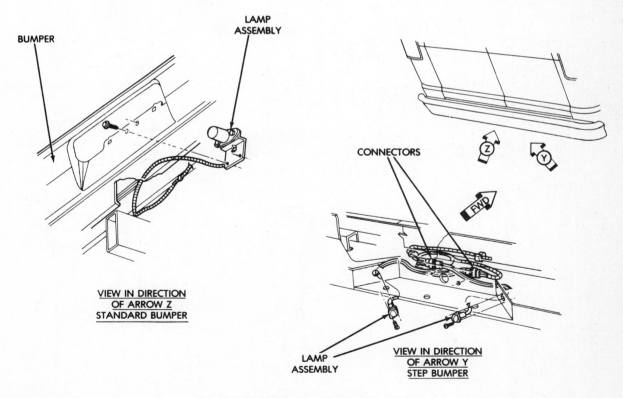

1991 Bumper wiring

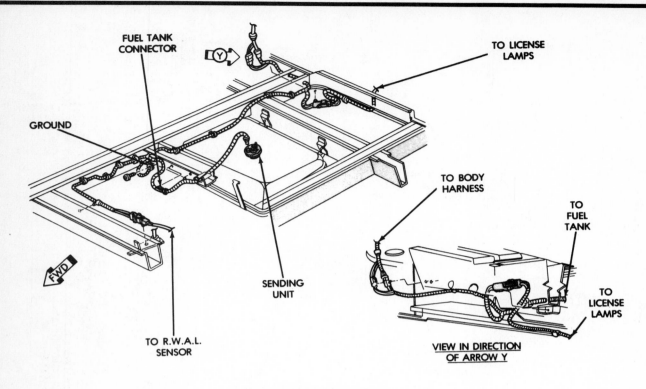

1991 Underbody wiring

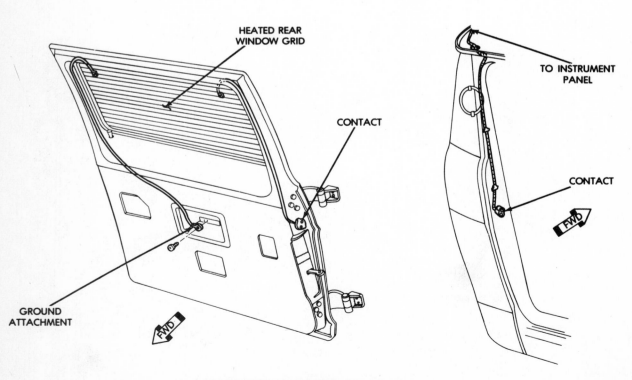

1991 Heated rear window wiring

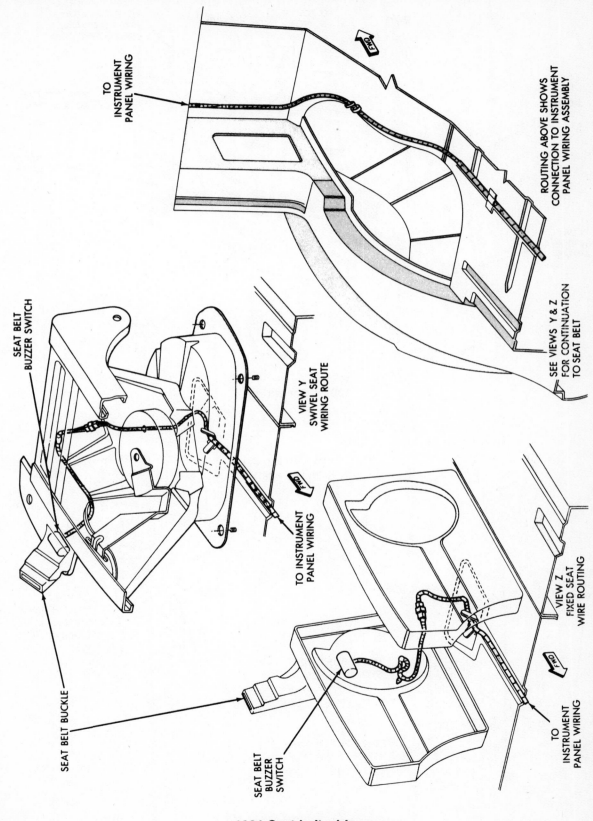

1991 Seat belt wiring

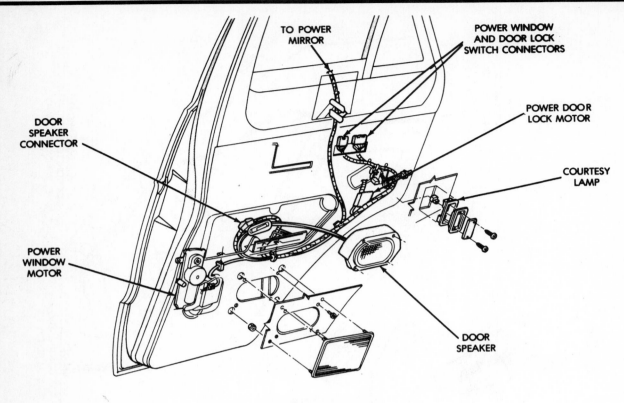

1991 Door wiring

TO POWER MIRROR

POWER WINDOW AND DOOR LOCK SWITCH CONNECTORS

POWER DOOR LOCK MOTOR

COURTESY LAMP

DOOR SPEAKER CONNECTOR

POWER WINDOW MOTOR

DOOR SPEAKER

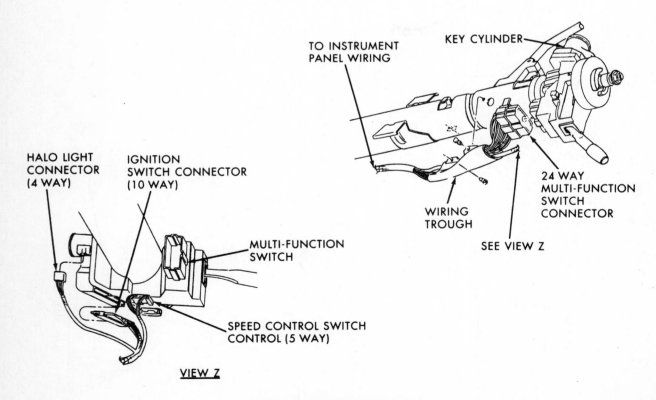

TO INSTRUMENT PANEL WIRING

KEY CYLINDER

HALO LIGHT CONNECTOR (4 WAY)

IGNITION SWITCH CONNECTOR (10 WAY)

MULTI-FUNCTION SWITCH

SPEED CONTROL SWITCH CONTROL (5 WAY)

VIEW Z

24 WAY MULTI-FUNCTION SWITCH CONNECTOR

WIRING TROUGH

SEE VIEW Z

1991 Steering column wiring

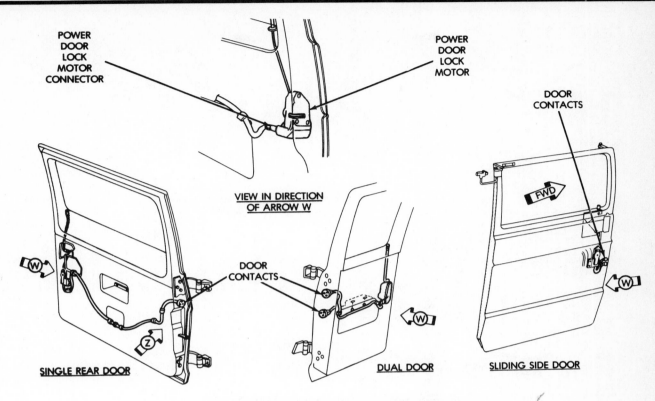

1991 Power door lock and window wiring

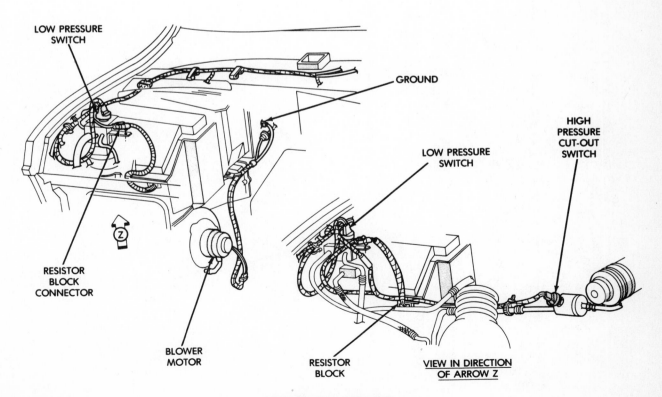

1991 Heater A/C wiring

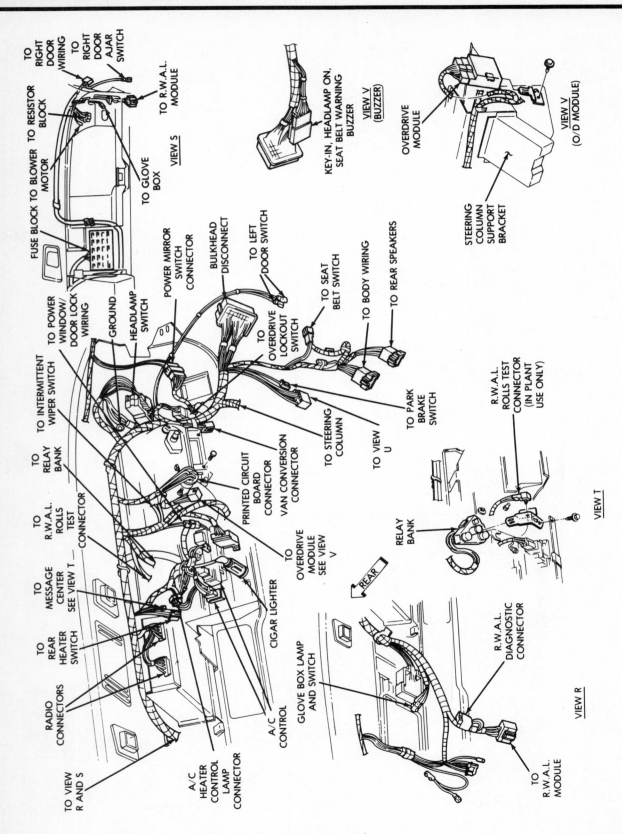

1991 Instrument panel wiring

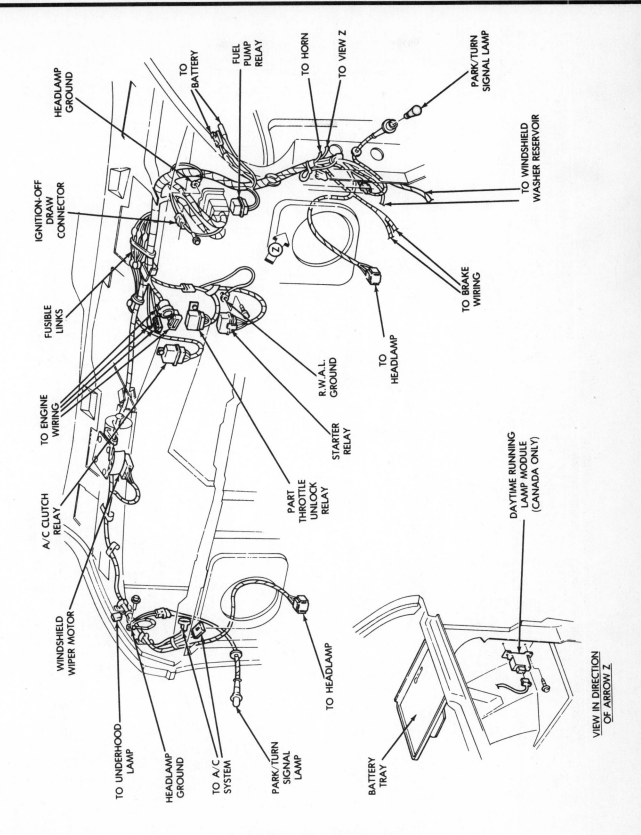

HEADLAMP GROUND

TO BATTERY

FUEL PUMP RELAY

TO HORN

TO VIEW Z

PARK/TURN SIGNAL LAMP

TO WINDSHIELD WASHER RESERVOIR

IGNITION-OFF DRAW CONNECTOR

TO BRAKE WIRING

FUSIBLE LINKS

TO ENGINE WIRING

R.W.A.L. GROUND

TO HEADLAMP

STARTER RELAY

A/C CLUTCH RELAY

PART THROTTLE UNLOCK RELAY

DAYTIME RUNNING LAMP MODULE (CANADA ONLY)

WINDSHIELD WIPER MOTOR

TO UNDERHOOD LAMP

HEADLAMP GROUND

TO A/C SYSTEM

PARK/TURN SIGNAL LAMP

TO HEADLAMP

BATTERY TRAY

VIEW IN DIRECTION OF ARROW Z

1991 Engine compartment wiring

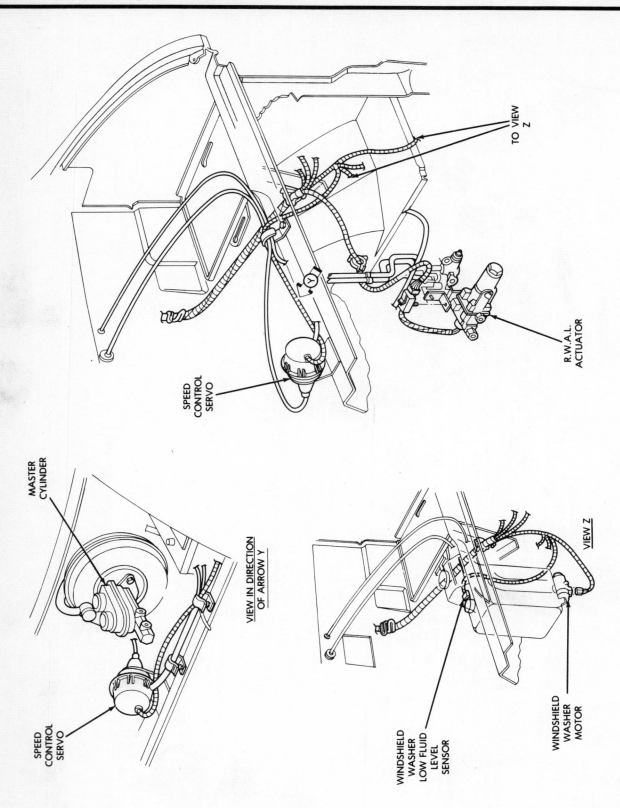

1991 Engine compartment wiring

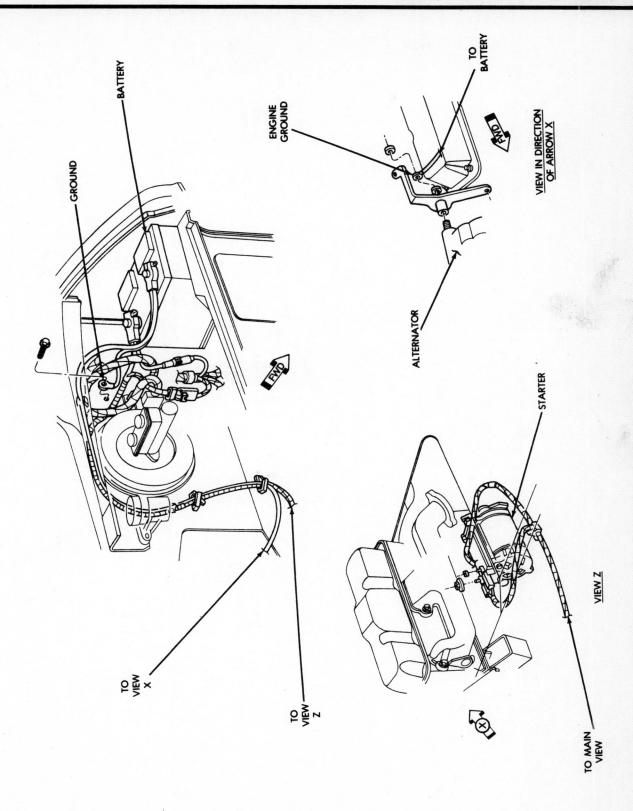

1991 Battery wiring

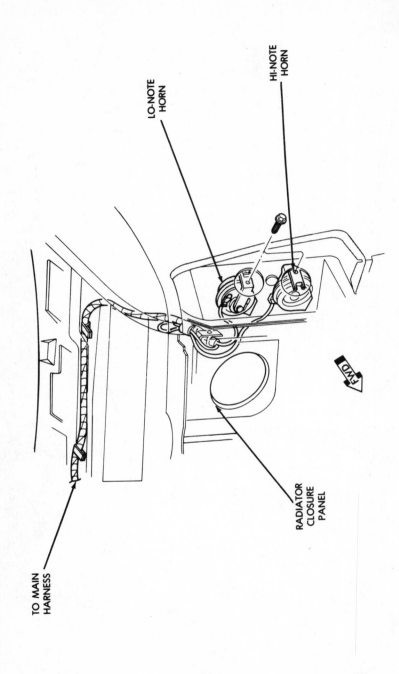

1991 Horn wiring

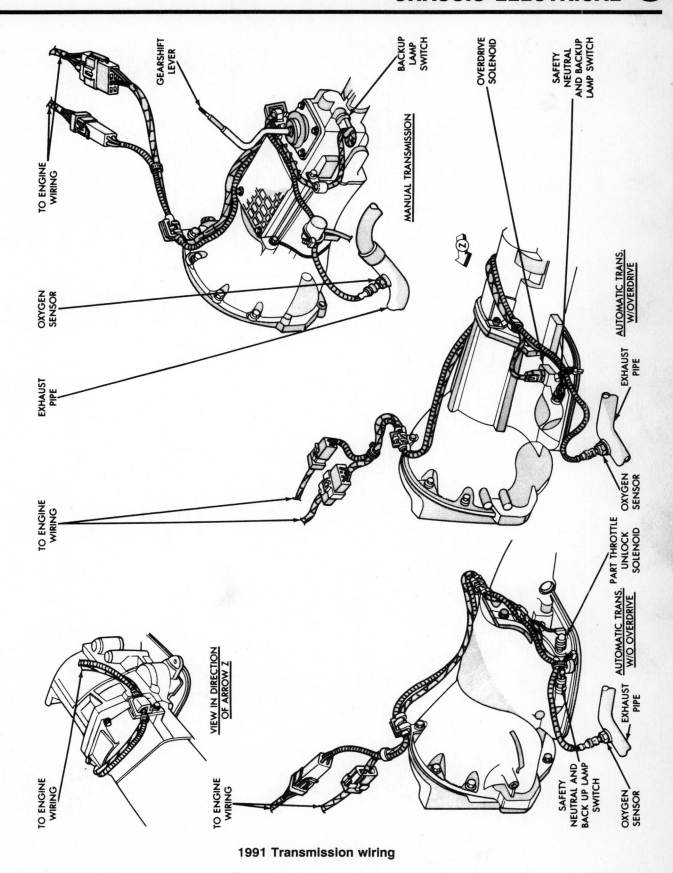

TO ENGINE WIRING

GEARSHIFT LEVER

BACKUP LAMP SWITCH

OVERDRIVE SOLENOID

SAFETY NEUTRAL AND BACKUP LAMP SWITCH

MANUAL TRANSMISSION

OXYGEN SENSOR

EXHAUST PIPE

TO ENGINE WIRING

AUTOMATIC TRANS. W/OVERDRIVE

EXHAUST PIPE

OXYGEN SENSOR

PART THROTTLE UNLOCK SOLENOID

AUTOMATIC TRANS. W/O OVERDRIVE

EXHAUST PIPE

VIEW IN DIRECTION OF ARROW Z

TO ENGINE WIRING

TO ENGINE WIRING

SAFETY NEUTRAL AND BACK UP LAMP SWITCH

OXYGEN SENSOR

1991 Transmission wiring

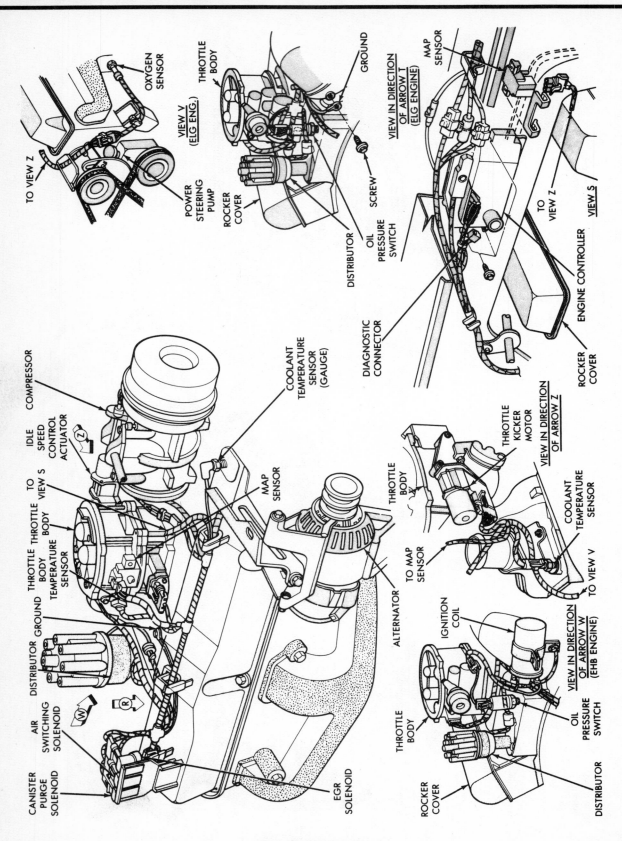

1991 Engine wiring

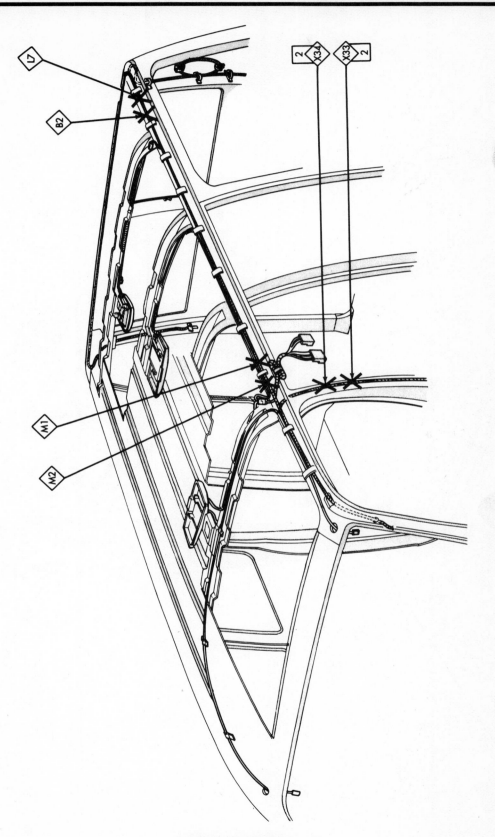

1989 Body splices

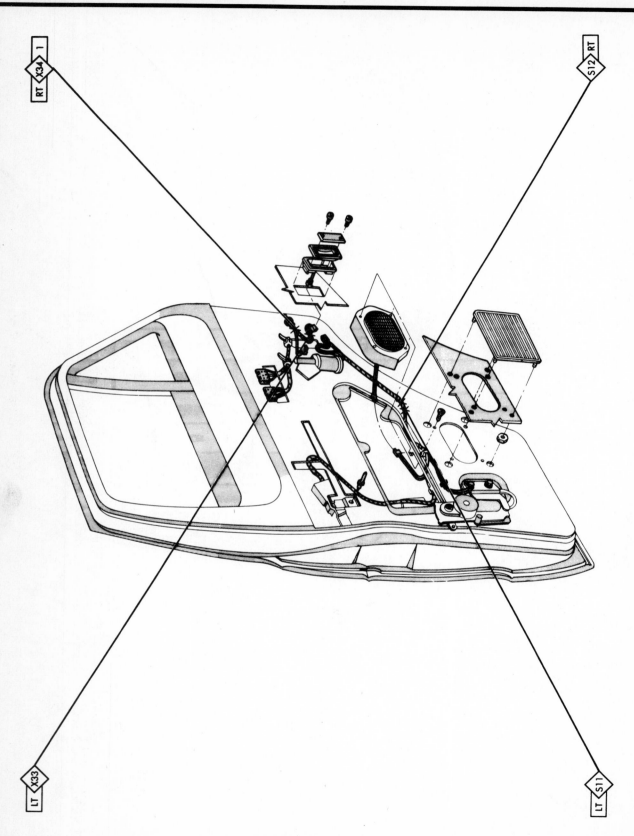

1989 Front door splices

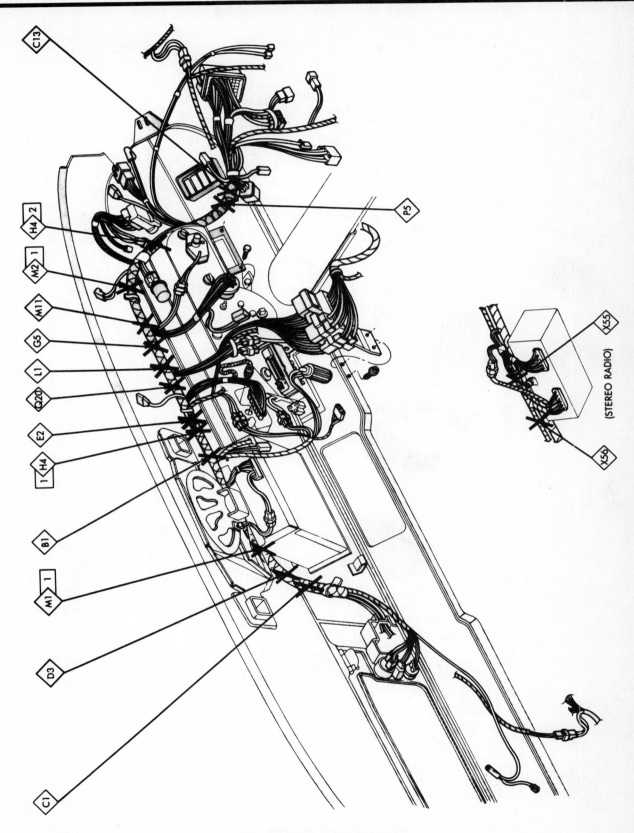

1989 Instrument panel splices

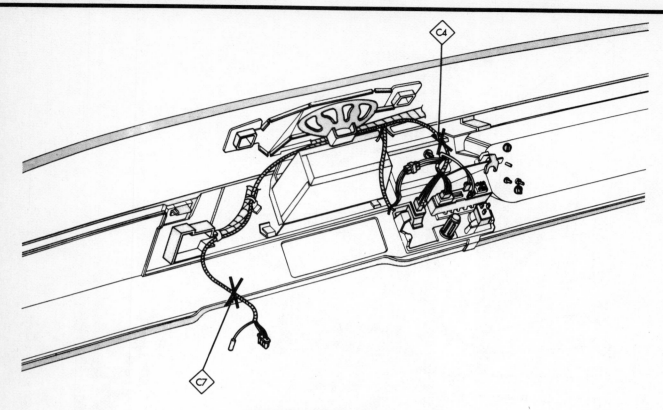

1989 A/C and heater splices

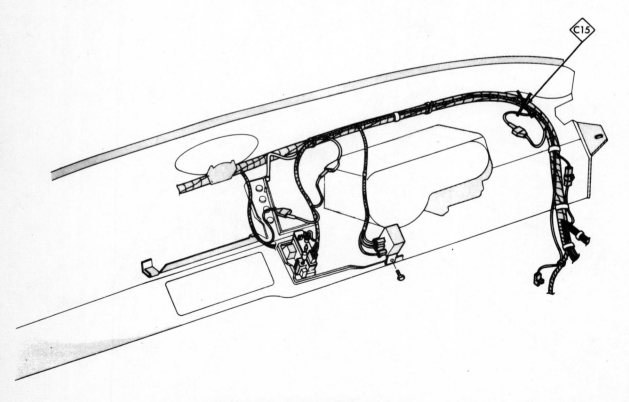

1989 Heated rear window splices

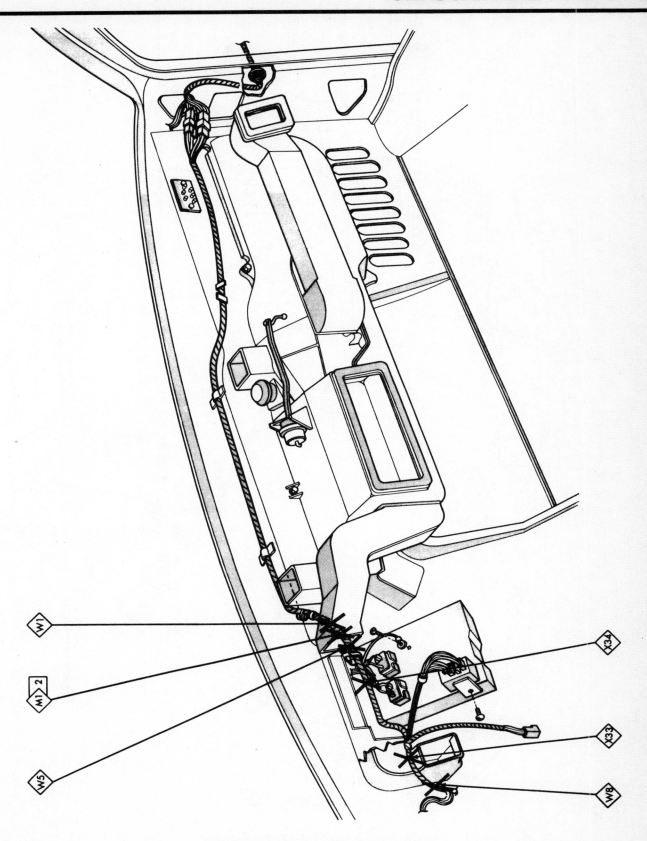

1989 Window lift and door lock splices

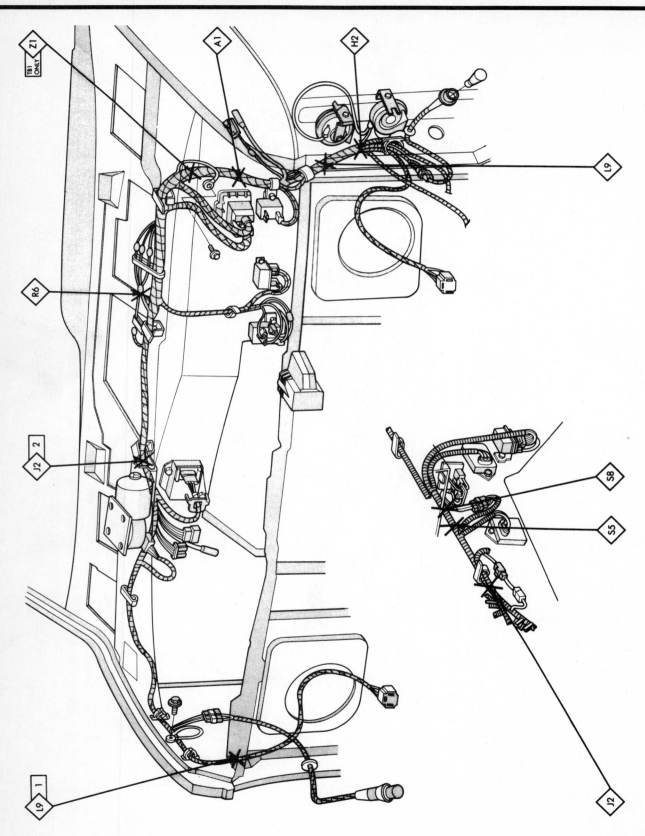

1989 Cowl panel splices

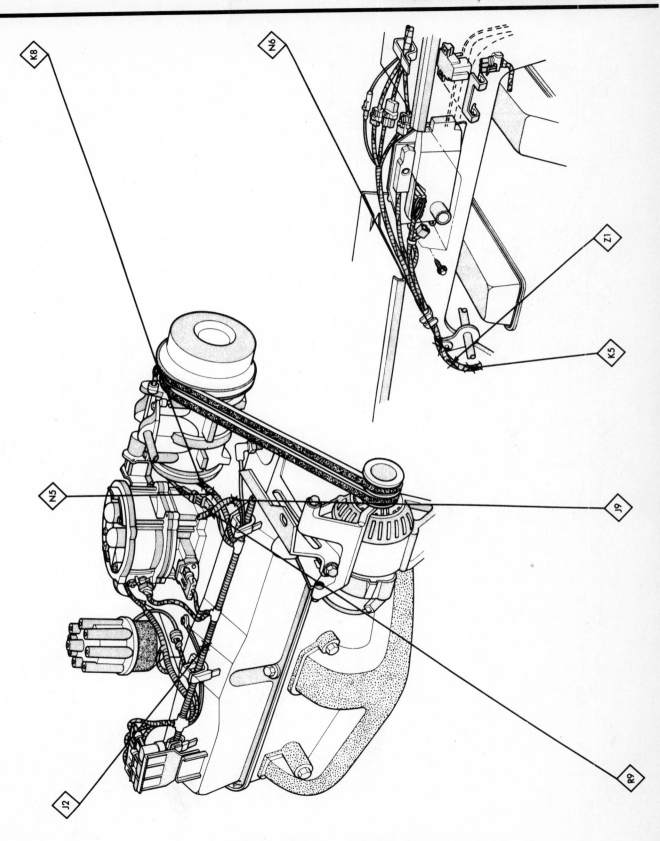

1989 Engine splices 3.9L and 5.2L EFI engine

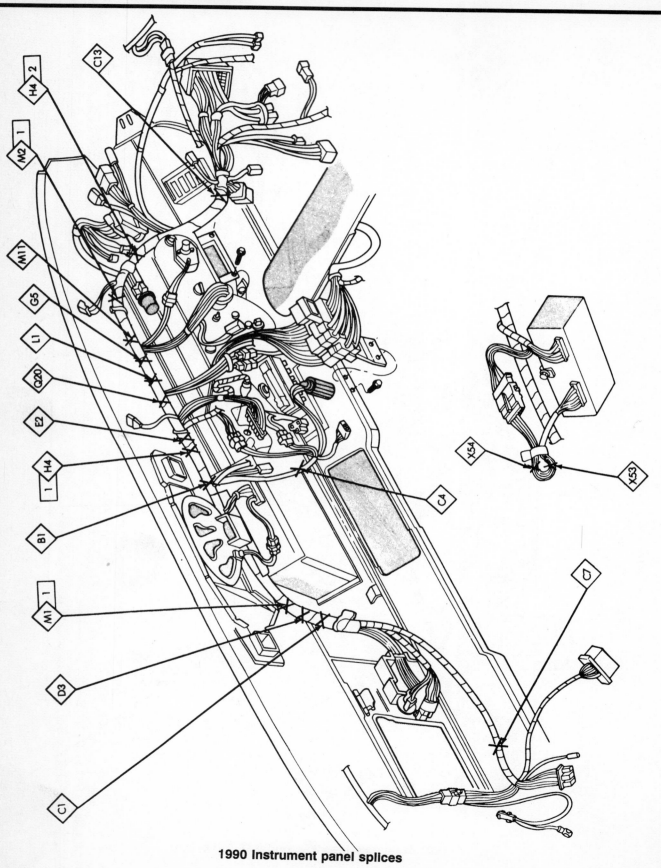

1990 Instrument panel splices

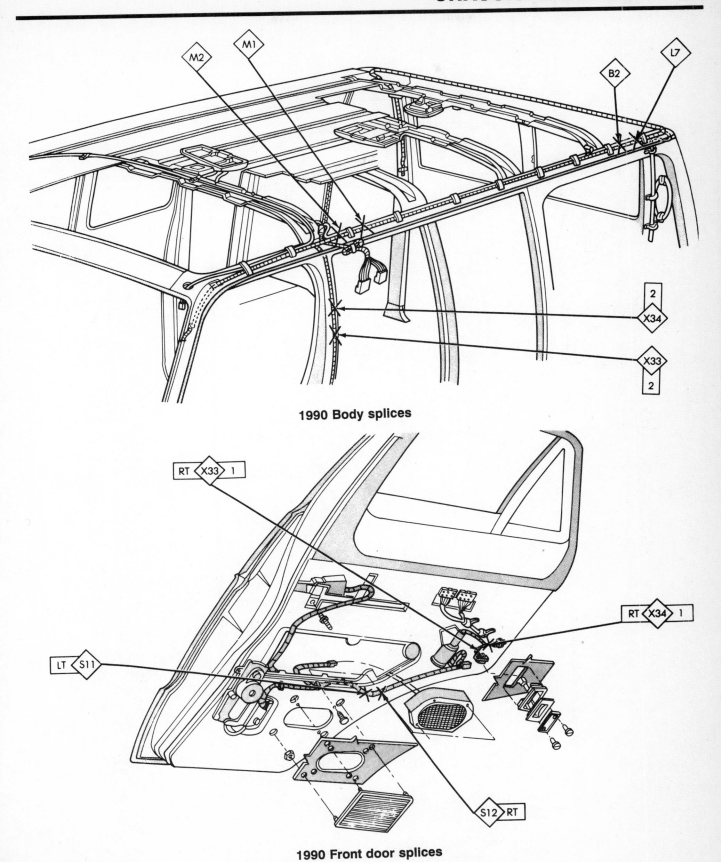

1990 Body splices

1990 Front door splices

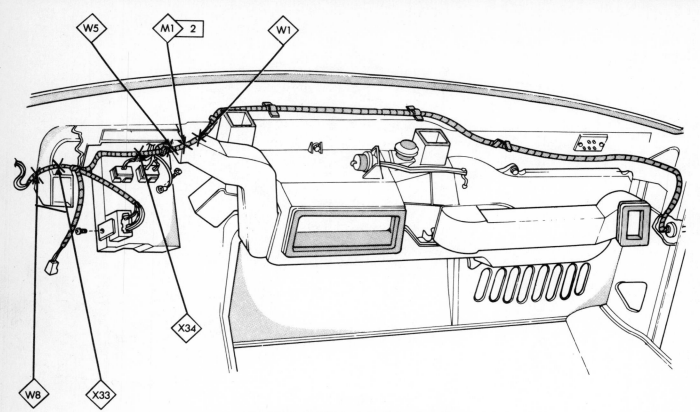

1990 Window lift and door lock splices

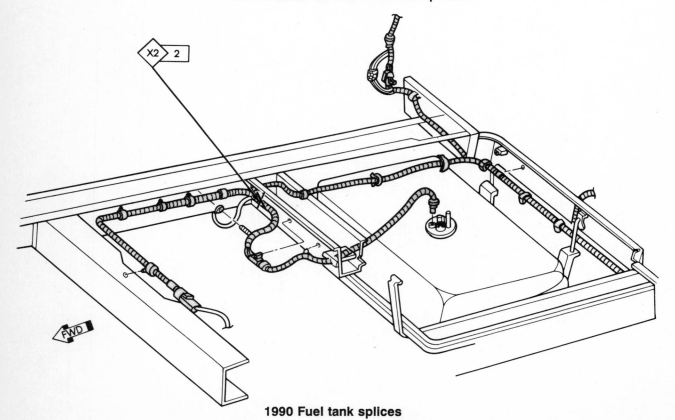

1990 Fuel tank splices

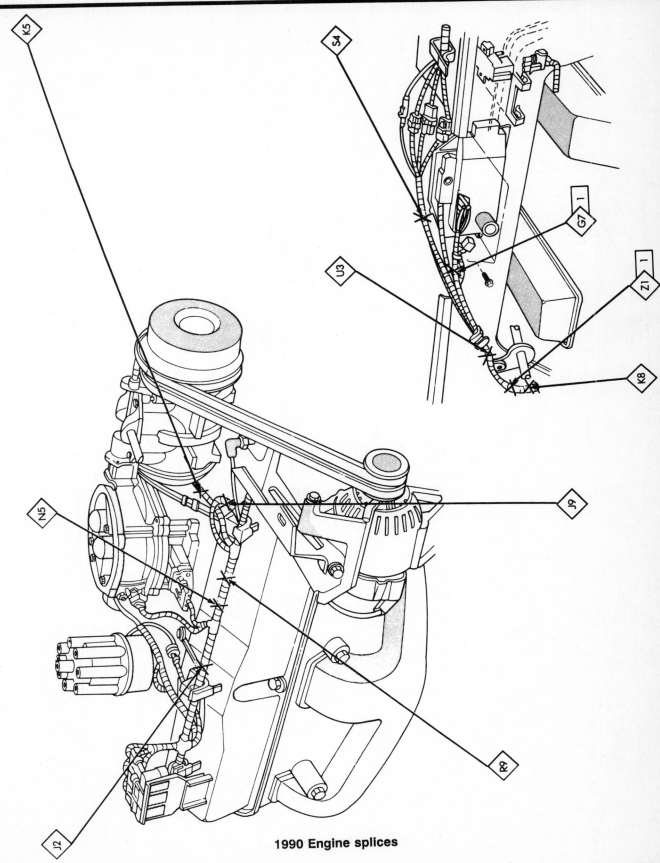

1990 Engine splices

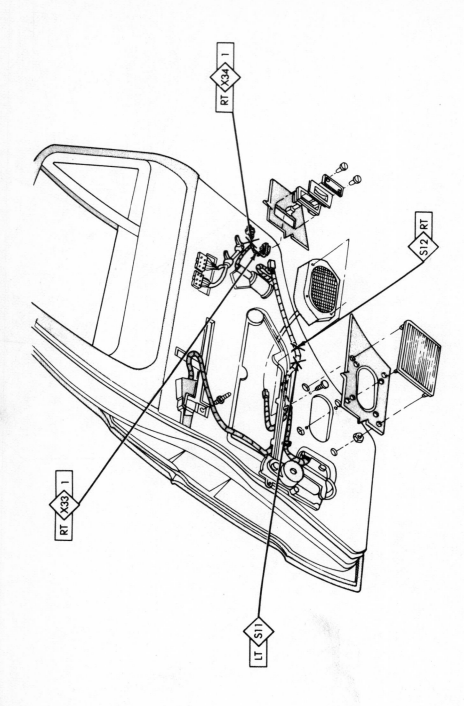

1991 Front door splices

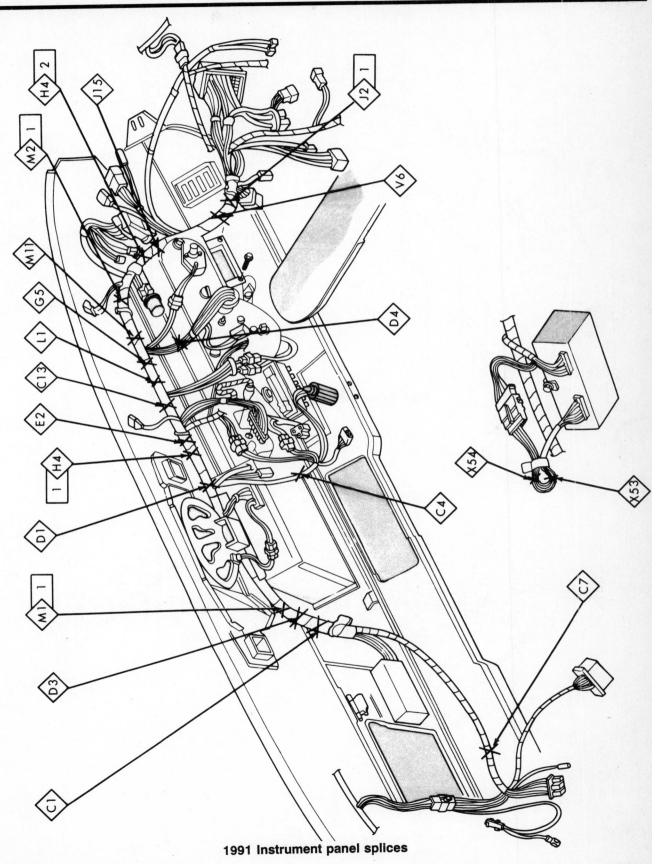

1991 Instrument panel splices

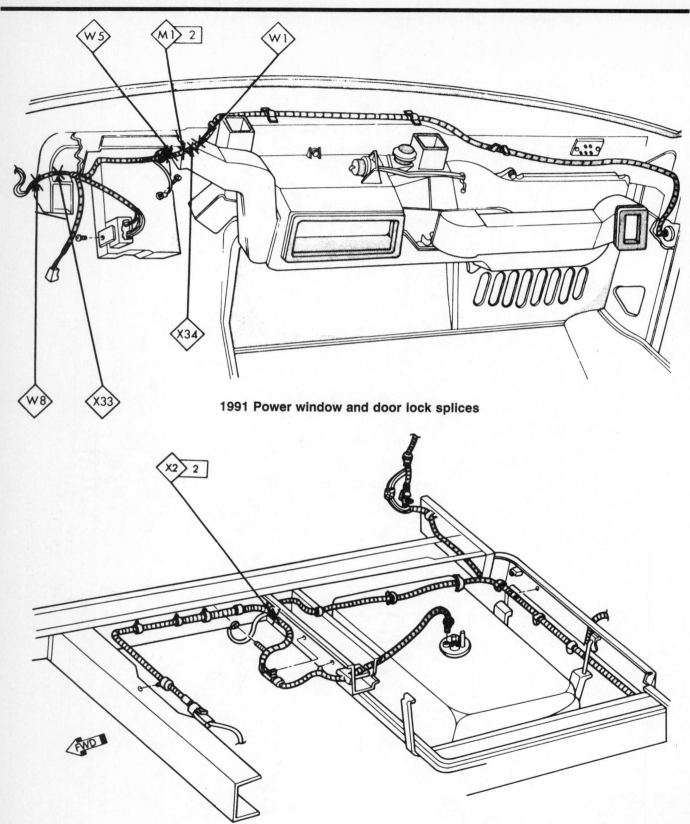

1991 Power window and door lock splices

1991 Fuel tank splices

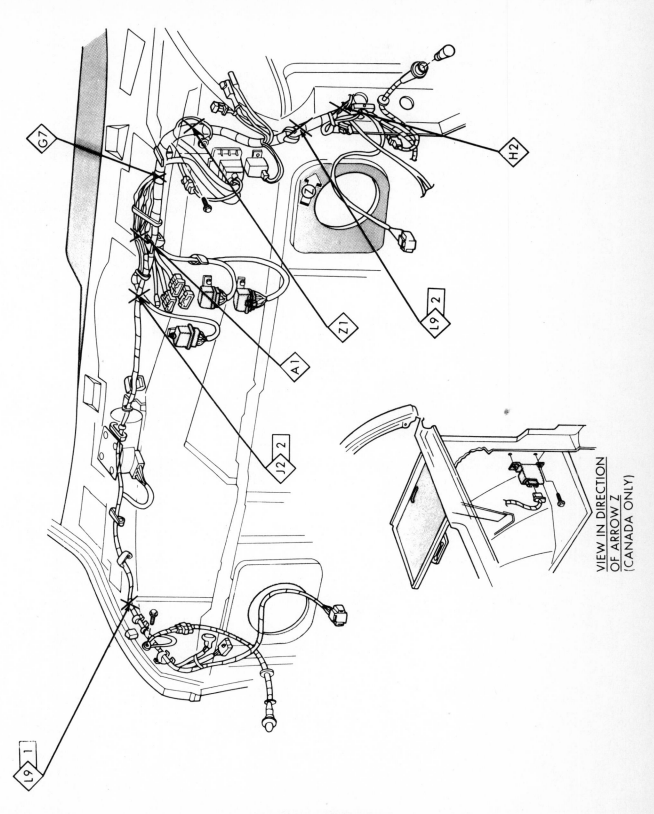

1991 Cowl panel splices

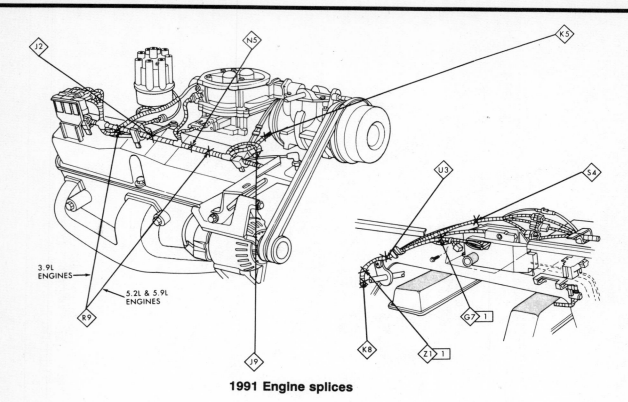

J2

N5

K5

U3

S4

3.9L
ENGINES

5.2L & 5.9L
ENGINES

R9

G7 1

J9

K8

Z1 1

1991 Engine splices

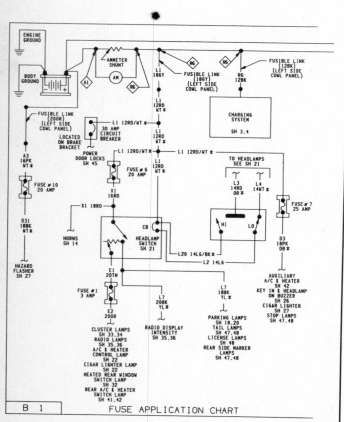

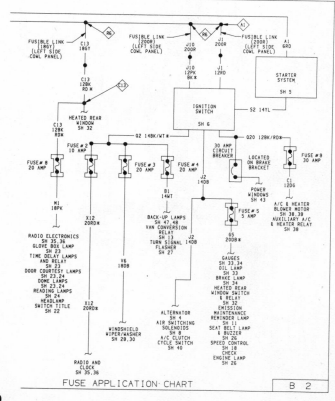

ENGINE GROUND

BODY GROUND

AMMETER SHUNT

AM

A1

R6

R6

R6

L1 18GY

FUSIBLE LINK
(18GY)
(LEFT SIDE
COWL PANEL)

FUSIBLE LINK
(12BK)
(LEFT SIDE COWL PANEL)

R6
12BK

FUSIBLE LINK
(20OR)
(LEFT SIDE
COWL PANEL)

A3
16PK
WT*

LOCATED
ON BRAKE
BRACKET

30 AMP
CIRCUIT
BREAKER

L1 12RD
WT*

L1 12RD
WT*

POWER
DOOR LOCKS
SH 45

L1 12RD/WT*

FUSE # 6
20 AMP

L1
12RD
WT*

L1 12RD/WT*

L1 12RD/WT*

TO HEADLAMPS
SEE SH 21

CHARGING
SYSTEM
SH 3,4

FUSE #10
20 AMP

X1
16RD

L3
14RD
OR*

L4
14WT*

FUSE # 7
25 AMP

D31
18BK
WT*

X1 18RD

HI

LO

D3
18PK
DB*

HORNS
SH 14

CB

HEADLAMP
SWITCH
SH 21

AUXILIARY
A/C & HEATER
SH 42

FUSE # 1
3 AMP

E1
20TN

L20 14LG/BK*

L2 14LG

KEY IN & HEADLAMP
ON BUZZER
SH 26

CIGAR LIGHTER
SH 27

STOP LAMPS
SH 47,48

HAZARD
FLASHER
SH 27

E2
20OR

L7
20BK
YL *

L7
20BK
YL *

PARKING LAMPS
SH 18,20
TAIL LAMPS
SH 47,48
LICENSE LAMPS
SH 48
REAR SIDE MARKER
LAMPS
SH 47,48

CLUSTER LAMPS
SH 33,34
RADIO LAMPS
SH 35,36
A/C & HEATER
CONTROL
LAMP
SH 22
CIGAR LIGHTER LAMP
SH 27
HEATED REAR WINDOW
SWITCH LAMP
SH 32
REAR A/C & HEATER
SWITCH LAMP
SH 41,42

RADIO DISPLAY
INTENSITY
SH 35,36

B 1

FUSE APPLICATION CHART

FUSIBLE LINK
(18GY)
(LEFT SIDE
COWL PANEL)

R6

C13
18GY

FUSIBLE LINK
(20OR)
(LEFT SIDE
COWL PANEL)

R6

A1

FUSIBLE LINK
(20OR)
(LEFT SIDE
COWL PANEL)

A1
6RD

J10
200R

J1
200R

C13
12BK
RD*

C13

C13
12BK
RD*

HEATED REAR
WINDOW
SH 32

J10
12PK
BK *

J1
12RD

IGNITION
SWITCH
SH 6

STARTER
SYSTEM
SH 5

S2 14YL

O2 14BK/WT*

O20 12BK/RD*

FUSE # 8
20 AMP

FUSE # 2
10 AMP

FUSE # 3
20 AMP

FUSE # 4
20 AMP

30 AMP
CIRCUIT
BREAKER

LOCATED
ON BRAKE
BRACKET

FUSE # 8
30 AMP

M1
18PK

X12
20RD*

B1
14WT

J2
14DB

C1
12DG

POWER
WINDOWS

A/C & HEATER
BLOWER MOTOR
SH 38,39
AUXILIARY A/C
& HEATER RELAY
SH 38

RADIO ELECTRONICS
SH 35,36
GLOVE BOX LAMP
SH 23
TIME DELAY LAMPS
AND RELAY
SH 23
DOOR COURTESY LAMPS
SH 23,24
DOME LAMPS
SH 23,24
READING LAMPS
SH 24
HEADLAMP
SWITCH TITLE
SH 22

V6
18DB

BACK-UP LAMPS
SH 47,48
VAN CONVERSION
RELAY
SH 13
TURN SIGNAL 14DB
FLASHER
SH 27

FUSE # 5
5 AMP

GS
20DB*

FUSE # 5
5 AMP

GAUGES
SH 33,34
OIL LAMP
SH 33
BRAKE LAMP
SH 34
HEATED REAR
WINDOW SWITCH
SH 32
EMISSION
MAINTENANCE
REMINDER LAMP
SH 11
SEAT BELT LAMP
& BUZZER
SH 26
SPEED CONTROL
SH 18
CHECK
ENGINE LAMP
SH 26

X12
20RD*

WINDSHIELD
WIPER/WASHER
SH 28,30

ALTERNATOR
SH 4
AIR SWITCHING
SOLENOIDS
SH 8
A/C CLUTCH
CYCLE SWITCH
SH 40

RADIO AND
CLOCK
SH 35,36

FUSE APPLICATION CHART

B 2

1989

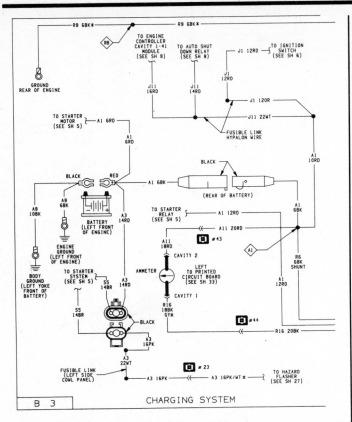

CHARGING SYSTEM

B 3

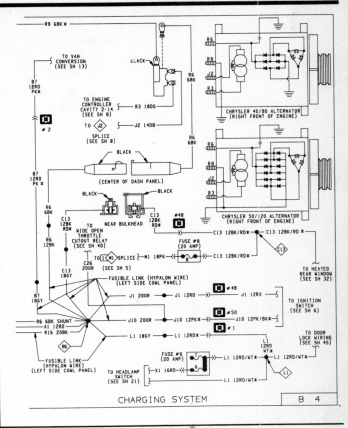

CHARGING SYSTEM

B 4

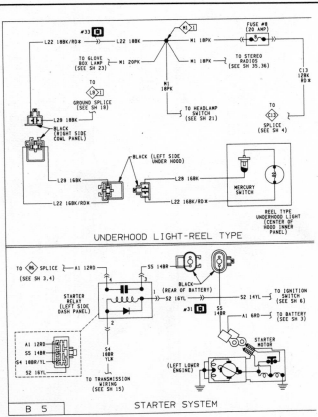

UNDERHOOD LIGHT-REEL TYPE

STARTER SYSTEM

B 5

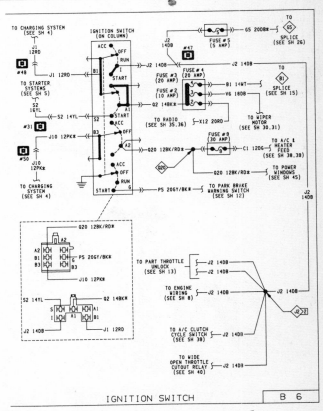

IGNITION SWITCH

B 6

1989

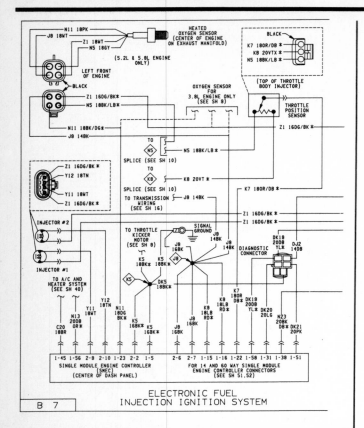

ELECTRONIC FUEL
INJECTION IGNITION SYSTEM

B 7

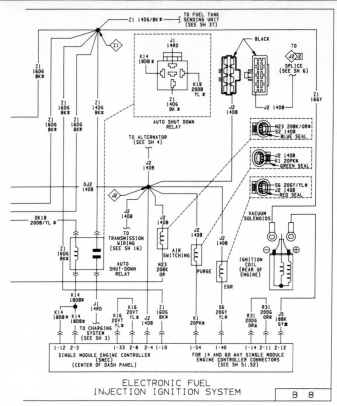

ELECTRONIC FUEL
INJECTION IGNITION SYSTEM

B 8

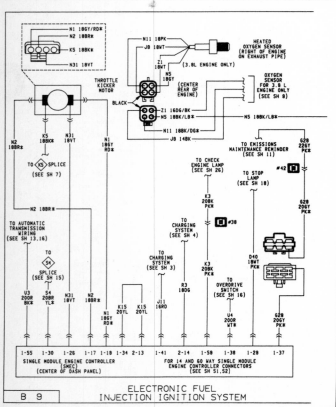

ELECTRONIC FUEL
INJECTION IGNITION SYSTEM

B 9

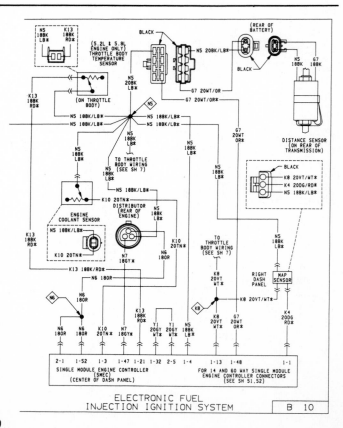

ELECTRONIC FUEL
INJECTION IGNITION SYSTEM

B 10

1989

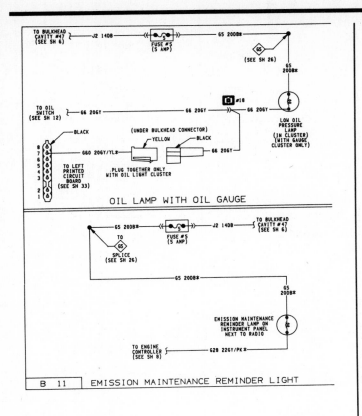

OIL LAMP WITH OIL GAUGE

B 11   EMISSION MAINTENANCE REMINDER LIGHT

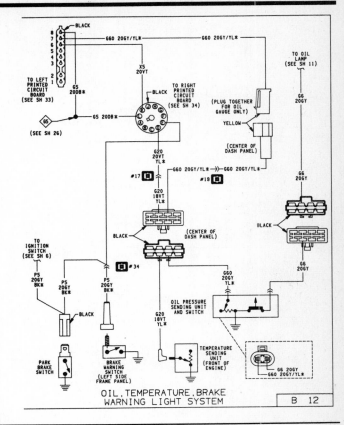

OIL, TEMPERATURE, BRAKE WARNING LIGHT SYSTEM

B 12

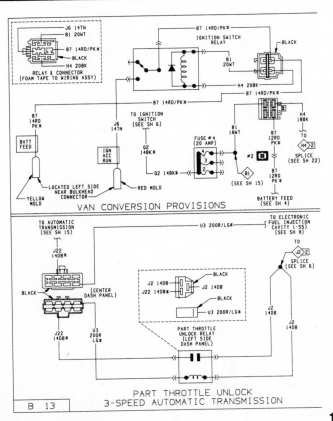

VAN CONVERSION PROVISIONS

PART THROTTLE UNLOCK 3-SPEED AUTOMATIC TRANSMISSION

B 13

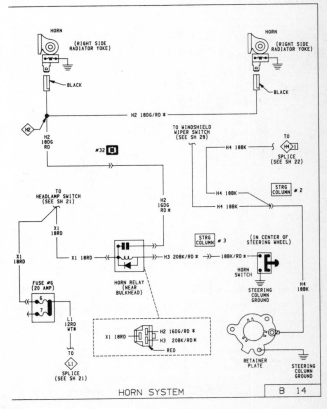

HORN SYSTEM

B 14

1989

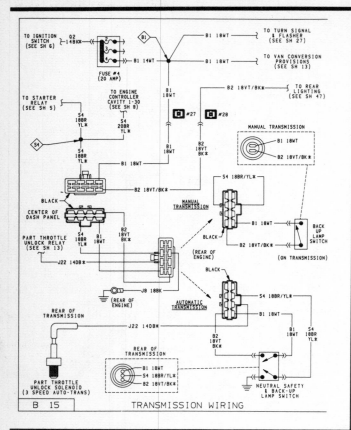

TRANSMISSION WIRING   B 15

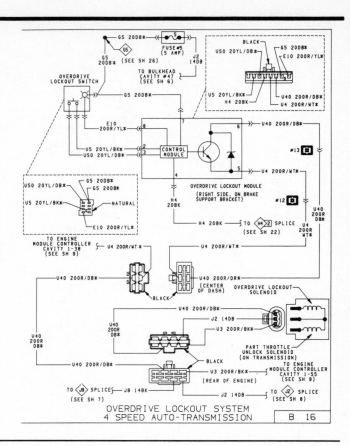

OVERDRIVE LOCKOUT SYSTEM
4 SPEED AUTO-TRANSMISSION   B 16

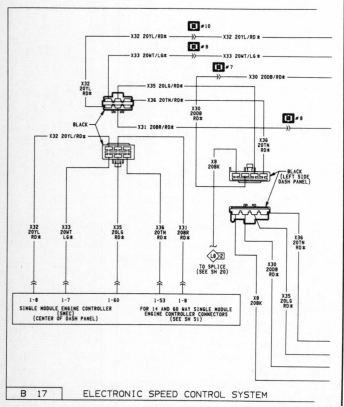

ELECTRONIC SPEED CONTROL SYSTEM   B 17

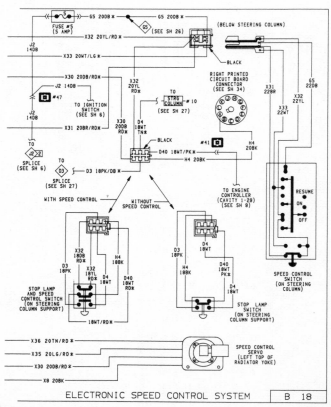

ELECTRONIC SPEED CONTROL SYSTEM   B 18

1989

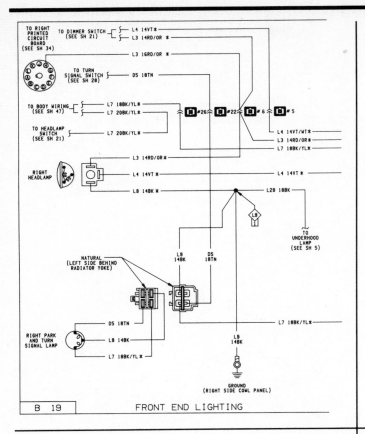

FRONT END LIGHTING

B 19

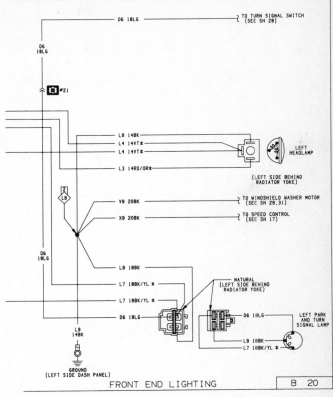

FRONT END LIGHTING

B 20

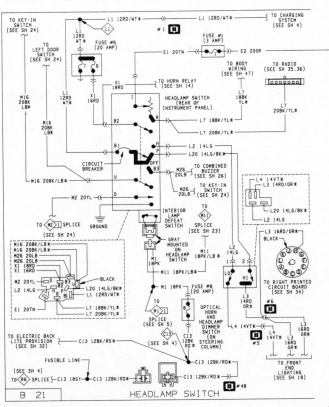

HEADLAMP SWITCH

B 21

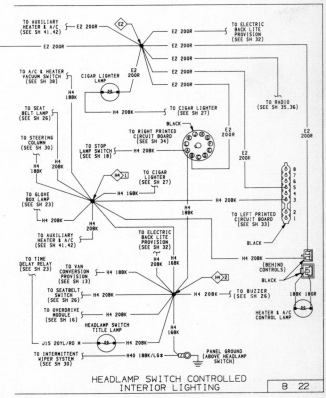

HEADLAMP SWITCH CONTROLLED
INTERIOR LIGHTING

B 22

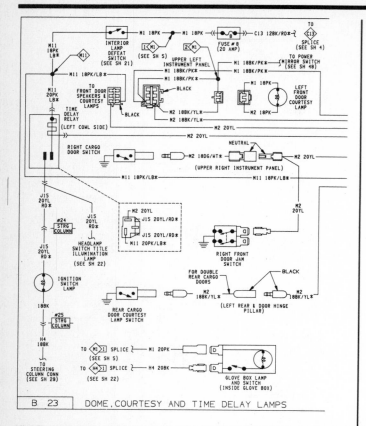

B 23   DOME, COURTESY AND TIME DELAY LAMPS

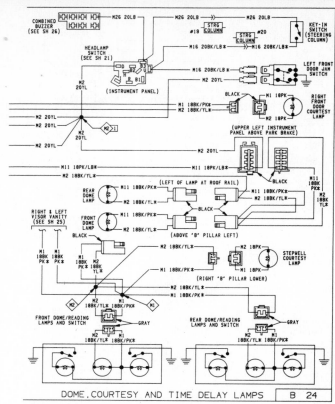

DOME, COURTESY AND TIME DELAY LAMPS   B 24

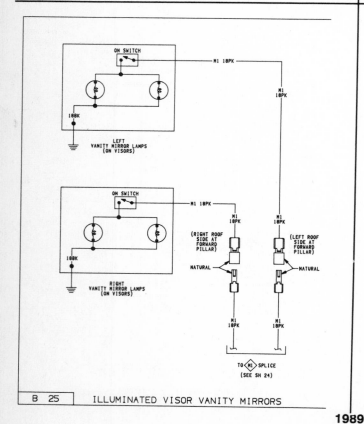

B 25   ILLUMINATED VISOR VANITY MIRRORS

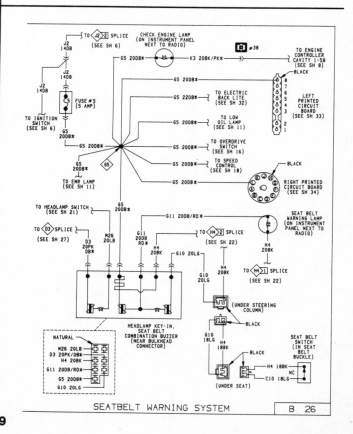

SEATBELT WARNING SYSTEM   B 26

1989

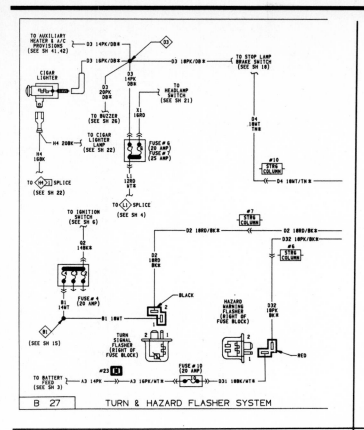

B 27   TURN & HAZARD FLASHER SYSTEM

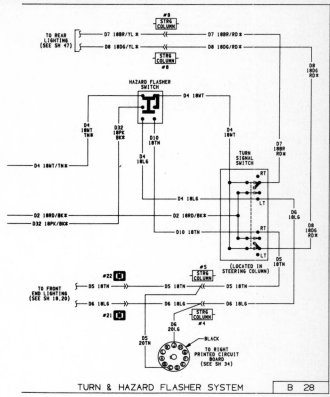

TURN & HAZARD FLASHER SYSTEM   B 28

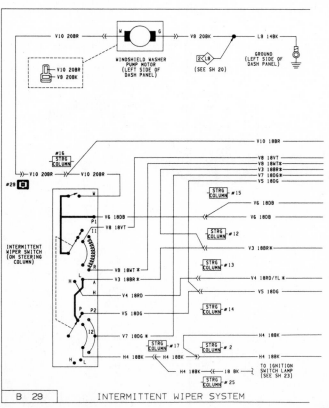

B 29   INTERMITTENT WIPER SYSTEM

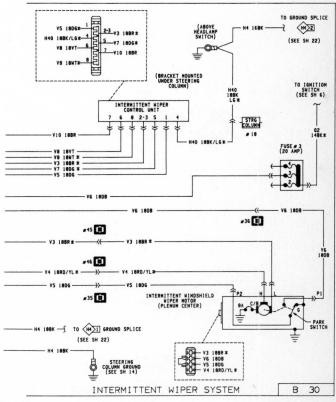

INTERMITTENT WIPER SYSTEM   B 30

1989

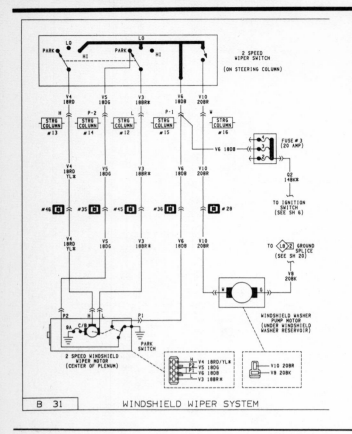

B 31 — WINDSHIELD WIPER SYSTEM

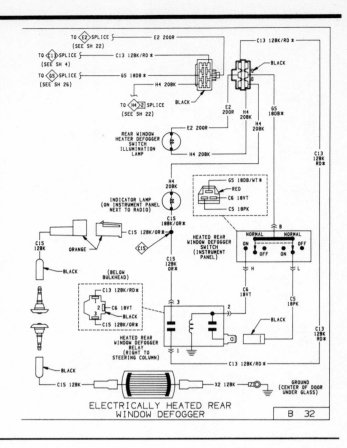

B 32 — ELECTRICALLY HEATED REAR WINDOW DEFOGGER

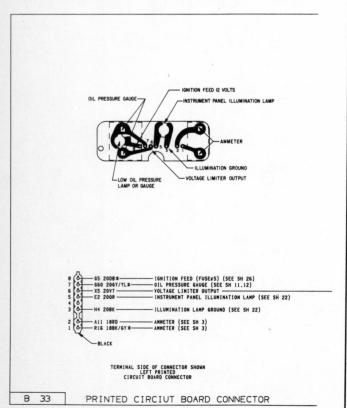

B 33 — PRINTED CIRCIUT BOARD CONNECTOR

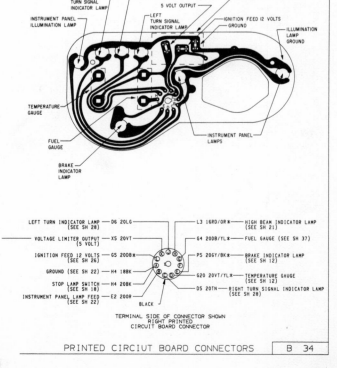

B 34 — PRINTED CIRCIUT BOARD CONNECTORS

1989

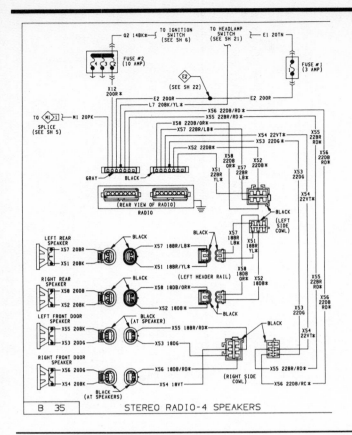

STEREO RADIO-4 SPEAKERS

B 35

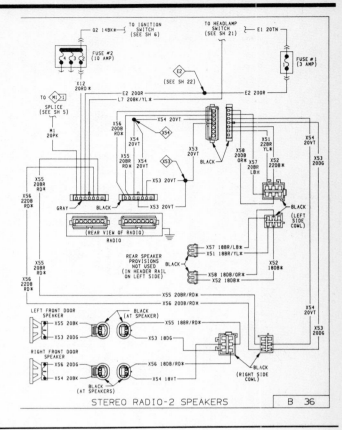

STEREO RADIO-2 SPEAKERS

B 36

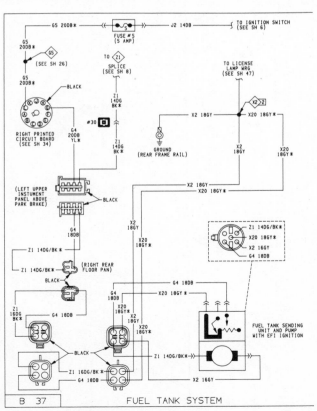

FUEL TANK SYSTEM

B 37

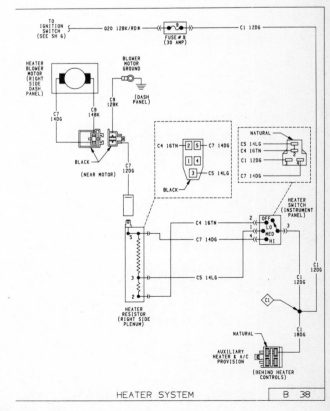

HEATER SYSTEM

B 38

1989

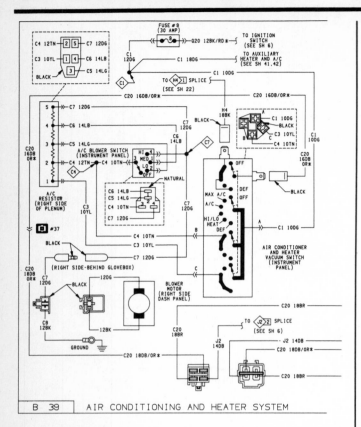

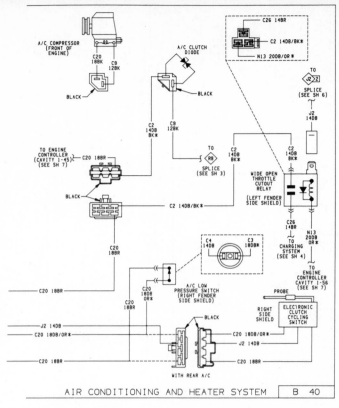

AIR CONDITIONING AND HEATER SYSTEM | B 39

AIR CONDITIONING AND HEATER SYSTEM | B 40

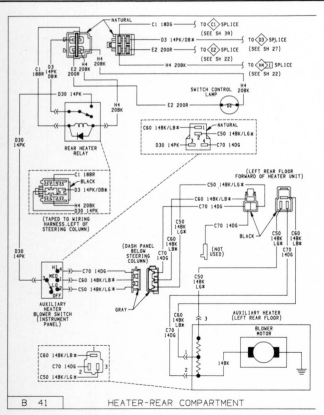

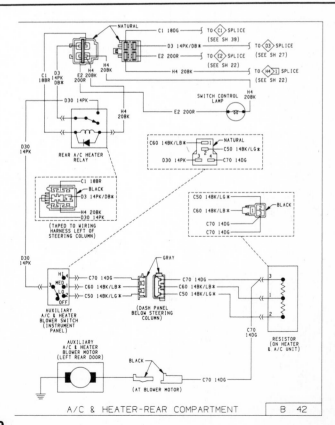

B 41 | HEATER-REAR COMPARTMENT

A/C & HEATER-REAR COMPARTMENT | B 42

1989

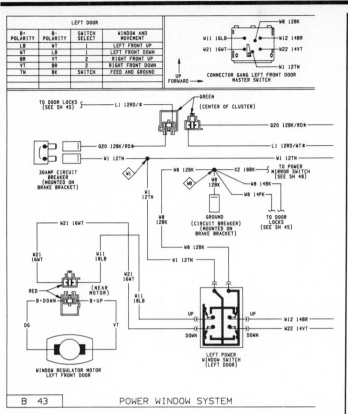

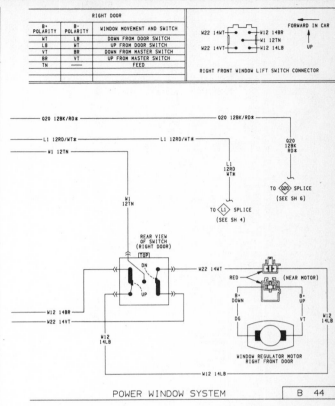

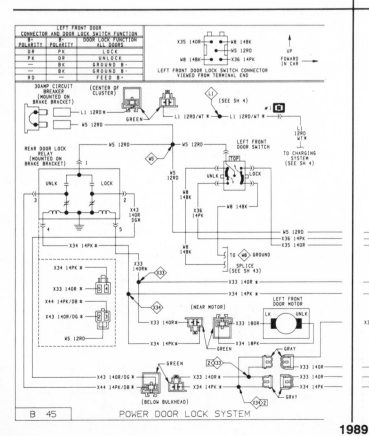

B 45   POWER DOOR LOCK SYSTEM

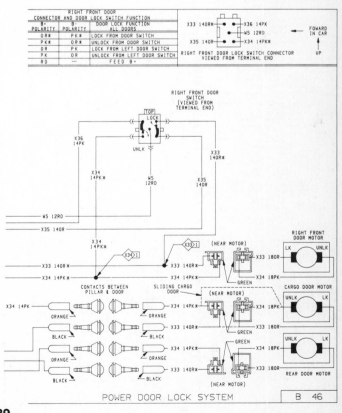

POWER DOOR LOCK SYSTEM   B 46

1989

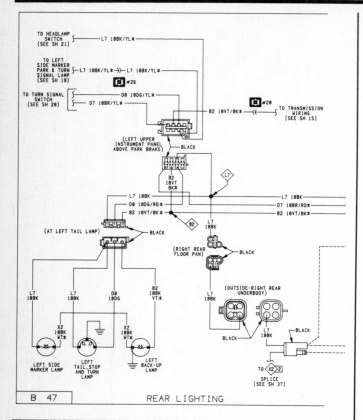

B 47   REAR LIGHTING

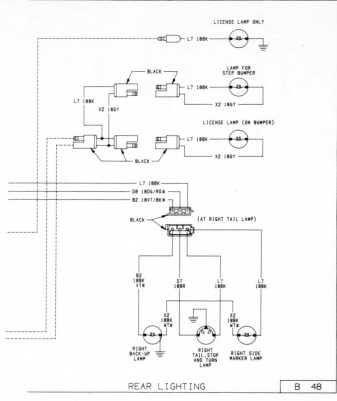

REAR LIGHTING   B 48

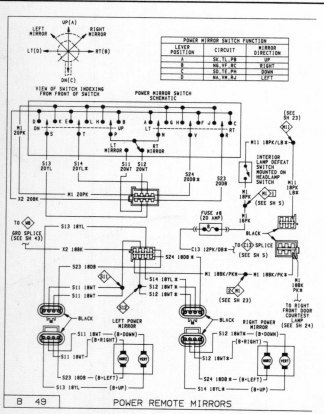

B 49   POWER REMOTE MIRRORS

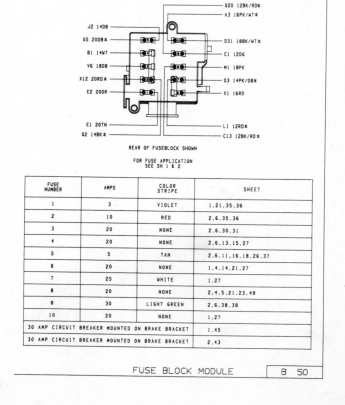

FUSE BLOCK MODULE   B 50

FUSE NUMBER	AMPS	COLOR STRIPE	SHEET
1	3	VIOLET	1,21,35,36
2	10	RED	2,6,35,36
3	20	NONE	2,6,30,31
4	20	NONE	2,6,13,15,27
5	5	TAN	2,6,11,16,18,26,37
6	20	NONE	1,4,14,21,27
7	25	WHITE	1,27
8	20	NONE	2,4,5,21,23,48
9	30	LIGHT GREEN	2,6,38,39
10	20	NONE	1,27
30 AMP CIRCUIT BREAKER MOUNTED ON BRAKE BRACKET			1,45
30 AMP CIRCUIT BREAKER MOUNTED ON BRAKE BRACKET			2,43

1989

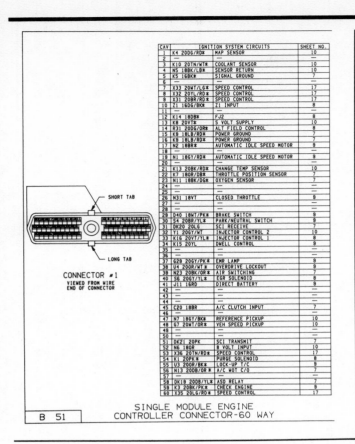

CAV		IGNITION SYSTEM CIRCUITS	SHEET NO.
1	K4 20DG/RD*	MAP SENSOR	10
2	—		—
3	K10 20TN/WT*	COOLANT SENSOR	10
4	N5 18BK/LB*	SENSOR RETURN	10
5	K5 16BK*	SIGNAL GROUND	—
6	—		—
7	X33 20WT/LG*	SPEED CONTROL	17
8	X32 20YL/RD*	SPEED CONTROL	17
9	X31 20BR/RD*	SPEED CONTROL	17
10	Z1 16DG/BK*	Z1 INPUT	8
11	—		—
12	K14 18DB*	FJ2	8
13	K8 20VT*	5 VOLT SUPPLY	10
14	R31 20DG/OR*	ALT FIELD CONTROL	8
15	K9 18LB*	POWER GROUND	7
16	K9 18LB*	POWER GROUND	7
17	N2 18BR*	AUTOMATIC IDLE SPEED MOTOR	9
18	—		—
19	N1 18GY/RD*	AUTOMATIC IDLE SPEED MOTOR	9
20	—		—
21	K13 20BK/RD*	CHANGE TEMP SENSOR	10
22	K7 18OR/DB*	THROTTLE POSITION SENSOR	7
23	N11 18BK/DG*	OXYGEN SENSOR	7
24	—		—
25	—		—
26	N31 18VT	CLOSED THROTTLE	9
27	—		—
28	—		—
29	D40 18WT/PK*	BRAKE SWITCH	8
30	S4 20BR/YL*	PARK/NEUTRAL SWITCH	9
31	DK20 20LG	SCI RECEIVE	7
32	Y1 20GY/WT	INJECTOR CONTROL 2	10
33	K16 20VT/YL*	INJECTOR CONTROL 1	8
34	K15 20YL	DWELL CONTROL	9
35	—		—
36	—		—
37	G29 20GY/PK*	EMR LAMP	9
38	U4 200R/WT*	OVERDRIVE LOCKOUT	9
39	N23 20BK/OR*	AIR SWITCHING	7
40	S6 20GY/YL*	EGR SOLENOID	8
41	J11 16RD	DIRECT BATTERY	9
42	—		—
43	—		—
44	—		—
45	C20 18BR	A/C CLUTCH INPUT	7
46	—		—
47	N7 18GY/BK*	REFERENCE PICKUP	10
48	G7 20WT/OR*	VEH SPEED PICKUP	10
49	—		—
50	—		—
51	DK21 20PK	SCI TRANSMIT	7
52	N6 18OR	8 VOLT INPUT	10
53	X36 20TN/RD*	SPEED CONTROL	17
54	K1 20PK*	PURGE SOLENOID	8
55	U3 200R/BK*	LOCK-UP T/C	9
56	N13 200B/OR*	A/C WOT C/O	7
57	—		—
58	DK19 20DB/YL*	ASD RELAY	7
59	K3 20BK/PK*	CHECK ENGINE	9
60	X35 20LG/RD*	SPEED CONTROL	17

SINGLE MODULE ENGINE CONTROLLER CONNECTOR-60 WAY

B 51

CONNECTOR #1
VIEWED FROM WIRE END OF CONNECTOR

SHORT TAB
LONG TAB

CONNECTOR #2

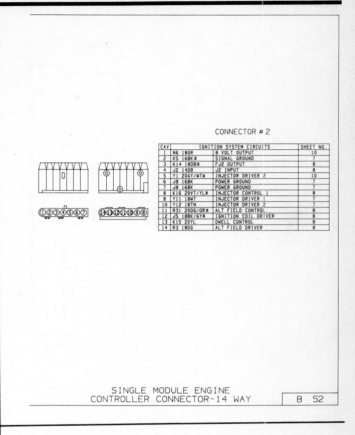

CAV		IGNITION SYSTEM CIRCUITS	SHEET NO.
1	N6 18OR	8 VOLT OUTPUT	10
2	K5 16BK*	SIGNAL GROUND	7
3	K14 18DB*	FJ2 OUTPUT	8
4	J2 14DB	J2 INPUT	8
5	Y1 20GY/WT*	INJECTOR DRIVER 2	10
6	J9 16BK	POWER GROUND	7
7	J9 16BK	POWER GROUND	7
8	K16 20VT/YL*	INJECTOR CONTROL 1	8
9	Y11 18WT	INJECTOR DRIVER 1	7
10	Y12 18TN	INJECTOR DRIVER 2	7
11	R31 20DG/OR*	ALT FIELD CONTROL	8
12	J5 18BK/GY*	IGNITION COIL DRIVER	7
13	K15 20YL	DWELL CONTROL	9
14	R3 18DG	ALT FIELD DRIVER	7

SINGLE MODULE ENGINE CONTROLLER CONNECTOR-14 WAY

B 52

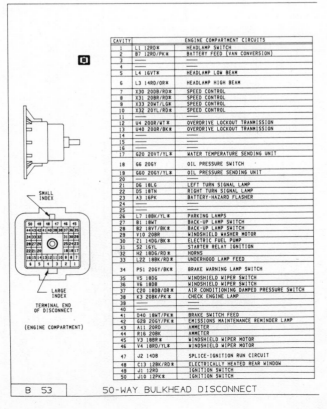

CAVITY		ENGINE COMPARTMENT CIRCUITS
1	L1 12RD*	HEADLAMP SWITCH
2	B7 12RD/PK*	BATTERY FEED (VAN CONVERSION)
3	—	
4	—	
5	L4 16VT*	HEADLAMP LOW BEAM
6	L3 14RD/OR*	HEADLAMP HIGH BEAM
7	X30 20DB/RD*	SPEED CONTROL
8	X31 20BR/RD*	SPEED CONTROL
9	X33 20WT/LG*	SPEED CONTROL
10	X32 20YL/RD*	SPEED CONTROL
11	—	
12	U4 200R/WT*	OVERDRIVE LOCKOUT TRANMISSION
13	U40 200R/BK*	OVERDRIVE LOCKOUT TRANMISSION
14	—	
15	—	
16	—	
17	G20 20VT/YL*	WATER TEMPERATURE SENDING UNIT
18	G6 20GY	OIL PRESSURE SWITCH
19	G60 20GY/YL*	OIL PRESSURE SENDING UNIT
20	—	
21	D6 18LG	LEFT TURN SIGNAL LAMP
22	D5 18TN	RIGHT TURN SIGNAL LAMP
23	A3 16PK	BATTERY-HAZARD FLASHER
24	—	
25	—	
26	L7 18BK/YL*	PARKING LAMPS
27	B1 18WT	BACK-UP LAMP SWITCH
28	B2 18VT/BK*	BACK-UP LAMP SWITCH
29	V10 20BR	WINDSHIELD WASHER MOTOR
30	Z1 14DG/BK*	ELECTRIC FUEL PUMP
31	S2 16YL	STARTER RELAY IGNITION
32	H2 18DG/RD*	HORNS
33	L22 18BK/RD*	UNDERHOOD LAMP FEED
34	P51 20GY/BK*	BRAKE WARNING LAMP SWITCH
35	V5 18DG	WINDSHIELD WIPER SWITCH
36	V6 18DB	WINDSHIELD WIPER SWITCH
37	C20 18DB/OR*	AIR CONDITIONING DAMPED PRESSURE SWITCH
38	K3 20BK/PK*	CHECK ENGINE LAMP
39	—	
40	—	
41	D40 18WT/PK*	BRAKE SWITCH FEED
42	G29 20GY/PK*	EMISSIONS MAINTENANCE REMINDER LAMP
43	A11 20RD	AMMETER
44	R16 20BK	AMMETER
45	V3 18BR*	WINDSHIELD WIPER MOTOR
46	V4 18RD/YL*	WINDSHIELD WIPER MOTOR
47	J2 14DB	SPLICE-IGNITION RUN CIRCUIT
48	C13 12BK/RD*	ELECTRICALLY HEATED REAR WINDOW
49	J1 12RD	IGNITION SWITCH
50	J10 12PK*	IGNITION SWITCH

SMALL INDEX
LARGE INDEX
TERMINAL END OF DISCONNECT
(ENGINE COMPARTMENT)

B 53   **50-WAY BULKHEAD DISCONNECT**

CAVITY		INSTRUMENT PANEL CIRCUITS	SHEET
1	L1 12RD/WT*	BATTERY FEED	4,21,45
2	B7 12RD/PK*	BATTERY FEED (VAN CONVERSION)	4,13
3	—		
4	—		
5	L4 14VT*	HEADLAMPS-LOW BEAM	19,21
6	L3 14RD/OR*	HEADLAMPS-HIGH BEAMS	19,21
	L3 16RD/OR*	HEADLAMPS-HIGH BEAMS	19,21
7	X30 20DB/RD*	SPEED CONTROL	17
8	X31 20BR/RD*	SPEED CONTROL	17
9	X33 20WT/LG*	SPEED CONTROL	17
10	X32 20YL/RD*	SPEED CONTROL	17
11	—		
12	U4 200R/WT*	OVERDRIVE LOCKOUT TRANSMISSION	16
13	U40 200R/DB*	OVERDRIVE LOCKOUT TRANSMISSION	16
14	—		
15	—		
16	—		
17	G20 20VT/YL*	TEMPERATURE-SEND/UNIT OR SWITCH	12
18	G6 20GY	OIL PRESSURE LAMP-SWITCH	11
	G6 20GY	OIL PRESSURE LAMP-SWITCH	11
19	G20 20GY/YL*	OIL PRESSURE GAUGE-SEND/UNIT	12
20	—		
21	D6 18LG	LEFT FRONT TURN SIGNAL LAMP	20,28
22	D5 18TN	RIGHT FRONT TURN SIGNAL LAMP	19,28
23	A3 16PK/WT*	HAZARD FLASHER	3,27
24	—		
25	—		
26	L7 18BK/YL*	PARKING LAMPS	18,47
27	B1 18WT	BACK-UP LAMP FEED	15
28	B2 18VT/BK*	BACK-UP LAMPS	15,47
29	V10 20BR	WINDSHIELD WIPER WASHER MOTOR	31
30	Z1 14DG/BK*	ELECTRIC FUEL PUMP	37
31	S2 14YL	IGNITION START	5,6
32	H2 16DG/RD*	HORNS	14
33	L22 18PK	UNDERHOOD LAMP	5
34	P5 20GY/BK*	BRAKE WARNING LAMP	12
	P5 20GY/BK*	PARK BRAKE SWITCH	12
35	V5 18DG	WINDSHIELD WIPER SWITCH	30,31
36	V6 18DB	WINDSHIELD WIPER SWITCH	30,31
37	C2 16DB/OR*	A/C CLUTCH	39
38	K3 20BK/PK*	CHECK ENGINE LAMP	9,26
39	—		
40	—		
41	D40 18WT/PK*	BRAKE SWITCH FEED	18
42	G29 20GY/BK*	EMR LAMP	9
43	A11 18RD	AMMETER	3
44	R16 18BK/GY*	AMMETER	3
45	V3 18BR*	WINDSHIELD WIPER LOW SPEED	30,31
46	V4 18RD/YL*	WINDSHIELD WIPER HIGH SPEED	30,31
47	J2 14DB	IGNITION-RUN & START	6
	J2 14DB	IGNITION-RUN & START	6
48	C13 12BK/DB*	ELECTRIC BACK LITE (EBL)	4,21
49	J1 12RD	IGNITION SWITCH (B1)	6
50	J10 12PK*	IGNITION SWITCH (B3)	4,6

TERMINAL END OF DISCONNECT
SMALL INDEX
LARGE INDEX
(INSTRUMENT PANEL)

B 54   **50-WAY BULKHEAD DISCONNECT**

1989

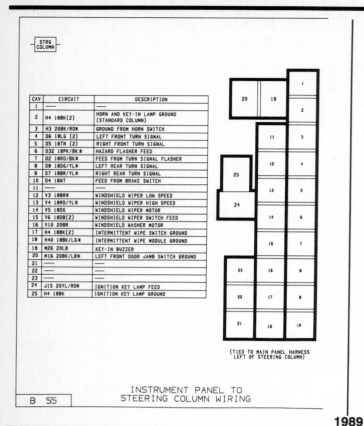

CAV	CIRCUIT	DESCRIPTION
1	—	
2	H4 18BK(2)	HORN AND KEY-IN LAMP GROUND (STANDARD COLUMN)
3	H3 20BK/RD*	GROUND FROM HORN SWITCH
4	D6 18LG (2)	LEFT FRONT TURN SIGNAL
5	D5 18TN (2)	RIGHT FRONT TURN SIGNAL
6	D32 18PK/BK*	HAZARD FLASHER FEED
7	D2 18RD/BK*	FEED FROM TURN SIGNAL FLASHER
8	D8 18DG/YL*	LEFT REAR TURN SIGNAL
9	D7 18BR/YL*	RIGHT REAR TURN SIGNAL
10	D4 18WT	FEED FROM BRAKE SWITCH
11	—	
12	V3 18BR*	WINDSHIELD WIPER LOW SPEED
13	V4 18RD/YL*	WINDSHIELD WIPER HIGH SPEED
14	V5 18DG	WINDSHIELD WIPER MOTOR
15	V6 18DB(2)	WINDSHIELD WIPER SWITCH FEED
16	V10 20BR	WINDSHIELD WASHER MOTOR
17	H4 18BK(2)	INTERMITTENT WIPE SWITCH GROUND
18	H40 18BK/LG*	INTERMITTENT WIPE MODULE GROUND
19	M26 20LB	KEY-IN BUZZER
20	M16 20BK/LB*	LEFT FRONT DOOR JAMB SWITCH GROUND
21	—	
22	—	
23	—	
24	J15 20YL/RD*	IGNITION KEY LAMP FEED
25	H4 18BK	IGNITION KEY LAMP GROUND

(TIED TO MAIN PANEL HARNESS LEFT OF STEERING COLUMN)

INSTRUMENT PANEL TO STEERING COLUMN WIRING

B 55

CAV	CIRCUIT	DESCRIPTION	SHEET
1			
2	H4 18BK(2)	HORN AND KEY-IN LAMP GROUND	14,29
3	H3 20BK/RD*	HORN SWITCH GROUND	14
4	D6 18LG	LEFT FRONT TURN SIGNAL	28
5	D5 18TN	RIGHT FRONT TURN SIGNAL	28
6	D32 18PK/BK*	HAZARD FLASHER FEED	27
7	D2 18RD/BK*	TURN SIGNAL FLASHER FEED	27
8	D8 18DG/YL*	LEFT REAR TURN SIGNAL	28
9	D7 18BR/YL*	RIGHT REAR TURN SIGNAL	28
10	D4 16WT/TN*	BRAKE SWITCH FEED	27
11			
12	V3 18BR*	INTERMITTENT WIPE MODULE	28,31
	V3 18BR*	WINDSHIELD WIPER LOW SPEED	
13	V4 18RD/YL*	WINDSHIELD WIPER HIGH SPEED	28,31
14	V5 18DG	INTERMITTENT WIPE MODULE	
	V5 18DG	WINDSHIELD WIPER MOTOR	28,31
15	V6 18DB(2)	WINDSHIELD WIPER SWITCH FEED	28,31
16	V10 20BR	INTERMITTENT WIPE MODULE	28,31
	V10 20BR	WINDSHIELD WASHER MOTOR	
17	H4 18BK(2)	INTERMITTENT WIPE SWITCH GROUND	28
18	H40 18BK/LG*	INTERMITTENT WIPE MODULE GROUND	30
19	M26 20LB	KEY-IN BUZZER OR CHIME	24
20	M16 20BK/LB*	LEFT FRONT DOOR JAMB SWITCH GROUND	24
21	—		
22	—		
23	—		
24	J15 20YL/RD*	IGNITION KEY LAMP FEED	23
25	H4 18BK	IGNITION KEY LAMP GROUND (TILT COLUMN)	23,28

TURN SIGNAL SWITCH CONNECTOR

WINDSHIELD WIPER SWITCH CONNECTOR

KEY-IN BUZZER SWITCH CONNECTOR

IGNITION KEY LAMP SWITCH CONNECTOR

(PLUGGED INTO STEERING COLUMN 25 WAY CONNECTOR)

STEERING COLUMN SWITCH TO PANEL WIRING

B 56

1989

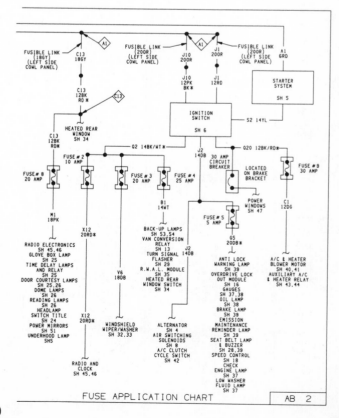

AB 1   FUSE APPLICATION CHART

FUSE APPLICATION CHART   AB 2

1990

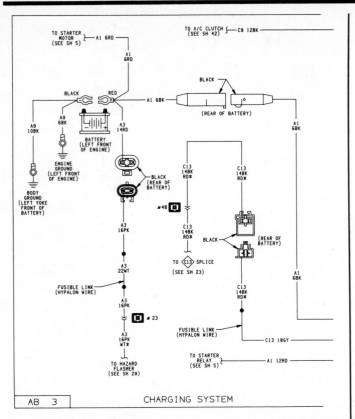

CHARGING SYSTEM

AB 3

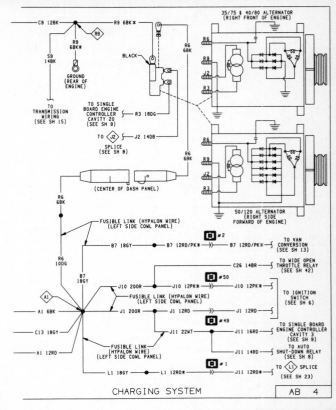

CHARGING SYSTEM

AB 4

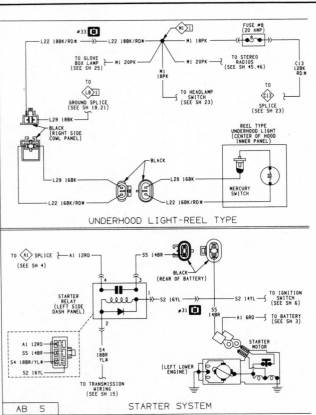

UNDERHOOD LIGHT-REEL TYPE

STARTER SYSTEM

AB 5

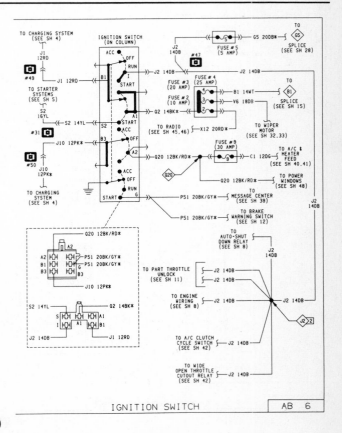

IGNITION SWITCH

AB 6

1990

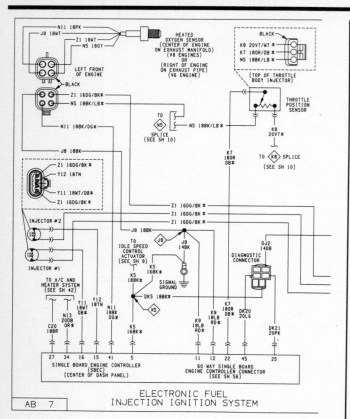

ELECTRONIC FUEL
INJECTION IGNITION SYSTEM

AB 7

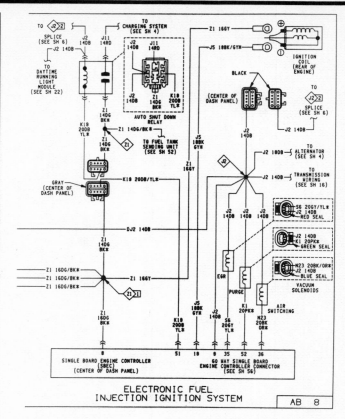

ELECTRONIC FUEL
INJECTION IGNITION SYSTEM

AB 8

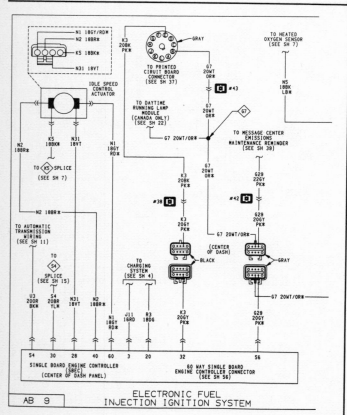

ELECTRONIC FUEL
INJECTION IGNITION SYSTEM

AB 9

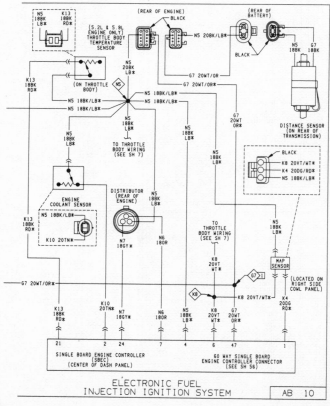

ELECTRONIC FUEL
INJECTION IGNITION SYSTEM

AB 10

1990

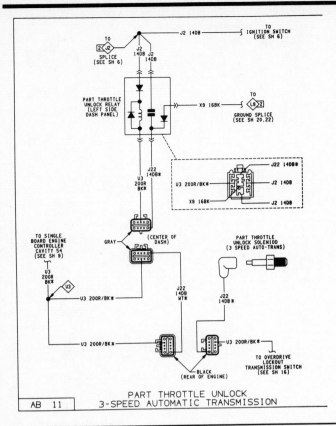

PART THROTTLE UNLOCK
3-SPEED AUTOMATIC TRANSMISSION

AB 11

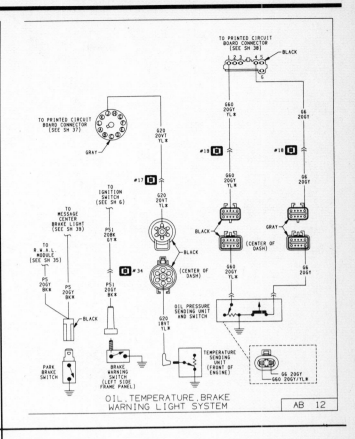

OIL, TEMPERATURE, BRAKE
WARNING LIGHT SYSTEM

AB 12

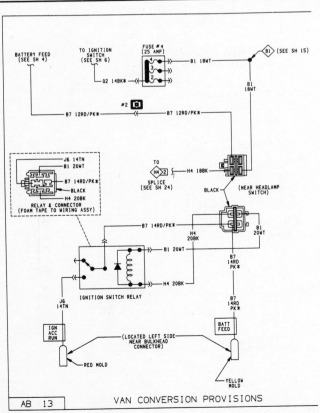

VAN CONVERSION PROVISIONS

AB 13

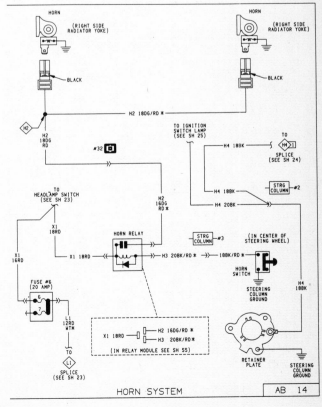

HORN SYSTEM

AB 14

1990

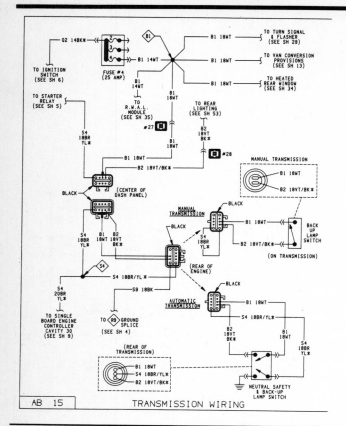

AB 15 · TRANSMISSION WIRING

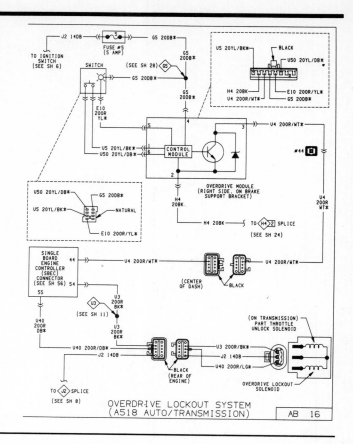

OVERDRIVE LOCKOUT SYSTEM
(A518 AUTO/TRANSMISSION) · AB 16

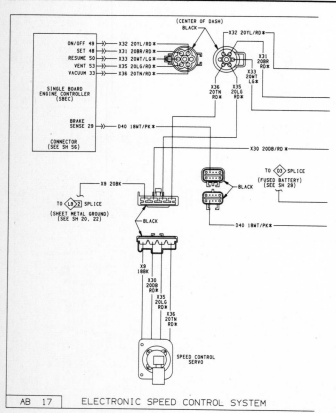

AB 17 · ELECTRONIC SPEED CONTROL SYSTEM

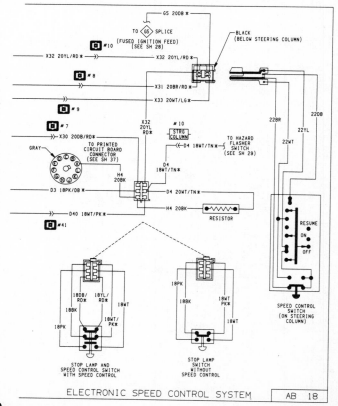

ELECTRONIC SPEED CONTROL SYSTEM · AB 18

1990

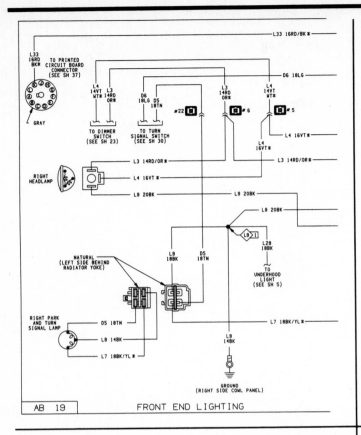

AB 19 — FRONT END LIGHTING

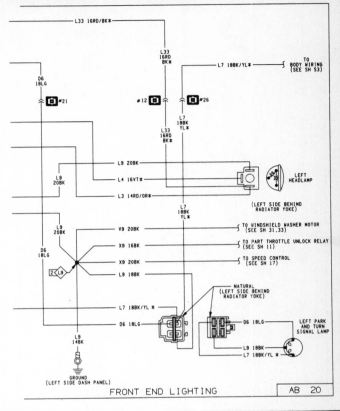

FRONT END LIGHTING — AB 20

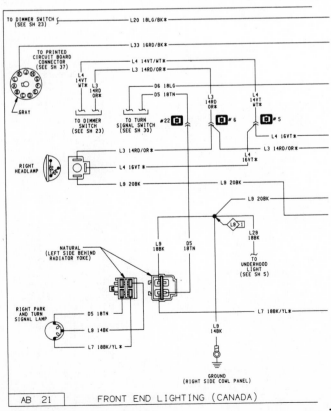

AB 21 — FRONT END LIGHTING (CANADA)

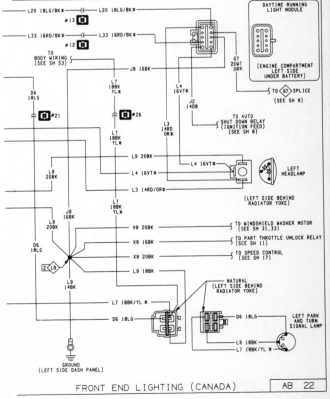

FRONT END LIGHTING (CANADA) — AB 22

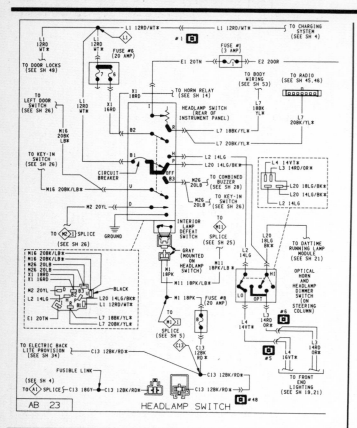

HEADLAMP SWITCH

AB 23

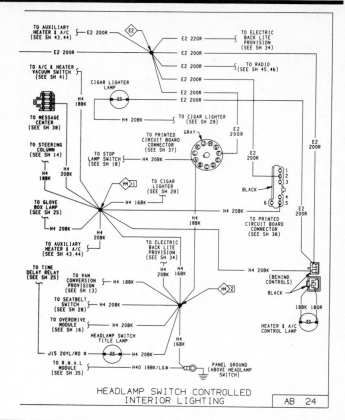

HEADLAMP SWITCH CONTROLLED
INTERIOR LIGHTING

AB 24

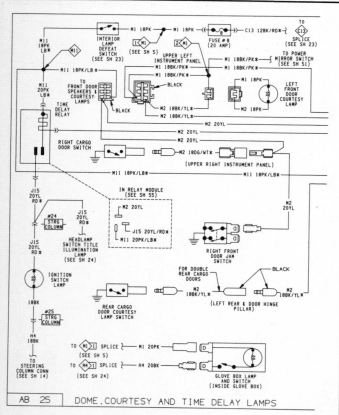

DOME, COURTESY AND TIME DELAY LAMPS

AB 25

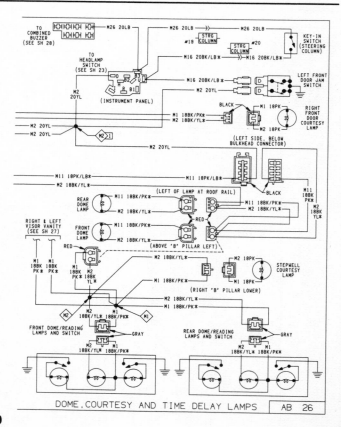

DOME, COURTESY AND TIME DELAY LAMPS

AB 26

1990

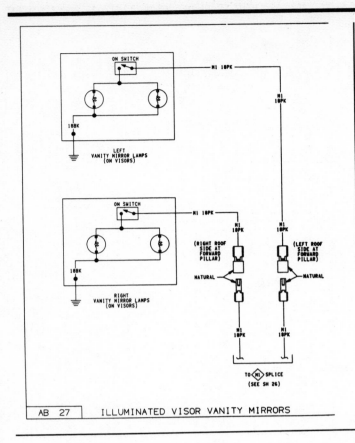

AB 27 — ILLUMINATED VISOR VANITY MIRRORS

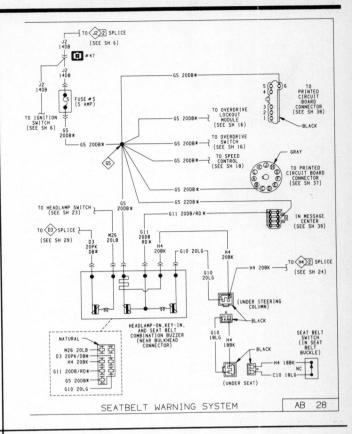

SEATBELT WARNING SYSTEM — AB 28

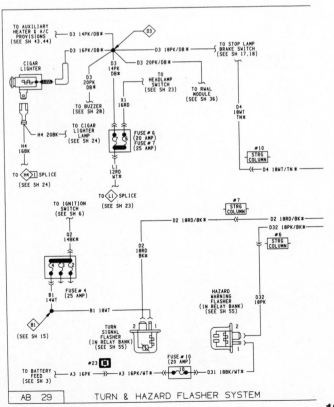

AB 29 — TURN & HAZARD FLASHER SYSTEM

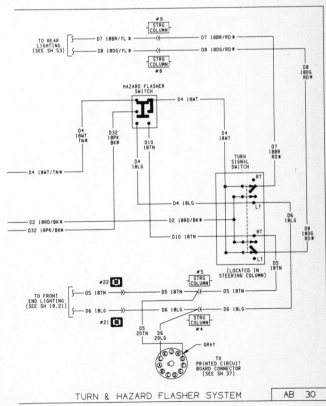

TURN & HAZARD FLASHER SYSTEM — AB 30

1990

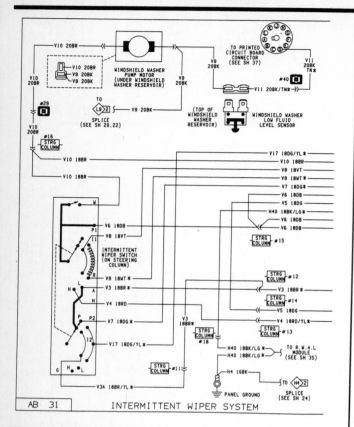

AB 31 — Intermittent Wiper System

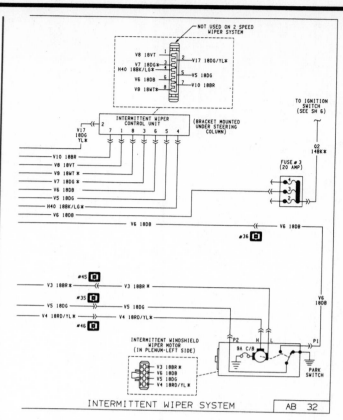

Intermittent Wiper System — AB 32

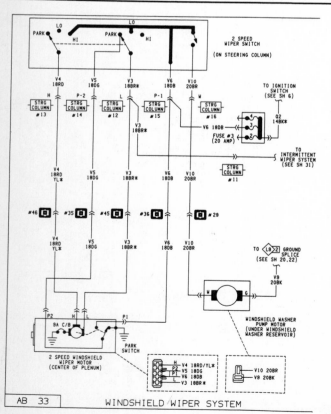

AB 33 — Windshield Wiper System

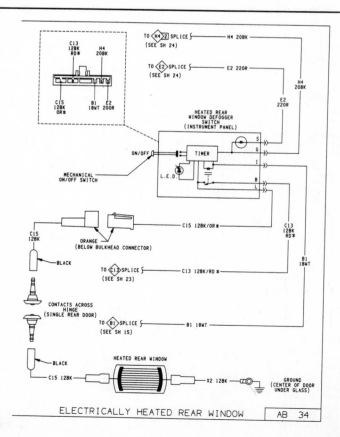

Electrically Heated Rear Window — AB 34

1990

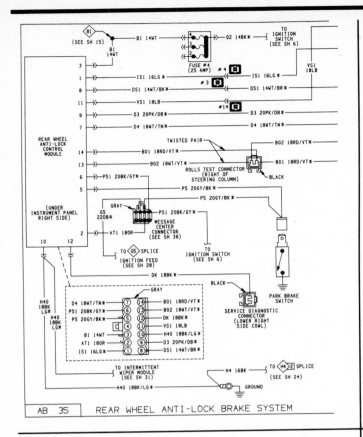

AB 35 — REAR WHEEL ANTI-LOCK BRAKE SYSTEM

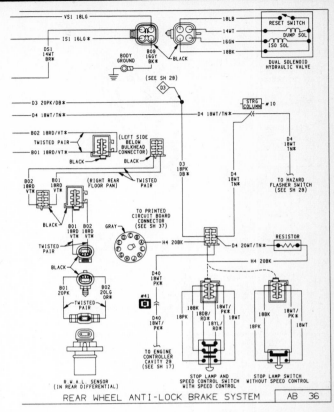

REAR WHEEL ANTI-LOCK BRAKE SYSTEM — AB 36

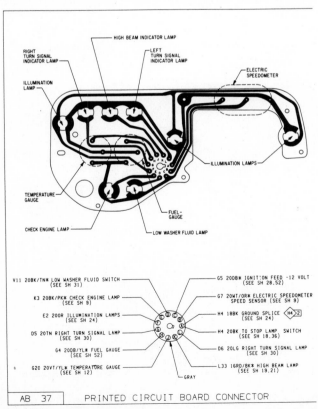

AB 37 — PRINTED CIRCUIT BOARD CONNECTOR

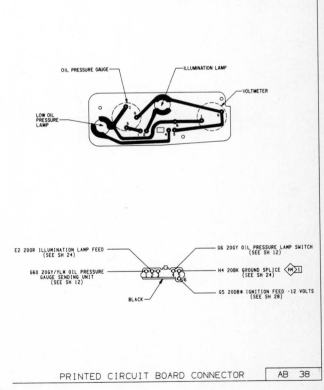

PRINTED CIRCUIT BOARD CONNECTOR — AB 38

1990

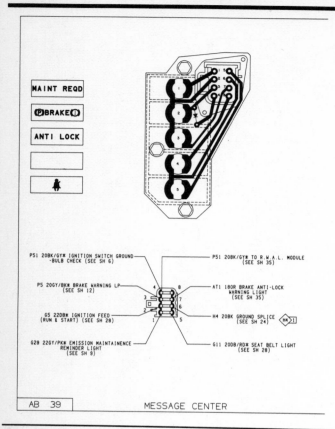

MESSAGE CENTER — AB 39

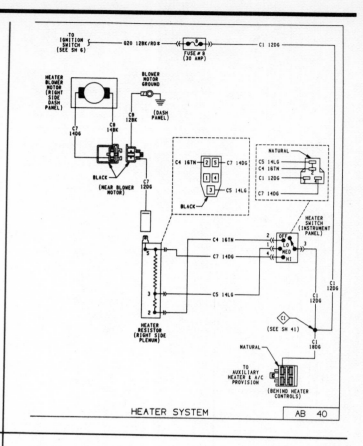

HEATER SYSTEM — AB 40

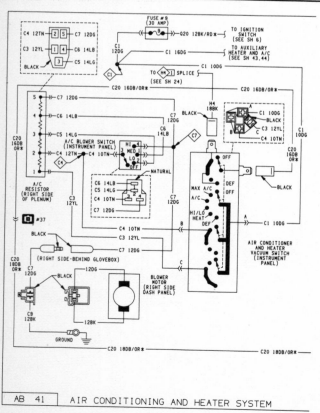

AB 41 — AIR CONDITIONING AND HEATER SYSTEM

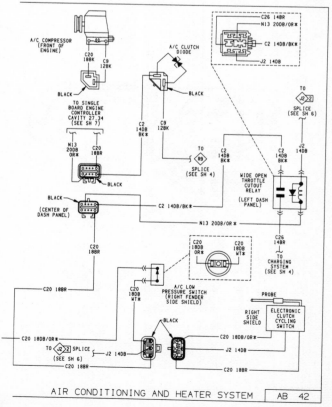

AIR CONDITIONING AND HEATER SYSTEM — AB 42

1990

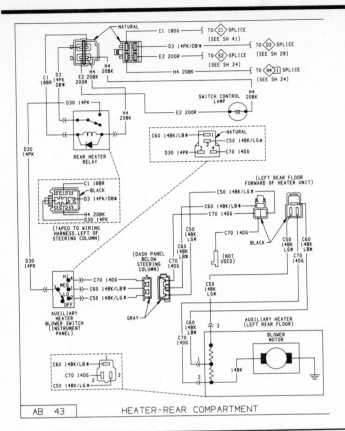

AB 43   HEATER-REAR COMPARTMENT

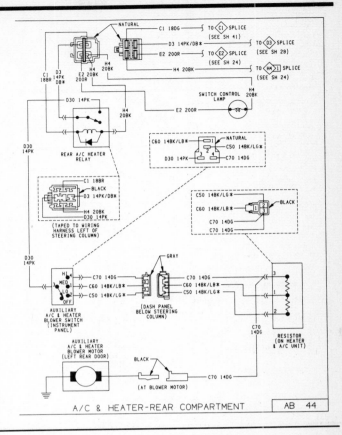

A/C & HEATER-REAR COMPARTMENT   AB 44

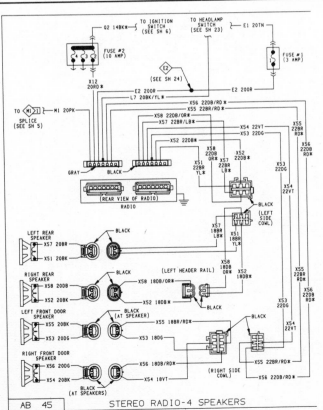

AB 45   STEREO RADIO-4 SPEAKERS

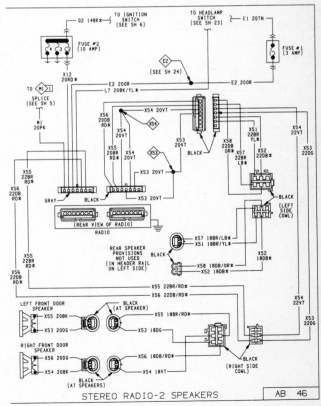

STEREO RADIO-2 SPEAKERS   AB 46

1990

6 CHASSIS ELECTRICAL

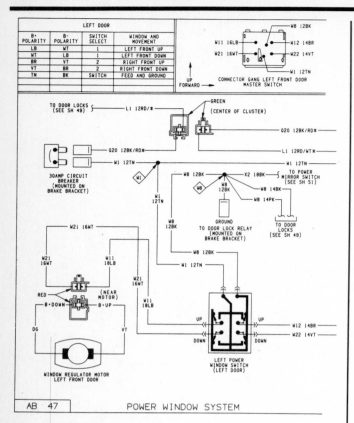

Power Window System — AB 47

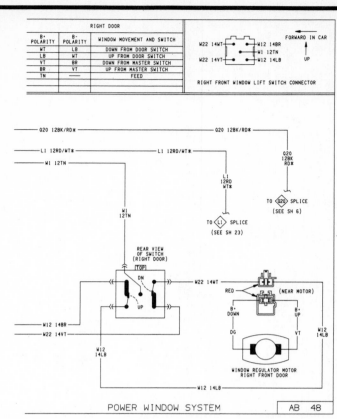

Power Window System — AB 48

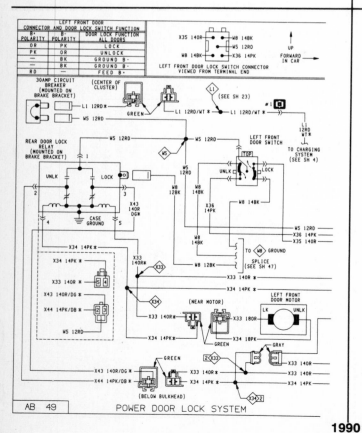

Power Door Lock System — AB 49

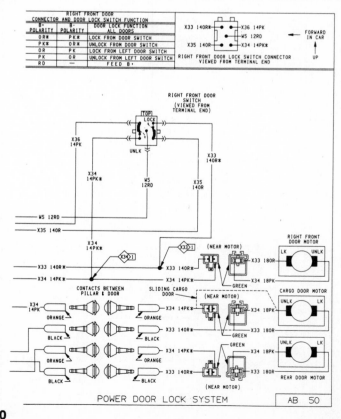

Power Door Lock System — AB 50

1990

6-132

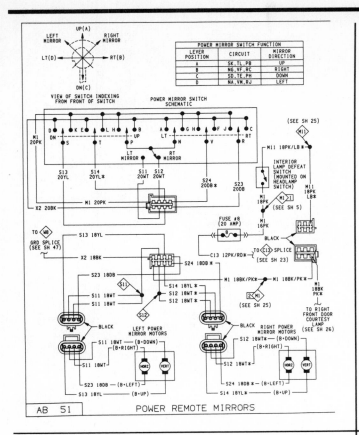

AB 51   POWER REMOTE MIRRORS

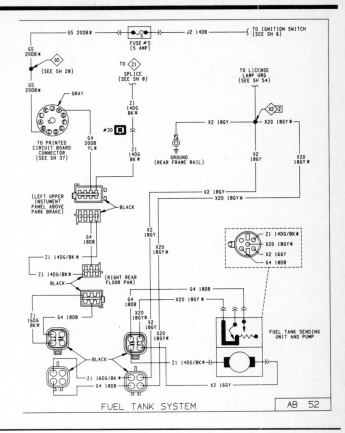

FUEL TANK SYSTEM   AB 52

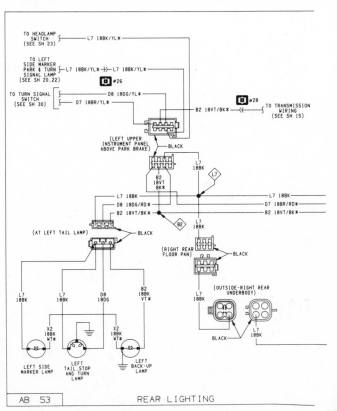

AB 53   REAR LIGHTING

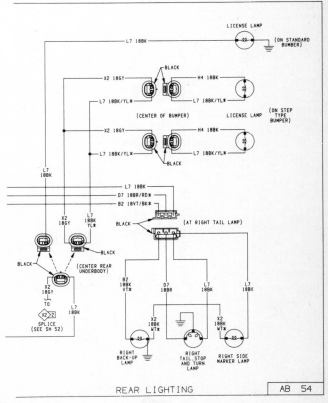

REAR LIGHTING   AB 54

1990

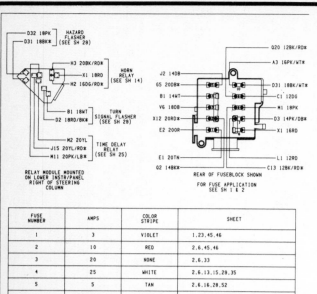

FUSE NUMBER	AMPS	COLOR STRIPE	SHEET
1	3	VIOLET	1,23,45,46
2	10	RED	2,6,45,46
3	20	NONE	2,6,33
4	25	WHITE	2,6,13,15,29,35
5	5	TAN	2,6,16,28,52
6	20	NONE	1,14,23,29
7	25	WHITE	1,29
8	20	NONE	2,5,23,25,51
9	30	LIGHT GREEN	2,6,40,41
10	20	NONE	1,29
30 AMP CIRCUIT BREAKER MOUNTED ON BRAKE BRACKET			1,49
30 AMP CIRCUIT BREAKER MOUNTED ON BRAKE BRACKET			2,47

AB 55   FUSEBLOCK MODULE AND RELAY BANK

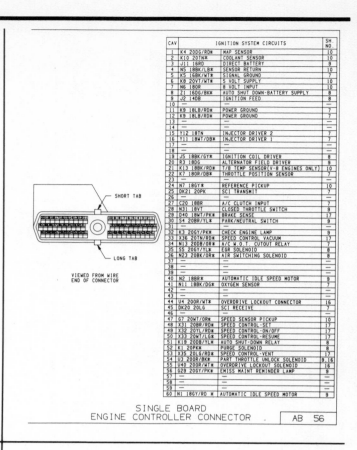

CAV	IGNITION SYSTEM CIRCUITS		SH. NO.
1	K4 20DG/RD*	MAP SENSOR	10
2	K10 20TN*	COOLANT SENSOR	10
3	J11 16RD	DIRECT BATTERY	9
4	N5 18BK/LB*	SENSOR RETURN	10
5	K5 18BK/WT*	SIGNAL GROUND	7
6	K8 20VT/WT*	5 VOLT SUPPLY	10
7	N6 180R	8 VOLT INPUT	10
8	Z1 16DG/BK*	AUTO SHUT DOWN-BATTERY SUPPLY	8
9	J2 14DB	IGNITION FEED	8
10			—
11	K9 18LB/RD*	POWER GROUND	7
12	K9 18LB/RD*	POWER GROUND	7
13			—
14			—
15	Y12 18TN	INJECTOR DRIVER 2	7
16	Y11 18WT/DB*	INJECTOR DRIVER 1	7
17			—
18			—
19	J5 18BK/GY*	IGNITION COIL DRIVER	8
20	R3 18DG	ALTERNATOR FIELD DRIVER	9
21	K13 18BK/RD*	T/B TEMP SENSOR (V-8 ENGINES ONLY)	10
22	K7 180R/DB*	THROTTLE POSITION SENSOR	7
23			—
24	N7 18GY*	REFERENCE PICKUP	10
25	DK21 20PK	SCI TRANSMIT	7
26			—
27	C20 18BR	A/C CLUTCH INPUT	8
28	N31 18VT	CLOSED THROTTLE SWITCH	9
29	D40 18WT/PK*	BRAKE SENSE	17
30	S4 20BR/YL*	PARK/NEUTRAL SWITCH	9
31			—
32	K3 20GY/PK*	CHECK ENGINE LAMP	9
33	X36 20TN/RD*	SPEED CONTROL VACUUM	17
34	N13 20DB/OR*	A/C W.O.T. CUTOUT RELAY	7
35	S5 20GY/YL*	EGR SOLENOID	8
36	N23 20BK/OR*	AIR SWITCHING SOLENOID	8
37			—
38			—
39			—
40	N2 18BR*	AUTOMATIC IDLE SPEED MOTOR	9
41	N11 18BK/DG*	OXYGEN SENSOR	7
42			—
43			—
44	U4 200R/WT*	OVERDRIVE LOCKOUT CONNECTOR	16
45	DK20 20LG	SCI RECEIVE	7
46			—
47	G7 20WT/OR*	SPEED SENSOR PICKUP	10
48	X31 20BR/RD*	SPEED CONTROL-SET	17
49	X32 20YL/RD*	SPEED CONTROL-ON/OFF	17
50	X33 20VT/RD*	SPEED CONTROL-RESUME	17
51	X19 20DB/YL*	AUTO SHUT-DOWN RELAY	8
52	K1 20PK*	PURGE SOLENOID	8
53	X35 20LG/RD*	SPEED CONTROL-VENT	17
54	U3 200R/BK*	PART THROTTLE UNLOCK SOLENOID	9,16
55	U40 200R/WT*	OVERDRIVE LOCKOUT SOLENOID	16
56	G29 20GY/PK*	EMISS MAINT REMINDER LAMP	9
57			—
58			—
59			—
60	N1 18GY/RD*	AUTOMATIC IDLE SPEED MOTOR	9

SINGLE BOARD
ENGINE CONTROLLER CONNECTOR   **AB 56**

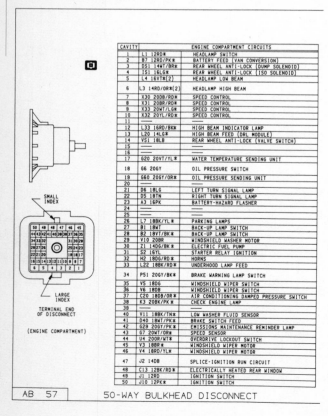

CAVITY		ENGINE COMPARTMENT CIRCUITS
1	L1 12RD*	HEADLAMP SWITCH
2	B7 12RD/PK*	BATTERY FEED (VAN CONVERSION)
3	DS1 14WT/BR*	REAR WHEEL ANTI-LOCK (DUMP SOLENOID)
4	IS1 16LG*	REAR WHEEL ANTI-LOCK (ISO SOLENOID)
5	L4 16VT*(2)	HEADLAMP LOW BEAM
6	L3 14RD/OR*(2)	HEADLAMP HIGH BEAM
7	X30 200B/RD*	SPEED CONTROL
8	X31 20BR/RD*	SPEED CONTROL
9	X33 20VT/LG*	SPEED CONTROL
10	X32 20YL/RD*	SPEED CONTROL
11		———
12	L33 16RD/BK*	HIGH BEAM INDICATOR LAMP
13	L20 18LG*	HIGH BEAM FEED (DRL MODULE)
14	VS1 18LB	REAR WHEEL ANTI-LOCK (VALVE SWITCH)
15		———
16		———
17	G20 20VT/YL*	WATER TEMPERATURE SENDING UNIT
18	G6 20GY	OIL PRESSURE SWITCH
19	G60 20GY/OR*	OIL PRESSURE SENDING UNIT
20		———
21	D6 18LG	LEFT TURN SIGNAL LAMP
22	D5 18TN	RIGHT TURN SIGNAL LAMP
23	A3 16PK	BATTERY-HAZARD FLASHER
24		———
25		———
26	L7 18BK/YL*	PARKING LAMPS
27	B1 18VT	BACK-UP LAMP SWITCH
28	B2 18VT/BK*	BACK-UP LAMP SWITCH
29	V10 20BR	WINDSHIELD WIPER WASHER MOTOR
30	Z1 14DG/BK*	ELECTRIC FUEL PUMP
31	S2 16YL	STARTER RELAY IGNITION
32	H2 16DG/RD*	HORNS
33	L22 18BK/RD*	UNDERHOOD LAMP FEED
34	P51 20GY/BK*	BRAKE WARNING LAMP SWITCH
35	V5 18DG	WINDSHIELD WIPER SWITCH
36	V6 18DB	WINDSHIELD WIPER SWITCH
37	C20 16DB/OR*	AIR CONDITIONING DAMPED PRESSURE SWITCH
38	K3 20BK/PK*	CHECK ENGINE LAMP
39		———
40	V11 18BK/TN*	LOW WASHER FLUID SENSOR
41	D40 18VT/PK*	BRAKE SWITCH FEED
42	G29 20GY/PK*	EMISSIONS MAINTENANCE REMINDER LAMP
43	G7 20VT/OR*	SPEED SENSOR
44	U4 200R/WT*	OVERDRIVE LOCKOUT SWITCH
45	V3 18BR*	WINDSHIELD WIPER MOTOR
46	V4 18RD/YL*	WINDSHIELD WIPER MOTOR
47	J2 14DB	SPLICE-IGNITION RUN CIRCUIT
48	C13 12BK/RD*	ELECTRICALLY HEATED REAR WINDOW
49	J1 12RD	IGNITION SWITCH
50	J10 12PK*	IGNITION SWITCH

AB 57   50-WAY BULKHEAD DISCONNECT

CAVITY		INSTRUMENT PANEL CIRCUITS	SHEET
1	L1 12RD/WT*	BATTERY FEED	4,23,49
2	B7 12RD/PK*	BATTERY FEED (VAN CONVERSION)	4,13
3	DS1 14WT/BR*	REAR WHEEL ANTI-LOCK (DUMP SOLENOID)	35
4	IS1 16LG*	REAR WHEEL ANTI-LOCK (ISO SOLENOID)	35
5	L4 14VT*	HEADLAMPS-LOW BEAM	19,21,23
6	L3 14RD/OR*	HEADLAMPS-HIGH BEAMS	19,21,23
7	X30 20DB/RD*	SPEED CONTROL	18
8	X31 20BR/RD*	SPEED CONTROL	18
9	X33 20WT/LG*	SPEED CONTROL	18
10	X32 20YL/RD*	SPEED CONTROL	18
11		———	
12	L33 16RD/BK*	HIGH BEAM INDICATOR LAMP	20,22
13	L20 18LG/BK*	HIGH BEAM FEED (DRL MODULE)	22
14	VS1 18LB	REAR WHEEL ANTI-LOCK (VALVE SWITCH)	35
15		———	
16		———	
17	G20 20VT/YL*	TEMPERATURE-SEND/UNIT OR SWITCH	12
18	G6 20GY	OIL PRESSURE LAMP-SWITCH	12
19	G20 20GY/YL*	OIL PRESSURE GAUGE-SEND/UNIT	12
20		———	
21	D6 18LG	LEFT FRONT TURN SIGNAL LAMP	20,22,30
22	D5 18TN	RIGHT FRONT TURN SIGNAL LAMP	19,21,30
23	A3 16PK/WT*	HAZARD FLASHER	3,29
24		———	
25		———	
26	L7 18BK/YL*	PARKING LAMPS	20,22,53
27	B1 18VT	BACK-UP LAMP FEED	15
28	B2 18VT/BK*	BACK-UP LAMPS	15,53
29	V10 20BR	WINDSHIELD WIPER WASHER MOTOR	31,33
30	Z1 14DG/BK*	ELECTRIC FUEL PUMP	52
31	S2 14YL	IGNITION START	5,6
32	H2 16DG/RD*	HORNS	14
33	L22 18BK/RD*	UNDERHOOD LAMP	5
34	P51 20BK/GY*	BRAKE WARNING LAMP SWITCH	12
35	V5 18DG	WINDSHIELD WIPER SWITCH	32,33
36	V6 18DB	WINDSHIELD WIPER SWITCH	32,33
37	C20 16DB/OR*	A/C CLUTCH	41
38	K3 20BK/PK*	CHECK ENGINE LAMP	9
39		———	
40	V11 20BK/TN*	LOW WASHER FLUID SENSOR	31,36
41	D40 18VT/PK*	BRAKE SWITCH FEED	18
42	G29 22GY/PK*	EMR LAMP	9
43	G7 20VT/OR*	SPEED SENSOR	9
44	U4 200R/WT*	OVERDRIVE LOCKOUT SWITCH	16
45	V3 18BR*	WINDSHIELD WIPER LOW SPEED	32,33
46	V4 18RD/YL*	WINDSHIELD WIPER HIGH SPEED	32,33
47	J2 14DB	IGNITION-RUN & START	6,28
48	J2 14DB	IGNITION-RUN & START	
49	J1 12RD	IGNITION SWITCH (B1)	4,6
50	J10 12PK	IGNITION SWITCH (B3)	6

AB 58   50-WAY BULKHEAD DISCONNECT

1990

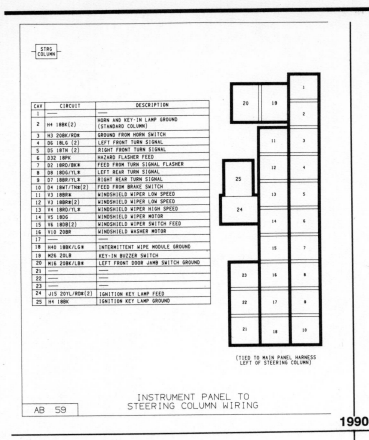

CAV	CIRCUIT	DESCRIPTION
1	—	—
2	H4 18BK(2)	HORN AND KEY-IN LAMP GROUND (STANDARD COLUMN)
3	H3 20BK/RD*	GROUND FROM HORN SWITCH
4	D6 18LG (2)	LEFT FRONT TURN SIGNAL
5	D5 18TN (2)	RIGHT FRONT TURN SIGNAL
6	D32 18PK	HAZARD FLASHER FEED
7	D2 18RD/BK*	FEED FROM TURN SIGNAL FLASHER
8	D8 18DG/YL*	LEFT REAR TURN SIGNAL
9	D7 18BR/YL*	RIGHT REAR TURN SIGNAL
10	D4 18WT/TN*(2)	FEED FROM BRAKE SWITCH
11	V3 18BR*	WINDSHIELD WIPER LOW SPEED
12	V3 18BR*(2)	WINDSHIELD WIPER LOW SPEED
13	V4 18RD/YL*	WINDSHIELD WIPER HIGH SPEED
14	V5 18DG	WINDSHIELD WIPER MOTOR
15	V6 18DB(2)	WINDSHIELD WIPER SWITCH FEED
16	V10 20BR	WINDSHIELD WASHER MOTOR
17	—	—
18	H40 18BK/LG*	INTERMITTENT WIPE MODULE GROUND
19	M26 20LB	KEY-IN BUZZER SWITCH
20	M16 20BK/LB*	LEFT FRONT DOOR JAMB SWITCH GROUND
21	—	—
22	—	—
23	—	—
24	J15 20YL/RD*(2)	IGNITION KEY LAMP FEED
25	H4 18BK	IGNITION KEY LAMP GROUND

(TIED TO MAIN PANEL HARNESS LEFT OF STEERING COLUMN)

INSTRUMENT PANEL TO
STEERING COLUMN WIRING

AB 59

1990

CAV	CIRCUIT	DESCRIPTION	SHEET
1	—	—	
2	H4 18BK(2)	HORN AND KEY-IN LAMP GROUND	14
3	H3 20BK/RD*	HORN SWITCH GROUND	14
4	D6 18LG	LEFT FRONT TURN SIGNAL	30
5	D5 18TN	RIGHT FRONT TURN SIGNAL	30
6	D32 18PK/BK*	HAZARD FLASHER FEED	29
7	D2 18RD/BK*	TURN SIGNAL FLASHER FEED	29
8	D8 18DG/YL*	LEFT REAR TURN SIGNAL	30
9	D7 18BR/YL*	RIGHT REAR TURN SIGNAL	30
10	D4 18WT/TN*	BRAKE SWITCH FEED	29

TURN SIGNAL
SWITCH
CONNECTOR

11	V3 18BR*	WINDSHIELD WIPER SWITCH	31,33
12	V3 18BR*	WINDSHIELD WIPER LOW SPEED	31,33
13	V4 18RD/YL*	WINDSHIELD WIPER HIGH SPEED	31,33
14	V5 18DG	INTERMITTENT WIPE MODULE	31,33
	V5 18DG	WINDSHIELD WIPER MOTOR	
15	V6 18DB(2)	WINDSHIELD WIPER SWITCH FEED	31,33
16	V10 20BR	INTERMITTENT WIPE MODULE	31,33
	V10 20BR	WINDSHIELD WASHER MOTOR	
17	—	—	
18	H40 18BK/LG*	INTERMITTENT WIPE MODULE GROUND	31

WINDSHIELD
WIPER SWITCH
CONNECTOR

19	M26 20LB	KEY-IN BUZZER SWITCH	26
20	M16 20BK/LB*	LEFT FRONT DOOR JAMB SWITCH GROUND	26

KEY-IN BUZZER
SWITCH CONNECTOR

21	—	—	
22	—	—	
23	—	—	

24	J15 20YL/RD*	IGNITION KEY LAMP FEED	25
25	H4 18BK	IGNITION KEY LAMP GROUND (TILT COLUMN)	25

IGNITION KEY LAMP
SWITCH CONNECTOR

(PLUGGED INTO STEERING COLUMN
25 WAY CONNECTOR)

STEERING COLUMN SWITCH
TO PANEL WIRING

AB 60

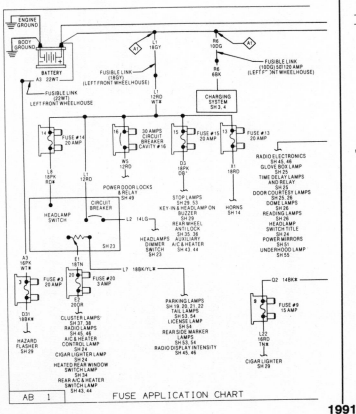

FUSE APPLICATION CHART

AB 1

1991

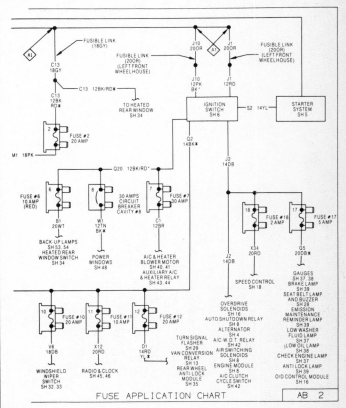

FUSE APPLICATION CHART

AB 2

6-135

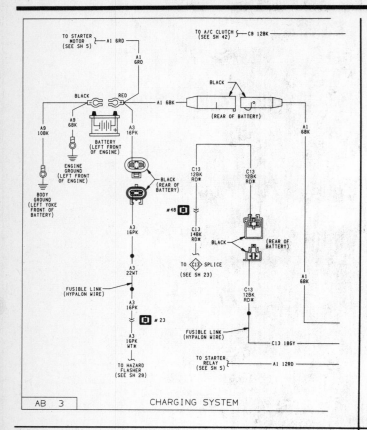

CHARGING SYSTEM | AB 3

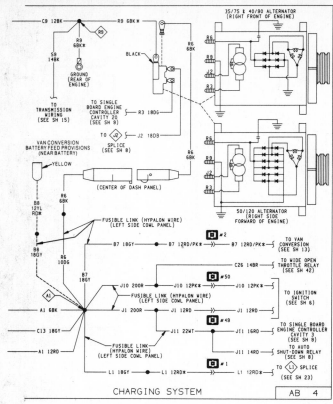

CHARGING SYSTEM   AB 4

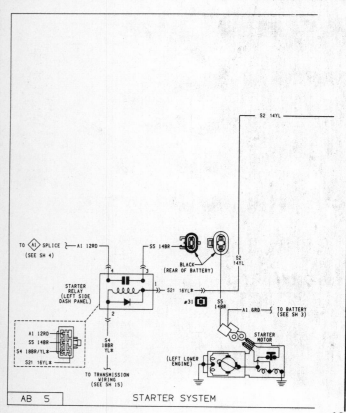

AB 5 | STARTER SYSTEM

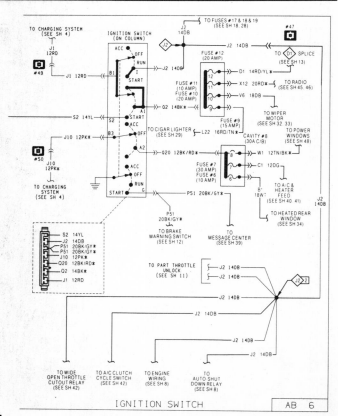

IGNITION SWITCH   AB 6

1991

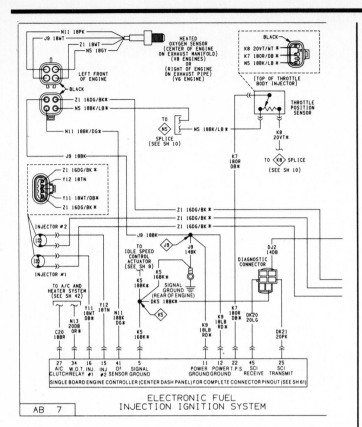

ELECTRONIC FUEL
Injection Ignition System

AB 7

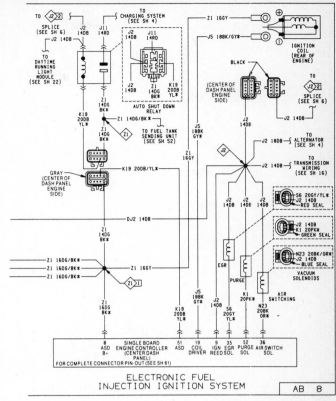

ELECTRONIC FUEL
Injection Ignition System

AB 8

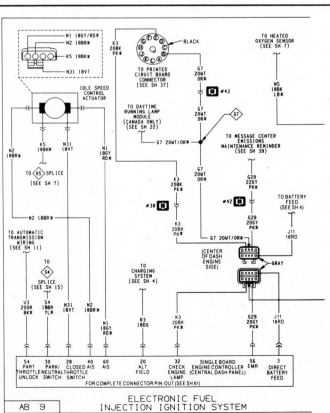

ELECTRONIC FUEL
Injection Ignition System

AB 9

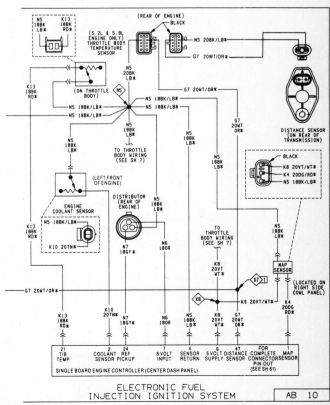

ELECTRONIC FUEL
Injection Ignition System

AB 10

1991

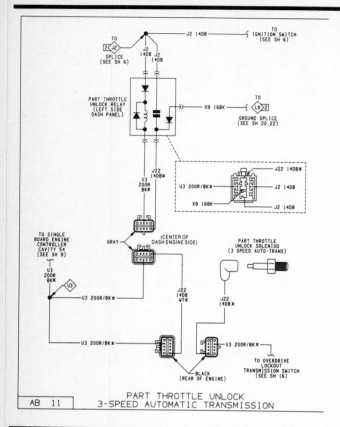

PART THROTTLE UNLOCK
3-SPEED AUTOMATIC TRANSMISSION

AB 11

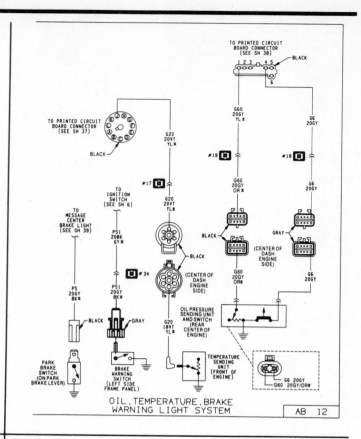

OIL, TEMPERATURE, BRAKE
WARNING LIGHT SYSTEM

AB 12

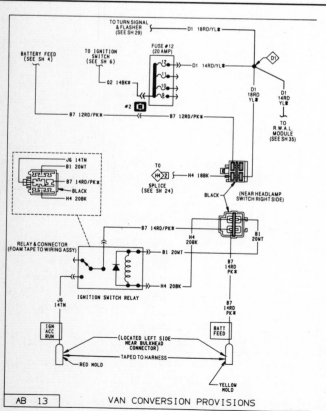

VAN CONVERSION PROVISIONS

AB 13

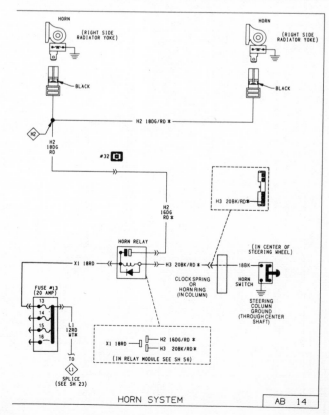

HORN SYSTEM

AB 14

1991

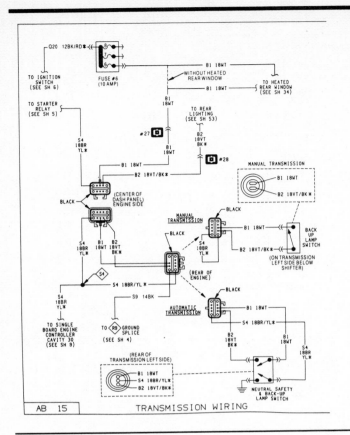

AB 15 — TRANSMISSION WIRING

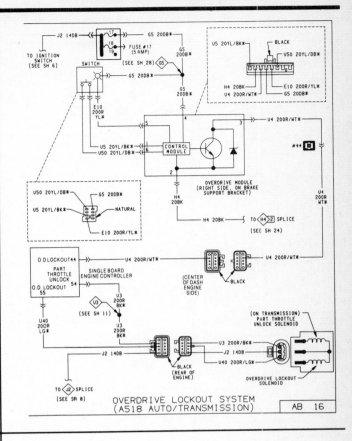

AB 16 — OVERDRIVE LOCKOUT SYSTEM (A518 AUTO/TRANSMISSION)

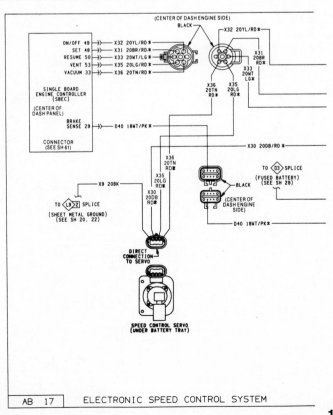

AB 17 — ELECTRONIC SPEED CONTROL SYSTEM

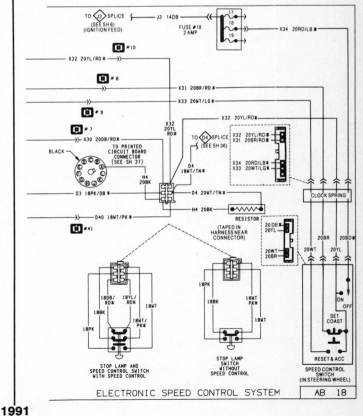

AB 18 — ELECTRONIC SPEED CONTROL SYSTEM

1991

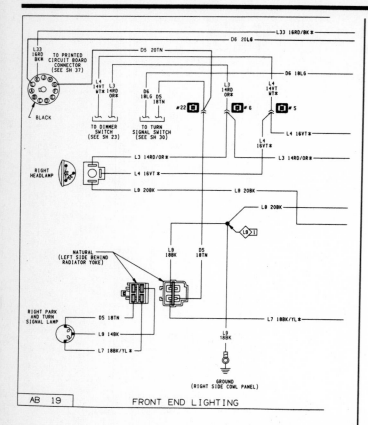

AB 19   FRONT END LIGHTING

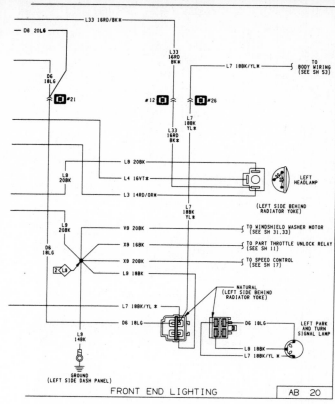

FRONT END LIGHTING   AB 20

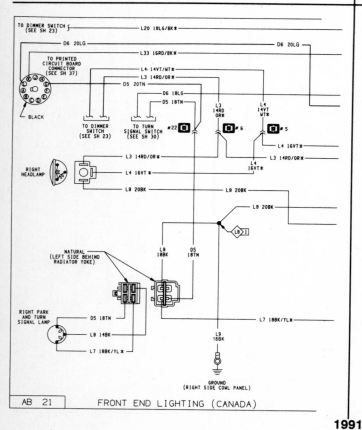

AB 21   FRONT END LIGHTING (CANADA)

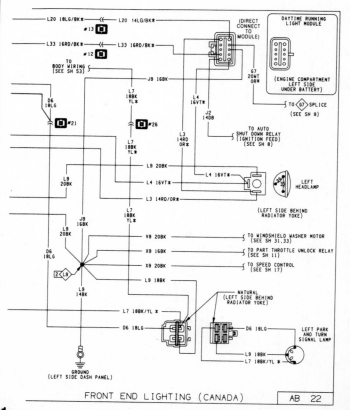

FRONT END LIGHTING (CANADA)   AB 22

1991

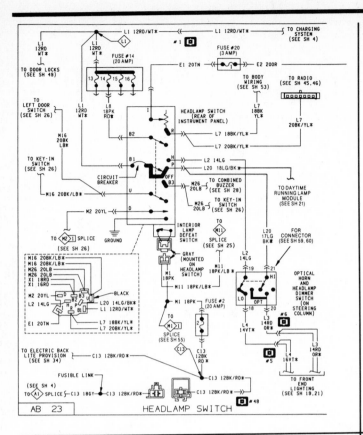

HEADLAMP SWITCH

AB 23

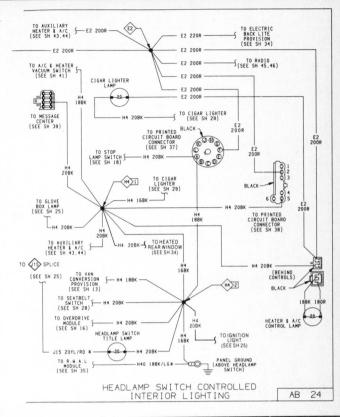

HEADLAMP SWITCH CONTROLLED
INTERIOR LIGHTING

AB 24

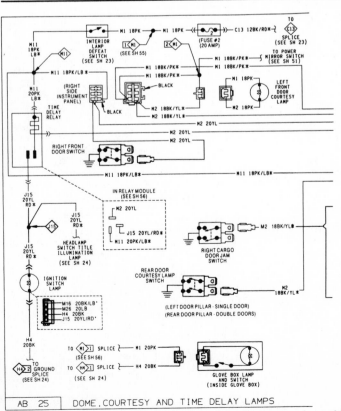

AB 25   DOME, COURTESY AND TIME DELAY LAMPS

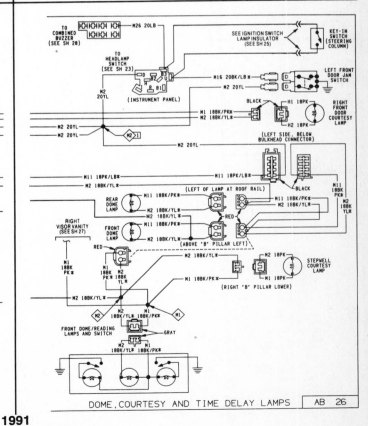

DOME, COURTESY AND TIME DELAY LAMPS   AB 26

1991

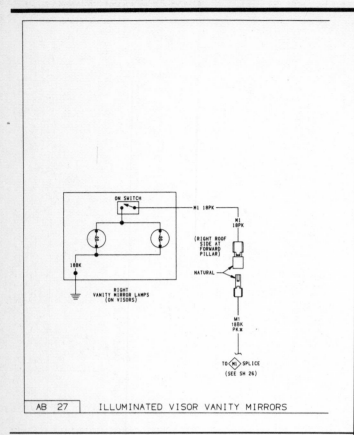

| AB 27 | ILLUMINATED VISOR VANITY MIRRORS |

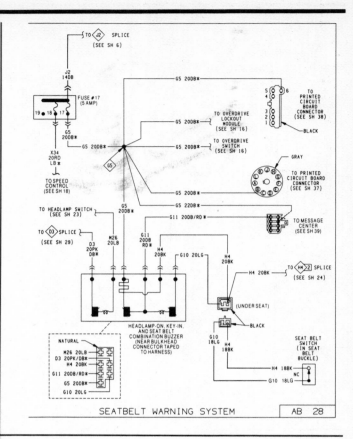

| SEATBELT WARNING SYSTEM | AB 28 |

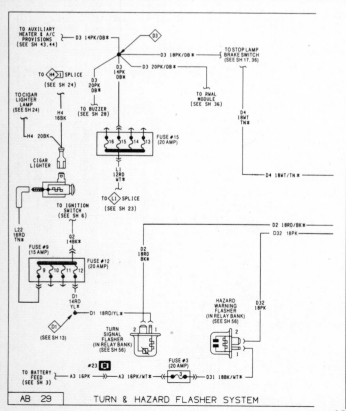

| AB 29 | TURN & HAZARD FLASHER SYSTEM |

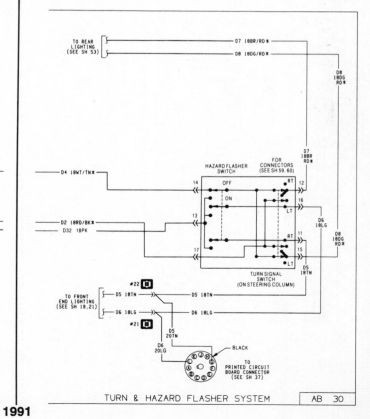

| TURN & HAZARD FLASHER SYSTEM | AB 30 |

1991

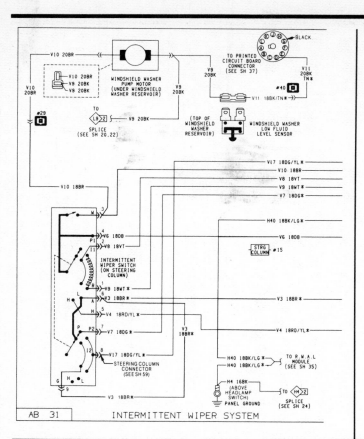

AB 31 — INTERMITTENT WIPER SYSTEM

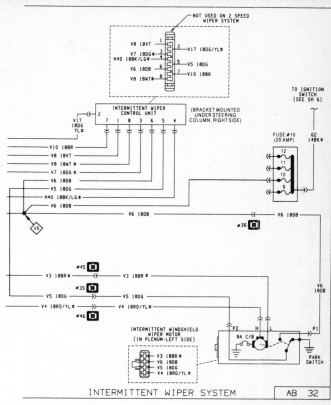

INTERMITTENT WIPER SYSTEM — AB 32

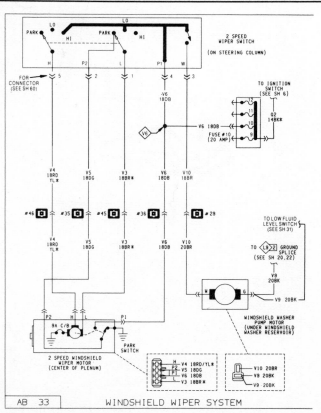

AB 33 — WINDSHIELD WIPER SYSTEM

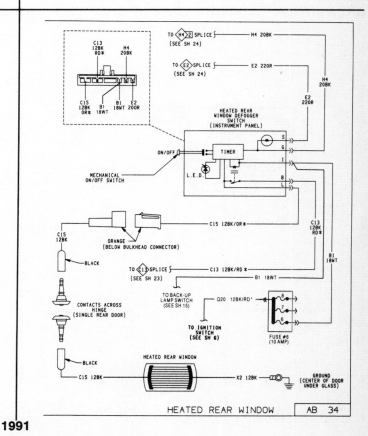

HEATED REAR WINDOW — AB 34

1991

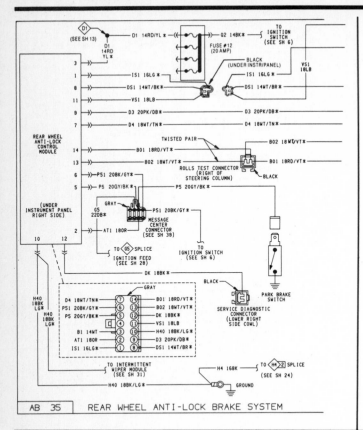

REAR WHEEL ANTI-LOCK BRAKE SYSTEM   AB 35

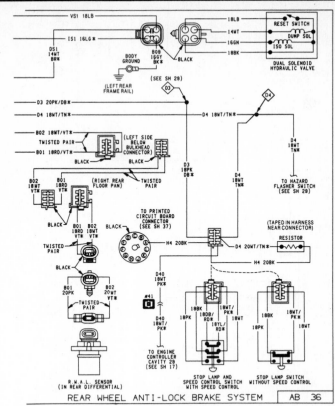

REAR WHEEL ANTI-LOCK BRAKE SYSTEM   AB 36

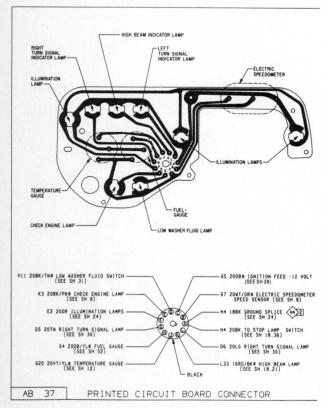

AB 37   PRINTED CIRCUIT BOARD CONNECTOR

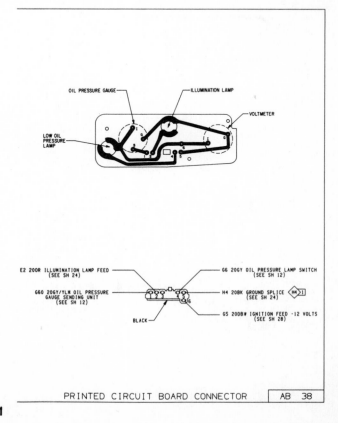

PRINTED CIRCUIT BOARD CONNECTOR   AB 38

1991

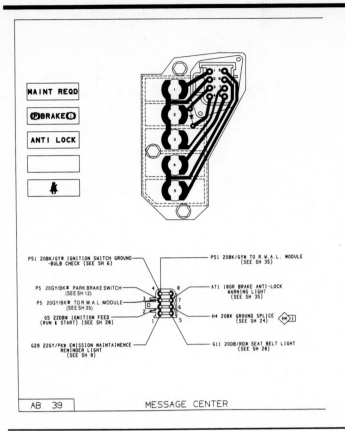

MESSAGE CENTER

AB 39

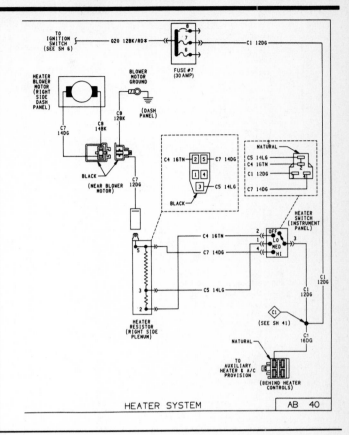

HEATER SYSTEM

AB 40

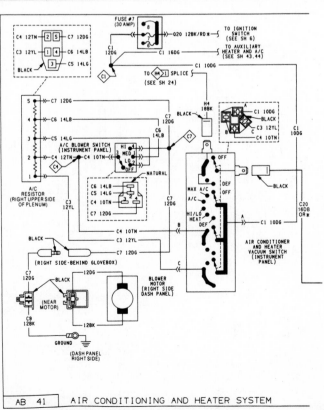

AB 41

AIR CONDITIONING AND HEATER SYSTEM

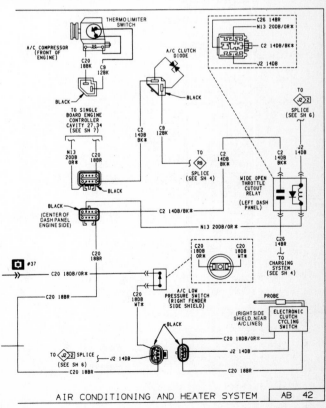

AIR CONDITIONING AND HEATER SYSTEM

AB 42

1991

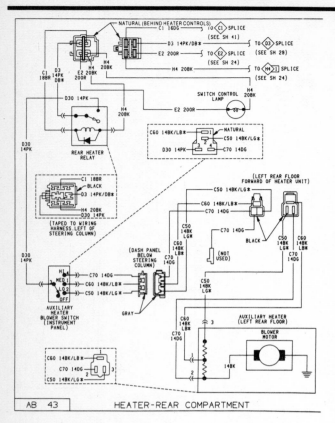

AB 43 — HEATER-REAR COMPARTMENT

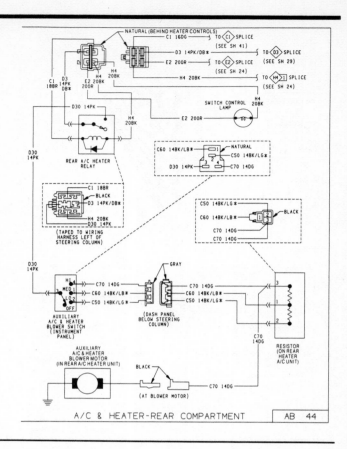

A/C & HEATER-REAR COMPARTMENT — AB 44

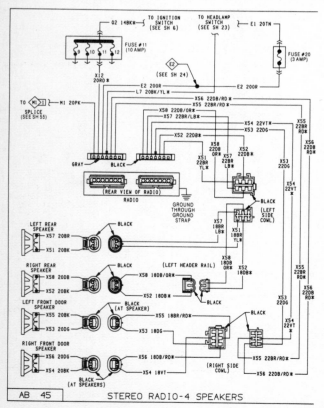

AB 45 — STEREO RADIO-4 SPEAKERS

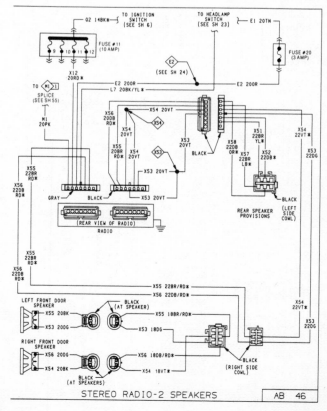

STEREO RADIO-2 SPEAKERS — AB 46

1991

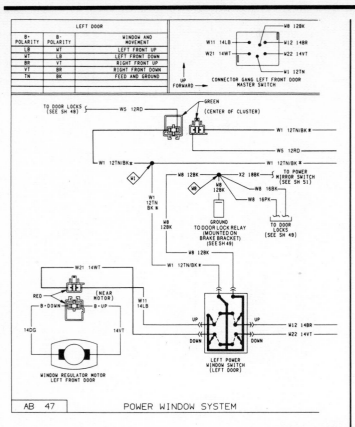

LEFT DOOR		
B+ POLARITY	B- POLARITY	WINDOW AND MOVEMENT
LB	WT	LEFT FRONT UP
WT	LB	LEFT FRONT DOWN
BR	VT	RIGHT FRONT UP
VT	BR	RIGHT FRONT DOWN
TN	BK	FEED AND GROUND

POWER WINDOW SYSTEM — AB 47

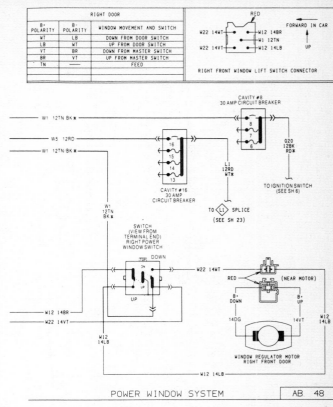

RIGHT DOOR		
B+ POLARITY	B- POLARITY	WINDOW MOVEMENT AND SWITCH
WT	LB	DOWN FROM DOOR SWITCH
LB	WT	UP FROM DOOR SWITCH
VT	BR	DOWN FROM MASTER SWITCH
BR	VT	UP FROM MASTER SWITCH
TN	—	FEED

POWER WINDOW SYSTEM — AB 48

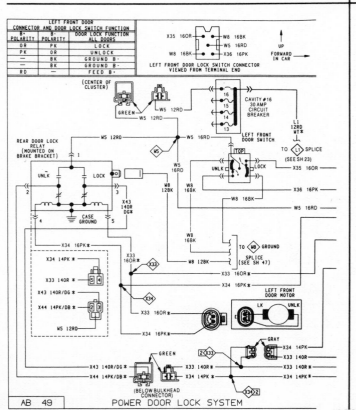

LEFT FRONT DOOR CONNECTOR AND DOOR LOCK SWITCH FUNCTION		
B+ POLARITY	B- POLARITY	DOOR LOCK FUNCTION ALL DOORS
OR	PK	LOCK
PK	OR	UNLOCK
—	BK	GROUND B-
—	BK	GROUND B-
RD	—	FEED B+

POWER DOOR LOCK SYSTEM — AB 49

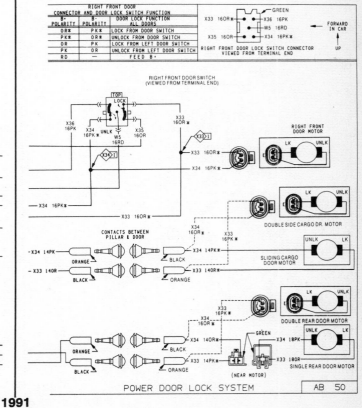

RIGHT FRONT DOOR CONNECTOR AND DOOR LOCK SWITCH FUNCTION		
B+ POLARITY	B- POLARITY	DOOR LOCK FUNCTION ALL DOORS
OR*	PK*	LOCK FROM DOOR SWITCH
PK*	OR*	UNLOCK FROM DOOR SWITCH
OR	PK	LOCK FROM LEFT DOOR SWITCH
PK	OR	UNLOCK FROM LEFT DOOR SWITCH
RD	—	FEED B+

POWER DOOR LOCK SYSTEM — AB 50

1991

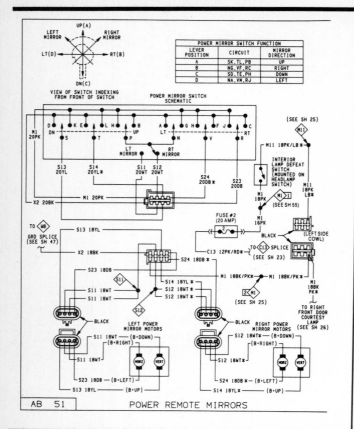

AB 51 — POWER REMOTE MIRRORS

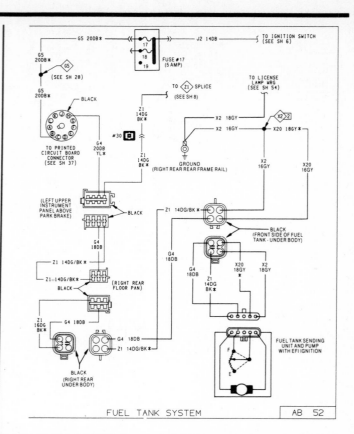

AB 52 — FUEL TANK SYSTEM

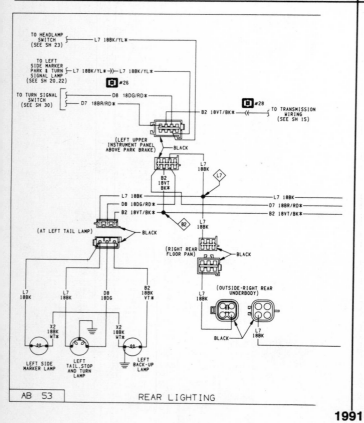

AB 53 — REAR LIGHTING

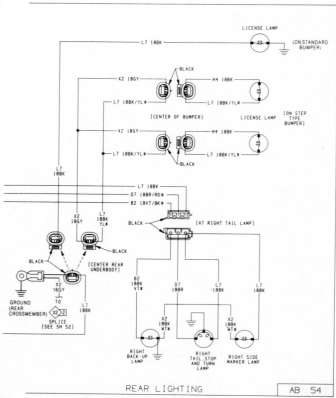

AB 54 — REAR LIGHTING

1991

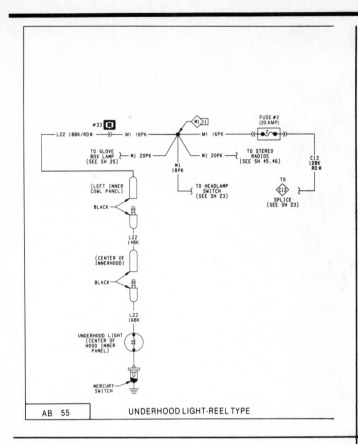

AB 55 · UNDERHOOD LIGHT-REEL TYPE

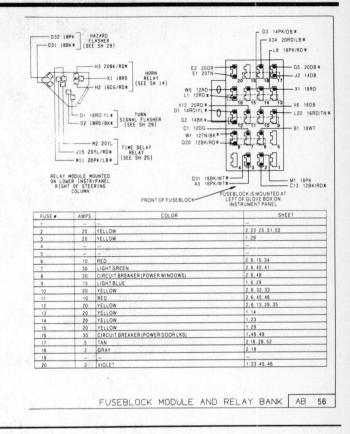

FUSE #	AMPS	COLOR	SHEET
1	—	—	—
2	20	YELLOW	2, 23, 25, 51, 55
3	20	YELLOW	1, 29
4	—	—	—
5	—	—	—
6	10	RED	2, 6, 15, 34
7	30	LIGHT GREEN	2, 6, 40, 41
8	30	CIRCUIT BREAKER (POWER WINDOWS)	2, 6, 48
9	15	LIGHT BLUE	1, 6, 29
10	20	YELLOW	2, 6, 32, 33
11	10	RED	2, 6, 45, 46
12	20	YELLOW	2, 6, 13, 29, 35
13	20	YELLOW	1, 14
14	20	YELLOW	1, 23
15	20	YELLOW	1, 29
16	30	CIRCUIT BREAKER (POWER DOOR LKS)	1, 48, 49
17	5	TAN	2, 16, 28, 52
18	2	GRAY	2, 18
19	—	—	—
20	3	VIOLET	1, 23, 45, 46

FUSEBLOCK MODULE AND RELAY BANK · AB 56

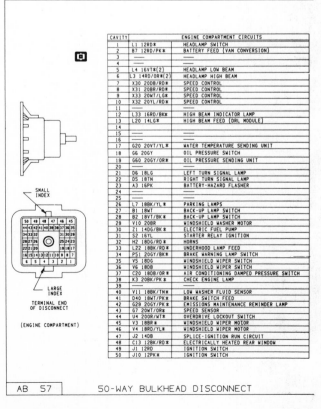

CAVITY		ENGINE COMPARTMENT CIRCUITS
1	L1 12RD*	HEADLAMP SWITCH
2	B7 12RD/PK*	BATTERY FEED (VAN CONVERSION)
3	—	—
4	—	—
5	L4 16VT*(2)	HEADLAMP LOW BEAM
6	L3 14RD/OR*(2)	HEADLAMP HIGH BEAM
7	X30 20DB/RD*	SPEED CONTROL
8	X31 20BR/RD*	SPEED CONTROL
9	X33 20WT/LG*	SPEED CONTROL
10	X32 20YL/RD*	SPEED CONTROL
11	—	—
12	L33 16RD/BK*	HIGH BEAM INDICATOR LAMP
13	L20 14LG*	HIGH BEAM FEED (DRL MODULE)
14	—	—
15	—	—
16	—	—
17	G20 20VT/YL*	WATER TEMPERATURE SENDING UNIT
18	G6 20GY	OIL PRESSURE SWITCH
19	G60 20GY/OR*	OIL PRESSURE SENDING UNIT
20	—	—
21	D6 18LG	LEFT TURN SIGNAL LAMP
22	D5 18TN	RIGHT TURN SIGNAL LAMP
23	A3 16PK	BATTERY-HAZARD FLASHER
24	—	—
25	—	—
26	L7 18BK/YL*	PARKING LAMPS
27	B1 18WT	BACK-UP LAMP SWITCH
28	B2 18VT/BK*	BACK-UP LAMP SWITCH
29	V10 20BR	WINDSHIELD WASHER MOTOR
30	Z1 14DG/BK*	ELECTRIC FUEL PUMP
31	S2 16YL	STARTER RELAY IGNITION
32	H2 18DG/RD*	HORNS
33	L22 18BK/RD*	UNDERHOOD LAMP FEED
34	PS1 20GY/BK*	BRAKE WARNING LAMP SWITCH
35	V5 18DG	WINDSHIELD WIPER SWITCH
36	V6 18DB	WINDSHIELD WIPER SWITCH
37	C20 18DB/OR*	AIR CONDITIONING DAMPED PRESSURE SWITCH
38	K3 20BK/PK*	CHECK ENGINE LAMP
39	—	—
40	V11 18BK/WT*	LOW WASHER FLUID SENSOR
41	D40 18WT/PK*	BRAKE SWITCH FEED
42	G29 20GY/PK*	EMISSIONS MAINTENANCE REMINDER LAMP
43	G7 20WT/OR*	SPEED SENSOR
44	U4 20OR/WT*	OVERDRIVE LOCKOUT SWITCH
45	V3 18BR*	WINDSHIELD WIPER MOTOR
46	V4 18RD/YL*	WINDSHIELD WIPER MOTOR
47	J2 14DB	SPLICE-IGNITION RUN CIRCUIT
48	C13 12BK/RD*	ELECTRICALLY HEATED REAR WINDOW
49	J1 12RD	IGNITION SWITCH
50	J10 12PK*	IGNITION SWITCH

AB 57 · 50-WAY BULKHEAD DISCONNECT

CAVITY		INSTRUMENT PANEL CIRCUITS	SHEET
1	L1 12RD/WT*	BATTERY FEED	4, 23
2	B7 12RD/PK*	BATTERY FEED (VAN CONVERSION)	4, 13
3	—	—	—
4	—	—	—
5	L4 14VT*	HEADLAMPS-LOW BEAM	19, 21, 23
6	L3 14RD/OR*	HEADLAMPS-HIGH BEAMS	19, 21, 23
7	X30 20DB/RD*	SPEED CONTROL	18
8	X31 20BR/RD*	SPEED CONTROL	18
9	X33 20WT/LG*	SPEED CONTROL	18
10	X32 20YL/RD*	SPEED CONTROL	18
11	—	—	—
12	L33 16RD/BK*	HIGH BEAM INDICATOR LAMP	20, 22
13	L20 18LG/BK*	HIGH BEAM FEED (DRL MODULE)	22
14	—	—	—
15	—	—	—
16	—	—	—
17	G20 20VT/YL*	TEMPERATURE-SEND/UNIT OR SWITCH	12
18	G6 20GY	OIL PRESSURE LAMP-SWITCH	12
19	G20 20GY/YL*	OIL PRESSURE GAUGE-SEND/UNIT	12
20	—	—	—
21	D6 18LG	LEFT FRONT TURN SIGNAL LAMP	20, 22, 30
22	D5 18TN	RIGHT FRONT TURN SIGNAL LAMP	19, 21, 30
23	A3 16PK/WT*	HAZARD FLASHER	3, 29
24	—	—	—
25	—	—	—
26	L7 18BK/YL*	PARKING LAMPS	20, 22, 53
27	B1 18WT	BACK-UP LAMP FEED	15
28	B2 18VT/BK*	BACK-UP LAMPS	15, 53
29	V10 18BR	WINDSHIELD WASHER MOTOR	31, 33
30	Z1 14DG/BK*	ELECTRIC FUEL PUMP	52
31	S2 14YL*	IGNITION START	5, 6
32	H2 16DG/RD*	HORNS	14
33	L22 18BK/RD*	UNDERHOOD LAMP	55
34	PS1 20BK/GY*	BRAKE WARNING LAMP SWITCH	12
35	V5 18DG	WINDSHIELD WIPER SWITCH	32, 33
36	V6 18DB	WINDSHIELD WIPER SWITCH	32, 33
37	C20 16DB/PK*	A/C CLUTCH	41
38	K3 20BK/PK*	CHECK ENGINE LAMP	9
39	—	—	—
40	V11 20BK/TN*	LOW WASHER FLIUD SENSOR	31, 36
41	D40 18WT/PK*	BRAKE SWITCH FEED	18, 36
42	G29 20GY/PK*	EMR LAMP	9
43	G7 20WT/OR*	SPEED SENSOR	9
44	U4 20OR/WT*	OVERDRIVE LOCKOUT SWITCH	16
45	V3 18BR*	WINDSHIELD WIPER LOW SPEED	32, 33
46	V4 18RD/YL*	WINDSHIELD WIPER HIGH SPEED	32, 33
47	J2 14DB	IGNITION-RUN & START	6, 28
48	C13 12BK/RD*	ELECTRIC BACK LITE (EBL)	3, 23
49	J1 12RD	IGNITION SWITCH (B1)	4, 6
50	J10 12PK*	IGNITION SWITCH (B3)	6

50-WAY BULKHEAD DISCONNECT · AB 58

STEERING COLUMN INSULATOR
AS VIEWED FROM THE WIRE END

CAV	CIRCUIT	DESCRIPTION
1	V9 18WT*	INTERMITTENT WIPE MODULE
2	V8 18VT	INTERMITTENT WIPE MODULE
3	V10 18BR	WINDSHIELD WASHER MOTOR
3	V10 18BR	INTERMITTENT WIPE MODULE
4	V6 18DB	WIPER SWITCH FEED
5	V4 18RD/YL*	WIPER MOTOR - HIGH SPEED
6	V3 18BR*	WIPER MOTOR - LOW SWITCH
6	V3 18BR*	WIPER SWITCH
7	V7 18DG*	INTERMITTENT WIPE MODULE
8	V17 18DG/YL*	INTERMITTENT WIPE MODULE
9	V3A 18BR*	WIPER SWITCH
10	—	—
11	D5 18TN	RIGHT FRONT TURN SIGNAL
12	D7 18BR/RD*	RIGHT REAR TURN SIGNAL
13	D32 18PK	HAZARD FLASHER FEED
14	D4 18WT/TN*	STOP LAMP FEED
14	D4 18WT/TN*	STOP LAMP FEED
15	D8 18DG/RD*	LEFT REAR TURN SIGNAL
16	D6 18LG	LEFT FRONT TURN SIGNAL
17	D2 18RD/BK*	TURN SIGNAL FLASHER FEED
18	L4 14VT*	LOW BEAM - HEADLAMPS
19	L2 14LG	HEADLAMP FEED
20	L3 14RD/OR*	HIGH BEAM - HEADLAMPS
21	L20 14LG/BK*	HEADLAMP - OPT HORN FEED
22	—	—
23	—	—
24	—	—

STEERING COLUMN CONNECTOR
WITH INTERMITTENT WIPERS (PULSE WIPE)

AB 59

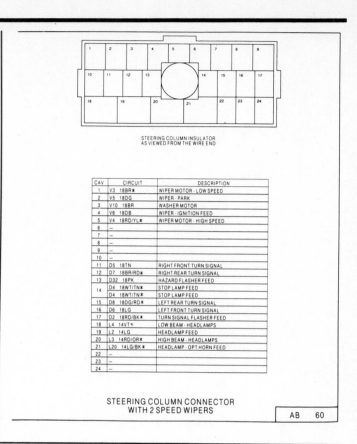

STEERING COLUMN INSULATOR
AS VIEWED FROM THE WIRE END

CAV	CIRCUIT	DESCRIPTION
1	V3 18BR*	WIPER MOTOR - LOW SPEED
2	V5 18DG	WIPER - PARK
3	V10 18BR	WASHER MOTOR
4	V6 18DB	WIPER - IGNITION FEED
5	V4 18RD/YL*	WIPER MOTOR - HIGH SPEED
6	—	—
7	—	—
8	—	—
9	—	—
10	—	—
11	D5 18TN	RIGHT FRONT TURN SIGNAL
12	D7 18BR/RD*	RIGHT REAR TURN SIGNAL
13	D32 18PK	HAZARD FLASHER FEED
14	D4 18WT/TN*	STOP LAMP FEED
14	D4 18WT/TN*	STOP LAMP FEED
15	D8 18DG/RD*	LEFT REAR TURN SIGNAL
16	D6 18LG	LEFT FRONT TURN SIGNAL
17	D2 18RD/BK*	TURN SIGNAL FLASHER FEED
18	L4 14VT*	LOW BEAM - HEADLAMPS
19	L2 14LG	HEADLAMP FEED
20	L3 14RD/OR*	HIGH BEAM - HEADLAMPS
21	L20 14LG/BK*	HEADLAMP - OPT HORN FEED
22	—	—
23	—	—
24	—	—

STEERING COLUMN CONNECTOR
WITH 2 SPEED WIPERS

AB 60

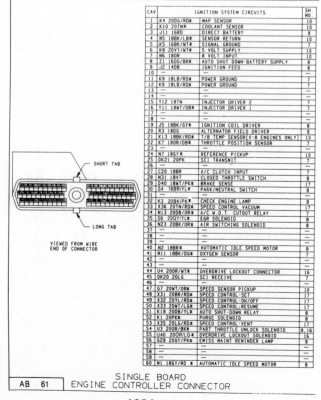

SHORT TAB

LONG TAB

VIEWED FROM WIRE
END OF CONNECTOR

CAV	IGNITION SYSTEM CIRCUITS		SH. NO.
1	K4 20DG/RD*	MAP SENSOR	10
2	K10 20TN*	COOLANT SENSOR	10
3	J11 16RD	DIRECT BATTERY	9
4	N5 18BK/LB*	SENSOR RETURN	10
5	K5 16BK/WT*	SIGNAL GROUND	7
6	K6 20VT/WT*	5 VOLT SUPPLY	10
7	N6 18OR	8 VOLT INPUT	10
8	Z1 16DG/BK*	AUTO SHUT DOWN - BATTERY SUPPLY	8
9	J2 14DB	IGNITION FEED	8
10	—	—	—
11	K9 18LB/RD*	POWER GROUND	7
12	K9 18LB/RD*	POWER GROUND	7
13	—	—	—
14	—	—	—
15	Y12 18TN	INJECTOR DRIVER 2	7
16	Y11 18WT/DB*	INJECTOR DRIVER 1	7
17	—	—	—
18	—	—	—
19	J5 18BK/GY*	IGNITION COIL DRIVER	8
20	R3 18DG	ALTERNATOR FIELD DRIVER	9
21	K13 18BK/RD*	T/B TEMP SENSOR (V-8 ENGINES ONLY)	10
22	K7 18OR/DB*	THROTTLE POSITION SENSOR	7
23	—	—	—
24	N7 18GY*	REFERENCE PICKUP	10
25	DK21 20PK	SCI TRANSMIT	7
26	—	—	—
27	C20 18BR	A/C CLUTCH INPUT	7
28	N31 18VT	CLOSED THROTTLE SWITCH	9
29	D40 18WT/*	BRAKE SENSE	17
30	S4 18BR/YL*	PARK/NEUTRAL SWITCH	9
31	—	—	—
32	K3 20BK/PK*	CHECK ENGINE LAMP	9
33	X36 20TN/RD*	SPEED CONTROL VACUUM	17
34	N13 20DB/OR*	A/C W.O.T. CUTOUT RELAY	7
35	S6 20GY/YL*	EGR SOLENOID	8
36	N23 20BK/OR*	AIR SWITCHING SOLENOID	8
37	—	—	—
38	—	—	—
39	—	—	—
40	N2 18BR*	AUTOMATIC IDLE SPEED MOTOR	9
41	N11 18BK/DG*	OXYGEN SENSOR	7
42	—	—	—
43	—	—	—
44	U4 20OR/WT*	OVERDRIVE LOCKOUT CONNECTOR	16
45	DK20 20LG	SCI RECEIVE	7
46	—	—	—
47	G7 20WT/OR*	SPEED SENSOR PICKUP	10
48	X31 20BR/RD*	SPEED CONTROL - SET	17
49	X32 20YL/RD*	SPEED CONTROL - ON/OFF	17
50	X33 20WT/LG*	SPEED CONTROL - RESUME	17
51	K18 20DB/YL*	AUTO SHUT - DOWN RELAY	8
52	K1 20PK	PURGE SOLENOID	8
53	X35 20LG/RD*	SPEED CONTROL - VENT	17
54	U3 20OR/BK*	PART THROTTLE UNLOCK SOLENOID	9,16
55	U40 20OR/LG*	OVERDRIVE LOCKOUT SOLENOID	16
56	G29 20GY/PK*	EMISS MAINT REMINDER LAMP	9
57	—	—	—
58	—	—	—
59	—	—	—
60	N1 18GY/RD *	AUTOMATIC IDLE SPEED MOTOR	9

AB 61

SINGLE BOARD
ENGINE CONTROLLER CONNECTOR

1991

Drive Train

QUICK REFERENCE INDEX

GENERAL INDEX

UNDERSTANDING THE MANUAL TRANSMISSION

Because of the way an internal combustion engine breathes, it can produce torque, or twisting force, only within a narrow speed range. Most modern, overhead valve engines must turn at about 2,500 rpm to produce their peak torque. By 4,500 rpm they are producing so little torque that continued increases in engine speed produce no power increases.

The torque peak on overhead camshaft engines is, generally, much higher, but much narrower.

The manual transmission and clutch are employed to vary the relationship between engine speed and the speed of the wheels so that adequate engine power can be produced under all circumstances. The clutch allows engine torque to be applied to the transmission input shaft gradually, due to mechanical slippage. The vehicle can, consequently, be started smoothly from a full stop.

The transmission changes the ratio between the rotating speeds of the engine and the wheels by the use of gears. 4-speed or 5-speed transmissions are most common. The lower gears allow full engine power to be applied to the rear wheels during acceleration at low speeds.

The clutch drive plate is a thin disc, the center of which is splined to the transmission input shaft. Both sides of the disc are covered with a layer of material which is similar to brake lining and which is capable of allowing slippage without roughness or excessive noise.

The clutch cover is bolted to the engine flywheel and incorporates a diaphragm spring which provides the pressure to engage the clutch. The cover also houses the pressure plate. The driven disc is sandwiched between the pressure plate and the smooth surface of the flywheel when the clutch pedal is released, thus forcing it to turn at the same speed as the engine crankshaft.

The transmission contains a mainshaft which passes all the way through the transmission, from the clutch to the driveshaft. This shaft is separated at one point, so that front and rear portions can turn at different speeds.

Power is transmitted by a countershaft in the lower gears and reverse. The gears of the countershaft mesh with gears on the mainshaft, allowing power to be carried from one to the other. All the countershaft gears are integral with that shaft, while several of the mainshaft gears can either rotate independently

Troubleshooting the Manual Transmission

Problem	Cause	Solution
Transmission shifts hard	• Clutch adjustment incorrect • Clutch linkage or cable binding • Shift rail binding	• Adjust clutch • Lubricate or repair as necessary • Check for mispositioned selector arm roll pin, loose cover bolts, worn shift rail bores, worn shift rail, distorted oil seal, or extension housing not aligned with case. Repair as necessary.
	• Internal bind in transmission caused by shift forks, selector plates, or synchronizer assemblies • Clutch housing misalignment • Incorrect lubricant • Block rings and/or cone seats worn	• Remove, dissemble and inspect transmission. Replace worn or damaged components as necessary. • Check runout at rear face of clutch housing • Drain and refill transmission • Blocking ring to gear clutch tooth face clearance must be 0.030 inch or greater. If clearance is correct it may still be necessary to inspect blocking rings and cone seats for excessive wear. Repair as necessary.
Gear clash when shifting from one gear to another	• Clutch adjustment incorrect • Clutch linkage or cable binding • Clutch housing misalignment • Lubricant level low or incorrect lubricant • Gearshift components, or synchronizer assemblies worn or damaged	• Adjust clutch • Lubricate or repair as necessary • Check runout at rear of clutch housing • Drain and refill transmission and check for lubricant leaks if level was low. Repair as necessary. • Remove, disassemble and inspect transmission. Replace worn or damaged components as necessary.

Troubleshooting the Manual Transmission

Problem	Cause	Solution
Transmission noisy	• Lubricant level low or incorrect lubricant	• Drain and refill transmission. If lubricant level was low, check for leaks and repair as necessary.
	• Clutch housing-to-engine, or transmission-to-clutch housing bolts loose	• Check and correct bolt torque as necessary
	• Dirt, chips, foreign material in transmission	• Drain, flush, and refill transmission
	• Gearshift mechanism, transmission gears, or bearing components worn or damaged	• Remove, disassemble and inspect transmission. Replace worn or damaged components as necessary.
	• Clutch housing misalignment	• Check runout at rear face of clutch housing
Jumps out of gear	• Clutch housing misalignment	• Check runout at rear face of clutch housing
	• Gearshift lever loose	• Check lever for worn fork. Tighten loose attaching bolts.
	• Offset lever nylon insert worn or lever attaching nut loose	• Remove gearshift lever and check for loose offset lever nut or worn insert. Repair or replace as necessary.
	• Gearshift mechanism, shift forks, selector plates, interlock plate, selector arm, shift rail, detent plugs, springs or shift cover worn or damaged	• Remove, disassemble and inspect transmission cover assembly. Replace worn or damaged components as necessary.
	• Clutch shaft or roller bearings worn or damaged	• Replace clutch shaft or roller bearings as necessary
Jumps out of gear (cont.)	• Gear teeth worn or tapered, synchronizer assemblies worn or damaged, excessive end play caused by worn thrust washers or output shaft gears	• Remove, disassemble, and inspect transmission. Replace worn or damaged components as necessary.
	• Pilot bushing worn	• Replace pilot bushing
Will not shift into one gear	• Gearshift selector plates, interlock plate, or selector arm, worn, damaged, or incorrectly assembled	• Remove, disassemble, and inspect transmission cover assembly. Repair or replace components as necessary.
	• Shift rail detent plunger worn, spring broken, or plug loose	• Tighten plug or replace worn or damaged components as necessary
	• Gearshift lever worn or damaged	• Replace gearshift lever
	• Synchronizer sleeves or hubs, damaged or worn	• Remove, disassemble and inspect transmission. Replace worn or damaged components.

Troubleshooting the Manual Transmission

Problem	Cause	Solution
Locked in one gear—cannot be shifted out	• Shift rail(s) worn or broken, shifter fork bent, setscrew loose, center detent plug missing or worn	• Inspect and replace worn or damaged parts
	• Broken gear teeth on countershaft gear, clutch shaft, or reverse idler gear	• Inspect and replace damaged part
	Gearshift lever broken or worn, shift mechanism in cover incorrectly assembled or broken, worn damaged gear train components	• Disassemble transmission. Replace damaged parts or assemble correctly.

of the shaft or be locked to it. Shifting from one gear to the next causes one of the gears to be freed from rotating with the shaft and locks another to it. Gears are locked and unlocked by internal dog clutches which slide between the center of the gear and the shaft. The forward gears usually employ synchronizers; friction members which smoothly bring gear and shaft to the same speed before the toothed dog clutches are engaged.

The clutch is operating properly if:

1. It will stall the engine when released with the vehicle held stationary.

2. The shift lever can be moved freely between 1st and reverse gears when the vehicle is stationary and the clutch disengaged.

A clutch pedal free-play adjustment is incorporated in the linkage. If there is about 1-2 in. (25-50mm) of motion before the pedal begins to release the clutch, it is adjusted properly. Inadequate free-play wears all parts of the clutch releasing mechanisms and may cause slippage. Excessive free-play may cause inadequate release and hard shifting of gears.

Some clutches use a hydraulic system in place of mechanical linkage. If the clutch fails to release, fill the clutch master cylinder with fluid to the proper level and pump the clutch pedal to fill the system with fluid. Bleed the system in the same way as a brake system. If leaks are located, tighten loose connections or overhaul the master or slave cylinder as necessary.

Manual Transmission

SHIFT LINKAGE ADJUSTMENT

Overdrive-4 (NP435)

1. Place the floorshift lever in Neutral. Insert a ¼ in. (6mm) drill bit through the bottom of the shifter to hold the levers in place.

2. Detach the shift rods. Make sure that the three transmission levers are in their Neutral detents.

3. Adjust the shift rods to make the length exactly right to fit into the transmission levers. Start with the 1st-2nd shift rod. It may be necessary to remove the clip at the shifter end of the rod to rotate this rod.

4. Replace the washers and the clips.

5. Remove the drill bit and check the shifting action.

NP2500 (5 Speed)

The NP2500 transmission shift mechanism consists of a single unit, top mounted shift lever and internal mounted shift rails. Since the shift mechanism is internally mounted, no linkage adjustment is necessary.

Shift Linkage and Shifter Handle

REMOVAL AND INSTALLATION

Overdrive-4 (NP435)

1. Disconnect the negative battery ground cable.

2. Remove the shift boot retaining screws and slide the boot up the shift lever.

3. Remove the shift lever by inserting a 0.010 in. (0.25mm) feeler gauge blade down the left (driver's) side of the shift lever mounting socket (just above the linkage pivot shift unit) to depress the pin detents that secure the lever.

4. Remove the retaining clips, washers and control rods from the shift unit levers.

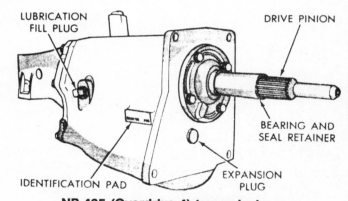

NP-435 (Overdrive 4) transmission

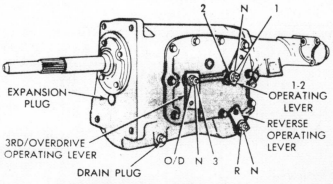

NP-435 (Overdrive 4) transmission

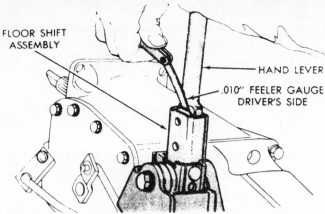

Removing the shift lever — NP-435

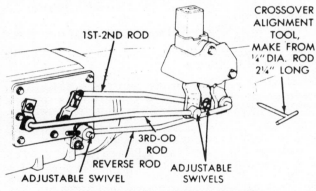

Shift linkage adjustment — NP-435

5. Remove the two bolts and washers that secure the shift unit to the side of the extension and remove the unit.

6. Service as required. Fasten the shift unit to the side of the extension housing and secure it with the two mounting bolts and washers. Tighten the mounting bolts to 30 ft.lbs. (41 N.m).

7. Install the shift rods, washers and clips.

8. Install the shift lever and secure the shifter boot.

NP2500 5 Speed

The gearshift lever is a two-piece design. The upper part of the lever is threaded to the transmission stub lever. The upper lever can be removed for service without having to remove the entire shift lever assembly or the transmission.

If the upper lever must be serviced, remove the boot mounting screws, the support and trim bezel for access to the lower end of the shifter. Unthread and remove the upper end of the shift lever. Service as required. Thread the upper lever to the stud end and secure the shift boot and components.

Back-Up Light Switch

REMOVAL AND INSTALLATION

NP-2500

The switch is located on the left side of the transmission just below and behind the shifter. It screws into place. Tighten it to 15 ft.lbs.

Overdrive-4 (NP435)

The switch is located on the left side of the case, just behind the shift linkage housing. It screws into place. Tighten it to 15 ft.lbs.

Extension Housing Seal

REMOVAL AND INSTALLATION

NOTE: Special seal removal/installing tools are required. Tool No. C-3985 Remover and C-3972 Installer, or their equivalents.

1. Raise and support the vehicle.

2. Place a drain pan under the end of the extension housing.

3. Mark the position of the driveshaft and differential yoke for correct installation. Disconnect the driveshaft at the rear universal joint and carefully slide the shaft out of the transmission extension housing.

4. Remove the extension housing seal with the special tool.

5. Clean the end and inside of the extension housing. Start the new seal into position, with the installing tool, tap the seal into place.

6. Check the condition of the driveshaft slip yoke. Clean and sand with crocus cloth to remove any minor nicks and rust.

7. Lubricate the lips of the seal and the slip yoke.

8. Align and install the driveshaft taking care when guiding the slip yoke into the extension housing.

9. Remove the drain pan. Check the transmission fluid level, add fluid if necessary. Lower the vehicle.

Transmission

REMOVAL AND INSTALLATION

4-Speed

1. Disconnect the negative battery ground cable. Remove the engine cover. Remove the gear shift lever and shift unit. Raise and support the vehicle. Drain the transmission fluid.

2. Mark the rear driveshaft yoke to differential for correct installation reference. Disconnect the rear universal joint and remove the driveshaft.

3. Disconnect the speedometer cable and backup lamp switch electrical harness.

4. Support the rear of the engine. Remove the rear mount to

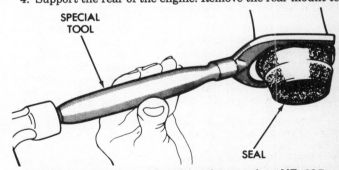

Removing the extension housing seal — NP-435

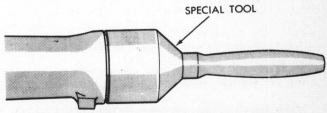

Install the extension housing seal — NP-435

crossmember bolts/nuts. Raise the rear of the engine slightly. Take care when raising the engine so nothing in the front of the engine, hoses lines, fan to radiator is strained or damaged.

5. Remove the rear transmission support crossmember. Position a suitable support jack under the transmission. Remove transmission to clutch bell housing retaining bolts and pull transmission rearward until drive pinion clears clutch, then remove transmission.

6. To install, place a small amount of MOPAR Multipurpose Grease 4318063, or equivalent into clutch pilot bushing and on the clutch release bearing sleeve. Do not get any grease on flywheel face. Do not lubricate the end of the transmission input shaft, clutch disc splines, or clutch release levers.

7. Align clutch disc and backing plate with a spare drive pinion shaft or clutch aligning tool.

8. Raise the transmission into position. Slide the transmission forward until the input shaft is aligned with and starts to enter the clutch disc splines. With the transmission in gear, turn the splines until they are aligned then push the transmission in until it mates to the bell housing.

9. Install transmission to bell housing bolts, tightening to 50 ft.lbs. (68 N.m). Install the crossmember and tighten the mounting bolts to 30 ft.lbs. Attach the rear transmission mount to the crossmember and tighten to 50 ft. lbs. (68 N.m).

10. Install gear shift unit and tighten the mounting bolts to 24 ft.lbs. (33 N.m). Install and adjust the shift linkage.

11. Install the driveshaft. Connect the speedometer cable and backup light.

12. Install the transmission drain plug and fill the transmission with lubricant.

13. Install the shift lever and engine cover. Lower the vehicle. Connect the battery cable.

5-Speed Models

1. Disconnect the negative battery ground cable. Remove the engine cover. Remove the upper shift lever. Raise and support the vehicle. Drain the transmission fluid.

2. Mark the rear driveshaft yoke to differential for correct installation reference. Disconnect the rear universal joint and remove the driveshaft.

3. Disconnect the backup lamp switch and distance sensor electrical harness. Separate the sensor from the speedometer adapter. Remove the speedometer components.

4. Support the rear of the engine. Remove the rear mount to crossmember bolts/nuts. Raise the rear of the engine slightly. Take care when raising the engine so nothing in the front of the engine, hoses lines, fan to radiator is strained or damaged.

5. Remove the rear transmission support crossmember. Position a suitable support jack under the transmission. Remove transmission to clutch bell housing retaining bolts and pull transmission rearward until drive pinion clears clutch, then remove transmission.

6. To install, place a small amount of MOPAR Multipurpose Grease 4318063, or equivalent into clutch pilot bushing and on the clutch release bearing sleeve. Do not get any grease on flywheel face. Do not lubricate the end of the transmission input shaft, clutch disc splines, or clutch release levers.

7. Align clutch disc and backing plate with a spare drive pinion shaft or clutch aligning tool.

8. Raise the transmission into position. Slide the transmission forward until the input shaft is aligned with and starts to enter the clutch disc splines. With the transmission in gear, turn the splines until they are aligned then push the transmission in until it mates to the bell housing.

9. Install transmission to bell housing bolts. Install the crossmember. Attach the rear transmission mount to the crossmember.

10. Install the driveshaft. Connect the speedometer components. Install the distance sensor. Connect the backup light.

12. Install the transmission drain plug and fill the transmission with lubricant.

13. Install the upper shift lever and engine cover. Lower the vehicle. Connect the battery cable.

Overdrive-4 (NP435) Overhaul

The Overdrive-4 Speed transmission is a four speed unit with all forward gears synchronized. 3rd gear is direct, while the 4th gear is the overdrive ratio.

DISASSEMBLY OF GEARSHIFT HOUSING AND MECHANISM

1. If available, mount the transmission on a repair stand.

2. Disconnect the gearshift control rods from the shift control levers and the transmission operating levers.

3. Remove the two gearshift control unit mounting bolts.

4. Remove the gearshift control unit from the transmission extension housing.

5. Remove gearshift housing-to-transmission case attaching bolts.

6. With all levers in the neutral detent position, pull housing out and away from the case.

NOTE: If 1st and 2nd, or 3rd and 4th shift forks remain in engagement with the synchronizer sleeves, move the sleeves and remove forks from the case.

7. Remove nuts, lock washers and flat washers that hold 1st/2nd, and 3rd/4th speed shift operating levers to the shafts.

8. Disengage shift levers from the flats on the shafts and remove levers. Remove the O-ring retainers and the O-rings from the housing. Remove the E-ring from the interlock lever pivot pin and remove the interlock levers and spring from the housing. Remove the reverse detent spring and ball from the bore at the side of the case.

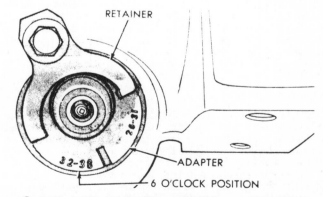

Speedometer pinion and adapter — NP-435

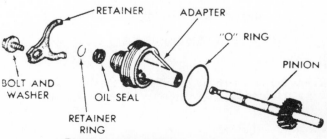

Speedometer drive — NP-435

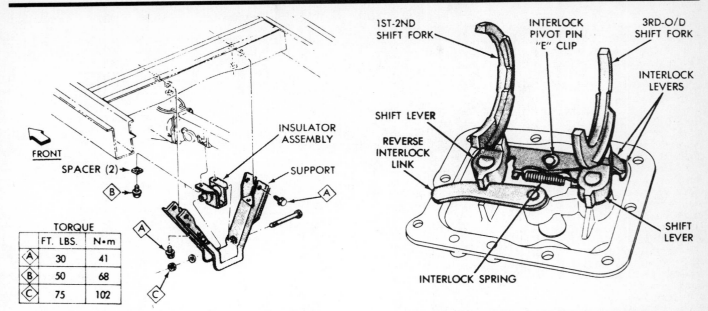

Crossmember and rear engine mount — NP-435

Gearshift housing assembly — NP-435

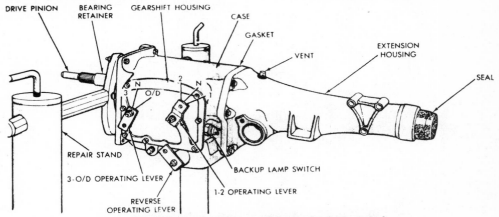

NP-435 mounted on a transmission stand

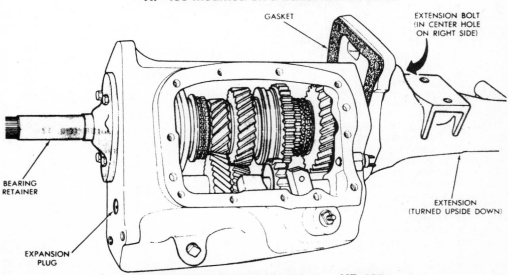

Rotating the extension housing — NP-435

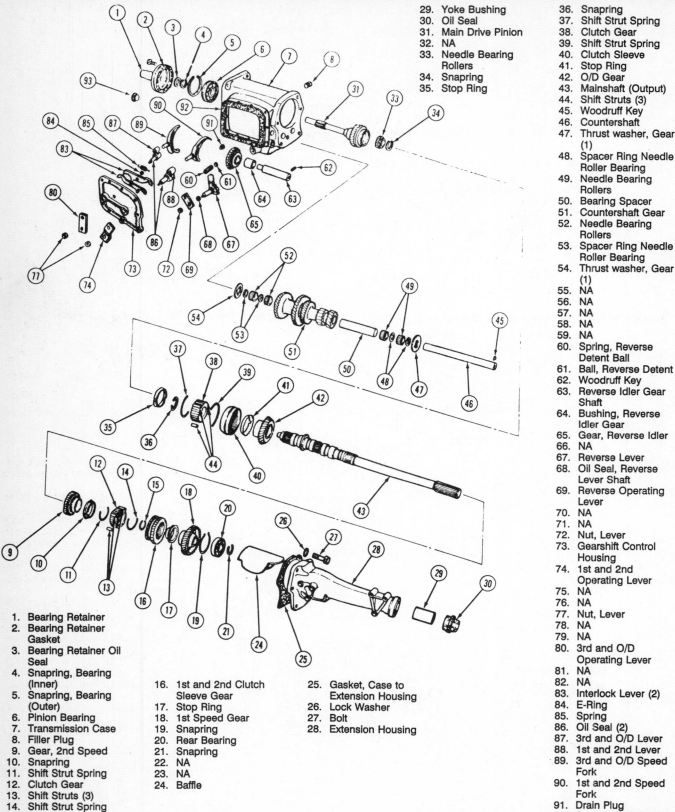

NP-435 (Overdrive 4) transmission components

29. Yoke Bushing
30. Oil Seal
31. Main Drive Pinion
32. NA
33. Needle Bearing Rollers
34. Snapring
35. Stop Ring

36. Snapring
37. Shift Strut Spring
38. Clutch Gear
39. Shift Strut Spring
40. Clutch Sleeve
41. Stop Ring
42. O/D Gear
43. Mainshaft (Output)
44. Shift Struts (3)
45. Woodruff Key
46. Countershaft
47. Thrust washer, Gear (1)
48. Spacer Ring Needle Roller Bearing
49. Needle Bearing Rollers
50. Bearing Spacer
51. Countershaft Gear
52. Needle Bearing Rollers
53. Spacer Ring Needle Roller Bearing
54. Thrust washer, Gear (1)
55. NA
56. NA
57. NA
58. NA
59. NA
60. Spring, Reverse Detent Ball
61. Ball, Reverse Detent
62. Woodruff Key
63. Reverse Idler Gear Shaft
64. Bushing, Reverse Idler Gear
65. Gear, Reverse Idler
66. NA
67. Reverse Lever
68. Oil Seal, Reverse Lever Shaft
69. Reverse Operating Lever
70. NA
71. NA
72. Nut, Lever
73. Gearshift Control Housing
74. 1st and 2nd Operating Lever
75. NA
76. NA
77. Nut, Lever
78. NA
79. NA
80. 3rd and O/D Operating Lever
81. NA
82. NA
83. Interlock Lever (2)
84. E-Ring
85. Spring
86. Oil Seal (2)
87. 3rd and O/D Lever
88. 1st and 2nd Lever
89. 3rd and O/D Speed Fork
90. 1st and 2nd Speed Fork
91. Drain Plug
92. Gasket, Shift Control Housing
93. Expansion Plug

1. Bearing Retainer
2. Bearing Retainer Gasket
3. Bearing Retainer Oil Seal
4. Snapring, Bearing (Inner)
5. Snapring, Bearing (Outer)
6. Pinion Bearing
7. Transmission Case
8. Filler Plug
9. Gear, 2nd Speed
10. Snapring
11. Shift Strut Spring
12. Clutch Gear
13. Shift Struts (3)
14. Shift Strut Spring
15. Snapring

16. 1st and 2nd Clutch Sleeve Gear
17. Stop Ring
18. 1st Speed Gear
19. Snapring
20. Rear Bearing
21. Snapring
22. NA
23. NA
24. Baffle

25. Gasket, Case to Extension Housing
26. Lock Washer
27. Bolt
28. Extension Housing

DISASSEMBLY OF THE EXTENSION HOUSING, MAINSHAFT & MAIN DRIVE PINION

1. Remove the bolt and retainer holding the speedometer pinion adapter in the extension housing, then remove the pinion adapter.

2. Remove the bolts attaching the extension housing to the transmission case.

3. Rotate the extension housing on the output shaft to expose the rear of the countershaft. Install one bolt to hold the extension in place.

4. Drill a hole in the countershaft extension plug at the front of the case.

5. Reaching through this hole, push the countershaft to the rear to expose the Woodruff key; when exposed, remove it. Push the countershaft forward against the expansion plug, and using a brass drift, tap the countershaft forward until the expansion plug is removed.

6. Using a countershaft arbor, push the countershaft out the rear of the case, but don't let the countershaft washers fall out of position. Lower the cluster gear to the bottom of the transmission case.

7. Remove the bolt and rotate the extension back to the normal position.

8. Remove the drive pinion attaching bolts and slide the retainer and gasket from the pinion shaft, then pry the pinion or

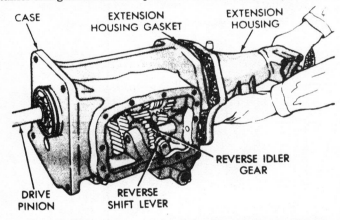

Remove/install the extension housing and mainshaft assembly — NP-435

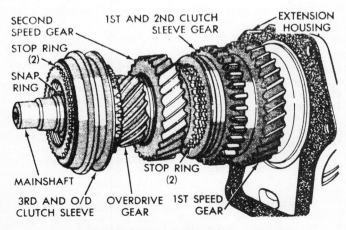

Mainshaft gear identification — NP-435

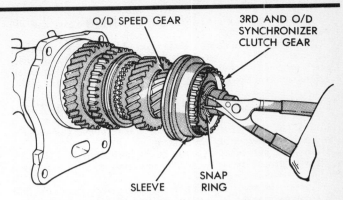

Disassembling the mainshaft — NP-435

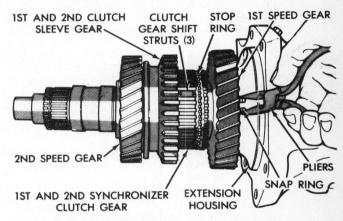

Mainshaft bearing snapring — NP-435

seal from the retainer. When installing the new seal, don't nick or scratch the seal bore in the retainer or the surface on which the seal bottoms.

9. Using a brass drift, tap the pinion and bearing assembly forward and remove through the front of the case.

10. Slide the 3rd and overdrive synchronizer sleeve slightly forward, slide the reverse idler gear to the center of its shaft, and tap the extension housing rearward. Slide the housing and mainshaft assembly out and away from the case.

11. Remove the snapring holding the 3rd and overdrive synchronizer clutch gear and sleeve assembly to the mainshaft, then remove the synchronizer assembly.

12. Slide the overdrive gear and stop ring off the mainshaft. Using pair of long nose pliers, compress the snapring holding the mainshaft bearing in the extension housing. With it compressed, pull the mainshaft assembly and bearing out of the extension housing.

13. Remove the snapring holding the mainshaft on the shaft. The bearing is removed by inserting steel plates on the front side of the 1st speed gear, then pressing the mainshaft through the bearing being careful not to damage the gear teeth.

14. Remove the bearing, retainer ring, 1st speed gear and stop ring from the shaft.

15. Remove the snapring. Remove the 1st and 2nd clutch gear and sleeve assembly from the mainshaft.

16. Remove the drive pinion bearing inner snapring, then using an arbor press, remove the bearing. Remove the snapring and bearing rollers from the cavity in the drive pinion.

17. Remove the countershaft gear from the bottom of the case, then remove the arbor, needle bearings, thrust washers and spacers from the center of the countershaft gear.

18. Remove the reverse gearshift lever detent spring retainer, gasket, plug, and detent ball spring from the rear of the case.

19. The reverse idler gear shaft is a tight fit in the case and will have to be pressed out.

20. If there is oil leakage visible around the reverse gearshift lever shaft, push the lever shaft in and remove it from the case. Remove the detent ball from the bottom of the transmission case and remove the shift fork from the shaft and detent plate.

ASSEMBLY OF THE REVERSE SHAFT

Follow the first four steps only if you removed the reverse shaft in the disassembly procedure.

1. Install a new oil seal O-ring on the lever shaft and coat the shaft with grease; insert it into its bore and install the reverse fork in the lever.

2. Install the reverse detent spring and gasket; insert the ball and spring and install the plug and gasket.

3. Place the reverse idler gear shaft in position in the end of the case and drive it in far enough to position the reverse idler gear on the protruding end of the shaft with the fork slot toward the rear. While doing this, engage the slot with the reverse shift fork.

4. With the reverse idler gear correctly positioned, drive the reverse gear shaft into the case far enough to install the Woodruff key. Drive the shaft in flush with the end of the transmission case. Install the back-up light switch and gasket.

ASSEMBLY OF THE COUNTERSHAFT GEAR AND DRIVE PINION

1. Coat the inside bore of the countershaft gear with a thin film of grease and install the roller bearing spacer with an arbor, into the gear; center the spacer and arbor.

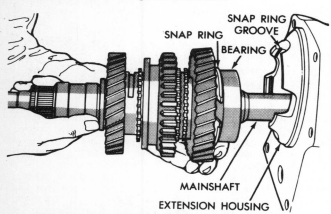

Remove/install the mainshaft — NP-435

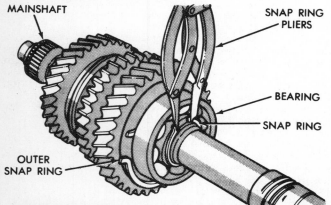

Remove/install the mainshaft bearing — NP-435

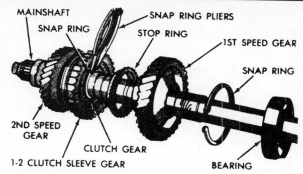

Remove/install the clutch gear snapring — NP-435

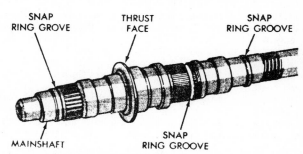

Mainshaft bearing surfaces — NP-435

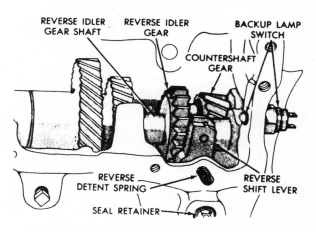

Gearshaft and lever identification — NP-435

2. Install the roller bearings and a spacer ring on each end.

3. Replace worn thrust washers; coat the new ones with grease and install them over the arbor with the tang side toward the case boss.

4. Install the countershaft assembly into the case and allow the gear assembly to sit on the bottom of the case so that the thrust washers won't come out of position.

5. Press the drive pinion bearing on the pinion shaft. Make sure the outer snapring groove is toward the front end and the bearing is seated against the shoulder on the gear.

6. Install a new snapring on the shaft to hold the bearing in place; make sure the snapring is seated and that there is minimum end play. There are several snapring thicknesses available for adjustment.

7. Place the pinion shaft in a soft-jawed vise and install the roller bearings in the cavity of the shaft. Coat them with grease and install the bearing retaining snapring.

8. Install a new oil seal in the bore.

ASSEMBLY OF THE MAINSHAFT

1. Place a stop ring flat on a bench followed by the clutch gear and sleeve; drop the struts in their slots and snap in a strut spring placing the tang inside one strut. Install the 2nd strut spring tang in a different strut after turning the assembly over.

2. Slide the 2nd speed gear over the mainshaft with the synchronizer cone toward the rear and down against the shoulder on the shaft.

3. Slide the 1st and 2nd gear synchronizer assembly including stop rings with lugs indexed in the hub slots, over the mainshaft down against the 2nd gear cone and hold it there with a new snapring. Slide the next snapring over the shaft and index the lugs into the clutch hub slots.

4. Slide the 1st speed gear with the synchronizer cone toward the clutch sleeve just installed over the mainshaft and into position against the clutch sleeve gear.

5. Install the mainshaft bearing retaining ring followed by the mainshaft rear bearing; press the bearing down into position and install a new snapring to secure it. There are several snapring thicknesses available for minimum end play.

6. Install the partially assembled mainshaft into the extension housing far enough to engage the bearing retaining ring in the slot in the extension housing. Compress the ring with pliers so that the mainshaft ball bearing can move in and bottom against its thrust shoulder in the extension housing. Release the ring and make sure that it is seated.

7. Slide the overdrive gear over the mainshaft with the synchronizer cone toward the front followed by the gear's snapring.

8. Install the 3rd-overdrive gear synchronizer clutch gear assembly on the mainshaft against the overdrive gear. Make sure to index the rear stop ring with the clutch gear struts.

9. Install the snapring and position the front stop ring over the clutch gear again lining up the ring lugs with the struts; coat a new extension gasket with grease and place it in position.

10. Slide the reverse idler gear to the center of its shaft and move the 3rd/overdrive synchronizer as far forward as possible without losing the struts.

11. Insert the mainshaft assembly in the case tilting it as necessary. Place the 3rd/overdrive sleeve in the neutral detent.

12. Rotate the extension on the mainshaft to expose the rear of the countershaft and install one bolt to hold it in position.

13. Install the drive pinion and bearing assembly through the front of the case and position it in the front bore. Install the outer snapring in the bearing groove and tap lightly into place. If it doesn't bottom easily, check to see if a strut, pinion roller or stop ring is out of position.

14. Turn the transmission upside down while holding the countershaft gear to prevent damage. Then lower the countershaft gear assembly into position making sure that the teeth mesh with the drive pinion gear.

15. Start the countershaft into the bore at the rear of the case and push until it is in about halfway; then install the Woodruff key and push it in until it is flush with the rear of the case.

16. Rotate the extension back to normal position and install the bolts; turn the transmission upright and install the drive pinion bearing retainer and gasket. Coat the threads with sealing compound and tighten the attaching bolts to 30 ft. lbs.

17. Install a new expansion plug in its bore.

ASSEMBLY OF THE GEARSHIFT HOUSING & MECHANISM

1. Install the interlock levers on the pivot pin and secure with the E-ring. Install the spring with a pair of pliers.

2. Grease and install new O-ring seals on both shift shafts; grease the housing bores and push the shafts through.

3. Install the operating levers and tighten the retaining nuts to 18 ft. lbs.; make sure the 3rd-overdrive lever points down.

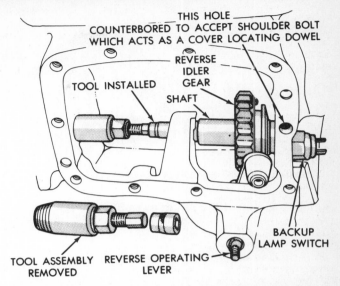

Removing the reverse idler gear shaft — NP-435

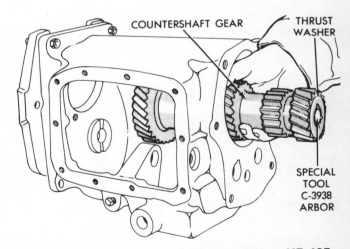

Countershaft gear and arbor assembly — NP-435

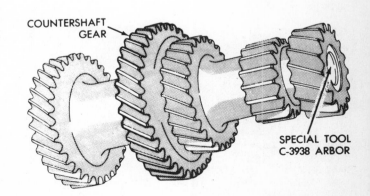

Countershaft gear assembly — NP-435

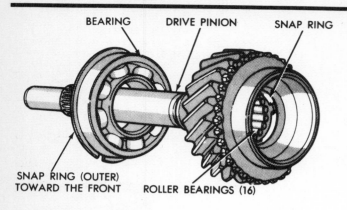

Drive pinion and bearing assembly — NP-435

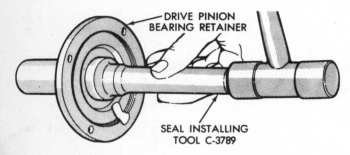

Replacing the drive pinion seal — NP-435

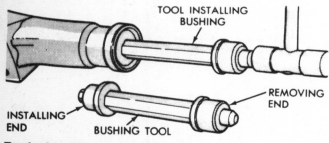

Replacing the extension housing bushing — NP-435

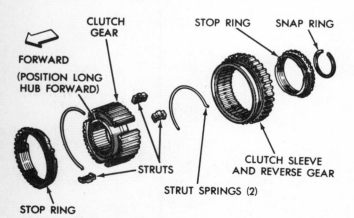

First/second synchronizer — NP-435

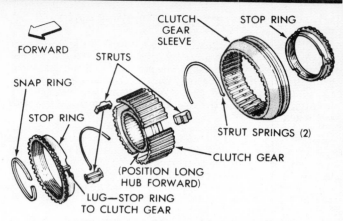

Third/OD synchronizer — NP-435

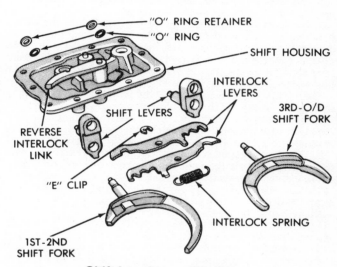

Shift housing — NP-435

4. Rotate each shift shaft fork bore straight up and install the 3rd/overdrive shift fork in its bore and under both interlock levers.

5. Position both synchronizer sleeves in neutral and place the 1st and 2nd gear shift fork in the groove of the 1st and 2nd gear synchronizer sleeve. Slide the reverse idler gear to neutral. Turn the transmission on its right side and place the gearshift housing gasket in place holding it there with grease. Install the reverse detent ball and spring into the case bore.

6. As the shift housing is lowered in place, guide the 3rd-overdrive shift fork into its synchronizer groove then lead the shaft of the 1st and 2nd shift lever.

7. Raise the interlock lever with a screwdriver to allow the 1st and 2nd shift fork to slip under the levers. The shift housing will now seat against the case.

8. Install the bolts lightly and shift through all the gears to check for proper operation.

9. The reverse shift lever and the 1st and 2nd gear shift lever have cam surfaces which mate in reverse position to lock the 1st and 2nd lever, the fork and synchronizer in the neutral position.

10. To check for proper operation, put the transmission in reverse, and, while turning the input shaft, move the 1st and 2nd lever in each direction. If it locks up or becomes harder to turn, select a new shift lever size with more or less clearance. If there is too little cam clearance, it will be difficult or impossible to shift into reverse.

11. Grease the reverse shaft, install the operating lever and nut, and install the speedometer drive pinion gear and adapter, making sure the range number is in the straight down position.

NP-2500 5-Speed Overhaul

MAJOR COMPONENT DISASSEMBLY

1. Remove the transmission from the van.
2. Remove the shifter assembly bolts.
3. The shifter assembly is sealed with a bead of RTV sealant so it must be pried loose before removal. Clean all RTV sealant from both mating surfaces.
4. Remove the access cover bolts.
5. The access cover is sealed with a bead of RTV sealant so it must be pried loose before removal. Clean all RTV sealant from both mating surfaces.
6. Remove the detent springs and bullets with a magnet. Keep track of which is which.
7. Remove the extension housing-to-case bolts.
8. The extension housing is sealed with a bead of RTV sealant so it must be pried loose before removal. When pulling off the housing, hold the Reverse/Overdrive shift rail as shown in the accompanying illustration. Clean all RTV sealant from both mating surfaces.
9. Remove the Reverse/Overdrive shift rail and blocker spring.
10. Remove the tapered pins from the shift forks.
11. Remove the 1st/2nd shift rail.
12. Remove the countershaft overdrive gear snapring.
13. Remove the countershaft overdrive gear using tool C-4982.
14. Remove the mainshaft overdrive gear snapring.
15. Remove the mainshaft overdrive gear thrust washer and anti-spin pin.
16. Remove the mainshaft overdrive gear.
17. Remove the Reverse/Overdrive hub snapring.
18. Remove the Reverse/Overdrive synchronizer, fork, and rail assembly.
19. Remove the 3rd/4th shift rail.
20. Remove the 1st/2nd and 3rd/4th shift forks.
21. Remove the 1st/2nd and 3rd/4th shift rails.
22. Remove the Reverse gear.
23. Remove the Reverse gear thrust washer.
24. Remove the center support plate bolts.
25. Remove the gear set/support plate assembly.

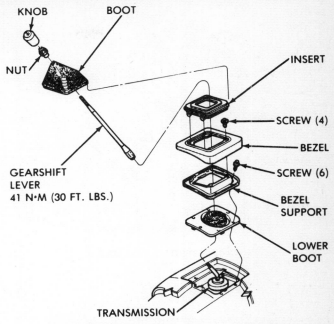

Gearshift lever components — NP-2500

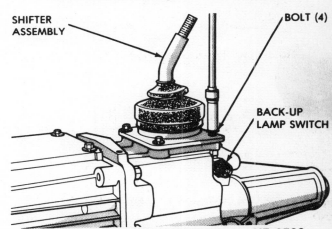

Removing/installing shifter bolts — NP-2500

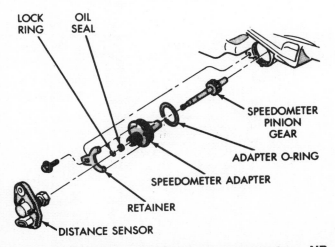

Speedometer adapter and pinion components — NP-2500

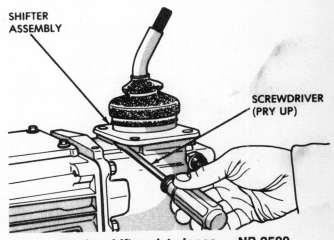

Prying the shifter plate loose — NP-2500

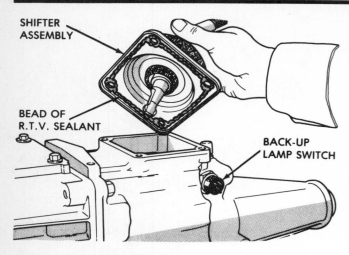

Shifter assembly — NP-2500

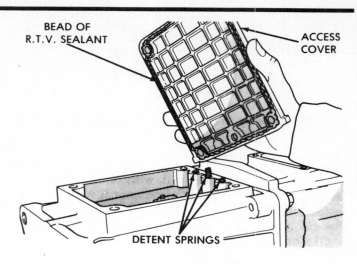

Access cover — NP-2500

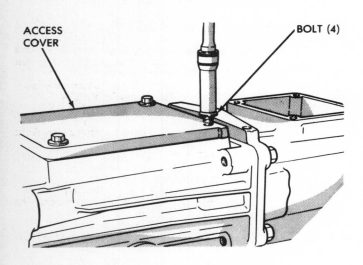

Access cover bolts — NP-2500

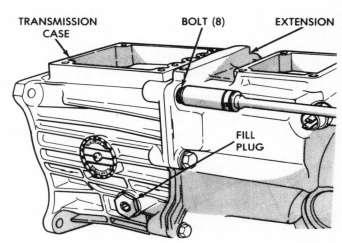

Extension housing bolts — NP-2500

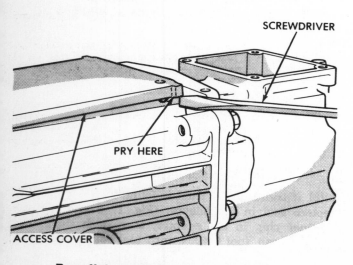

Pry off the access cover — NP-2500

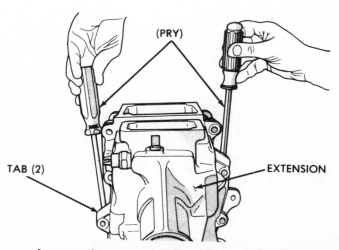

Loosen the extension housing — NP-2500

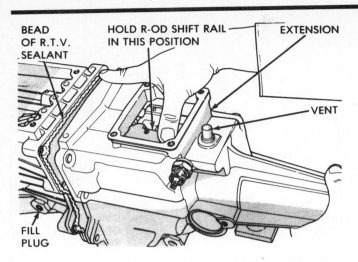

Remove/install the extension housing — NP-2500

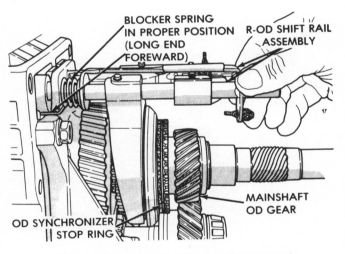

Blocker spring in position — NP-2500

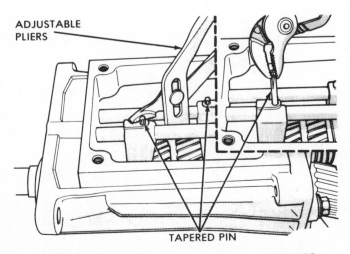

Shift fork tapered pin placement — NP-2500

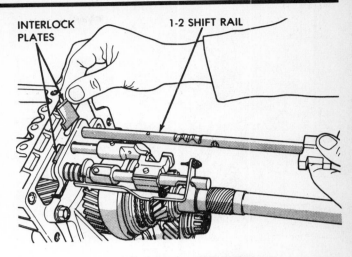

1-2 shift rail — NP-2500

GEAR SET DISASSEMBLY

1. Open the mainshaft center bearing snapring with a pliers and, using a plastic mallet on the center support, tap the mainshaft assembly out of the center support.
2. Remove the countershaft gear from the center support.
3. Remove the 3rd/4th synchronizer hub snapring.
4. Remove the 3rd/4th synchronizer assembly.
5. Remove the 3rd speed gear.
6. Remove the split thrust washer and retaining ring from the mainshaft.
7. Remove the 2nd speed gear.
8. Remove the 1st/2nd synchronizer hub snapring.
9. Remove the 1st/2nd synchronizer assembly.
10. Remove the 1st speed gear.
11. The synchronizers may be disassembled by removing the energizing springs.
12. Remove the mainshaft center bearing snapring.
13. Place the mainshaft in an arbor press and press off the mainshaft center bearing.
14. Remove the input shaft bearing retainer bolts.
15. Remove the retainer and discard the gasket.
16. Remove the large input bearing snapring.
17. Remove the input shaft from the case.

NOTE: There should be 16 roller bearings in the bore of the input shaft.

18. Remove the small input shaft bearing retaining snapring.
19. Using an arbor press, remove the input bearing.
20. Remove the reverse idler shaft snapring from the center support plate.
21. Remove the reverse idler gear and shaft.
22. Remove the countershaft center bearing snapring.
23. Using an arbor press, remove the countershaft center bearing.
24. Using tool C-4983-4 and adapter, remove the front bearing cone from the countershaft.
25. Using an arbor press remove the front bearing cup from the case.
26. Using tool C-4983-5 and adapter, remove he rear bearing cone from the countershaft.
27. Using tools L-4454 and L-4518, remove the countershaft rear bearing cup from the extension housing.
28. Remove the snapring and remove the extension housing ball bearing.
29. Remove the extension housing yoke seal.
30. Drive out the extension housing bushing.

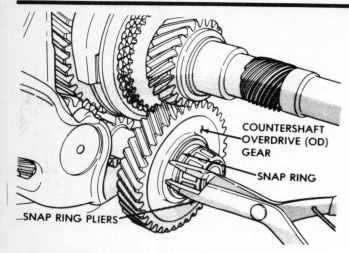

Remove/install the countershaft OD gear snapring — NP-2500

Labels: COUNTERSHAFT OVERDRIVE (OD) GEAR, SNAP RING, SNAP RING PLIERS

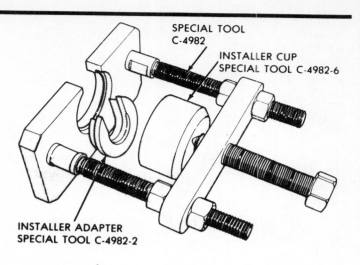

Assembly sequence for tool C-4982 — NP-2500

Labels: SPECIAL TOOL C-4982, INSTALLER CUP SPECIAL TOOL C-4982-6, INSTALLER ADAPTER SPECIAL TOOL C-4982-2

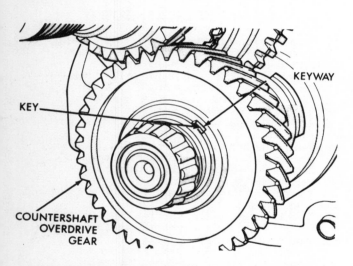

Overdrive gear keyway position — NP-2500

Labels: KEYWAY, KEY, COUNTERSHAFT OVERDRIVE GEAR

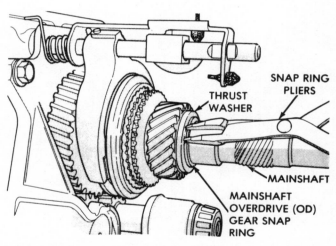

Remove/install the mainshaft OD gear snapring — NP-2500

Labels: THRUST WASHER, SNAP RING PLIERS, MAINSHAFT, MAINSHAFT OVERDRIVE (OD) GEAR SNAP RING

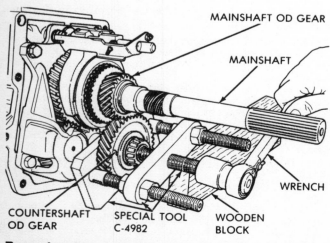

Removing the overdrive countershaft gear — NP-2500

Labels: MAINSHAFT OD GEAR, MAINSHAFT, WRENCH, WOODEN BLOCK, SPECIAL TOOL C-4982, COUNTERSHAFT OD GEAR

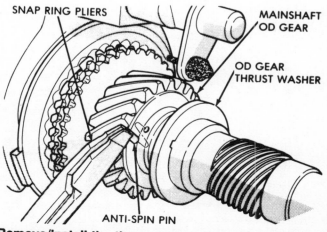

Remove/install the thrust washer and anti-spin pin — NP-2500

Labels: SNAP RING PLIERS, MAINSHAFT OD GEAR, OD GEAR THRUST WASHER, ANTI-SPIN PIN

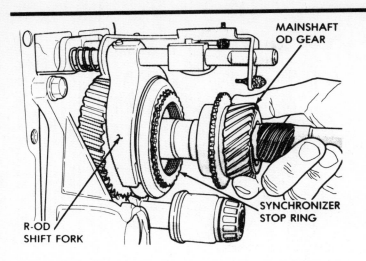

Remove/install the mainshaft OD gear — NP-2500

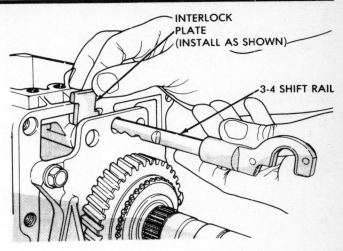

3-4 shift rail — NP-2500

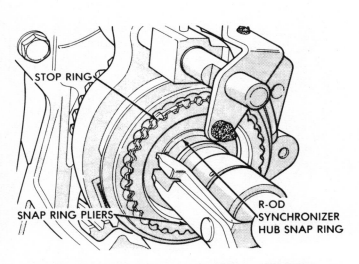

ROD synchronizer hub snapring — NP-2500

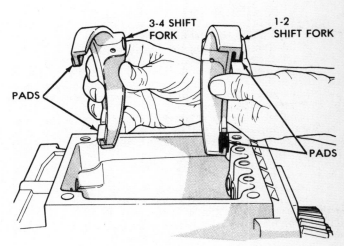

1-2 and 3-4 shift forks — NP-2500

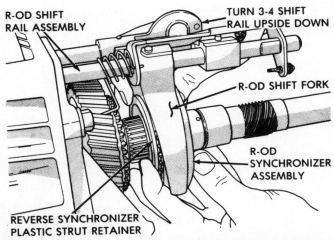

ROD synchronizer, fork and rail assembly — NP-2500

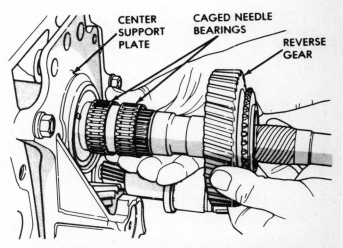

Remove/install reverse gear — NP-2500

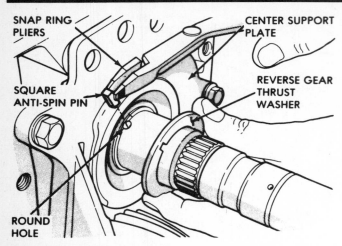

Remove/install the reverse gear thrust washer and anti-spin pin — NP-2500

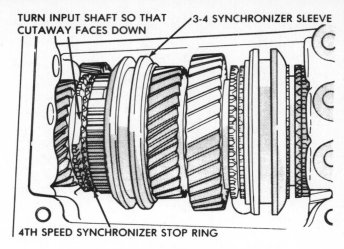

Gear set position — NP-2500

GEAR SET ASSEMBLY

1. Drive in a new extension housing bushing.
2. Install a new extension housing yoke seal.
3. Install the extension housing ball bearing and snapring.
4. Using an arbor press and tools C-4171 and C-4973, install the countershaft rear bearing cup in the extension housing.
5. Using and arbor press and tool C-4966, install the rear bearing cone on the countershaft.
6. Using an arbor press and tool C-4171, install the front bearing cup in the case.
7. Using an arbor press and tool C-4967, install the front bearing cone on the countershaft.
8. Using an arbor press and tool C-4171, install the countershaft center bearing.
9. Install the countershaft center bearing snapring.
10. Using a dial indicator, check the countershaft gear endplay with the assembly positioned in the case. Endplay should be 0.03-0.13mm. Select-fit shims are available in 0.02mm increments in sizes from 1.37mm to 2.29mm.
11. Assemble and install the reverse idler gear and shaft.
12. Install the reverse idler shaft snapring on the center support plate.
13. Install a new input shaft retainer oil seal.
14. Using an arbor press and tool C-4965, install the input bearing.
15. Install the small input shaft bearing retaining snapring.
16. Install the input shaft in the case.
17. Install the large input bearing snapring.
18. Install the retainer and new gasket. Torque the bolts to 21 ft. lbs.
19. Assemble the 16 roller bearings in the bore of the input shaft. Hold them in place by coating them with heavy grease.
20. Place the mainshaft in an arbor press and press on the mainshaft center bearing.
21. Install the mainshaft center bearing snapring.
22. Assemble the synchronizers.
23. Install the needle bearings and snapring with the chamfered side of the snapring facing 1st gear.
24. Install the 1st speed gear.
25. Install the 1st/2nd synchronizer assembly.
26. Install the 1st/2nd synchronizer hub snapring.
27. Install the 2nd speed gear.
28. Install the split thrust washer and retaining ring on the mainshaft.
29. Install the 3rd speed gear.
30. Install the 3rd/4th synchronizer assembly.

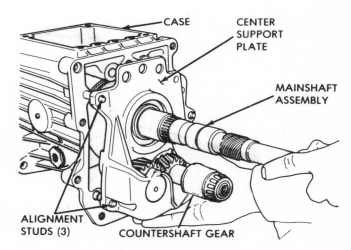

Remove/install the center support and gear set — NP-2500

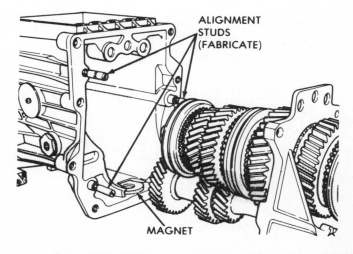

Gear case alignment studs — NP-2500

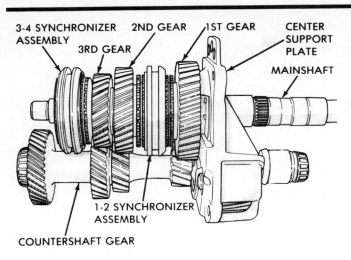

Gear identification — NP-2500

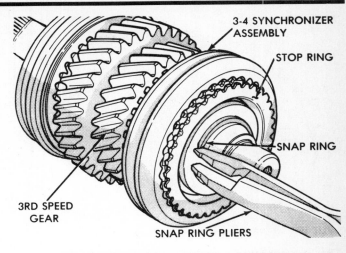

Synchronizer hub snapring — NP-2500

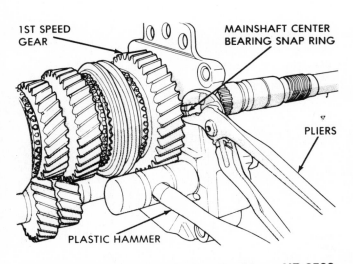

Remove/install the mainshaft assembly — NP-2500

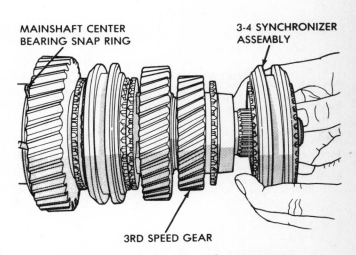

3-4 synchronizer assembly — NP-2500

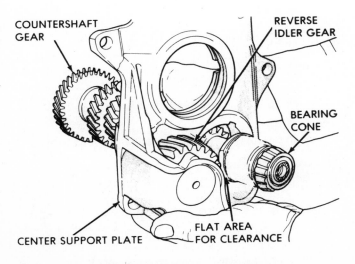

Countershaft gear — NP-2500

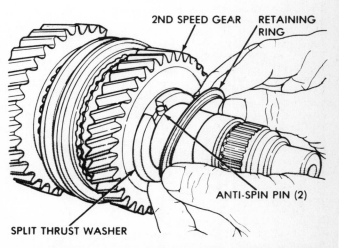

Remove/install the split thrust washer, retaining ring and anti-spin pins — NP-2500

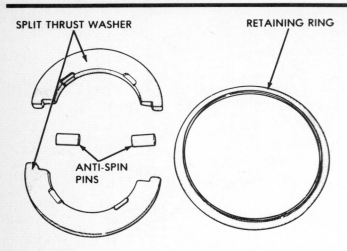

Thrust washer and retaining installation position — NP-2500

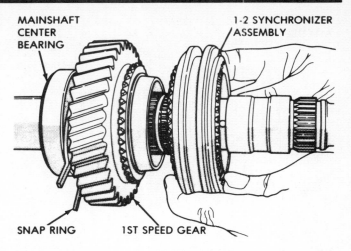

Remove/install the 1-2 synchronizer assembly — NP-2500

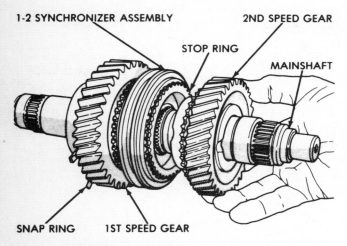

Remove/install the second speed gear — NP-2500

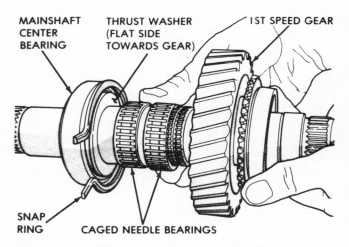

Remove/install the first gear and needle bearings — NP-2500

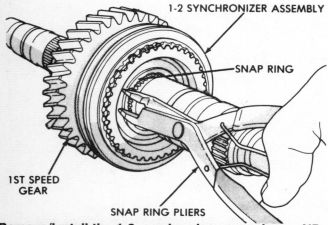

Remove/install the 1-2 synchronizer snap ring — NP-2500

31. Install the 3rd/4th synchronizer hub snapring.
32. Install the countershaft gear from the center support.
33. Open the mainshaft center bearing snapring with a pliers and, using a plastic mallet on the center support, tap the mainshaft assembly onto the center support.

MAIN COMPONENT ASSEMBLY

1. Install the gear set/support plate assembly.
2. Install the center support plate bolts. Torque the bolts to 40 ft. lbs.
3. Install the Reverse gear thrust washer.
4. Install the Reverse gear.
5. Install the 1st/2nd and 3rd/4th shift forks.
6. Install the 3rd/4th shift rail.
7. Install the Reverse/Overdrive synchronizer, fork, and rail assembly.
8. Install the Reverse/Overdrive hub snapring.
9. Install the mainshaft overdrive gear.
10. Install the mainshaft overdrive gear thrust washer and anti-spin pin.

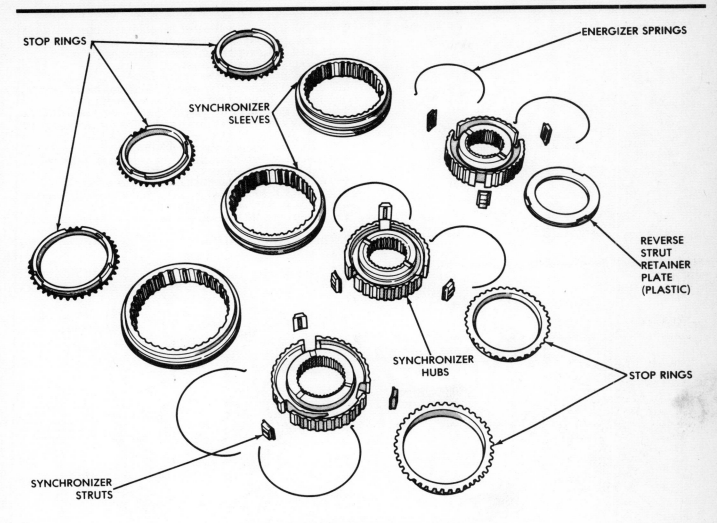

STOP RINGS

ENERGIZER SPRINGS

SYNCHRONIZER SLEEVES

REVERSE STRUT RETAINER PLATE (PLASTIC)

SYNCHRONIZER HUBS

STOP RINGS

SYNCHRONIZER STRUTS

Synchronizer identification — NP-2500

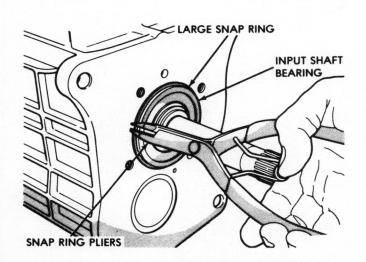

LARGE SNAP RING

INPUT SHAFT BEARING

SNAP RING PLIERS

Remove/install the front bearing locating snapring — NP-2500

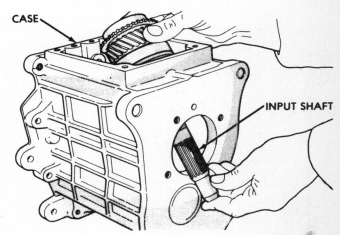

CASE

INPUT SHAFT

Remove/install the iNP-ut shaft — NP-2500

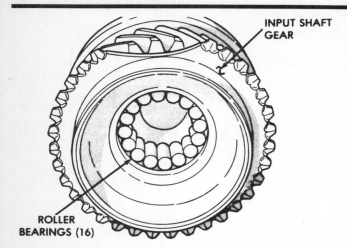

Remove/install the mainshaft pilot roller bearings — NP-2500

11. Install the mainshaft overdrive gear snapring.
12. Install the countershaft overdrive gear using tool C-4982.
13. Install the countershaft overdrive gear snapring.
14. Install the 1st/2nd shift rail.
15. Install the tapered pins on the shift forks.
16. Install the Reverse/Overdrive shift rail and blocker spring. The long end of the blocker spring faces forward.
17. Install the extension housing. The housing is sealed with a bead of RTV sealant rather than a gasket. Torque the bolts to 40 ft. lbs.
18. Install the detent springs and bullets.
19. Install the access cover. The access cover is sealed with a bead of RTV sealant rather than a gasket.
20. Install the access cover bolts. Torque the bolts to 21 ft. lbs.
21. Install the shifter assembly. The shifter assembly is sealed with a bead of RTV sealant rather than a gasket.
22. Install the shifter assembly bolts. Torque the bolts to 21 ft. lbs.
23. Install the transmission.

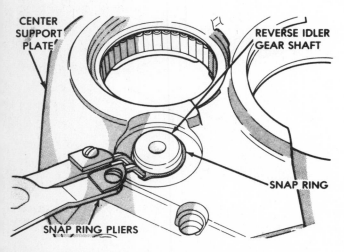

Remove/install the idler shaft snapring — NP-2500

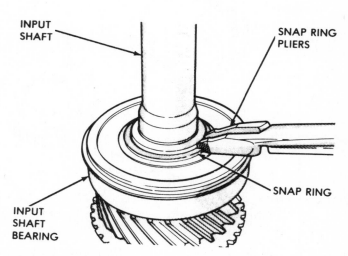

Remove/install the front bearing retaining snapring — NP-2500

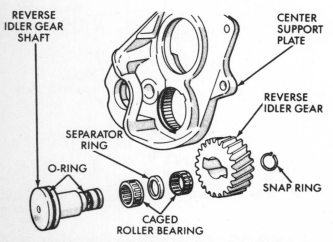

Remove/install the reverse idler gear components — NP-2500

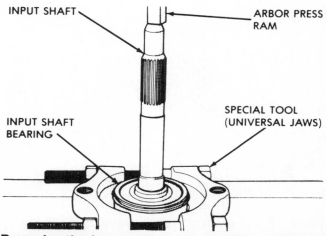

Removing the front bearing from the iNP-ut shaft — NP-2500

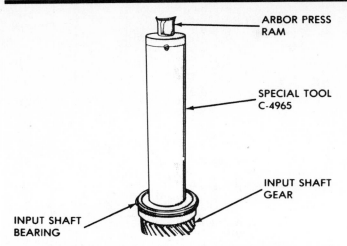

Installing the front bearing on the iNP-ut shaft — NP-2500

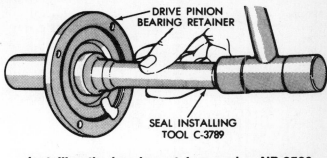

Installing the bearing retainer seal — NP-2500

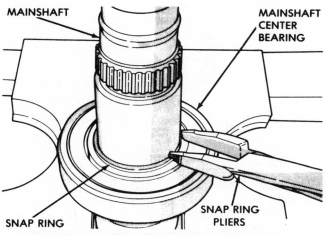

Remove/install the center bearing snapring — NP-2500

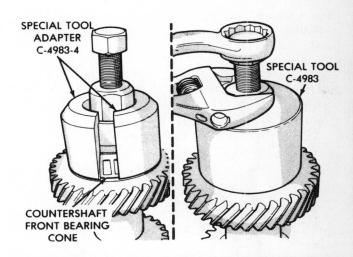

Removing the countershaft gear front bearing cone — NP-2500

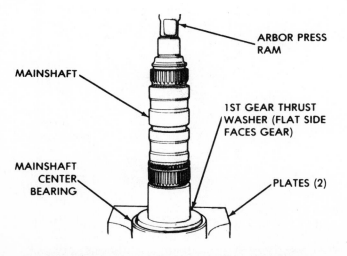

Installing the center bearing — NP-2500

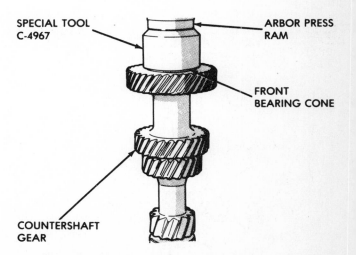

Installing the countershaft gear front bearing cone — NP-2500

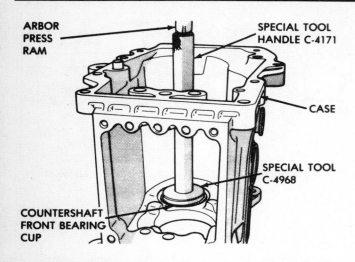

Installing the countershaft gear front bearing cup — NP-2500

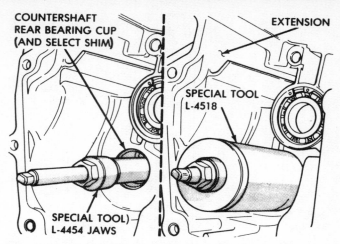

Removing the countershaft gear rear bearing cup — NP-2500

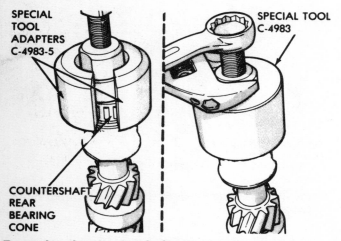

Removing the countershaft gear rear bearing cone — NP-2500

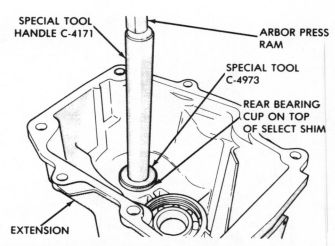

Installing the countershaft gear rear bearing cup — NP-2500

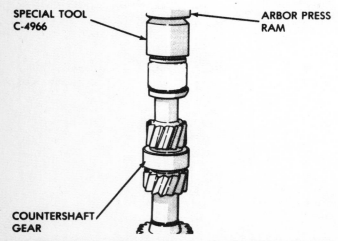

Installing the countershaft gear rear bearing cone — NP-2500

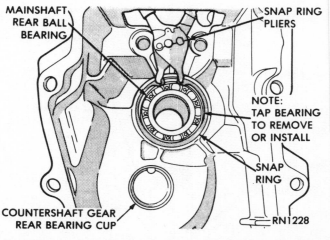

Snapring removal/installation countershaft bearing — NP-2500

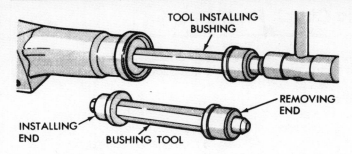

Replacing the extension housing bushing — NP-2500

Part No.	Thickness	
	Millimeters	**Inches**
4338275	1.37-1.39	.0539
4338276	1.46-1.48	.0579
4338277	1.55-1.57	.0614
4338278	1.64-1.66	.0650
4338279	1.73-1.75	.0685
4338280	1.82-1.84	.0720
4338281	1.91-1.93	.0756
4338282	2.00-2.02	.0791
4338283	2.09-2.11	.0827
4338284	2.18-2.20	.0862
4338285	2.27-2.29	.0898

Countershaft gear end play shim chart — NP-2500

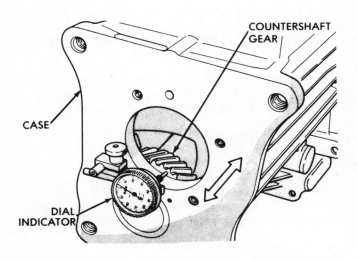

Measuring the countershaft gear end play — NP-2500

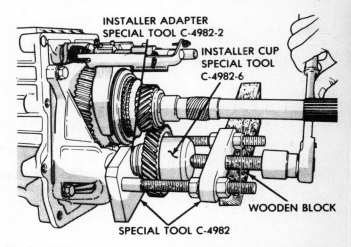

Installing the countershaft overdrive gear — NP-2500

Troubleshooting Basic Clutch Problems

Problem	Cause
Excessive clutch noise	Throwout bearing noises are more audible at the lower end of pedal travel. The usual causes are: • Riding the clutch • Too little pedal free-play • Lack of bearing lubrication A bad clutch shaft pilot bearing will make a high pitched squeal, when the clutch is disengaged and the transmission is in gear or within the first 2″ of pedal travel. The bearing must be replaced. Noise from the clutch linkage is a clicking or snapping that can be heard or felt as the pedal is moved completely up or down. This usually requires lubrication. Transmitted engine noises are amplified by the clutch housing and heard in the passenger compartment. They are usually the result of insufficient pedal free-play and can be changed by manipulating the clutch pedal.
Clutch slips (the car does not move as it should when the clutch is engaged)	This is usually most noticeable when pulling away from a standing start. A severe test is to start the engine, apply the brakes, shift into high gear and SLOWLY release the clutch pedal. A healthy clutch will stall the engine. If it slips it may be due to: • A worn pressure plate or clutch plate • Oil soaked clutch plate • Insufficient pedal free-play
Clutch drags or fails to release	The clutch disc and some transmission gears spin briefly after clutch disengagement. Under normal conditions in average temperatures, 3 seconds is maximum spin-time. Failure to release properly can be caused by: • Too light transmission lubricant or low lubricant level • Improperly adjusted clutch linkage
Low clutch life	Low clutch life is usually a result of poor driving habits or heavy duty use. Riding the clutch, pulling heavy loads, holding the car on a grade with the clutch instead of the brakes and rapid clutch engagement all contribute to low clutch life.

CLUTCH

The purpose of the clutch is to disconnect and connect engine power from the transmission. A vehicle at rest requires a lot of engine torque to get all that weight moving. An internal combustion engine does not develop a high starting torque (unlike steam engines), so it must be allowed to operate without any load until it builds up enough torque to move the vehicle. Torque increases with engine rpm. The clutch allows the engine to build up torque by physically disconnecting the engine from the transmission, relieving the engine of any load or resistance. The transfer of engine power to the transmission (the load) must be smooth and gradual; if it weren't, drive line components would wear out or break quickly. This gradual power transfer is made possible by gradually releasing the clutch ped-

al. The clutch disc and pressure plate are the connecting link between the engine and transmission. When the clutch pedal is released, the disc and plate contact each other (clutch engagement), physically joining the engine and transmission. When the pedal is pushed in, the disc and plate separate (the clutch is disengaged), disconnecting the engine from the transmission.

The clutch assembly consists of the flywheel, the clutch disc, the clutch pressure plate, the throwout bearing and fork, the actuating linkage and the pedal. The flywheel and clutch pressure plate (driving members) are connected to the engine crankshaft and rotate with it. The clutch disc is located between the flywheel and pressure plate, and splined to the transmission shaft. A driving member is one that is attached to the engine and transfers engine power to a driven member (clutch disc) on the transmission shaft. A driving member (pressure plate) rotates (drives) a driven member (clutch disc) on contact and, in so doing, turns the transmission shaft. There is a circular diaphragm spring within the pressure plate cover (transmission side). In a relaxed state (when the clutch pedal is fully released), this spring is convex; that is, it is dished outward toward the transmission. Pushing in the clutch pedal actuates an attached linkage rod. Connected to the other end of this rod is the throwout bearing fork. The throwout bearing is attached to the fork. When the clutch pedal is depressed, the clutch linkage pushes the fork and bearing forward to contact the diaphragm spring of the pressure plate. The outer edges of the spring are secured to the pressure plate and are pivoted on rings so that when the center of the spring is compressed by the throwout bearing, the outer edges bow outward and, by so doing, pull the pressure plate in the same direction - away from the clutch disc. This action separates the disc from the plate, disengaging the clutch and allowing the transmission to be shifted into another gear. A coil type clutch return spring attached to the clutch pedal arm permits full release of the pedal. Releasing the pedal pulls the throwout bearing away from the diaphragm spring resulting in a reversal of spring position. As bearing pressure is gradually released from the spring center, the outer edges of the spring bow outward, pushing the pressure plate into closer contact with the clutch disc. As the disc and plate move closer together, friction between the two increases and slippage is reduced until, when full spring pressure is applied (by fully releasing the pedal), The speed of the disc and plate are the same. This stops all slipping, creating a direct connection between the plate and disc which results in the transfer of power from the engine to the transmission. The clutch disc is now rotating with the pressure plate at engine speed and, because it is splined to the transmission shaft, the shaft now turns at the same engine speed. Understanding clutch operation can be rather difficult at first; if you're still confused after reading this, consider the following analogy. The action of the diaphragm spring can be compared to that of an oil can bottom. The bottom of an oil can is shaped very much like the clutch diaphragm spring and pushing in on the can bottom and then releasing it produces a similar effect. As mentioned earlier, the clutch pedal return spring permits full release of the pedal and reduces linkage slack due to wear. As the linkage wears, clutch free-pedal travel will increase and free-travel will decrease as the clutch wears. Free-travel is actually throwout bearing lash.

The diaphragm spring type clutches used are available in two different designs: flat diaphragm springs or bent spring. The bent fingers are bent back to create a centrifugal boost ensuring quick re-engagement at higher engine speeds. This design enables pressure plate load to increase as the clutch disc wears and makes low pedal effort possible even with a heavy-duty clutch. The throwout bearing used with the bent finger design is 1¼ in. (31.75mm) long and is shorter than the bearing used with the flat finger design. These bearings are not interchangeable. If the longer bearing is used with the bent finger clutch, free-pedal travel will not exist. This results in clutch slippage and rapid wear.

The transmission varies the gear ratio between the engine and rear wheels. It can be shifted to change engine speed as driving conditions and loads change. The transmission allows disengaging and reversing power from the engine to the wheels.

Adjustment and Linkage

The vehicle is equipped with an hydraulic operating linkage system, no adjustment is required. The hydraulic linkage system is serviced as an assembly only. The individual components that form the system cannot be overhauled or serviced separately.

Clutch Pedal

REMOVAL AND INSTALLATION

1. Remove the cotter pin, nut and washer that attach the lever to the clutch pedal.
2. Remove the parking brake apply pedal from its mounting to gain more working clearance if needed.
3. Press the clutch pedal shaft and shaft bushing to the left and out of the pedal support and brake pedal.
4. Inspect the pedal and shaft. Replace the pedal if bent, cracked or damaged in any way.
5. Lubricate the pedal shaft, bushing and bore with multipurpose grease.
6. Insert the clutch pedal shaft through the support and brake pedal.
7. Secure the shaft with the washer, nut and new cotter pin.
8. Attach the parking brake pedal if removed.

Driven Disc and Pressure Plate

REMOVAL AND INSTALLATION

—————————— CAUTION ——————————
The clutch driven disc contains asbestos, which has been determined to be a cancer causing agent. Never clean clutch surfaces with compressed air! Avoid inhaling any dust from any clutch surface! When cleaning clutch surfaces, use commercially available brake cleaning fluid.

1. Remove the transmission.
2. Remove the slave cylinder from the bell housing. Remove the clutch bell housing.
3. Remove the clutch fork and release bearing assembly.
4. If the pressure plate is to be reused, mark the clutch cover and flywheel, with a suitable tool to assure correct reassembly.
5. Remove the pressure plate retaining bolts, loosening them evenly so the clutch cover will not be distorted.
6. Pull the pressure plate assembly clear of the flywheel and, while supporting pressure plate, slide the clutch disc from between flywheel and pressure plate.
7. Thoroughly clean all the working surfaces of the flywheel and the pressure plate. Inspect the flywheel. Resurface or replace if required.
8. Lubricate the pilot bearing with high temperature grease.

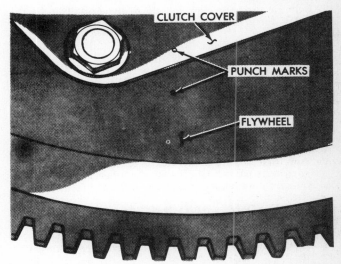

Mark the clutch pressure plate cover and flywheel position

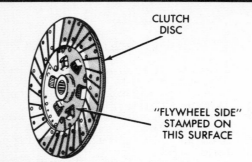

Flywheel side marked on clutch disc

CLUTCH
DISC

"FLYWHEEL SIDE"
STAMPED ON
THIS SURFACE

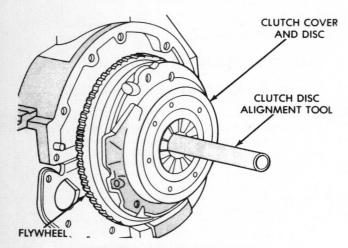

CLUTCH COVER
AND DISC

CLUTCH DISC
ALIGNMENT TOOL

FLYWHEEL

Aligning the clutch disc

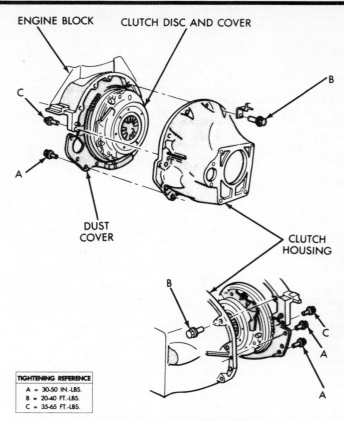

ENGINE BLOCK   CLUTCH DISC AND COVER

C

A

DUST
COVER

B

CLUTCH
HOUSING

B

C

A

A

TIGHTENING REFERENCE
A = 30-50 IN.-LBS.
B = 20-40 FT.-LBS.
C = 35-65 FT.-LBS.

Clutch bell housing installation

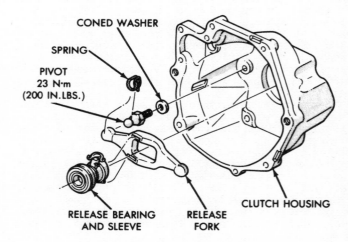

CONED WASHER

SPRING

PIVOT
23 N·m
(200 IN.LBS.)

RELEASE BEARING
AND SLEEVE

RELEASE
FORK

CLUTCH HOUSING

Clutch release bearing mounting

9. Position clutch disc and plate against flywheel and insert spare transmission main drive gear shaft or clutch installing tool through clutch disc hub and into main drive pilot bearing. Be sure the clutch disc hub marked flywheel side is facing the flywheel.

10. Rotate clutch cover until the punch marks on cover and flywheel line up.

11. Bolt the pressure plate loosely to flywheel. Tighten the bolts a few turns at a time, in progression, until tight. Then tighten bolts to:
- $5/16$ in. bolts: 17 ft.lbs. (23 N.m)
- $3/8$ in. bolts: 30 ft.lbs. (41 N.m)

12. Install the bell housing and transmission. Install the clutch slave cylinder. Lower the vehicle.

Clutch Master Cylinder and Slave Cylinder

The clutch master cylinder, remote reservoir, slave cylinder and connecting lines are serviced as an assembly only. The linkage components cannot be overhauled or serviced separately. The cylinders and connecting lines are filled and factory sealed.

REMOVAL AND INSTALLATION

1. Raise and safely support the vehicle.
2. Remove the nuts attaching the slave cylinder to the clutch housing.
3. Remove the hydraulic fluid line from the body retaining clips.
4. Lower the vehicle.

5. Remove the clutch master cylinder locating clip from the mounting bracket.

6. Remove the attaching ring, flat washer and wave washer that attach the clutch master cylinder pushrod to the clutch pedal.

7. Slide the pushrod off of the clutch pedal pin.

8. Make sure that the fluid remote reservoir is tight. Remove the screws that attach the reservoir to the mounting panel. Pull the clutch master cylinder rubber seal from the dash panel.

9. Rotate the clutch master cylinder 45° counterclockwise to unlock it.

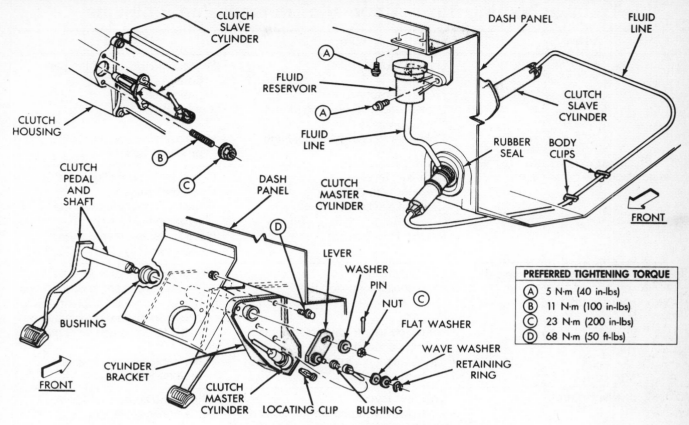

Clutch pedal and clutch hydraulic linkage

10. Remove the master cylinder from the dash panel. Remove the clutch cylinders, reservoir and connecting lines from the vehicle.

11. Position the cylinders, reservoir and connecting lines in the vehicle.

12. Insert the clutch master cylinder into the dash panel and rotate it 45° clockwise to lock it in position. Install the locating clip into the
master cylinder mounting bracket.

13. Lubricate the master cylinder rubber seal with liquid soap to ease installation. Seat the seal into position.

14. Install the remote reservoir. Tighten the mounting screws to 40 in. lbs. (5 N.m).

15. Connect the clutch master cylinder pushrod to the clutch pedal and secure it with the wave washer, flat washer and retainer ring.

16. Raise and safely support the vehicle.

17. Install the slave cylinder to the bell housing. Make sure the rod engages the clutch release fork. Secure the mounting nuts to 200 in. lbs. (23 N.m).

18. Secure the connecting line in the body clips. Lower the vehicle.

Troubleshooting Basic Automatic Transmission Problems

Problem	Cause	Solution
Fluid leakage	• Defective pan gasket	• Replace gasket or tighten pan bolts
	• Loose filler tube	• Tighten tube nut
	• Loose extension housing to transmission case	• Tighten bolts
	• Converter housing area leakage	• Have transmission checked professionally
Fluid flows out the oil filler tube	• High fluid level	• Check and correct fluid level
	• Breather vent clogged	• Open breather vent
	• Clogged oil filter or screen	• Replace filter or clean screen (change fluid also)
	• Internal fluid leakage	• Have transmission checked professionally
Transmission overheats (this is usually accompanied by a strong burned odor to the fluid)	• Low fluid level	• Check and correct fluid level
	• Fluid cooler lines clogged	• Drain and refill transmission. If this doesn't cure the problem, have cooler lines cleared or replaced.
	• Heavy pulling or hauling with insufficient cooling	• Install a transmission oil cooler
	• Faulty oil pump, internal slippage	• Have transmission checked professionally
Buzzing or whining noise	• Low fluid level	• Check and correct fluid level
	• Defective torque converter, scored gears	• Have transmission checked professionally
No forward or reverse gears or slippage in one or more gears	• Low fluid level	• Check and correct fluid level
	• Defective vacuum or linkage controls, internal clutch or band failure	• Have unit checked professionally
Delayed or erratic shift	• Low fluid level	• Check and correct fluid level
	• Broken vacuum lines	• Repair or replace lines
	• Internal malfunction	• Have transmission checked professionally

Lockup Torque Converter Service Diagnosis

Problem	Cause	Solution

No lockup	• Faulty oil pump
	• Sticking governor valve
	• Valve body malfunction
	   (a) Stuck switch valve
	   (b) Stuck lockup valve
	   (c) Stuck fail-safe valve
	• Failed locking clutch
	• Leaking turbine hub seal
	• Faulty input shaft or seal ring
		• Replace oil pump
		• Repair or replace as necessary
		• Repair or replace valve body or its internal components as necessary
		
		• Replace torque converter
		• Replace torque converter
		• Repair or replace as necessary

Will not unlock	• Sticking governor valve
	• Valve body malfunction
	   (a) Stuck switch valve
	   (b) Stuck lockup valve
	   (c) Stuck fail-safe valve
		• Repair or replace as necessary
		• Repair or replace valve body or its internal components as necessary

Stays locked up at too low a speed in direct	• Sticking governor valve
	• Valve body malfunction
	   (a) Stuck switch valve
	   (b) Stuck lockup valve
	   (c) Stuck fail-safe valve
		• Repair or replace as necessary
		• Repair or replace valve body or its internal components as necessary

Locks up or drags in low or second	• Faulty oil pump
	• Valve body malfunction
	   (a) Stuck switch valve
	   (b) Stuck fail-safe valve
		• Replace oil pump
		• Repair or replace valve body or its internal components as necessary

Sluggish or stalls in reverse	• Faulty oil pump
	• Plugged cooler, cooler lines or fittings
	• Valve body malfunction
	   (a) Stuck switch valve
	   (b) Faulty input shaft or seal ring
		• Replace oil pump as necessary
		• Flush or replace cooler and flush lines and fittings
		• Repair or replace valve body or its internal components as necessary

Loud chatter during lockup engagement (cold)	• Faulty torque converter • Failed locking clutch • Leaking turbine hub seal	• Replace torque converter • Replace torque converter • Replace torque converter
Vibration or shudder during lockup engagement	• Faulty oil pump • Valve body malfunction • Faulty torque converter • Engine needs tune-up	• Repair or replace oil pump as necessary • Repair or replace valve body or its internal components as necessary • Replace torque converter • Tune engine
Vibration after lockup engagement	• Faulty torque converter • Exhaust system strikes underbody • Engine needs tune-up • Throttle linkage misadjusted	• Replace torque converter • Align exhaust system • Tune engine • Adjust throttle linkage

Lockup Torque Converter Service Diagnosis

Problem	Cause	Solution
Vibration when revved in neutral Overheating: oil blows out of dip stick tube or pump seal	• Torque converter out of balance • Plugged cooler, cooler lines or fittings • Stuck switch valve	• Replace torque converter • Flush or replace cooler and flush lines and fittings • Repair switch valve in valve body or replace valve body
Shudder after lockup engagement	• Faulty oil pump • Plugged cooler, cooler lines or fittings • Valve body malfunction • Faulty torque converter • Fail locking clutch • Exhaust system strikes underbody • Engine needs tune-up • Throttle linkage misadjusted	• Replace oil pump • Flush or replace cooler and flush lines and fittings • Repair or replace valve body or its internal components as necessary • Replace torque converter • Replace torque converter • Align exhaust system • Tune engine • Adjust throttle linkage

Transmission Fluid Indications

The appearance and odor of the transmission fluid can give valuable clues to the overall condition of the transmission. Always note the appearance of the fluid when you check the fluid level or change the fluid. Rub a small amount of fluid between your fingers to feel for grit and smell the fluid on the dipstick.

If the fluid appears:	It indicates:
Clear and red colored	• Normal operation
Discolored (extremely dark red or brownish) or smells burned	• Band or clutch pack failure, usually caused by an overheated transmission. Hauling very heavy loads with insufficient power or failure to change the fluid, often result in overheating. Do not confuse this appearance with newer fluids that have a darker red color and a strong odor (though not a burned odor).
Foamy or aerated (light in color and full of bubbles)	• The level is too high (gear train is churning oil) • An internal air leak (air is mixing with the fluid). Have the transmission checked professionally.
Solid residue in the fluid	• Defective bands, clutch pack or bearings. Bits of band material or metal abrasives are clinging to the dipstick. Have the transmission checked professionally.
Varnish coating on the dipstick	• The transmission fluid is overheating

AUTOMATIC TRANSMISSION

Understanding Automatic Transmissions

The automatic transmission allows engine torque and power to be transmitted to the rear wheels within a narrow range of engine operating speeds. The transmission will allow the engine to turn fast enough to produce plenty of power and torque at very low speeds, while keeping it at a sensible rpm at high vehicle speeds. The transmission performs this job entirely without driver assistance. The transmission uses a light fluid as the medium for the transmission of power. This fluid also works in the operation of various hydraulic control circuits and as a lubricant. Because the transmission fluid performs all of these three functions, trouble within the unit can easily travel from one part to another. For this reason, and because of the complexity and unusual operating principles of the transmission, a very sound understanding of the basic principles of operation will simplify troubleshooting.

THE TORQUE CONVERTER

The torque converter replaces the conventional clutch. It has three functions:

1. It allows the engine to idle with the vehicle at a standstill, even with the transmission in gear.

2. It allows the transmission to shift from range to range smoothly, without requiring that the driver close the throttle during the shift.

3. It multiplies engine torque to an increasing extent as vehicle speed drops and throttle opening is increased. This has the effect of making the transmission more responsive and reduces the amount of shifting required.

The torque converter is a metal case which is shaped like a sphere that has been flattened on opposite sides. It is bolted to the rear end of the engine's crankshaft. Generally, the entire metal case rotates at engine speed and serves as the engine's flywheel.

The case contains three sets of blades. One set is attached directly to the case. This set forms the torus or pump. Another set is directly connected to the output shaft, and forms the turbine. The third set is mounted on a hub which, in turn, is mounted on a stationary shaft through a one-way clutch. This third set is known as the stator.

A pump, which is driven by the converter hub at engine speed, keeps the torque converter full of transmission fluid at all times. Fluid flows continuously through the unit to provide cooling.

Under low speed acceleration, the torque converter functions as follows:

The torus is turning faster than the turbine. It picks up fluid at the center of the converter and, through centrifugal force, slings it outward. Since the outer edge of the converter moves faster than the portions at the center, the fluid picks up speed.

The fluid then enters the outer edge of the turbine blades. It then travels back toward the center of the converter case along the turbine blades. In impinging upon the turbine blades, the fluid loses the energy picked up in the torus.

If the fluid were now to immediately be returned directly into the torus, both halves of the converter would have to turn at approximately the same speed at all times, and torque input and output would both be the same.

In flowing through the torus and turbine, the fluid picks up two types of flow, or flow in two separate directions. It flows through the turbine blades, and it spins with the engine. The stator, whose blades are stationary when the vehicle is being accelerated at low speeds, converts one type of flow into another. Instead of allowing the fluid to flow straight back into the torus, the stator's curved blades turn the fluid almost 90° toward the direction of rotation of the engine. Thus the fluid does not flow as fast toward the torus, but is already spinning when the torus picks it up. This has the effect of allowing the torus to turn much faster than the turbine. This difference in speed may be compared to the difference in speed between the smaller and larger gears in any gear train. The result is that engine power output is higher, and engine torque is multiplied.

As the speed of the turbine increases, the fluid spins faster and faster in the direction of engine rotation. As a result, the ability of the stator to redirect the fluid flow is reduced. Under cruising conditions, the stator is eventually forced to rotate on its one-way clutch in the direction of engine rotation. Under these conditions, the torque converter begins to behave almost like a solid shaft, with the torus and turbine speeds being almost equal.

THE PLANETARY GEARBOX

The ability of the torque converter to multiply engine torque is limited. Also, the unit tends to be more efficient when the turbine is rotating at relatively high speeds. Therefore, a planetary gearbox is used to carry the power output of the turbine to the driveshaft.

Planetary gears function very similarly to conventional transmission gears. However, their construction is different in that three elements make up one gear system, and, in that all three elements are different from one another. The three elements are: an outer gear that is shaped like a hoop, with teeth cut into the inner surface; a sun gear, mounted on a shaft and located at the very center of the outer gear; and a set of three planet gears, held by pins in a ring-like planet carrier, meshing with both the sun gear and the outer gear. Either the outer gear or the sun gear may be held stationary, providing more than one possible torque multiplication factor for each set of gears. Also, if all three gears are forced to rotate at the same speed, the gearset forms, in effect, a solid shaft.

Most modern automatics use the planetary gears to provide either a single reduction ratio of about 1.8:1, or two reduction gears: a low of about 2.5:1, and an intermediate of about 1.5:1. Bands and clutches are used to hold various portions of the gearsets to the transmission case or to the shaft on which they are mounted. Shifting is accomplished, then, by changing the portion of each planetary gearset which is held to the transmission case or to the shaft.

THE SERVOS AND ACCUMULATORS

The servos are hydraulic pistons and cylinders. They resemble the hydraulic actuators used on many familiar machines, such as bulldozers. Hydraulic fluid enters the cylinder, under pressure, and forces the piston to move to engage the band or clutches.

The accumulators are used to cushion the engagement of the servos. The transmission fluid must pass through the accumulator on the way to the servo. The accumulator housing contains a thin piston which is sprung away from the discharge passage of the accumulator. When fluid passes through the accumulator on the way to the servo, it must move the piston against spring pressure, and this action smooths out the action of the servo.

THE HYDRAULIC CONTROL SYSTEM

The hydraulic pressure used to operate the servos comes from the main transmission oil pump. This fluid is channeled to the various servos through the shift valves. There is generally a manual shift valve which is operated by the transmission selector lever and an automatic shift valve for each automatic upshift the transmission provides: i.e., 2-speed automatics have a low/high shift valve, while 3-speeds have a 1-2 valve, and a 2-3 valve.

There are two pressures which effect the operation of these valves. One is the governor pressure which is affected by vehicle speed. The other is the modulator pressure which is affected by intake manifold vacuum or throttle position. Governor pressure rises with an increase in vehicle speed, and modulator pressure rises as the throttle is opened wider. By responding to these two pressures, the shift valves cause the upshift points to be delayed with increased throttle opening to make the best use of the engine's power output.

Most transmissions also make use of an auxiliary circuit for downshifting. This circuit may be actuated by the throttle linkage or the vacuum line which actuates the modulator, or by a cable or solenoid. It applies pressure to a special downshift surface on the shift valve or valves.

The transmission modulator also governs the line pressure, used to actuate the servos. In this way, the clutches and bands will be actuated with a force matching the torque output of the engine.

The LoadFlite® automatic transmission consists of a torque converter and a fully automatic three-speed gear system, housed in an integral aluminum casing. The transmission uses two multiple disc clutches, an overrunning clutch, two servos, two bands, and two planetary gearsets to provide three forward, and one reverse gear. The transmission hydraulic system is composed of a fluid pump and a single valve body that contains all of the control valves, except for the governor valve. The torque converter is a sealed unit and cannot be disassembled.

The transmission fluid is filtered through an internal dacron filter which is attached to the lower side of the valve body. The fluid is cooled by circulating it through a cooler in the lower radiator tank.

Fluid Pan

REMOVAL AND INSTALLATION

1. Raise the front of the van and support it on jackstands. Place a large drain pan under the transmission.
2. Loosen the pan attaching bolts and tap the pan at one corner to break it loose.
3. Allow the fluid to drain into the drain pan.
4. After most of the fluid has drained, carefully remove the attaching bolts, lower the pan and drain the rest of the fluid.
5. Remove the filter attaching screws and remove the filter.
6. Install a new filter. Tighten the screws to 35 in. lbs (4 N.m).
7. Thoroughly clean the fluid pan with safe solvent and allow it to dry.
8. Using a new gasket, install the pan to the transmission. Tighten the attaching bolts to 150 in. lbs. (17 N.m).
9. Lower the vehicle. Pour three quarts of MOPAR ATF Plus, Type 7176, or the equivalent such as Dexron®II automatic transmission fluid in through the dipstick tube.
10. Start the engine and allow it to run for a few minutes. With the parking brake set, slowly move the gear selector to each position. Return it to the Neutral position.
11. Check the fluid level. Add more fluid as necessary to bring it up to **ADD** level.
12. Drive the van to bring the transmission up to normal operating temperature. Check the level again. It should be between the **ADD** and **FULL** marks.

FILTER SERVICE

Always change the filter when changing the transmission fluid. Refer to the previous Fluid Pan servicing for instructions.

Adjustments

FRONT (KICKDOWN) BAND

The kickdown band adjusting screw is located on the left-hand side of the transmission case near the throttle lever shaft.

1. Loosen the locknut and back it off about five turns. Be sure that the adjusting screw turns freely in the case.
2. Torque the adjusting screw to 72 in. lbs. (8 N.m).
3. Back off the adjusting screw 2½ turns.
4. Hold the adjuster screw in position and tighten the locknut to 30 ft.lbs. (41 N.m).

REAR (LOW AND REVERSE) BAND

The pan must be removed from the transmission to gain access to the low and reverse band adjusting screw.

1. Drain the transmission fluid and remove the pan.
2. Loosen the band adjusting screw locknut and back it off about five turns. Be sure that the adjusting screw turns freely in the lever.
3. Torque the adjusting screw to 72 in. lbs. (8 N.m)
4. Back off the adjusting screw as follows:
● A-727 & A-518: 2 turns
● A-998 & A-999: 4 turns
● A-500: 4 turns
Keep the adjusting screw from turning, tighten and torque the locknut to 25 ft.lbs. (34 N.m).
5. Use a new gasket and install the transmission pan.

SHIFT LINKAGE

NOTE: To insure proper adjustment, it is suggested that new linkage grommets be installed.

1. Place the gearshift lever in Park position.
2. Move the shift control lever on the transmission all the way to the rear (in the Park detent).
3. Set the adjustable rod to the proper length and install it with no load in either direction. Tighten the swivel bolt.
4. The shift linkage must be free of binding and be positive in all positions. Make sure that the engine can start only when the gearshift lever is in the Park or Neutral position. Be sure that the gearshift lever will not jump into an unwanted gear.

Neutral Start Switch

ADJUSTMENT/REMOVAL AND INSTALLATION

The neutral safety switch is thread mounted into the transmission case. When the gearshift lever is placed in either the Park or Neutral position, a cam, which is attached to the transmission throttle lever inside the transmission, contact the neutral safety switch and provides a ground to complete the starter solenoid circuit.

The back-up light switch is incorporated into the neutral safety switch. The center terminal is for the neutral safety switch and the two outer terminals are for the back-up lamps.

There is no adjustment for the switch. If a malfunction occurs, the switch must be removed and replaced. To remove the switch:

1. Disconnect the electrical leads and unscrew the switch. Use a drain pan to catch the transmission fluid.
2. Using a new seal, install the new switch and torque it to 25 ft.lbs. (34 N.m).
3. Check the transmission fluid level as follows:
4. Start the engine and idle it for at least 2 minutes.
5. Set the parking brake and move the selector through each position, ending in Park.

6. Add sufficient fluid to bring the level to the FULL mark on the dipstick. The level should be checked in Park, with the engine idling at normal operating temperature.

Extension Housing Seal

REMOVAL AND INSTALLATION

NOTE: Special tool C-3985 Seal Remover and C-3995 or C-3972 Seal Installer, or their equivalents are required.

1. Raise and safely support the vehicle.
2. Place a suitable container under the extension housing end.
3. Mark the rear universal driveshaft yoke and differential for installation reference. Disconnect the rear universal joint and remove the driveshaft.
4. Remove the old extension housing seal using the special tool.
5. Clean the mounting surfaces of the extension housing.
6. Place the new seal into position and install it with the special tool by tapping it into place.
7. Lubricate the seal lips and slip yoke spline on the front of the driveshaft.

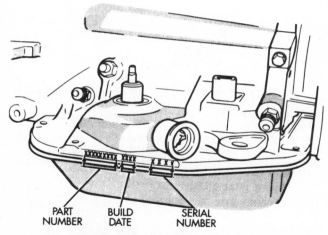

Transmission ID number and code location

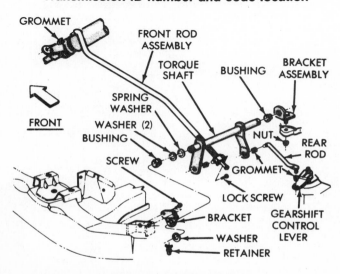

Column gearshift linkage

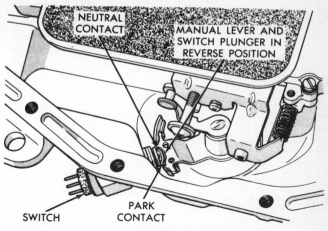

Neutral switch contacts

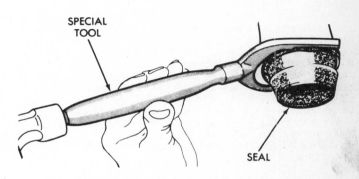

Removing the extension housing seal

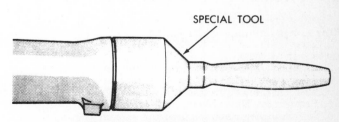

Installing the extension housing seal

8. Install the driveshaft. Lower the vehicle and check the transmission fluid. Add fluid as necessary.

Transmission

REMOVAL AND INSTALLATION

1. Remove the transmission and converter as an assembly; otherwise the converter drive plate pump bushing and oil seal will be damaged. The drive plate will not support a load. Therefore, none of the weight of transmission should be allowed to rest on the plate during removal.
2. Disconnect the negative battery ground cable. Remove the engine cover. Raise and safely support the vehicle.
3. Remove engine to transmission struts, if necessary. You may have to drop the exhaust system on some models.
4. Remove the starter. Remove wire from the neutral starting switch and converter solenoid (if equipped).

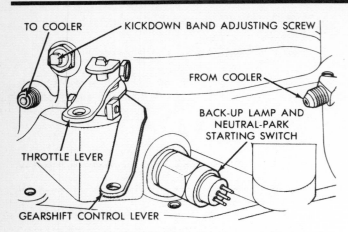

Front band adjusting screw location

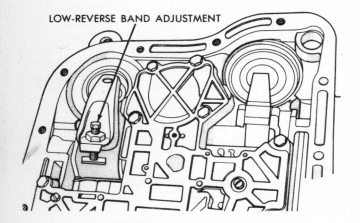

Rear band adjusting screw location

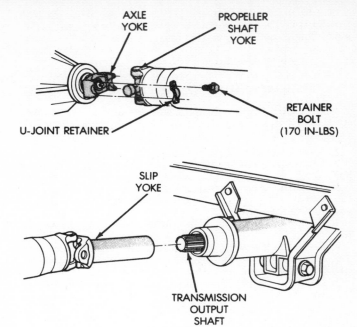

Driveshaft mounting

5. Remove gearshift cable or rod from the transmission and the lever.

6. Disconnect the throttle rod from left side of transmission. Disconnect the distance sensor and speedometer.

7. Drain the transmission fluid and reinstall the pan. Disconnect the oil cooler lines at transmission and remove the oil filler tube. Disconnect the speedometer cable.

8. Remove the converter front cover. Mark the converter and flexplate for installation reference. Remove the flexplate to converter mounting bolts/nuts. Turn the engine in a clockwise direction to gain access to the blots. Mark for reference and remove the driveshaft.

9. Install engine support fixture or position a jack to hold up the rear of the engine. Care should be taken not to place strain on the radiator hoses, etc. Watch fan to radiator clearance.

10. Position a suitable transmission jack under the transmission. Raise the transmission slightly with the jack to relieve load and remove support bracket or crossmember. Remove all of the bell housing bolts and carefully work transmission and converter rearward off engine dowels and disengage converter hub from end of crankshaft.

WARNING: Attach a small C-clamp to edge of bell housing to hold converter in place during transmission removal; otherwise the front pump bushing might be damaged.

Install transmission and converter as an assembly. The drive plate will not support a load. Do not allow weight of transmission to rest on the plate during installation.

11. Using a jack, position transmission and converter assembly in alignment with engine.

12. Rotate converter so mark on converter (made during removal) will align with the mark on the flexplate. Carefully work the transmission assembly forward over engine block dowels with converter hub entering the crankshaft opening and the flexplate to converter mounts lined up.

13. Install converter housing to engine bolts and tighten them.

14. Install the converter to flexplate bolts.

15. Install the rear crossmember and connect the rear transmission mount if loosened. Install the driveshaft.

16. Connect oil cooler lines, install oil filler tube and connect the speedometer and distance sensor.

17. Connect gearshift cable or rod and torque shaft assembly to the transmission case and to the lever.

18. Connect throttle rod to the lever at left side of transmission bell housing.

19. Connect wire to back-up and neutral starting switch. Connect the converter solenoid (if equipped).

20. Install cover plate in front of the converter assembly. Install the starter motor and engine struts.

19. Lower the vehicle and refill transmission with fluid.

20. Adjust linkage.

DRIVELINE

Troubleshooting Basic Driveshaft and Rear Axle Problems

When abnormal vibrations or noises are detected in the driveshaft area, this chart can be used to help diagnose possible causes. Remember that other components such as wheels, tires, rear axle and suspension can also produce similar conditions.

BASIC DRIVESHAFT PROBLEMS

Problem	Cause	Solution
Shudder as car accelerates from stop or low speed	• Loose U-joint • Defective center bearing	• Replace U-joint • Replace center bearing
Loud clunk in driveshaft when shifting gears	• Worn U-joints	• Replace U-joints
Roughness or vibration at any speed	• Out-of-balance, bent or dented driveshaft • Worn U-joints • U-joint clamp bolts loose	• Balance or replace driveshaft • Replace U-joints • Tighten U-joint clamp bolts
Squeaking noise at low speeds	• Lack of U-joint lubrication	• Lubricate U-joint; if problem persists, replace U-joint
Knock or clicking noise	• U-joint or driveshaft hitting frame tunnel • Worn CV joint	• Correct overloaded condition • Replace CV joint

BASIC REAR AXLE PROBLEMS

First, determine when the noise is most noticeable.

Drive Noise—Produced under vehicle acceleration.

Coast Noise—Produced while the car coast with a closed throttle.

Float Noise—Occurs while maintaining constant car speed (just enough to keep speed constant) on a level road.

Road Noise

Brick or rough surfaced concrete roads produce noises that seem to come from the rear axle. Road noise is usually identical in Drive or Coast and driving on a different type of road will tell whether the road is the problem.

Tire Noise

Tire noises are often mistaken for rear axle problems. Snow treads or unevenly worn tires produce vibrations seeming to originate elsewhere. Temporarily inflating the tire to 40 lbs will significantly alter tire noise, but will have no effect on rear axle noises (which normally cease below about 30 mph).

Engine/Transmission Noise

Determine at what speed the noise is more pronounced, then stop the car in a quiet place. With the transmission in Neutral, run the engine through speeds corresponding to road speeds where the noise was noticed. Noises produced with the car standing still are coming from the engine or transmission.

Front Wheel Bearings

While holding the car speed steady, lightly apply the foot brake; this will often decease bearing noise, as some of the load is taken from the bearing.

Rear Axle Noises

Eliminating other possible sources can narrow the cause to the rear axle, which normally produces noise from worn gears or bearings. Gear noises tend to peak in a narrow speed range, while bearing noises will usually vary in pitch with engine speeds.

NOISE DIAGNOSIS

The Noise Is	Most Probably Produced By
· Identical under Drive or Coast	· Road surface, tires or front wheel bearings
· Different depending on road surface	· Road surface or tires
· Lower as the car speed is lowered	· Tires
· Similar with car standing or moving	· Engine or transmission
· A vibration	· Unbalanced tires, rear wheel bearing, unbalanced driveshaft or worn U-joint
· A knock or click about every 2 tire revolutions	· Rear wheel bearing
· Most pronounced on turns	· Damaged differential gears
· A steady low-pitched whirring or scraping, starting at low speeds	· Damaged or worn pinion bearing
· A chattering vibration on turns	· Wrong differential lubricant or worn clutch plates (limited slip rear axle)
· Noticed only in Drive, Coast or Float conditions	· Worn ring gear and/or pinion gear

DRIVELINE

Driveshaft and U-Joints

REMOVAL AND INSTALLATION

1. Raise and support the rear of the vehicle. If both ends of the vehicle are raised, keep the front slightly lower than the rear. This will help prevent transmission fluid from spilling when the driveshaft is removed.

2. Matchmark the shaft and pinion flange to assure proper balance at installation.

3. Remove both rear U-joint roller and bushing clamps from the rear axle pinion flange. Do not disturb the retaining strap that holds the bushing assemblies on the U-Joint cross (if equipped).

NOTE: Do not allow the driveshaft to hang during removal. Suspend it from the frame with a piece of wire. Before removing the driveshaft, lower the front end to prevent loss of fluid.

4. Slide the driveshaft from the van.
5. Service as required.
6. Position the driveshaft under the vehicle. Guide the slip yoke into the transmission extension housing. Align the rear joint with the pinion yoke and secure the universal joint. Lower the vehicle.

U-JOINT OVERHAUL

1. If replacing the front sliding yoke assembly u-joint, paint

Driveshaft and U-joints

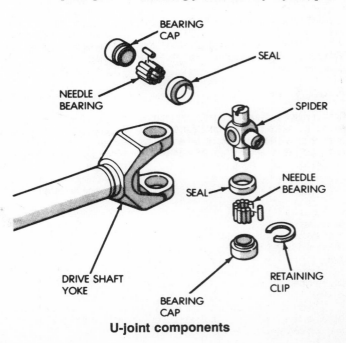

U-joint components

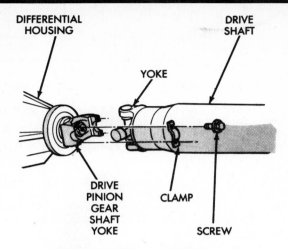

Driveshaft to pinion yoke

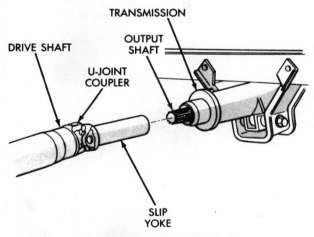

Slip yoke to output shaft

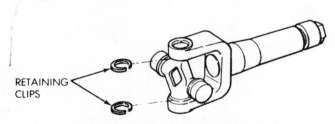

Bearing cap retaining clips

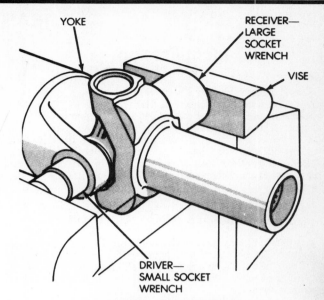

Bearing cap removal

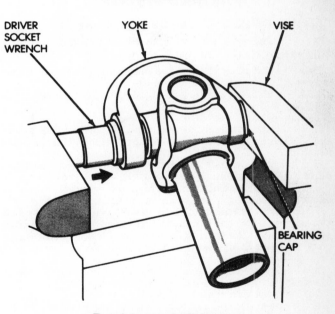

Bearing cap installation

or score alignment marks on the yoke and driveshaft for installation reference.

2. Position the yoke in a soft jawed vise. Support the free end of the driveshaft. Remove the retaining clips from the bearing caps.

3. Loosen the vise and place a socket bigger than the bearing cap on one side and one smaller than the bearing cap on the other side.

4. Tighten the vise, forcing one cap out into the larger socket. Reverse the vise and socket arrangement to remove the other bearing cap and the cross.

5. Clean any burrs or sharp edges on the yoke. Apply lithium base lubricant to the yoke bores, bearing caps and needles.

6. Place a bearing cap in position in the yoke bore and press in slightly. Position the cross into the yoke with one end in the bearing caps. Place the other side bearing cap in position on the yoke, align the cross and push the cap in slightly. Tap the bearing caps in until the cross is retained.

7. Locate a socket that is smaller than the bearing cap. Place the socket against one bearing cap. Position the yoke, with the socket on one bearing cap, in between the jaws of the vise. Tighten the vise to drive the caps in position on the cross.

8. Press the socket end in far enough to expose the retainer slot. Loosen the vise and install the retainer. Reverse the socket and install the other side retainer.

REAR AXLE

Identification

1988-91
- Chrysler semi-floating 8⅜ in. ring gear
- Chrysler semi-floating 9¼ in. HD ring gear
- Spicer (Dana) 60 full-floating 9¾ in. ring gear
- Spicer (Dana 60HD) full-floating 9¾ in. ring gear

The Chrysler 8⅜ in. and 9¼ in. axles can be fitted with a Sure-Grip differential. The Spicer axles can be fitted with a Trac-Lok differential. Both of the limited-slip differentials require a friction modifier to be added to the lubricant.

Axle Shafts and Bearings

Before servicing any axle shafts, be sure to jack and support the truck so that both rear wheels are off the ground. This will ensure that the vehicle will not roll off the supports if the vehicle is equipped with a limited slip rear axle and one wheel is turned inadvertently.

REMOVAL AND INSTALLATION

Chrysler 8⅜ in., 9¼ in. and 9¼ in. HD Axles

NOTE: There is no provision for axle shaft end-play adjustment on these axles.

1. Raise and safely support the rear of the vehicle with both wheels off of the ground. Remove the rear wheels.
2. Clean all dirt from the housing cover and remove the housing cover to drain the lubricant.
3. Remove the brake drum.
4. Rotate the differential case until the differential pinion shaft lockscrew can be removed. Remove the lockscrew and pinion shaft.
5. Push the axle shaft toward the center of the vehicle and remove the C-locks from the groove at the end of the axle shaft.
6. Pull the axle shaft from the housing, being careful not to damage the bearing which remains in the housing.
7. Remove the seal from the housing with a small pry bar. Use a appropriate slide hammer puller and attachment, if necessary, to remove the bearing from the axle housing. Always replace the bearing and seal when they have been removed from the axle housing.
8. Check the bearing shoulder in the axle housing for imperfections. These should be corrected with a file.
9. Clean the axle shaft bearing cavity.
10. Grease and install the axle shaft bearing in the cavity. An installer tool is recommended, although a driver can be used. Be sure that the bearing is not cocked and that it is seated firmly against the shoulder.
11. Install the axle shaft bearing seal. It should be seated beyond the end of the flange face.
12. Insert the axle shaft, making sure that the splines do not damage the seal. Be sure that the splines are properly engaged with the differential side gear splines.
13. Install the C-locks in the grooves on the axle shafts. Pull the shafts outward so that the C-locks seat in the counterbore of the differential side gears.
14. Install the differential pinion shaft through the case and pinions. Install the lockscrew and secure it in position.
15. Clean the housing and gasket surfaces. Install the cover and a new gasket. Refill the axle with the specified lubricant.

NOTE: Replacement differential cover gaskets may not be available. The use of gel type nonsticking sealant is recommended.

Spicer 60
AXLE

1. Remove the axle shaft flange locknuts and washers, or bolts.
2. On axles with locknuts, rap the axle shafts sharply in the center of the flange with a hammer to free the dowels.
3. Remove the tapered dowels and/or axle shafts.
4. Clean the gasket area with solvent and install a new flange gasket.
5. Install the axle shaft into the housing.
6. If the axle has an outer wheel bearing seal, install new gaskets on each side of the seal mounting flange.
7. Install the tapered dowels, lockwashers, and nuts or Durlock® bolts. Torque the nuts to 70 ft.lbs. (95 N.m).

REAR AXLE BEARINGS

For Removal, Repacking and Installation, see Section 1.

ADJUSTMENT

1. Raise and support the rear end on jackstands.
2. Remove the axle shaft.
3. Remove the lock ring and loosen the adjusting nut.
4. While rotating the wheel, tighten the adjusting nut to 120-140 ft.lbs. (163-190 N.m).

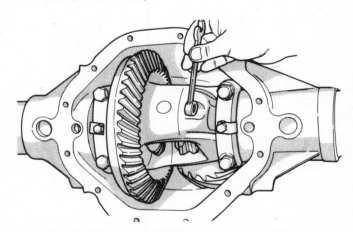

Remove/install the differential pinion shaft lock pin

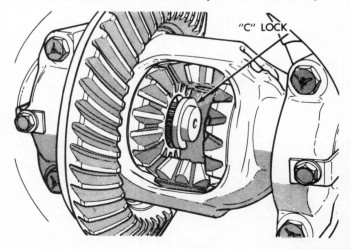

Remove/install the axle shaft C-lock retainers

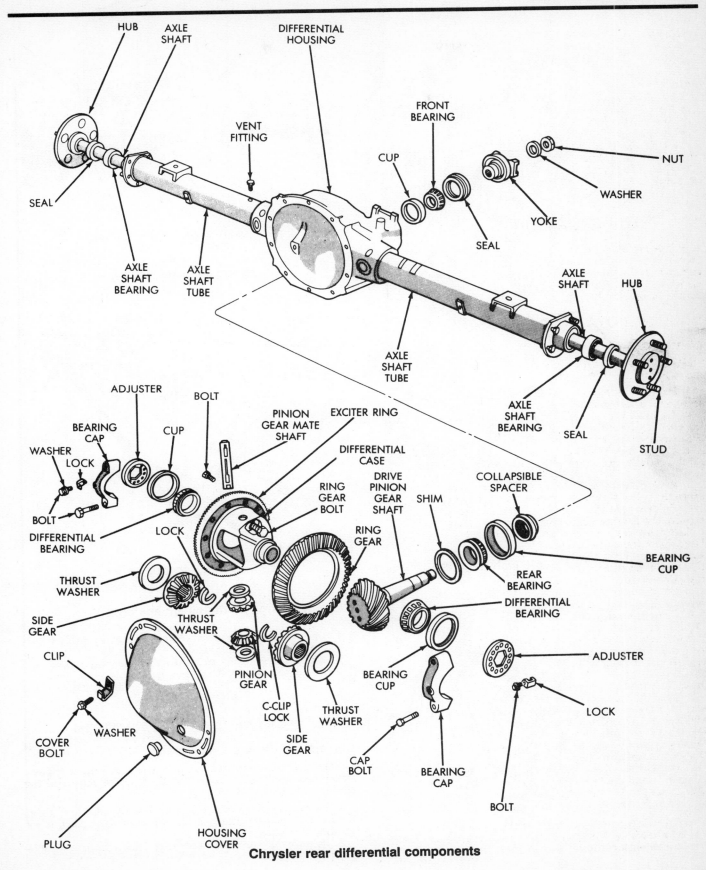

Chrysler rear differential components

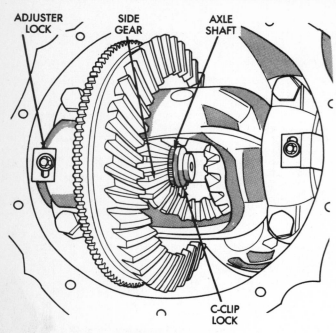

Axle shaft C-lock

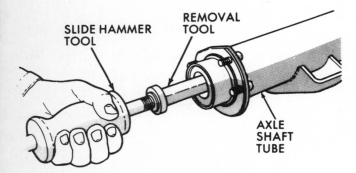

Bearing removal — 8⅜ inch ring gear

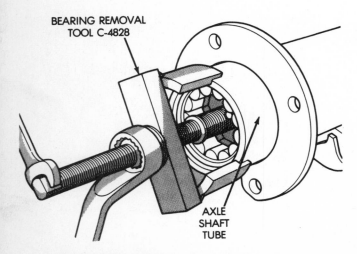

Bearing removal — 9¼ inch ring gear

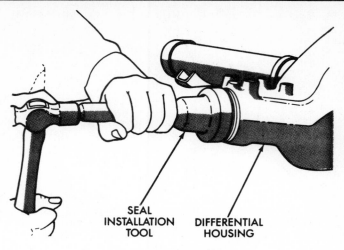

Pinion seal installation

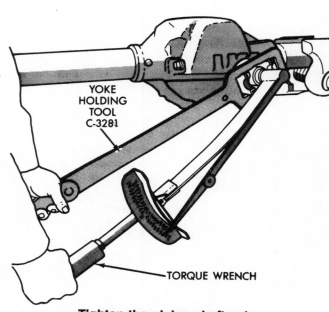

Tighten the pinion shaft nut

5. Back off the nut ⅓ turn (120°). This will provide 0.001-0.008 in. (0.025–0.203mm) endplay.
6. Install the lock ring onto the spindle keyway.
7. Install the axle shaft and new gasket.

Pinion Seal

REMOVAL AND INSTALLATION

Chrysler 8⅜ in. and 9¼ in. Axles

NOTE: An inch-pound torque wrench and a torque wrench capable of at least 250 ft. lbs. are required for pinion seal installation.

1. Raise and safely support the vehicle with jackstands under the frame rails. Allow the axle to drop to rebound position for working clearance.
2. Remove the rear wheels and brake drums. No drag must be present on the axle.

3. Mark the companion flanges and U-joints for correct reinstallation position.

4. Remove the driveshaft.

5. Using an inch pound torque wrench and socket on the pinion yoke nut measure the amount of torque needed to maintain differential rotation through several clockwise revolutions. Record the measurement.

6. Use a suitable tool to hold the companion flange. Remove the pinion nut and washer.

7. Place a drain pan under the differential, clean the area around the seal, and mark the yoke-to-pinion relation.

8. Use a 2-jawed puller to remove the pinion.

9. Remove the seal with a small prybar.

10. Thoroughly clean the oil seal bore.

NOTE: If you are not absolutely certain of the proper seal installation depth, the proper seal driver must be used. If the seal is misaligned or damaged during installation, it must be removed and a new seal installed.

11. Drive the new seal into place with a seal driver. Coat the seal lip with clean, waterproof wheel bearing grease.

12. Coat the splines with a small amount of wheel bearing grease and install the yoke, aligning the matchmarks. Never hammer the yoke onto the pinion!

13. Install a NEW nut on the pinion.

14. Hold the yoke with a holding tool. Tighten the pinion nut to 210 ft. lbs. Using the inch-pound torque wrench, take several readings. Continue tightening the nut until the original recorded preload reading is achieved.

NOTE: Under no circumstances should the preload be more than 5 in. lbs. higher than the original reading.

15. Bearing preload should be uniform through several complete revolutions. If binding exists, the condition must be diagnosed and corrected. The assembly is unacceptable if the final pinion nut torque is below 210 ft. lbs. or pinion bearing preload is not correct.

WARNING: Under no circumstances should the nut be backed off to reduce the preload reading! If the preload is exceeded, the yoke and bearing must be removed and a new collapsible spacer must be installed. The entire process of preload adjustment must be repeated.

16. Install the driveshaft using the matchmarks. Torque the nuts to 15 ft. lbs.

Dana 60

NOTE: A torque wrench capable of at least 300 ft.lbs. is required for pinion seal installation.

1. Raise and support the truck on jackstands.

2. Allow the axle to hang freely.

3. Matchmark and remove the driveshaft from the axle.

4. Using a holding tool, hold the pinion flange while removing the pinion nut.

5. Using a puller, remove the pinion flange.

6. Use a puller to remove the seal, or punch the seal out using a pin punch.

7. Thoroughly clean the seal bore and make sure that it is not damaged in any way. Coat the sealing edge of the new seal with a small amount of 80W/90 oil and drive the seal into the housing using a seal driver.

8. Coat the inside of the pinion flange with clean 80W/90 oil and install the flange onto the pinion shaft.

9. Install the nut on the pinion shaft and tighten it to 250-300 ft. lbs.

10. Install the driveshaft. Lower the vehicle.

Axle Housing

REMOVAL AND INSTALLATION

Chrysler 8⅜ in. and 9¼ in. Axles

1. Raise and support the rear end on jackstands placed under the frame.

2. Prop the brake pedal in the UP position.

3. Remove the wheels. Do not remove the brake drums.

4. Disconnect the brake lines.

5. Disconnect the parking brake cables.

6. Matchmark the driveshaft and flange.

7. Disconnect the driveshaft at the flange and position it out of the way.

8. Disconnect the shock absorbers at the axle.

9. Position a floor jack under the axle to take up the weight.

10. Remove the spring U-bolt nuts and lower the axle.

11. Raise the axle into position.

12. Install the spring U-bolt nuts.

13. Connect the shock absorbers at the axle. Torque the nuts to 55 ft. lbs.

14. Connect the driveshaft at the flange.

15. Connect the parking brake cables.

16. Connect the brake lines.

17. Install the wheels.

18. Bleed the brakes.

Observe the following torques for U-bolt nuts:

- ½-20: 62-70 ft. lbs.
- ¾-16: 175-225 ft. lbs.
- ⁹⁄₁₆-18: 120-130 ft. lbs.
- ⅝-18: 175-200 ft. lbs.

Spicer 60

1. Raise and support the rear end on jackstands placed under the frame.

2. Prop the brake pedal in the UP position.

3. Remove the axle shafts.

4. Remove the wheels and drums.

5. Disconnect the brake line and the flexible line connector.

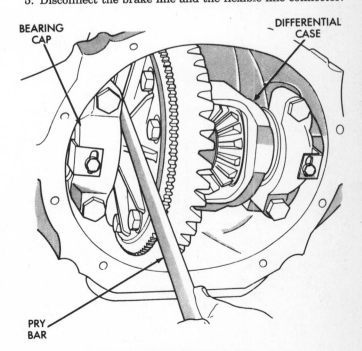

Differential side play test

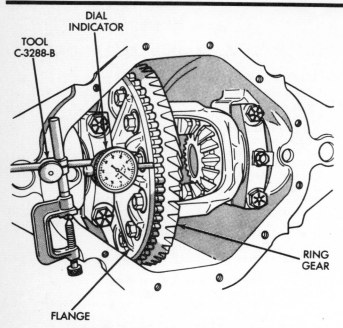

Ring gear runout measurement

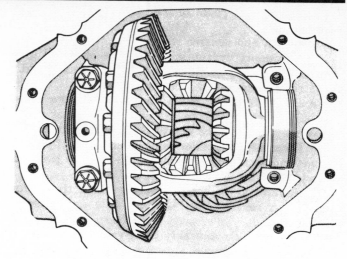

Carrier with one bearing removed

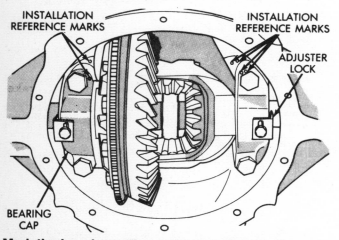

Mark the housing and caps for installation reference

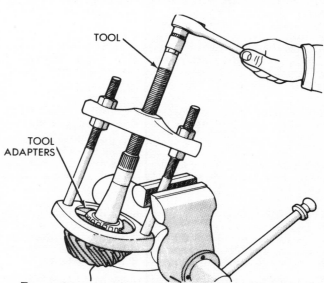

Removing the drive bearing rear bearing cone

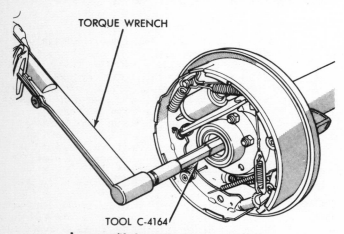

Loosen/tighten the hex adjuster

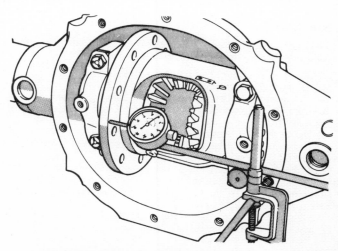

Checking the mounting flange runout

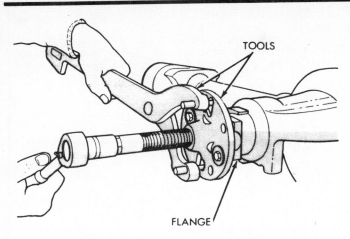

Removing the drive pinion companion flange

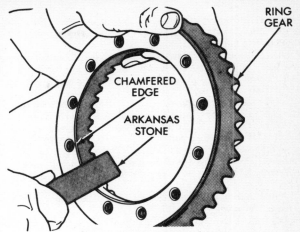

Sharp edge removal

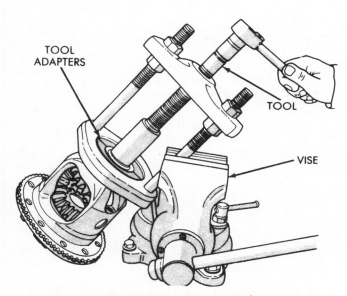

Differential bearing removal

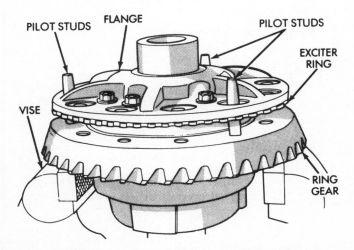

Differential case to heated ring gear alignment

6. Disconnect the parking brake cables.
7. Matchmark the driveshaft and flange.
8. Disconnect the driveshaft at the flange and position it out of the way.
9. Disconnect the shock absorbers at the axle.
10. Position a floor jack under the axle to take up the weight.
11. Remove the spring U-bolt nuts and lower the axle.
12. Raise the axle into position.
13. Install the spring U-bolt nuts. Torque the nuts to 200 ft. lbs.
14. Connect the shock absorbers at the axle. Torque the nuts to 55 ft. lbs.
15. Connect the driveshaft at the flange.
16. Connect the parking brake cables.
17. Connect the brake lines.
18. Install the drums and wheels. On 60 series axles, torque the lug nuts to 225 ft. lbs for cone-shaped nuts; 325 ft. lbs. for flanged nuts. Bleed the brakes.
20. Install the axle shafts.
21. Refill the housing with fluid.

Differential

OVERHAUL

Special tools are required throughout the following procedures. If the tools are not on hand, or obtainable, do not attempt overhauling the differential.

Chrysler 8⅜ in. and 9¼ in. Axles

It is not necessary to remove the complete axle assembly from the vehicle for differential service.

1. Raise and safely support the rear of the vehicle on the frame rails in front of rear spring. This will allow the axle assembly to hang free for more working room. Remove the axles and driveshaft. Thoroughly clean the outer area of the housing and axle tubes with a safe cleaning solvent and wipe dry.
2. Place a drain pan under the differential. Remove the cover and drain the lubricant.
3. If equipped with a Sure-Grip differential do not perform this step. Wash and flush the internal components of the differential. Allow them to dry.
4. Side play and runout checks should be made at this time. They will be very useful in reassembly.
5. To measure for side play: Position a small pry bar between the left side of the axle housing and case flange. Use a prying motion to determine if any side play exists. There should be no side play.

6. Side play resulting from bearing races being loose on the case hubs requires replacement of the differential case. Otherwise, use the threaded adjuster to remove the side play before measuring the ring gear runout.

7. Eliminate any side play. Attach a dial indicator to a pilot stud. Place the indicator at a right angle to the brake exciter ring or ring gear. The plunger should exert a slight pressure. Zero the dial.

8. Rotate the ring gear several revolutions. Observe the dial indicator pointer. Mark the differential case and ring gear at the maximum runout. Runout should not exceed 0.005 in. (0.127mm). If runout exceeds 0.005 in. (0.127mm) a damaged differential case could be the cause. The marking of the ring gear and differential case will be use as reference later on.

9. Remove the rear wheel anti-lock (RWAL) sensor, if equipped.

10. Mark the housing and differential case bearing caps for installation reference.

11. Remove the threaded bearing adjuster lock from each bearing cap. Loosen, but do not remove the bearing caps.

12. Insert the special wrench C-4164 through the axle tube and loosen the threaded bearing adjusters on each side. Use extreme caution to avoid damage to the RWAL brake exciter ring, if equipped.

13. Remove the bearing caps, adjusters and the differential case assembly. Each differential bearing cup and threaded adjuster must be kept with their respective bearing. They are a matched set.

14. Use an inch-pound torque wrench to measure the pinion gear preload. Rotate the pinion yoke with the torque wrench and record the maximum amount of torque necessary. Remove the pinion yoke.

15. Remove the pinion seal. Force the pinion shaft out of the front bearing and remove the shaft assembly. This will damage the front bearing rollers and bearing cup. They must be replaced at assembly time. Discard the collapsible spacer.

16. Use the appropriate puller and remove the pinion front and rear bearing cups. Remove the pinion gear depth shims. Keep the shims together.

17. Press the rear pinion bearing from the shaft.

18. Do not remove the ring gear from the case unless gear replacement or flange runout must be measured.

19. To measure flange runout: If the ring gear runout exceed 0.005 in. (0.127mm) the case flange runout should be checked. Clamp the case assembly in a soft jawed vise. Remove the ring gear mounting bolts. The bolts have LEFT-HAND threads. Remove the ring gear using a hammer and brass drift.

20. Detach the RWAL brake exciter, if equipped, with a brass drift.

21. Install the case into the housing with the bearing cups and threaded adjusters to their original positions.

22. Install the bearing caps and bolts. Use the special wrench to tighten both adjusters through the axle tubes and remove all side play.

23. Mount a dial indicator and position the plunger against the right side of the case flange between the outer edge and bolt holes. Zero the indicator.

24. Rotate the case several times. Mark the area of maximum runout. The flange runout should not exceed 0.003 in. (0.08mm). If the runout exceeds 0.003 in. (0.08mm), the differential case must be replaced.

25. It is possible to reduce the combined ring gear/flange runout by positioning the marked area of maximum runout 180° opposite the other marked maximum area.

26. Remove the differential case from the axle housing.

27. Clamp the case assembly in a soft jawed vise.

28. Remove the pinion gear mate shaft lock screw and the shaft. Rotate the differential side gears until the differential pinion gears are located at a case opening and remove them.

29. Remove the differential side gears and thrust washers.

30. Remove the case from the vise. Press off the carrier side bearings.

31. Clean all of the components in a safe solvent and allow them to dry. Inspect all components and replace as necessary. Inspect the RWAL brake exciter ring, if equipped, for damaged or broken teeth. Replace it if necessary.

Exciter Ring Replacement

If installed, the ring gear must be removed before the RWAL brake exciter ring can be replaced.

1. If RWAL brake exciter ring replacement is necessary, remove with a hammer and drift.

2. Heat the replacement exciter ring with a heat lamp or by immersing in a hot fluid. The temperature should not exceed (149°C) 300°F. Do not use a torch to heat the ring.

3. After heating, quickly position the exciter ring on the differential case adjacent to the flange.

Differential Case Assembly

1. Lubricate all the differential case components with gear lubricant. Place the thrust washers on the differential side gears. Position the gears in the differential case counterbores.

NOTE: If replacement side gears or thrust washers are used, refer to Side Gear Clearance Measurement and Adjustment.

2. Position the thrust washers on the differential pinion gears. Mesh the pinion gears with the side gears. Ensure the pinion gears are exactly 180° opposite each other. Rotate the side gears to align the pinion gears and thrust washers. Align these components with the mate shaft bores in the case.

3. Install the lock screw and tighten it with (10 N.m) 90 in. lbs. If the ring gear was removed, clean all contact surfaces.

4. Use an Arkansas stone to remove any sharp areas from the chamfered inside diameter. If removed, install the exciter ring.

5. Heat the ring gear with a heat lamp or by immersing in a hot fluid. The temperature should not exceed (149°C) 300°F. Do not use a torch to heat the ring.

6. Position the heated gear on to the case adjacent to the exciter ring. Use three equally spaced pilot studs to align the gear with the flange holes.

7. Insert new ring gear bolts (with left hand threads) through the differential case flange and thread them into the ring gear. Alternately tighten each bolt to (95 N.m) 70 ft.lbs.

8. Place a differential bearing on each hub (bearing taper facing outward). Carefully seat the bearings on the hubs. An arbor press should be used with the installation tools. When installing a differential bearing, never apply force to the bearing cage because bearing damage will result.

Pinion Gear Assembly/Installation

1. Select the correct pinion gear adjustment gauges. 8⅜ in. axles use Tool Set C-3715-B or C-758-D6 for 9¼ in. axles (or equivalents).

2. Insert both bearing cups into the differential housing bores. On 8⅜ in. diameter, position Locating Spacer SP-6030 over the shaft of SP-5385 and follow it with the rear bearing. Position the tools (with bearing) in the differential housing. Install Shaft Locating Sleeve SP-5382. Install the pinion front bearing. Install Washer SP-6022 followed by Compression Sleeve SP-3194-B. Install Centralizing Washer SP-534 followed by Compression Nut SP-3193.

3. On 9¼ in. diameter, position Locating Spacer SP-6017 over the shaft of SP-526 and follow it with the pinion rear bearing. Position the tools (with the bearing) in the differential housing. Install Shaft Locating Sleeve SP-1730. Install the pinion front bearing. Install Washer SP-6022 followed by Compression Sleeve SP-535-A. Install Centralizing Washer SP-534 followed by Compression Nut SP-533.

3. Prevent compression sleeve tool from turning with Wrench C-3281. Tighten the nut to seat the pinion bearing cups in the housing. Allow the sleeve to turn several times during the tightening to prevent brinelling the bearing cups or the bearings.

NOTE: Depth Shim(s) are positioned between the pinion gear rear bearing and pinion gear to provide separation distance. The required thickness of the depth shim(s) is determined according to the following information.

4. Loosen the compression nut tool. Lubricate the pinion gear front and rear bearings with gear lubricant. Tighten the compression nut tool to (1 to 3 N.m) 15 to 25 in. lbs. Rotate the pinion gear several complete revolution to align the bearing rollers.

5. On 8⅜ in. diameter ring gear axles. Install Gauge Block SP-5383 at the end of SP-5385. Install Cap Screw SP-536. Tighten securely with Wrench SP-531.

6. On 9¼ in. diameter ring gear axles. Install Gauge Block SP-6020 at the end of SP-526. Install Cap Screw SP-536. Tighten securely with Wrench SP-531.

7. On 8⅜ in. axles, position Crossbore Arbor SP-6029 in the differential housing.

8. On 9¼ in. axles, position Crossbore Arbor SP-6018 in the differential housing.

9. Center the tool. Place a piece of 0.002 in. (0.05mm) shim stock at each end of the arbor tool. Position the bearing caps on the arbor tool. Install the attaching bolts. Tighten the cap bolts to (14 N.m) 10 ft.lbs.

10. Trial fit depth shim(s) between the crossbore arbor tool and gauge block tool. The depth shim(s) fit must be snug but not tight (drag friction of a feeler gauge blade). Depth shims are available in 0.001 in. (0.0254mm) increments from 0.020 in. (0.5mm) to 0.038 in. (0.965mm).

11. Note the etched number on the face of the drive pinion gear (e.g., $-0, -1, -2, +1, +2$, etc.). The numbers represent thousands-of-an-inch deviation from the standard. If the number is $(-)$ negative, add that value to the required thickness of the depth shim(s). If the number is $(+)$ positive, subtract that value from the thickness of the depth shim(s). If the number is **0**, no change is necessary.

12. Remove the tools from the differential housing.

13. Position the depth shim selected above on the pinion shaft followed by the rear bearing (ensure the Contact surface, bearing and depth shim are clean and without foreign particles). Use Bearing Installer C-4040 for 8⅜ in. diameter. Use Sleeve C-3095-A for 9¼ in. diameter ring gear axles. Force the rear bearing onto the pinion shaft. An arbor press can be used with the installation tool.

14. Lubricate the pinion gear front and rear bearings with gear lubricant. Install the pinion with rear bearing in the housing. Position the bearing replacement collapsible spacer at the end of the pinion gear. Position the pinion gear front bearing at the end of the pinion gear.

15. Install the yoke. On 8⅜ in. diameter use Installer C-3718 and Wrench C-3281. On 9¼ in. diameter use Installer C-496 and Wrench C-3281.

NOTE: Because of the front bearing interference fit on the drive pinion gear shaft, it is necessary to use the tools to correctly seat the front bearing on the drive pinion gear shaft. Use care to prevent collapsing preload spacer during installation of the yoke and seating the front bearing.

16. Remove the yoke and tools from the pinion gear. Install the pinion seal with Seal Installer. On 8⅜ in. diameter use Installer C-4076. On 9¼ in. diameter use Installer C-3980 or C-4109. The seal is correctly installed when the seal flange contacts the face of the differential housing flange. The outer perimeter of

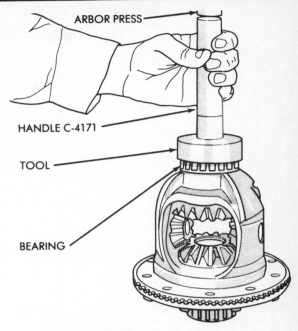

Differential bearing installation

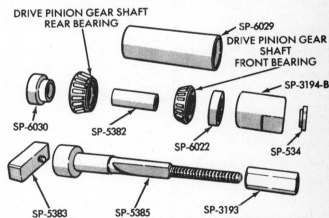

8⅜ inch ring gear/pinion gear adjusting tools

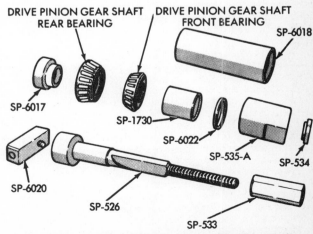

9¼ inch ring gear/pinion gear adjusting tools

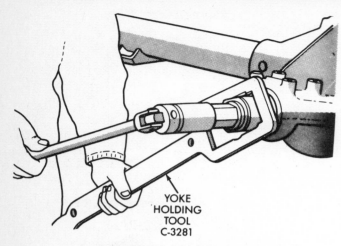

Bearing cup installation

YOKE HOLDING TOOL C-3281

TOOL

Installing the pinion oil seal

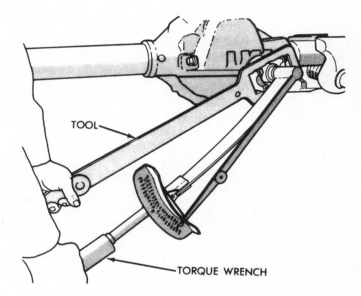

ARBOR

GAUGE BLOCK

SPACER (SELECTIVE)

Measuring the housing for pinion shim thickness

TOOL

TORQUE WRENCH

Tightening the pinion nut

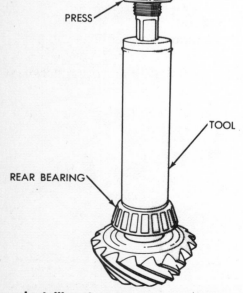

PRESS

TOOL

REAR BEARING

Installing the pinion rear bearing cone

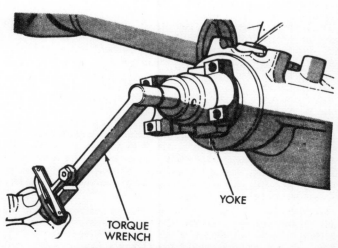

YOKE

TORQUE WRENCH

Measuring bearing preload

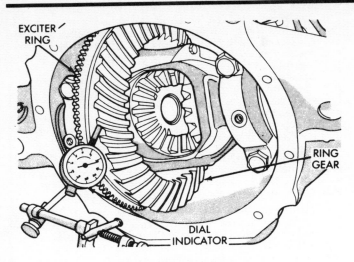

Measuring backlash

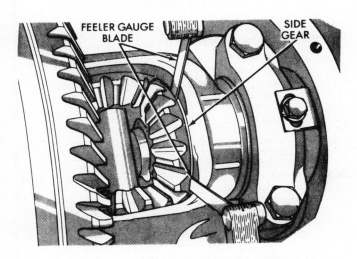

Side gear clearance measurement

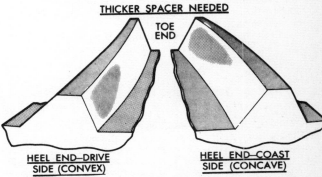

Desired tooth contact under light load

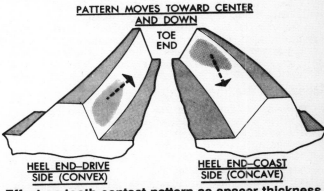

Incorrect tooth contact — increase spacer thickness

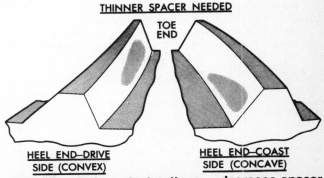

Effect on tooth contact pattern as spacer thickness is increased

Incorrect tooth contact pattern — decrease spacer thickness

the seal is pre-coated with a special sealant. An additional application of sealant is not required.

17. Install the yoke. Remove the tools. Install the Belleville washer. The convex side of the washer must face outward. Install the pinion nut.

18. Hold the pinion yoke with Wrench C-3281. Initially tighten the drive pinion gear shaft nut enough to remove the bearing end play. While tightening, rotate the pinion shaft to ensure the bearing rollers are correctly seated.

19. Tighten the pinion nut to (285 N.m) 210 ft.lbs. (minimum).

20. Remove the tools from the shaft. Rotate the pinion several complete revolutions (both directions) to additionally seat the bearing rollers. **Never loosen the pinion gear nut to decrease the pinion gear bearing preload torque.** If the specified preload torque is exceeded, a replacement collapsible spacer must be installed. The torque sequence will have to be repeated.

21. Measure the bearing preload torque by rotating pinion gear with a Newton-meter or an inch-pound torque wrench. The correct replacement bearing preload torque is (2 to 4 N.m) 20 to 35 in. lbs. for both 8⅜ in. and 9¼ in. axles. Tighten the pinion gear nut a minimum of (285 N.m) 210 ft.lbs.

22. When using the original pinion rear bearing and a replacement front bearing. The correct preload torque is (1 N.m) 10 in. lbs. in addition to the torque measured and recorded during disassembly. The bearing preload torque should be constant during a complete revolution of the drive pinion gear shaft. If the preload torque varies during rotation of the shaft, there is an internal binding that must be corrected before final assembly.

23. If the specified torque is not obtained with (285 N.m) 210 ft.lbs., continue tightening the nut in small increments until the recorded bearing preload torque is obtained. The differential will be unacceptable for use if the final nut torque is less than (285 N.m) 210 ft.lbs. If the preload torque is not within the specified range this is also unacceptable.

Differential Case Installation

1. Apply a coating of hypoid gear lubricant to the differential bearings, bearing cups and the threaded adjusters. Carefully position the assembled differential case in the housing.

2. Observe the reference marks and install the differential bearing caps at their original locations.

3. Install the bearing cap bolts. Tighten the upper bolts to (14 N.m) 10 ft.lbs. Tighten the lower bolts finger-tight until the bolt head is lightly seated.

Bearing Preload Torque and Ring Gear Backlash Adjustment

The following limitations must be considered when adjusting the differential bearing preload torque and ring gear backlash: The maximum permissible ring gear backlash variation is 0.003 in. (0.076mm). For example, if the minimum backlash is 0.006 in. (0.152mm) at one location, the maximum can be no more than 0.009 in. (0.229mm) at another location.

This variation represents the maximum permissible runout and it is important to mark the gears so the some teeth are engaged (meshed) during all backlash measurements.

It is also important to maintain the specified threaded-adjuster torque while adjusting the differential bearing preload torque and the ring gear backlash; Excessive adjuster torque will introduce a high bearing load and cause premature bearing failure. Insufficient adjuster torque can result in excessive differential case free-play and ring gear noise.

The differential bearing cups will not always immediately follow the threaded adjusters as they are moved during adjustment. Ensure accurate bearing cup responses to the adjustments. Maintain the gear teeth engaged (meshed) as marked. The bearings must be seated by rapidly rotating the pinion gear a half turn back and forth. Do this five to ten times each time the threaded adjusters are adjusted.

1. Use Wrench C-4164 to adjust each threaded adjuster inward until the differential bearing free-play is eliminated. Allow some ring gear backlash (approximately 0.01 in./0.25mm) between the ring and pinion gear. Seat the bearing cups with the procedure described previously.

2. Install Deal Indicator C-3339. Position the plunger against the drive side of a ring gear tooth. Measure the backlash at 4 positions around the perimeter of the ring gear (with an arc interval of approximately 90° between each position) to locate the area of minimum backlash.

3. Rotate the ring gear to the position of the least backlash (as indicated by the dial indicator pointer). Mark the gear so that all future backlash measurements will be taken with the same gear teeth meshed.

4. Loosen the right-side, tighten the left-side threaded adjuster. Obtain backlash of 0.003 to 0.004 in. (0.076 to 0.102mm) with each adjuster tightened to (14 N.m) 10 ft.lbs. Seat the bearing cups with the procedure described previously.

5. Tighten the differential bearing cap bolts. On 8⅜ in. diameter to (95 N.m) 70 ft.lbs. On 9¼ in. diameter with 136 (N.m) 100 ft.lbs.

6. Use Wrench C-4164 to tighten the right-side threaded adjuster to (102 N.m) 75 ft.lbs. Seat the bearing rollers as previously described. Continue to tighten the right-side adjuster and seat bearing rollers until the torque remains constant at (102 N.m) 75 ft.lbs.

7. Measure the ring gear backlash. If the backlash is not 0.006 to 0.008 in. (0.15 to 0.20mm), continue increasing the torque at the right-side adjuster until obtained. Continue increasing the torque at the right-side threaded adjuster (seat the bearing cups with the procedure described above) until the specified backlash is obtained.

NOTE: If all the instructions have been correctly completed, the left-side threaded adjuster should be approximately (102 N.m) 75 ft.lbs. If the torque is substantially less, the complete procedure must be repeated.

8. Tighten the left-side threaded adjuster until (102 N.m) 75 ft.lbs. is indicated. Seat the bearing rollers with the procedure previously described. Do this until the torque remains constant.

9. After the adjustments are completed, install the threaded adjuster locks. On 9¼ in. diameter, the lock teeth are engaged with the adjuster threads. On 8⅜ in. diameter, the lock finger is engaged with an adjuster hole. Tighten the lock screws to (10 N.m) 90 in.lbs.

Side Gear Clearance Measurement and Adjustment

The correct differential side gear clearance is obtained by selection of a thrust washer that has the correct thickness. Refer to the replacement parts catalog for the required side gear thrust washer package.

When measuring side gear clearance, check each gear independently. One side hear can have an acceptable clearance and the other side gear to have an unacceptable clearance. If it necessary to replace a side gear, replace both gears as a matched set.

1. Install the axle shafts and C-clips locks.

2. Measure each side gear clearance. Insert a matched pair of feeler gauge blades between the gear and differential housing on opposite sides of the hub.

3. If side gear clearances is no more than 0.005 in. (0.127mm). Determine if the shaft is contacting the pinion gear mate shaft. Do not remove the feeler gauges, inspect the axle shaft with the feeler gauge inserted behind the side gear. If the end of the axle shaft is not contacting the pinion gear mate shaft, the side gear clearance is acceptable.

4. If clearance is more than 0.005 (axle shaft not contacting mate shaft), record the side gear clearance. Remove the thrust washer and measure its thickness with a micrometer. Add the washer thickness to the recorded side gear clearance. The sum of gear clearance and washer thickness will determine required thickness of replacement thrust washer. For example, if the side gear clearance is 0.007 in. (0.178mm) and the thrust washer thickness is 0.033 in. (0.84mm), the sum is 0.04 in. (1mm). Install the thickest thrust washer from the service package that does not exceed the sum above. In the example situation, 0.037 in. (0.94mm) thick washer from the service package should be installed because the next larger washer size is 0.042 in. (1.06mm) thick, which would be to thick. When the replacement thrust washer is installed, the side gear clearance should be 0.003 in. (0.076mm). In some cases, the end of the axle shaft will move and contact the mate shaft when the feeler gauge is inserted. The C-clip lock is preventing the side gear from sliding on the axle shaft.

5. To determine the total side gear clearance with this situation (above), re-measure the clearance with the C-clip lock removed. Compare this measurement with the measurement when the C-clip lock was installed.

6. If there is no clearance, remove the C-clip from the side with no end play. With the differential case disassembled, use a micrometer to measure the thrust washer thickness. Record the thickness and return the thrust washer behind the side gear.

Assemble the differential case without the C-clip lock, installed and re-measure the side gear clearance.

7. Compare both clearance measurements. If the difference is less than 0.012 in. (0.305mm), add the side gear clearance recorded when the C-clip lock was installed to the thrust washer thickness (measured with the micrometer). The sum will determine the required thickness of the replacement thrust washer. For example, side gear clearance; 0.006 in. (0.152mm) with C-clip installed, 0.015 in. (0.481mm) with C-clip removed. Clearance difference in 0.009 in. (0.329). Add 0.006 in. (0.152mm) to the thrust washer thickness (e.g., 0.032 in./0.813mm). The sum is 0.038 in. (0.965mm). The closest thrust washer not exceeding 0.038 in. (0.965mm) is 0.037 in. (0.940mm).

8. If the clearance difference is 0.012 in. (0.305mm) or greater, both side gears must be replaced (matched set) and the clearance measurements repeated.

9. If the side gear clearance difference continues to be 0.012 in. (0.305mm) or greater with replacement side gears and the thickest thrust washers from the service package installed, the differential case must be replaced.

Ring Gear Teeth Contact Pattern

The ring gear teeth contact patterns will show if the pinion gear depth shim(s) have the correct thickness. It will also show if the ring gear backlash has been correctly adjusted. The backlash must be maintained within specified limits until the correct teeth contact patterns are obtained.

1. Apply a thin coat of hydrated ferric oxide, or equivalent, to both the drive side and coast side of the ring gear teeth.

2. Rotate the ring gear one complete revolution in both directions while applying a load force with a pry bar placed between the differential housing and the case flange. A distinct contact pattern on both the drive side and coast side of the ring gear teeth will be produced.

3. Compare the pattern on the teeth with the illustrations shown in the section. Adjust as required.

4. After all required adjustments are completed: Install the driveshaft. Tighten the U-joint mounting bolts to (19 N.m) 14 ft.lbs.

5. Install new axle shaft oil seals. Install the axles.

6. Make sure the cover plate mounting surfaces are clean and install the cover, and or the RWAL brake speed sensor. Fill the

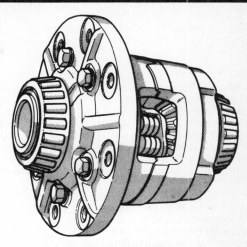

Sure-Grip differential

differential with the proper type and amount of lubricant. Install the wheels and lower the vehicle.

SURE-GRIP SERVICE

Limited slip Sure-Grip is an option with the 8⅜ in. and 9¼ in. axles. The Sure-Grip differential use the same internal components as the standard axles, except that it has a two piece case carrier containing the limited slip components.

With the exception of the ring gear, pinion, pinion bearings and differential side bearings, the differential case and internal parts are serviced as an assembly only and should not be disassembled.

Should the differential be submerged in water, the axle lubricant must be replaced immediately.

Fluid change is described in Section 1. 4 ounces of MOPAR Hypoid Gear Lubricant additive, or equivalent should be added. The additive contains a friction modifier for the limited slip clutches.

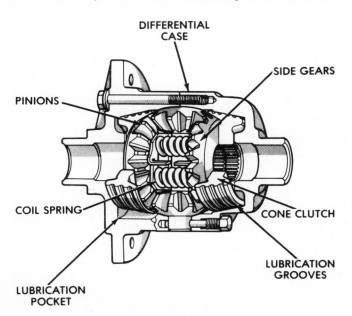

Sure-Grip components

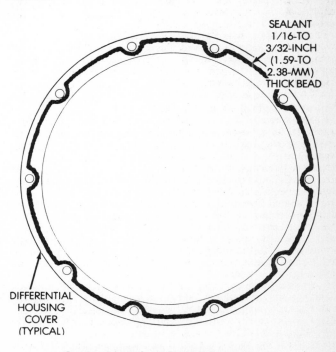

Sealant bead on housing cover

After a fluid change, drive the vehicle and make several slow figure eight turns. This maneuver will pump the lubricant through the clutch discs.

Spicer 60 Axle

It is not necessary to remove the complete axle assembly from the vehicle for differential service.

1. Raise and safely support the rear of the vehicle on the frame rails in front of rear spring. This will allow the axle assembly to hang free for more working room. Remove the axles and driveshaft. Thoroughly clean the outer area of the housing and axle tubes with a safe cleaning solvent and wipe dry.

2. Place a drain pan under the differential. Remove the cover and drain the lubricant.

3. If equipped with a Trac-Lok differential do not perform this step. Wash and flush the internal components of the differential. Allow them to dry.

4. Install a pilot stud and attach a dial indicator. Locate the plunger slightly at a right angle to the back of the ring gear. Zero the dial indicator.

5. Rotate the ring gear several times. Mark the ring gear and differential case flange at the maximum runout point. The mark on the case flange will be very useful when measuring differential case runout. If the combined ring gear runout exceeds 0.006 in. (0.15mm), possible causes could be a loose ring gear or damaged case.

6. Check the clearance between the differential bearing cap and bearing cup by trying to insert a 0.003 in. (0.076mm) thick feeler gauge between them. The 0.003 in. (0.076mm) blade should not enter between the two surfaces. If the clearance is more than 0.003 in. (0.076mm) the cause could be the bearing cup rotated in the housing seat and caused excessive wear.

7. Note the position reference marks stamped on the bearing caps and machined housing sealing surface.

8. Remove the bearing caps. Install a position spreader on the differential housing with the tool dowel pins in the proper locating holes. Install the holddown clamps and tighten the turnbuckle finger-tight.

9. Install a pilot stud at the left side of the differential housing. Attach a dial indicator to the housing pilot stud. Load the indicator plunger against the opposite housing and zero the indicator.

10. Separate the housing enough to remove the case from the housing. DO NOT separate the case more than 0.015 in. (0.38mm) with the spreader tool. Measure the distance with the dial indicator.

11. When the separating distance is reached, remove the dial indicator. Pry the differential case loose from the housing. To prevent damage, pivot on the housing with the end of the pry bar against the case.

12. Remove the case from the housing. Retain the differential bearing cups and bearings as matched sets if they are to be used over again.

13. Clamp the differential in a soft jawed vise. Remove and discard the ring gear mounting bolts. Tap the ring gear with a soft mallet to remove it.

14. If the ring gear runout exceeded 0.006 in. (0.15mm), the case flange runout should be measured at this time.

15. Install the case, with the original bearings, into the housing. Remove the spreader. Install the bearing caps and tighten the mounting bolts finger tight.

16. Set up the dial indicator with the plunger on the inner side of the case flange. Zero the plunger.

17. Rotate the flange several times an observe the reading. the flange maximum runout should not exceed 0.003 in. (0.076mm). It is possible to reduce excessive ring gear runout by positioning maximum runout mark located 180° opposite the flange maximum runout. Remove the case from the housing.

18. Remove the pinion yoke nut and washer. Use a puller and remove the yoke from the pinion shaft.

19. Remove the pinion seal. Remove the oil slinger, front bearing and shims. Record the thickness of the shims. This will save time should the shims be misplaced.

20. Remove the pinion gear and rear bearing from the housing.

21. Use an appropriate puller and remove the front bearing cup from the housing.

22. Use an appropriate puller and remove the rear bearing cup from the housing.

23. Remove the depth shims from the housing. Record the thickness of the shims.

24. Remove the rear bearing from the pinion shaft, using the appropriate puller.

25. Clamp the differential case in a soft jawed vise. Use a pin punch to remove the pinion gear mate shaft lock pin from the case.

26. Remove the pinion gear mate shaft. Remove the differential pinion gears and thrust washers.

27. Remove the differential side gears and thrust washers.

28. Remove the case from the vise.

29. Remove the end bearings from the case using the appropriate puller. Take care not to damage the bearing cage.

30. Remove the bearing shims from the case hubs. Mark them for assembly reference. Record the shim thicknesses.

31. Wash and clean all of the differential components in a safe solvent. Do not steam clean.

32. Clean the axle shaft tubes. Pull a rag attached to wire through the tubes. Inspect all of the component parts.

33. Replace any worn parts. Cup and bearings must be replaced as matched sets only.

To assemble:

34. Lubricate all of the differential components with hypoid gear lubricant.

35. Install the differential side gears and thrust washers, the pinion gears and thrust washers, and the gear mate shaft into the differential case.

36. Insert and seat the lock pin in the case and shaft. If replacement gears and thrust washers were installed, it is not necessary to measure the gear backlash. Correct fit is due to close machining tolerances during manufacture.

37. Position and align the ring gear threaded holes to the case flange. Install but do not tighten new ring gear bolts. After the bolts are threaded into the ring gear bolt holes, tap the ring gear with a non-metallic mallet. Make sure the gear is flush against the carrier flange.

38. Clamp the differential case in a soft jawed vise. Alternately tighten each ring gear bolt to 81 ft.lbs. (110 N.m).

39. Place a set of dummy bearing special tools on each case bearing hub.

40. Install the differential spreader and dial indicator on the housing. Spread the housing. Take care not to exceed the maximum of 0.015 in. (0.38mm).

41. Remove the dial indicator. Install the ring gear case assembly in the housing.

42. Remove the spreader. Observe the assembly reference marks and locate the bearing caps in their original positions.

43. Set up the dial indicator to against the back side of the ring gear. Insert a small pry bar between the bearing cap and the left side of the differential case. Pry the case to the left and record the travel distance.

44. The measurement is the shim thickness required for case zero end-play. The total thickness will be determined during the ring gear backlash adjustment.

45. Remove the dial indicator. Remove the bearing caps.

46. Spread the housing and remove the differential case.

47. If the original ring and pinion gear are not reusable, the pinion gear depth shim thickness must be determined before installing the differential case.

48. To determine the gear depth a special tool set such as D-271 is used. The following is the procedure using the tool set.

49. Insert Master Pinion Block D-120 into the pinion bore.

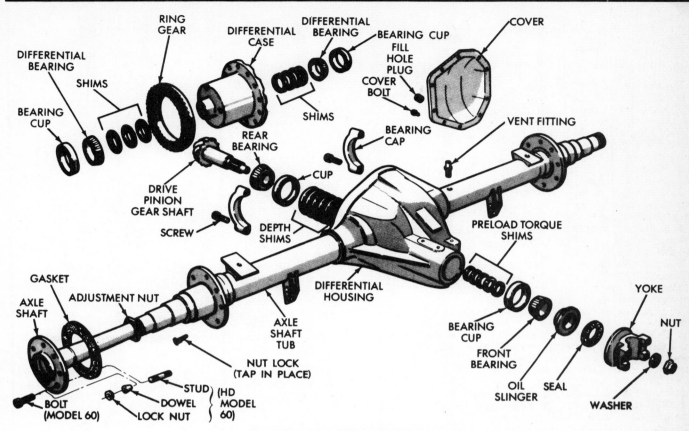

Components of the Model 60 axle

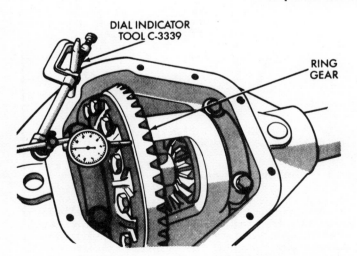

Measuring gear runout — Model 60

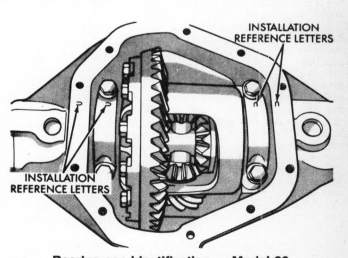

Bearing cap identification — Model 60

Place Disc D-116-2 on Arbor D-115-3 and position in the bearing cradles. This is the centerline of the ring gear/axle shaft. Place Pinion Height Block D-116-1 on top of master pinion block tool and against arbor tool. Firmly place Gauge Block D-115-2 and dial indicator D-106-5 on the lowest step of the pinion height block tool. Zero the dial indicator point. Move the gauge block toward the arbor until the indicator contacts the arbor tool. Slide the gauge block across along the arbor while observing the indicator. Record the longest travel distance, whether inward

(−) or outward (+) as indicated by the pointer. The plunger distance indicated, plus or minus the variance etched in the gear is the required thickness for the depth shims. Measure the thickness of each depth shim with a micrometer and combine the shims necessary for total required pack thickness. Include the oil slinger thickness with the total shim pack thickness. Remove the measurement tools from the differential housing. Place the depth shims in the pinion gear rear bearing bore. Install the rear pinion bearing cup. Install the front pinion bearing cup. In-

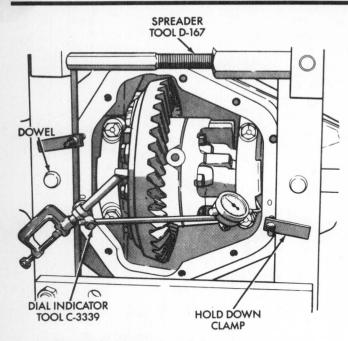

Differential housing separation — Model 60

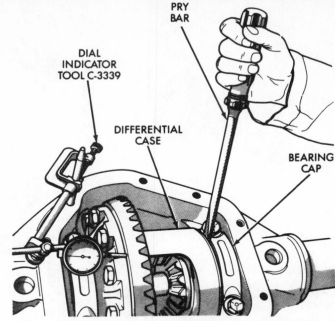

Differential case end play measurement — Model 60

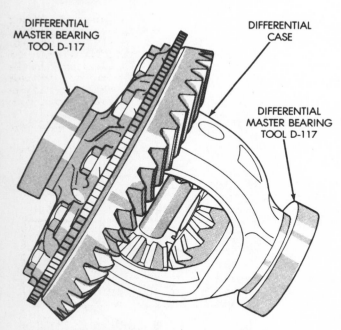

Master bearing tools on hubs — Model 60

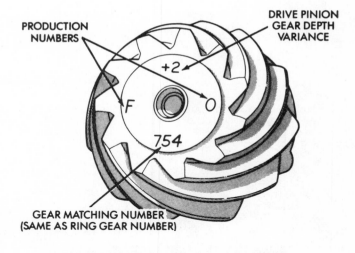

Pinion gear ID numbers — Model 60

stall the rear pinion bearing completely seated on the pinion shaft using an appropriate press.

50. Install the pinion gear into the housing. Install the front pinion bearing, oil slinger, yoke, washer and nut. Tighten the yoke nut until it requires 10 in. lbs. (1 N.m) torque to rotate the pinion gear. This will seat the bearings.

51. Measure the pinion gear depth of mesh. Place the height block on the face of the pinion gear. Place the arbor and discs into the bearing cradles. Firmly place the gauge block with the indicator on the lowest step of the height block. Zero the dial indicator.

52. Move the gauge block toward the arbor until the indicator contacts the arbor tool. Slide the gauge block across along the arbor while observing the indicator. Record the longest travel distance, whether inward (−) or outward (+), indicated by the pointer.

53. If the indicator reading actual depth is within 0.002 in. (0.05mm) of variance etched, it is acceptable. If not within specs, correct the shim pack thickness.

54. Remove the measuring tools. Remove the pinion gear nut, washer, yoke, oil slinger and front bearing.

55. Install the shims that were removed that were removed during disassembly. Install the front bearing and oil slinger.

56. Lubricate the pinion bearing and install it.

57. Install the pinion yoke, washer and new pinion nut. Tighten the nut to 250-270 ft.lbs. (339-366 N.m).

58. Check the preload torque. The torque necessary to turn the pinion shaft should be between 10-20 in. lbs. (1 to 3 N.m).

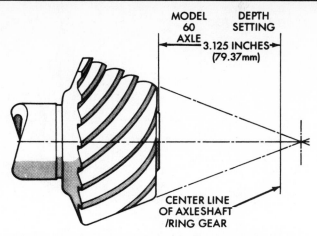

Pinion gear standard depth/distance — Model 60

59. If the preload torque is not within specs, correct the shim thickness accordingly. To increase torque, decrease the shim thickness. To decrease the torque, increase the shim thickness.
60. To check the ring gear backlash:
61. Spread the housing. Install the dummy bearing special tools on the case hubs. Position the case in the differential housing and install the bearing caps finger tight. Remove the spreader.
62. Install a pilot stud and dial indicator on the housing. Load the indicator against the back of the ring gear. Ensure that the ring and pinion gears are tightly meshed. Zero the dial indicator.
63. Insert a small pry bar between the bearing cap and the left side of the case. Pry the ring gear and case toward the right side away from the pinion gear. Observe the dial indicator reading. Repeat the measurement several times to check consistency. Record the travel distance.
64. The measurement shows shim thickness necessary to eliminate ring gear backlash. Subtract this thickness from the case zero endplay shim thickness. The shims must be placed on the ring gear side between the case and the bearing.

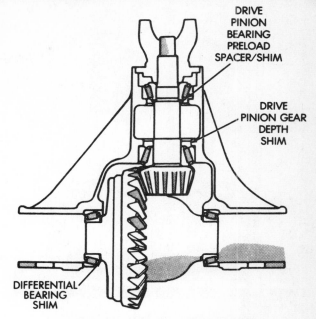

Shim locations — Model 60

65. Remove the dial indicator and stud. Remove the bearing caps. Spread the housing and remove the case assembly. Remove the dummy bearing special tools from the case hub ends.
66. Position the determined backlash shims in position on the ring gear side of the case hub and install the bearing. Place the remaining zero end-play shims on the other case hub, include a 0.015 in. (0.38mm) additional shim to provide the required preload, and install the bearing. Match the bearing to cups and install the cups over the bearings.
67. Spread the housing and install the differential case assembly.
68. Remove the spreader. Install the bearing caps in their

Original Pinion Gear Depth Variance	Replacement Pinion Gear Depth Variance								
	− 4	− 3	− 2	− 1	0	+ 1	+ 2	+ 3	+ 4
+ 4	+ 0.008	+ 0.007	+ 0.006	+ 0.005	+ 0.004	+ 0.003	+ 0.002	+ 0.001	0
+ 3	+ 0.007	+ 0.006	+ 0.005	+ 0.004	+ 0.003	+ 0.002	+ 0.001	0	− 0.001
+ 2	+ 0.006	+ 0.005	+ 0.004	+ 0.003	+ 0.002	+ 0.001	0	− 0.001	− 0.002
+ 1	+ 0.005	+ 0.004	+ 0.003	+ 0.002	+ 0.001	0	− 0.001	− 0.002	− 0.003
0	+ 0.004	+ 0.003	+ 0.002	+ 0.001	0	− 0.001	− 0.002	− 0.003	− 0.004
− 1	+ 0.003	+ 0.002	+ 0.001	0	− 0.001	− 0.002	− 0.003	− 0.004	− 0.005
− 2	+ 0.002	+ 0.001	0	− 0.001	− 0.002	− 0.003	− 0.004	− 0.005	− 0.006
− 3	+ 0.001	0	− 0.001	− 0.002	− 0.003	− 0.004	− 0.005	− 0.006	− 0.007
− 4	0	− 0.001	− 0.002	− 0.003	− 0.004	− 0.005	− 0.006	− 0.007	− 0.008

Pinion gear depth variance — Model 60

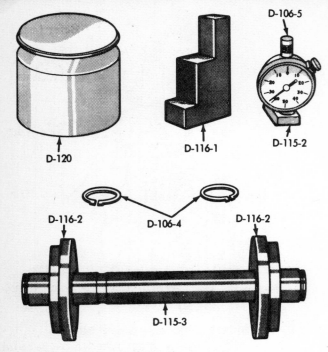

Pinion gear depth gauge tool set (D116-60) — Model 60

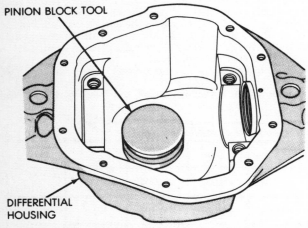

Pinion block tool inserted in the shaft bore — Model 60

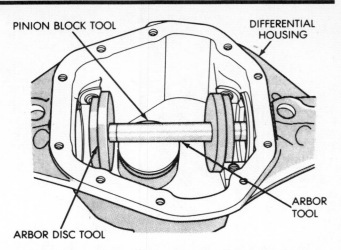

Arbor disc and arbor tools in the housing — Model 60

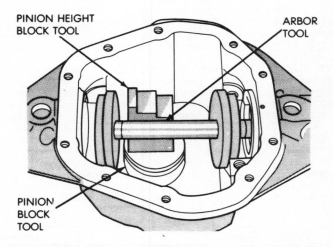

Pinion height tool against the arbor tool — Model 60

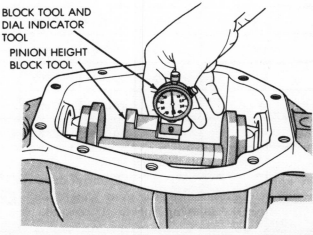

Drive pinion gear depth measurement — Model 60

proper locations by aligning the reference marks. Tighten the mounting bolts to 70-90 ft.lbs. (95-122 N.m).

69. Measure the backlash at three equally spaced locations. The gear backlash must be within 0.004-0.009 in. (0.10–0.23mm). It cannot vary more than 0.002 in. (0.05mm) between the points checked.

70. Excessive backlash is corrected by moving the ring gear teeth closer to the pinion. Insufficient backlash is corrected by moving the ring gear away from the pinion gear. Backlash correction is accomplished by transferring shims from one side to the other.

71. Install the axles. Install the housing cover with a silicone sealer bead as a gasket. Fill the differential with the correct

amount of lubricant and install the filler plug. Lower the vehicle. Test the brakes and axle for proper operation.

TRAC-LOK SERVICE

Limited slip Trac-Lok is an option with the Spicer 60 axle. The Trac-Lok differential use the same internal components as the standard Spicer 60 plus two clutch disc packs.

With the exception of the ring gear, pinion, pinion bearings and differential side bearings, the differential case and internal parts are serviced as an assembly only and should not be disassembled.

Should the differential be submerged in water, the axle lubricant must be replaced immediately.

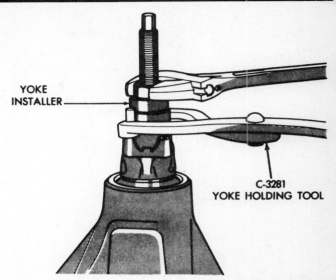

Drive pinion gear shaft yoke installation — Model 60

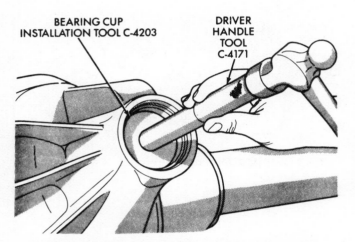

Shaft front bearing cup installation — Model 60

Fluid change is described in Section 1. 4 ounces of MOPAR Hypoid Gear Lubricant additive, or equivalent should be added. The additive contains a friction modifier for the limited slip clutches.

After a fluid change, drive the vehicle and make several slow figure eight turns. This maneuver will pump the lubricant through the clutch discs.

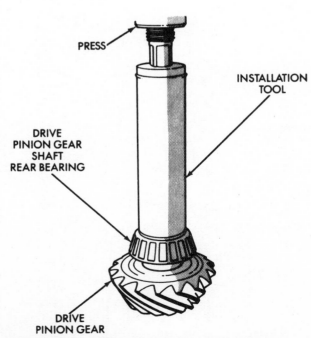

Shaft rear bearing installation — Model 60

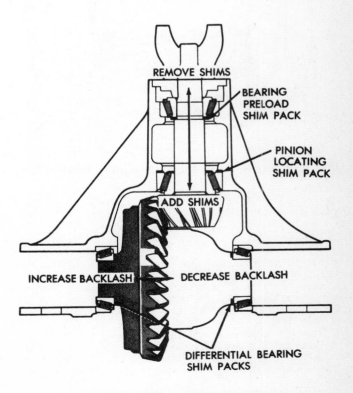

Shim locations — Model 60

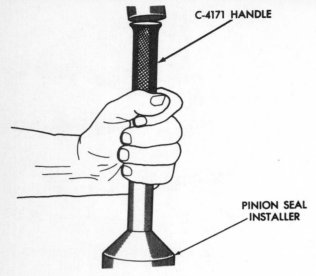

Pinion seal installation

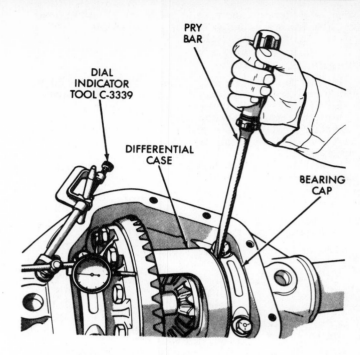

Ring gear backlash measurement — Model 60

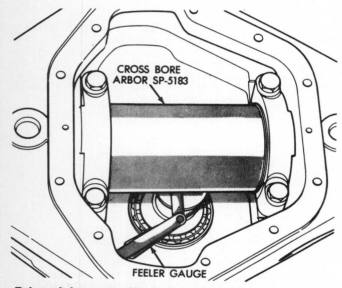

Drive pinion gear depth measurement — Model 60

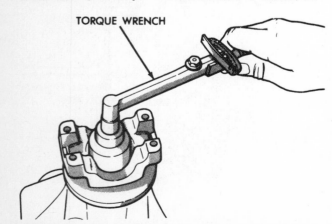

Rotating the drive pinion gear shaft — Model 60

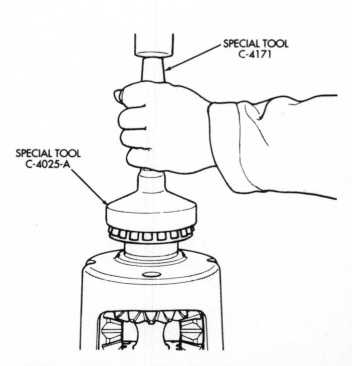

Differential bearing installation — Model 60

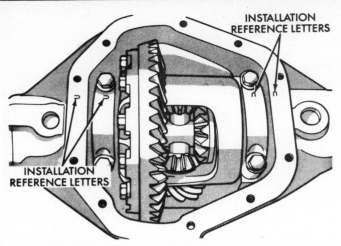

Differential bearing cap reference letters — Model 650

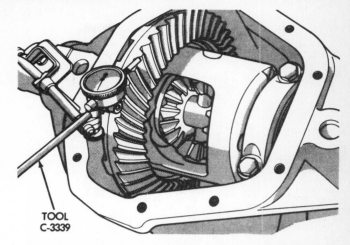

Ring gear backlash measurement — Model 60

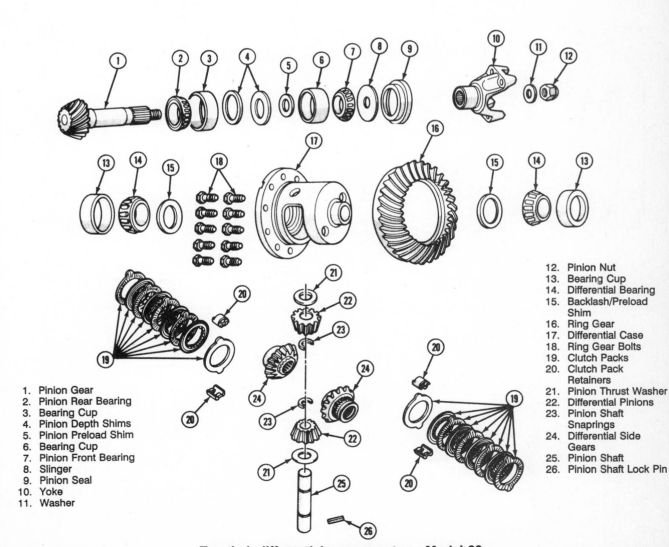

1. Pinion Gear
2. Pinion Rear Bearing
3. Bearing Cup
4. Pinion Depth Shims
5. Pinion Preload Shim
6. Bearing Cup
7. Pinion Front Bearing
8. Slinger
9. Pinion Seal
10. Yoke
11. Washer

12. Pinion Nut
13. Bearing Cup
14. Differential Bearing
15. Backlash/Preload Shim
16. Ring Gear
17. Differential Case
18. Ring Gear Bolts
19. Clutch Packs
20. Clutch Pack Retainers
21. Pinion Thrust Washer
22. Differential Pinions
23. Pinion Shaft Snaprings
24. Differential Side Gears
25. Pinion Shaft
26. Pinion Shaft Lock Pin

Trac-Lok differential components — Model 60

TORQUE SPECIFICATIONS

Component	English	Metric
Manual Transmission		
NP-435		
Shift unit-to-extension housing:	30 ft. lbs.	41 Nm
Back-up light switch:	15 ft. lbs.	20 Nm
Transmission-to-clutch housing bolts:	50 ft. lbs.	68 Nm
Crossmember mounting bolts:	30 ft. lbs.	41 Nm
Rear transmission mount-to-cross-member:	50 ft. lbs.	68 Nm
Extension housing attaching bolts:	30 ft. lbs.	41 Nm
Operating levers retaining nuts:	18 ft. lbs.	25 Nm
Shift cover bolts:	30 ft. lbs.	41 Nm
Drive gear retainer bolt:	20 ft. lbs.	28 Nm
Countershaft rear bearing retainer bolts:	20 ft. lbs.	28 Nm
Yoke nut:	125 ft. lbs.	169 Nm
Mainshaft rear bearing retainer bolt:	20 ft. lbs.	28 Nm
Fill & drain plugs:	35 ft. lbs.	47 Nm
Reverse idel shaft lock bolt:	30 ft. lbs.	41 Nm
NP-2500		
Access cover bolts:	21 ft. lbs.	28 Nm
Back-up light switch:	15 ft. lbs.	20 Nm
Center support-to-case bolts:	40 ft. lbs.	54 Nm
Input bearing retainer bolts:	21 ft. lbs.	29 Nm
Center support plate bolts:	40 ft. lbs.	54 Nm
Extension housing bolts:	40 ft. lbs.	54 Nm
Access cover bolts:	21 ft. lbs.	28 Nm
Shifter assembly bolts:	21 ft. lbs.	28 Nm
Shift lever bolt:	30 ft. lbs.	41 Nm
Speedometer adapter screw:	8 ft. lbs.	11 Nm
Transmission-to-clutch housing bolts:	50 ft. lbs.	68 Nm
Drain & fill plugs:	30 ft. lbs.	41 Nm
Clutch		
Pressure plate-to-flywheel		
⁵⁄₁₆ in. bolts:	17 ft. lbs.	23 Nm
⅜ in. bolts:	30 ft. lbs.	41 Nm
Remote reservoir mounting screws:	40 inch lbs.	5 Nm
Slave cylinder-to-the bell housing:	200 inch lbs.	23 Nm
Automatic Transmission		
Cooler line fittings:	155 inch lbs.	18 Nm
Cooler line nuts:	85 inch lbs.	10 Nm
Filter screws:	35 inch lbs.	4 Nm
Flexplate bolts:	55 ft. lbs.	75 Nm
Pan attaching bolts:	150 inch lbs.	17 Nm
Front band adjusting screw:	72 inch. lbs.	8 Nm
Front band adjuster screw locknut:	30 ft. lbs.	41 Nm
Rear band adjusting screw:	72 inch lbs.	8 Nm
Rear band locknut:	25 ft. lbs.	34 Nm
Neutral start switch:	25 ft. lbs.	34 Nm
Speedometer adapter screw:	100 inch lbs.	11 Nm
Torque converter bolts:	23 ft. lbs.	31 Nm
Transmission-to-engine bolts:	30 ft. lbs.	41 Nm
Driveshaft		
Center bearing bolts:	50 ft. lbs.	68 Nm
U-bolts nuts:	15 ft. lbs.	20 Nm
Rear Axle		
Dana 60 Axle		
Axle shaft retaining nuts:	70 ft. lbs.	95 Nm
Pinion shaft nut:	250-300 ft. lbs.	340-408 Nm
U-bolt nuts:	200 ft. lbs.	272 Nm
Shock absorbers-to-axle nuts:	55 ft. lbs.	75 Nm
Ring gear bolts:	113 ft. lbs.	153 Nm
Pinion nut:	250-270 ft. lbs.	339-366 Nm
Pinion preload torque:	10-20 inch lbs.	1-3 Nm
Differential case mounting bolts:	70-90 ft. lbs.	95-122 Nm
Differential bearing cap bolts:	85 ft. lbs.	115 Nm
Differential cover bolts:	35 ft. lbs.	47 Nm
Brake backing plate nuts:	85 ft. lbs.	115 Nm
RWAL brakes sensor cover screw:	17 ft. lbs.	24 Nm
Chrysler 8⅜ in. and 9¼ in. Axles		
Brake backing plate nuts:	35 ft. lbs.	47 Nm
Differential bearing cap bolts		
8⅜ in. axle:	70 ft. lbs.	90 Nm
9¼ in. axle:	100 ft. lbs.	136 Nm
Differential cover bolts:	35 ft. lbs.	47 Nm
Pinion nut:	210 ft. lbs.	286 Nm
Shock absorbers-to-axle nuts:	55 ft. lbs.	75 Nm
U-bolt nuts		
½-20:	62-70 ft. lbs.	84-95 Nm
¾-16:	175-225 ft. lbs.	238-306 Nm
⁹⁄₁₆-18:	120-130 ft. lbs.	163-177 Nm
⅝-18:	175-200 ft. lbs.	238-272 Nm
Side gears-to-pinion gears lock screw:	90 inch lbs.	10 Nm
Ring gear bolts:	70 ft. lbs.	95 Nm
Pinion bearing preload torque:		
Used bearings:	20-35 inch lbs.	2-4 Nm
New bearings (added to used bearing torque):	10 inch lbs.	1 Nm
RWAL brakes sensor cover screw:	17 ft. lbs.	24 Nm

Suspension and Steering
8

QUICK REFERENCE INDEX

GENERAL INDEX

WHEELS

Wheels

Standard equipment wheels are drop center (depending on year-type), steel wheels with safety rims. Optional wheels include styled steel and cast aluminum. The steel wheels are the two-piece type that consist of a rim and center section. The two sections are welded together to form a seamless, airtight unit.

A wheel safety rim has a ridge (raised edge) located inboard of each rim flange and at the top of the rim well. Initial inflation of the tire forces the tire bead over the ridge sections. In case of tire failure the raised sections help hold the tire in position on the wheel until the vehicle can be brought to a safe stop.

The wheels should be inspected on a frequent basis. Replace any wheel that is either cracked, bent, severely dented, has excessive runout or has broken welds. The tire inflation valve should also be inspected frequently for wear, leakage, cuts and looseness. The valve should be replaced if defective or its condition is doubtful.

CLEANING

Clean the wheels with a mild soap and water solution only and rinse thoroughly with water. Never use abrasive or caustic materials, especially on aluminum or chrome-plated wheels because the surface could be etched or the plating severely damaged. After cleaning aluminum or chrome-plated wheels, apply a coat of protective wax to preserve the finish and luster.

REMOVAL AND INSTALLATION

Always block the wheels, apply the parking brake, raise and safely support the vehicle after removing the wheel cover and slightly loosening the lugs on the wheel to be removed. After the vehicle is raised and supported, remove the wheel assembly. When mounting the wheel assembly, start all of the lug nuts on the studs, and then tighten them in a crisscross pattern. All wheel lugs have right-handed threads. After the lug nuts have been snugged up, lower the vehicle enough to prevent the wheel from turning and finish tightening the lug nuts to the specified torque. Once again torque the lugs in the same crisscross pattern.

Wheel Lug Studs

Wheel lug studs on the front disc rotors usually require the services of a machine shop when replacement is required. Rear axle flange mounted wheel studs can sometimes be pushed out and the new stud pulled in (clearance providing) when replacement is required. Once again, axle flange (axle removed from vehicle) or rear drum mounted studs may require the services of a machine shop for replacement.

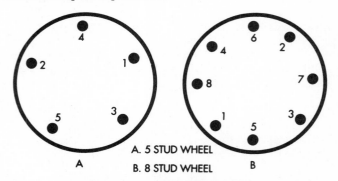

A. 5 STUD WHEEL
B. 8 STUD WHEEL

Wheel lug nut tightening pattern

DESCRIPTION			TORQUE	
Vehicle	Type Lug Nut	Stud Size	N-m	Ft -Lbs
150, 250	60° Cone	1/2-20	115-149	85-110
150, 250	Styled — 60° Cone	1/2-20	115-149	85-110
350	90° Cone	1/2-20	115-149	85-110
350 H.D. Axle	90° Cone	5/8-18	217-305	175-225
350 H.D. Axle	Flanged	5/8-18	407-475	300-350

Wheel lug, stud and torque specifications

FRONT SUSPENSION

Coil Springs

REMOVAL AND INSTALLATION

1. Raise and support the front end on jackstands placed under the frame.

2. Remove the wheels.
3. Remove the brake calipers and suspend them out of the way. DO NOT DISCONNECT THE BRAKE LINE! Remove the inner pad.
4. Remove the shock absorbers.
5. Disconnect the sway bar, if equipped.
6. Remove the lower control arm strut bar.

Troubleshooting Basic Steering and Suspension Problems

Problem	Cause	Solution
Hard steering (steering wheel is hard to turn)	• Low or uneven tire pressure • Loose power steering pump drive belt • Low or incorrect power steering fluid • Incorrect front end alignment • Defective power steering pump • Bent or poorly lubricated front end parts	• Inflate tires to correct pressure • Adjust belt • Add fluid as necessary • Have front end alignment checked/adjusted • Check pump • Lubricate and/or replace defective parts
Loose steering (too much play in the steering wheel)	• Loose wheel bearings • Loose or worn steering linkage • Faulty shocks • Worn ball joints	• Adjust wheel bearings • Replace worn parts • Replace shocks • Replace ball joints
Car veers or wanders (car pulls to one side with hands off the steering wheel)	• Incorrect tire pressure • Improper front end alignment • Loose wheel bearings • Loose or bent front end components • Faulty shocks	• Inflate tires to correct pressure • Have front end alignment checked/adjusted • Adjust wheel bearings • Replace worn components • Replace shocks
Wheel oscillation or vibration transmitted through steering wheel	• Improper tire pressures • Tires out of balance • Loose wheel bearings • Improper front end alignment • Worn or bent front end components	• Inflate tires to correct pressure • Have tires balanced • Adjust wheel bearings • Have front end alignment checked/adjusted • Replace worn parts
Uneven tire wear	• Incorrect tire pressure • Front end out of alignment • Tires out of balance	• Inflate tires to correct pressure • Have front end alignment checked/adjusted • Have tires balanced

7. Install a spring compressor, such as tool DD-1278, tighten the nut finger tight and back it off ½ turn.

8. Remove the ball joint nuts.

9. Using ball joint separator C-3564-A, or equivalent, spread the tool against the lower joint just enough to exert pressure then strike the knuckle sharply with a hammer to free the joint. NEVER ATTEMPT TO FORCE THE BALL JOINT OUT WITH TOOL PRESSURE ALONE! Remove the tool.

10. Slowly loosen the spring compressor until all tension is relieved from the coil.

11. Remove the compressor and spring.

To install:

12. Position the spring on the control arm and install the compressor.

13. Compress the spring until the ball joint is properly positioned.

14. Install the ball joint nuts and tighten them to 135 ft.lbs. (183 N.m). On ¾–16, tighten to 175 ft.lbs. (237 N.m).

15. Install the strut. Tighten the mounting bolts to 100 ft.lbs. (136 N.m); the retainer nut to 52 ft.lbs. (71 N.m).

16. Connect the sway bar and tighten the link to 100 inch lbs.

17. Remove the spring compressor.

18. Install the shock absorber. Tighten the upper end to 25 ft.lbs. (34 N.m); the lower end to 200 inch lbs. (23 N.m).

19. Install the inboard brake pad and caliper. Tighten the retaining clips to 180 inch lbs. (20 N.m).

20. Install the wheels and lower the van.

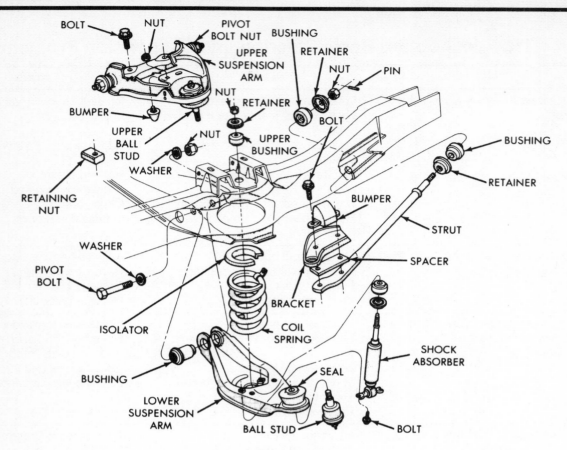

Front suspension components

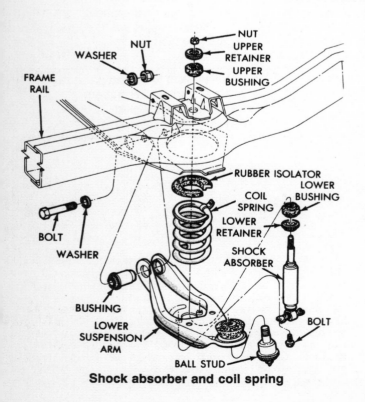

Shock absorber and coil spring

Lower suspension arm strut bar

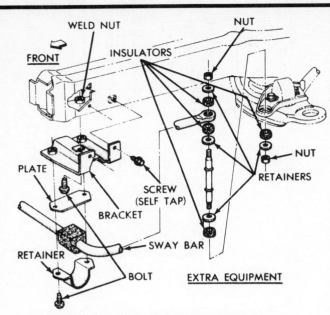

Stabilizer bar removal/installation

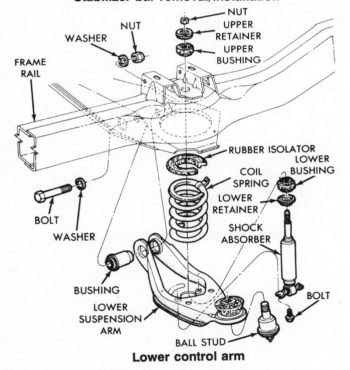

Lower control arm

Shock Absorber

REMOVAL AND INSTALLATION

1. Raise and support the vehicle with jackstands positioned at the extreme front ends of the frame rails.
2. Remove the wheel.
3. Remove the upper nut and retainer.
4. Remove the two lower mounting bolts.
5. Remove the shock absorber.
6. When installing the shock absorber, make sure the upper bushings are in the correct position. Replace any worn or

cracked bushing. Tighten the top nut to 25 ft.lbs. (31 N.m). Then, tighten the lower bolts to 200 inch lbs. (23 N.m).

Upper Ball Joint

Front end noise, front wheel shimmy and difficult steering could be caused by a worn upper ball joint. Jack the vehicle up under the lower control arm. Position a suitable pry bar under the tire and pry upward. Observe any free travel and determine what piece of the suspension has excessive play.

REMOVAL AND INSTALLATION

1. Install a jack under the outer end of the lower control arm and raise the vehicle.
2. Remove the wheel.
3. Remove the ball joint to knuckle nut. Using a ball joint breaker, loosen the upper ball joint.
4. Unscrew the ball joint from the control arm. Special tool C-3561, or the equivalent is required.
To install:
5. Screw a new ball joint into the control arm and tighten to 125 ft.lbs. (170 N.m).
6. Install the new ball joint seal, using a 2 in. (51mm) socket. Be sure that the seal is seated on the ball joint housing.
7. Insert the ball joint into the steering knuckle and install the ball joint nuts. Tighten the nut to 135 ft.lbs. (183 N.m) and install the cotter pins. On ¾–16, the torque is 175 ft.lbs. (237 N.m).
8. Install the wheel and lower the van.

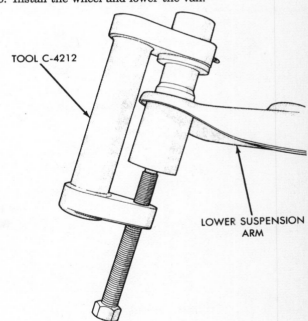

Lower ball joint removal

Lower Ball Joint

Front end noise, front wheel shimmy and difficult steering could be caused by a worn lower ball joint. Jack the vehicle up under the lower control arm. Position a suitable pry bar under the tire and pry upward. Observe any free travel and determine what piece of the suspension has excessive play. The lower ball joint is "loaded". Replace the ball joint when the endplay is 0.020 in. (0.5mm) or more.

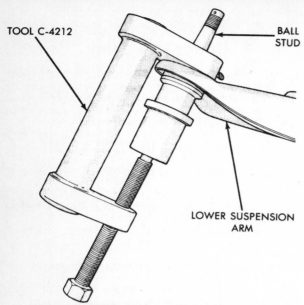

Lower ball joint installation

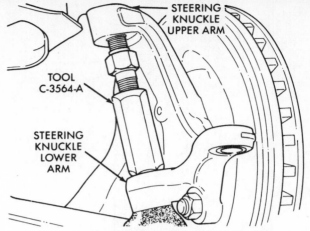

Loosening tool installed on upper ball joint

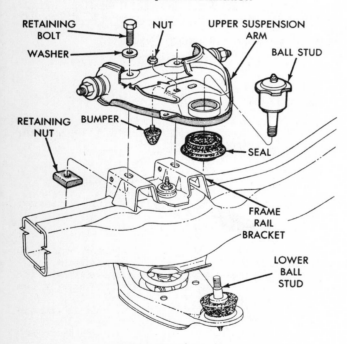

Upper control arm

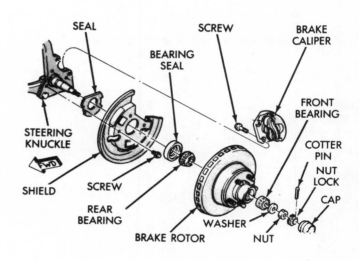

Rotor and splash shield removal

REMOVAL AND INSTALLATION

1. Remove the lower control arm.
2. Remove the ball joint seal.
3. Using tool C-4212 or an arbor press and a sleeve, press the ball joint from the control arm.
4. Installation is the reverse of removal. Be sure that the ball joint is fully seated. Install a new ball joint seal.
5. Install the lower control arm. Be sure to install the ball joint cotter pins.

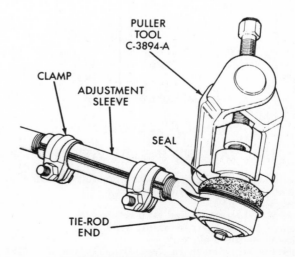

Tie rod end removal

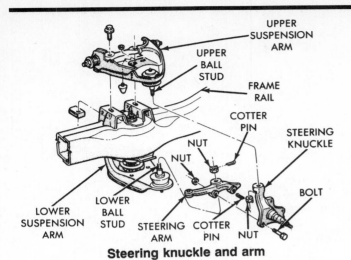

Steering knuckle and arm

Upper Arm Control
REMOVAL AND INSTALLATION

NOTE: Any time the control arm is removed, it is necessary to align the front end.

1. Raise and support the vehicle with jackstands under the frame rails.
2. Remove the wheel.
3. Remove the shock absorber and shock absorber upper bushing and sleeve.
4. Install a spring compressor and tighten it finger-tight.
5. Remove the cotter pins and ball joint nuts.
6. Install a ball joint breaker and turn the threaded portion of the tool, locking it securely against the upper stud. Spread the tool enough to place the upper ball joint under pressure and strike the steering knuckle sharply to loosen the stud. Do not attempt to remove the stud from the steering knuckle with the tool.
7. Remove the tool.
8. Remove the eccentric pivot bolts, after marking their relative positions in the control arm.
9. Remove the upper control arm.
To install:
10. Install the upper control arm.
11. Install the pivot bolts and finger-tighten them for now.
12. Position the spring on the control arm and install the compressor.
13. Compress the spring until the ball joint is properly positioned.
14. Install the ball joint nut and tighten to 135 ft. lbs. (183 N.m). For 350 models, the torque is 175 ft. lbs. (237 N.m).
15. Install new cotter pins.
16. Remove the spring compressor.
17. Install the shock absorber. Tighten the upper end to 25 ft. lbs.; the lower end to 15 ft. lbs.
18. Install the wheels.
19. Lower the truck to the ground.
20. Tighten the pivot bolts to 195 ft. lbs. (264 N.m).
21. Have the front end alignment checked.

Lower Control Arm
REMOVAL AND INSTALLATION

1. Follow the procedure outlined under Coil Spring Removal and Installation.
2. Remove the inner mounting bolt from the crossmember.

3. Remove the lower control arm from the vehicle.
4. When installing the control arm, install the crossmember bolt finger tight. After the vehicle has been lowered to the ground, tighten the mounting bolt to 175 ft. lbs. 237 N.m).

Lower Control Arm Strut Bar
REMOVAL AND INSTALLATION

1. Raise and support the front end on jackstands.
2. Using a small drift and hammer, drive out the spring pin from the front end of the strut.
3. Remove the nut, retainer and bushing.
4. Remove the rear mounting bolts along with the jounce bumper and bracket.
5. Installation is the reverse of removal. Torque the mounting bracket bolts to 100 ft. lbs. (136 N.m). Torque the retainer nut to 52 ft. lbs. (71 N.m). Install the lock pin.

Sway Bar
REMOVAL AND INSTALLATION

1. Disconnect the bar at each end link.
2. Remove the bolts from the frame mounting brackets.
3. Remove the sway bar.
4. Installation is the reverse of removal. Tighten the frame bracket bolts to 23 ft. lbs.; the end links to 100 inch lbs.

Knuckle and Spindle
REMOVAL AND INSTALLATION

1. Support the brake pedal in the UP position.
2. Raise and support the front end on jackstands.
3. Remove the wheels.
4. Remove the brake calipers and suspend them out of the way. DO NOT DISCONNECT THE BRAKE LINE! Remove the inner pad.
5. Remove the hub/rotor assembly.
6. Remove the brake splash shield.
7. Support the lower control arm with a floor jack.
8. Disconnect the tie rod from the knuckle.
9. Disconnect the ball joints from the knuckle, and remove the knuckle.
10. Unbolt the brake adapter from the knuckle.
To install:
11. Install the adapter on the knuckle. Torque the bolts to 100 ft. lbs. (136 N.m).
12. Align the knuckle arm and the knuckle. Torque the mounting bolts to 215 ft. lbs. (290 N.m). On ¾-16 bolts, tighten to 225 ft. lbs. (330 N.m).
13. Install the knuckle assembly on the control arms and connect the ball joints. Torque the nuts to 135 ft. lbs. (183 N.m). On 350 models, the torque is 175 ft. lbs. (237 N.m). Always use new cotter pins.
14. Connect the tie rod end and torque the nut to 55 ft. lbs. (75 N.m). Install a new cotter pin.
15. Install the splash shield and new dust seal. Torque the bolts to 16 ft. lbs.
16. Install the hub/rotor assembly and adjust the wheel bearings.
17. Install the caliper and pads.
18. Install the wheels.

Front Wheel Bearings
REPLACEMENT

1. Block the brake pedal in the up, non-applied position. Raise and safely support the front of the vehicle.

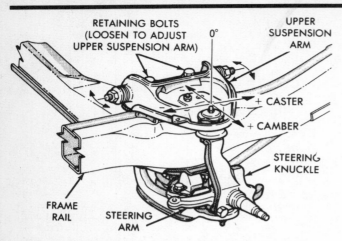

Caster and camber adjustment location

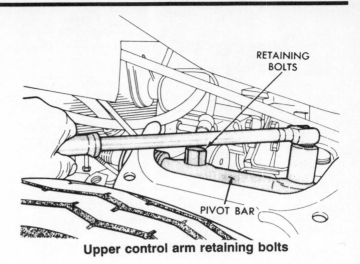

Upper control arm retaining bolts

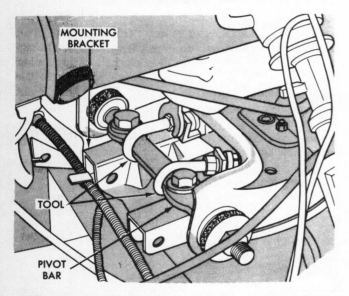

Alignment tools attached to frame and pivot bar

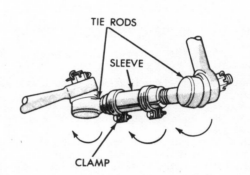

Tie rod adjustment

2. Remove the wheel and tire assembly.

3. Remove the caliper mounting bolts and slide the caliper assembly up and off of the rotor. Insert a block of wood between the brake pads to prevent caliper piston application. Suspend the caliper out of the way. Do not allow the caliper to hang by the brake hose.

4. Remove the brake rotor. Take care not to drop the outer bearing.

5. Remove the grease seal and the inner bearing.

6. Remove the bearing cups, mounted in the rotor hub, by inserting a brass drift through the opposite side of the hub. Position the drift on the inner edge of the bearing cup and drive the cup from the hub.

To install:

7. Clean the rotor hub. Install the new bearing cups by carefully driving or pressing them into position until they are firmly seated in the hub. Take care to drive or press them in evenly. Use a soft tool against the edge of the cup to avoid damage to the bearing contact surface, or tapered edges.

8. Pack the bearings. Install the inner bearing and new grease retainer. Install the rotor and remaining components. Lower the vehicle.

WHEEL ALIGNMENT SPECIFICATIONS

Years	Models	Caster (deg.) Range	Pref.	Camber (deg.) Range	Pref.	Toe-in (in.)	Steering Axis Inclination (deg.)
1989-91	B100-B350	+1.25 to +3.75	+2.5	−0.6 to +0.6	0	0①	NA

FRONT END ALIGNMENT

Steering Axis Inclination

Steering axis inclination is the number of degrees that the spindle support centerline is tilted from the true vertical as viewed from the front. It has a fixed relationship with camber and does not change except in the event of damage to a spindle or ball joint. The angle is not adjustable and damaged parts must be replaced.

Camber

Camber is expressed as the number of degrees that the top of the wheel is tilted outward or inward from the true vertical when viewed from the front. Inward tilt is negative camber and outward tilt is positive camber. Excessive camber causes premature tire wear; negative camber causes wear on the inside of the tire and positive camber causes the tire to wear on the outside edge.

Camber is adjusted by upper control arm position. Camber cannot be accurately measured without professional equipment.

Caster

Caster is the backward or forward tilt from the vertical of the steering knuckle centerline at the top, measured in degrees. A steering knuckle centerline tilted backward has positive (+) caster, while one tilted forward has negative (−) caster. Positive caster produces greater directional stability and requires great-er steering effort, since it increases the self-centering effect at the steering wheel.

Caster is adjusted by upper control arm position. Caster can-not be measured accurately without professional equipment.

Toe-In

Toe-in is the amount measured in inches, that the centerlines of the wheels are closer together at the front than at the rear. Toe-in must be checked after caster and camber have been ad-justed., but it can be adjusted without disturbing the other two settings. You can make this adjustment without special equip-ment, if you make careful measurements. The adjustment is made at the tie rod sleeves. The wheels must be straight-ahead.

1. Toe-in can be determined by measuring the distance be-tween the centers of the tire treads, front and rear. If the tread pattern makes this impossible, you can measure between the edges of the wheel rims, but make sure to move the truck for-ward and measure in a couple of places to avoid errors caused by bent rims and wheel runout.

2. Loosen the clamp bolts on the tie rod sleeves.

3. Rotate the sleeves equally (in opposite directions) to obtain the correct measurement. If the sleeves are not adjusted equal-ly, the steering will be crooked. If the steering wheel is already crooked, it can be straightened by turning the sleeves equally in the same direction.

4. When the adjustment is complete, tighten the clamps.

REAR SUSPENSION

Springs

REMOVAL AND INSTALLATION

1. Raise the truck and support the rear with jackstands un-der the frame. Be sure that the front wheels are chocked and that the parking brake is set. The wheels should be touching the floor, but the weight must be off of the springs.

2. Remove the nuts, lockwashers and U-bolts that hold the axle to the springs.

3. Remove the front pivot bolt.

4. Remove the rear shackle bolt nuts and the rear shackle plate.

NOTE: Some models are equipped with a one piece shackle which is to be removed with the spring. On in-stallation; attach the shackle to the spring first.

5. Remove the spring.
To install:

6. Position the spring in place and install the eye bolt and nut. Do not tighten the bolt yet.

7. Install the shackles and bolts. Tighten them just enough to make them snug.

8. Make sure that the spring center bolt enters the locating hole in the axle pad. On headless-type spring bolts, install the bolts with the lock groove lined up with the lock bolt hole in the bracket. Install the lock bolt and tighten the lock bolt nut. In-stall the lubrication fittings.

9. Install the U-bolts, lockwashers and nuts. Make them just snug for now. Align the auxiliary spring parallel with the main spring.

10. Lower the vehicle to its normal position with the weight back on the springs. Now tighten all bolts and nuts as follows:
- U-bolt nuts:
  exc. 350 — 45 ft. lbs. (61 N.m)
  350 model — 110 ft. lbs. (149 N.m)
- Shackle nuts:
  exc. 350 — 35 ft. lbs. (47 N.m)
  350 model — 155 ft. lbs. (210 N.m)
- Fixed end:
  All — 100 ft. lbs. (135 N.m)

Shock Absorbers

REMOVAL AND INSTALLATION

1. Jack and support the vehicle.

2. Remove the nut from the stud or bolt at the upper end. Re-move the stud or bolt from the upper end.

3. Remove the lower nut at the bushing end.

4. Pivot the shock absorber and washers from the lower stud.

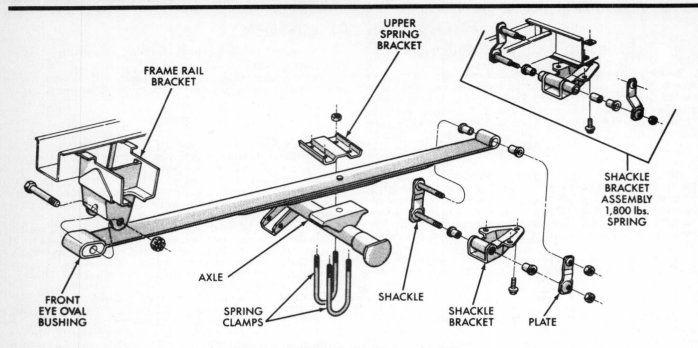

Rear spring components — exc. 350

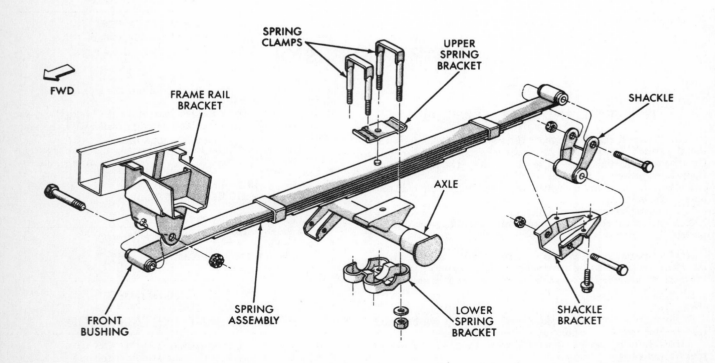

Rear spring components — 350

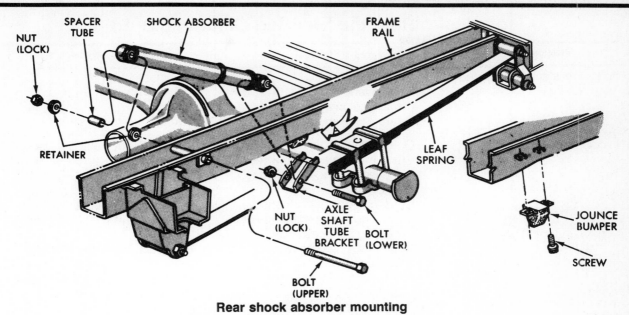

Rear shock absorber mounting

5. Remove the shock absorber and washers from the lower stud.

6. Purge the new shock of air by extending it in its normal position and compressing it while inverted. Do this several times.

It is normal for there to be more resistance to extension than to compression.

7. Installation is the reverse of removal. Torque the nuts to 60 ft. lbs. (82 N.m).

Troubleshooting the Steering Column

Problem	Cause	Solution
Will not lock	• Lockbolt spring broken or defective	• Replace lock bolt spring
High effort (required to turn ignition key and lock cylinder)	• Lock cylinder defective	• Replace lock cylinder
	• Ignition switch defective	• Replace ignition switch
	• Rack preload spring broken or deformed	• Replace preload spring
	• Burr on lock sector, lock rack, housing, support or remote rod coupling	• Remove burr
	• Bent sector shaft	• Replace shaft
	• Defective lock rack	• Replace lock rack
	• Remote rod bent, deformed	• Replace rod
	• Ignition switch mounting bracket bent	• Straighten or replace
	• Distorted coupling slot in lock rack (tilt column)	• Replace lock rack
Will stick in "start"	• Remote rod deformed	• Straighten or replace
	• Ignition switch mounting bracket bent	• Straighten or replace
Key cannot be removed in "off-lock"	• Ignition switch is not adjusted correctly	• Adjust switch
	• Defective lock cylinder	• Replace lock cylinder
Lock cylinder can be removed without depressing retainer	• Lock cylinder with defective retainer	• Replace lock cylinder
	• Burr over retainer slot in housing cover or on cylinder retainer	• Remove burr

Troubleshooting the Steering Column (cont.)

Problem	Cause	Solution
High effort on lock cylinder between "off" and "off-lock"	• Distorted lock rack • Burr on tang of shift gate (automatic column) • Gearshift linkage not adjusted	• Replace lock rack • Remove burr • Adjust linkage
Noise in column	• One click when in "off-lock" position and the steering wheel is moved (all except automatic column) • Coupling bolts not tightened • Lack of grease on bearings or bearing surfaces • Upper shaft bearing worn or broken • Lower shaft bearing worn or broken • Column not correctly aligned • Coupling pulled apart • Broken coupling lower joint • Steering shaft snap ring not seated • Shroud loose on shift bowl. Housing loose on jacket—will be noticed with ignition in "off-lock" and when torque is applied to steering wheel.	• Normal—lock bolt is seating • Tighten pinch bolts • Lubricate with chassis grease • Replace bearing assembly • Replace bearing. Check shaft and replace if scored. • Align column • Replace coupling • Repair or replace joint and align column • Replace ring. Check for proper seating in groove. • Position shroud over lugs on shift bowl. Tighten mounting screws.
High steering shaft effort	• Column misaligned • Defective upper or lower bearing • Tight steering shaft universal joint • Flash on I.D. of shift tube at plastic joint (tilt column only) • Upper or lower bearing seized	• Align column • Replace as required • Repair or replace • Replace shift tube • Replace bearings
Lash in mounted column assembly	• Column mounting bracket bolts loose • Broken weld nuts on column jacket • Column capsule bracket sheared	• Tighten bolts • Replace column jacket • Replace bracket assembly
Lash in mounted column assembly (cont.)	• Column bracket to column jacket mounting bolts loose • Loose lock shoes in housing (tilt column only) • Loose pivot pins (tilt column only) • Loose lock shoe pin (tilt column only) • Loose support screws (tilt column only)	• Tighten to specified torque • Replace shoes • Replace pivot pins and support • Replace pin and housing • Tighten screws
Housing loose (tilt column only)	• Excessive clearance between holes in support or housing and pivot pin diameters • Housing support-screws loose	• Replace pivot pins and support • Tighten screws

Troubleshooting the Steering Column (cont.)

Problem	Cause	Solution
Steering wheel loose—every other tilt position (tilt column only)	• Loose fit between lock shoe and lock shoe pivot pin	• Replace lock shoes and pivot pin
Steering column not locking in any tilt position (tilt column only)	• Lock shoe seized on pivot pin • Lock shoe grooves have burrs or are filled with foreign material • Lock shoe springs weak or broken	• Replace lock shoes and pin • Clean or replace lock shoes • Replace springs
Noise when tilting column (tilt column only)	• Upper tilt bumpers worn • Tilt spring rubbing in housing	• Replace tilt bumper • Lubricate with chassis grease
One click when in "off-lock" position and the steering wheel is moved	• Seating of lock bolt	• None. Click is normal characteristic sound produced by lock bolt as it seats.
High shift effort (automatic and tilt column only)	• Column not correctly aligned • Lower bearing not aligned correctly • Lack of grease on seal or lower bearing areas	• Align column • Assemble correctly • Lubricate with chassis grease
Improper transmission shifting— automatic and tilt column only	• Sheared shift tube joint • Improper transmission gearshift linkage adjustment • Loose lower shift lever	• Replace shift tube • Adjust linkage • Replace shift tube

Troubleshooting the Ignition Switch

Problem	Cause	Solution
Ignition switch electrically inoperative	• Loose or defective switch connector • Feed wire open (fusible link) • Defective ignition switch	• Tighten or replace connector • Repair or replace • Replace ignition switch
Engine will not crank	• Ignition switch not adjusted properly	• Adjust switch
Ignition switch wil not actuate mechanically	• Defective ignition switch • Defective lock sector • Defective remote rod	• Replace switch • Replace lock sector • Replace remote rod
Ignition switch cannot be adjusted correctly	• Remote rod deformed	• Repair, straighten or replace

Troubleshooting the Turn Signal Switch

Problem	Cause	Solution
Turn signal will not cancel	• Loose switch mounting screws • Switch or anchor bosses broken • Broken, missing or out of position detent, or cancelling spring	• Tighten screws • Replace switch • Reposition springs or replace switch as required
Turn signal difficult to operate	• Turn signal lever loose • Switch yoke broken or distorted • Loose or misplaced springs • Foreign parts and/or materials in switch • Switch mounted loosely	• Tighten mounting screws • Replace switch • Reposition springs or replace switch • Remove foreign parts and/or material • Tighten mounting screws
Turn signal will not indicate lane change	• Broken lane change pressure pad or spring hanger • Broken, missing or misplaced lane change spring • Jammed wires	• Replace switch • Replace or reposition as required • Loosen mounting screws, reposition wires and retighten screws
Turn signal will not stay in turn position	• Foreign material or loose parts impeding movement of switch yoke • Defective switch	• Remove material and/or parts • Replace switch
Hazard switch cannot be pulled out	• Foreign material between hazard support cancelling leg and yoke	• Remove foreign material. No foreign material impeding function of hazard switch—replace turn signal switch.
No turn signal lights	• Inoperative turn signal flasher • Defective or blown fuse • Loose chassis to column harness connector • Disconnect column to chassis connector. Connect new switch to chassis and operate switch by hand. If vehicle lights now operate normally, signal switch is inoperative • If vehicle lights do not operate, check chassis wiring for opens, grounds, etc.	• Replace turn signal flasher • Replace fuse • Connect securely • Replace signal switch • Repair chassis wiring as required

Troubleshooting the Turn Signal Switch (cont.)

Problem	Cause	Solution
Instrument panel turn indicator lights on but not flashing	• Burned out or damaged front or rear turn signal bulb • If vehicle lights do not operate, check light sockets for high resistance connections, the chassis wiring for opens, grounds, etc. • Inoperative flasher • Loose chassis to column harness connection • Inoperative turn signal switch • To determine if turn signal switch is defective, substitute new switch into circuit and operate switch by hand. If the vehicle's lights operate normally, signal switch is inoperative.	• Replace bulb • Repair chassis wiring as required • Replace flasher • Connect securely • Replace turn signal switch • Replace turn signal switch
Stop light not on when turn indicated	• Loose column to chassis connection • Disconnect column to chassis connector. Connect new switch into system without removing old.	• Connect securely • Replace signal switch
Stop light not on when turn indicated (cont.)	Operate switch by hand. If brake lights work with switch in the turn position, signal switch is defective. • If brake lights do not work, check connector to stop light sockets for grounds, opens, etc.	• Repair connector to stop light circuits using service manual as guide
Turn indicator panel lights not flashing	• Burned out bulbs • High resistance to ground at bulb socket • Opens, ground in wiring harness from front turn signal bulb socket to indicator lights	• Replace bulbs • Replace socket • Locate and repair as required
Turn signal lights flash very slowly	• High resistance ground at light sockets • Incorrect capacity turn signal flasher or bulb • If flashing rate is still extremely slow, check chassis wiring harness from the connector to light sockets for high resistance • Loose chassis to column harness connection • Disconnect column to chassis connector. Connect new switch into system without removing old. Operate switch by hand. If flashing occurs at normal rate, the signal switch is defective.	• Repair high resistance grounds at light sockets • Replace turn signal flasher or bulb • Locate and repair as required • Connect securely • Replace turn signal switch

Troubleshooting the Turn Signal Switch (cont.)

Problem	Cause	Solution
Hazard signal lights will not flash—turn signal functions normally	• Blow fuse • Inoperative hazard warning flasher • Loose chassis-to-column harness connection • Disconnect column to chassis connector. Connect new switch into system without removing old. Depress the hazard warning lights. If they now work normally, turn signal switch is defective. • If lights do not flash, check wiring harness "K" lead for open between hazard flasher and connector. If open, fuse block is defective	• Replace fuse • Replace hazard warning flasher in fuse panel • Conect securely • Replace turn signal switch • Repair or replace brown wire or connector as required

STEERING

Steering Wheel

REMOVAL AND INSTALLATION

1. Disconnect the battery negative ground cable.
2. Working through the access holes in the back of the wheel, push the horn pad off. DO NOT PRY THE PAD OFF!
3. Disconnect the horn wire.
4. Matchmark the steering wheel and shaft.
5. Remove the steering wheel retaining nut.
6. Using a puller, remove the steering wheel from the shaft. NEVER HAMMER THE SHAFT TO FREE THE WHEEL!
7. **To install:** place the steering wheel on to the column shaft with matchmarks aligned. Tighten the nut to 45 ft. lbs. (61 N.m).

Turn Signal/Combination Switch

The multi function (combination) switch contains electrical circuitry for turn signal, hazard warning, headlamp beam select, windshield wiper and windshield washer switching.

REMOVAL AND INSTALLATION

1989-90

STANDARD COLUMNS

1. Disconnect the negative battery cable. Remove the steering wheel. Remove the lower instrument panel bezel. Remove the wiring trough by prying out the retainer buttons.
2. Position the gearshift lever to its full clockwise position. Disconnect the wiring connections.
3. Remove the screw that attaches the wiper-washer switch to the turn signal pivot. Remove the three switch mounting screws and the switch from the steering column.
To install:
4. Position the switch on the column and install the attaching screws.
5. Install the wiper-washer switch.
6. Connect the wiring.
7. Install the wiring cover, lower bezel and steering wheel.
8. Connect the battery cable.

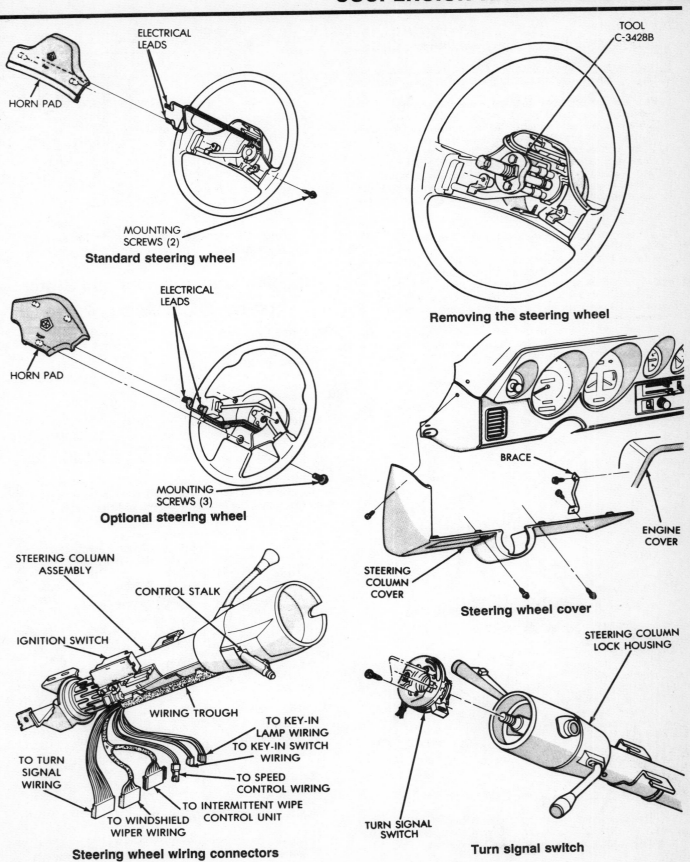

ELECTRICAL LEADS

HORN PAD

MOUNTING SCREWS (2)

Standard steering wheel

ELECTRICAL LEADS

HORN PAD

MOUNTING SCREWS (3)

Optional steering wheel

TOOL C-3428B

Removing the steering wheel

BRACE

ENGINE COVER

STEERING COLUMN COVER

Steering wheel cover

STEERING COLUMN ASSEMBLY

CONTROL STALK

IGNITION SWITCH

WIRING TROUGH

TO KEY-IN LAMP WIRING

TO KEY-IN SWITCH WIRING

TO TURN SIGNAL WIRING

TO SPEED CONTROL WIRING

TO INTERMITTENT WIPE CONTROL UNIT

TO WINDSHIELD WIPER WIRING

Steering wheel wiring connectors

STEERING COLUMN LOCK HOUSING

TURN SIGNAL SWITCH

Turn signal switch

TILT COLUMNS

1. Disconnect the negative battery cable. Remove the steering wheel.
2. Depress the lock plate with tool C-4156, or equivalent, just enough to remove the retaining ring, and pry the retaining ring out of the groove.
3. Remove the lock plate and upper bearing spring.
4. Place the turn signal switch in the right turn position.
5. Remove the screw which attaches the link between the turn signal switch and wiper/washer switch pivot.
6. Remove the screw which attaches the hazard switch knob.
7. Remove the 3 screws securing the turn signal switch.
8. Gently pull the switch and wiring from the column.
9. Assemble the switch to the steering in the reverse order. Connect the negative battery cable and check switch operation.

1991

1. Disconnect the negative battery cable. On tilt wheel models, remove the tilt lever.
2. Remove both the upper and lower steering column covers.
3. Remove the multi function switch tamper proof screws (Torx). Pull the switch away from the column carefully.
4. Loosen the connector screw (the screw will remain in the connector). Remove the wiring connector from the multi function switch.
5. Install the connector to the switch. Mount and secure the multi function switch to the steering column. Install the tilt lever. Connect the battery cable and check the switch functions.

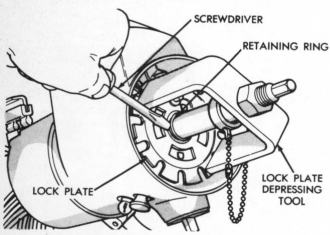

Removing the lock plate retaining ring

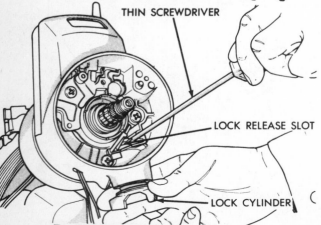

Removing the lock cylinder

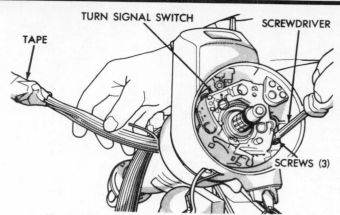

Tape the connectors and wires for ease in installation

Ignition Lock Cylinder and Switch

REMOVAL AND INSTALLATION

Standard Columns

1. Remove the steering wheel.
2. Remove the turn signal switch.
3. Remove the retaining screw and lift the ignition lock cylinder lamp out of the way.
4. Remove the bearing housing.
5. Remove the coil spring.
6. Remove the lock plate from the shaft.
7. Remove the 2 retaining screws and lift the lock lever guide plate to expose the lock cylinder release hole.
8. Insert the key and place the lock cylinder in the **LOCK** position. Remove the key.
9. Insert a thin punch into the lock cylinder release hole and push inward to release the spring-loaded lock retainer. At the same time, pull the lock cylinder out of the column.

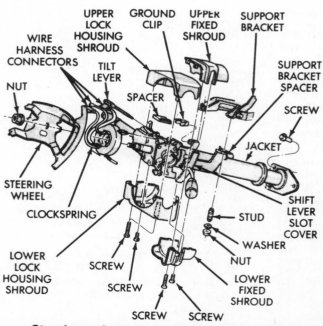

Steering column components (1991 shown)

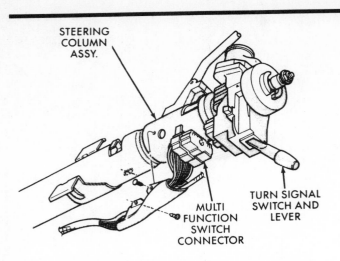

Multi-function switch wiring (1991 shown)

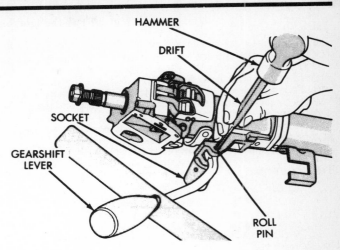

Gear shift removal

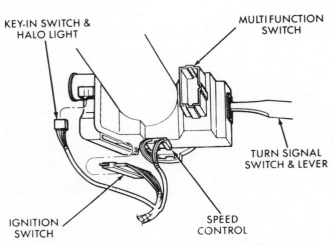

Steering column wiring (1991 shown)

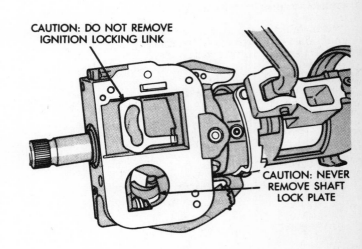

Observe cautions

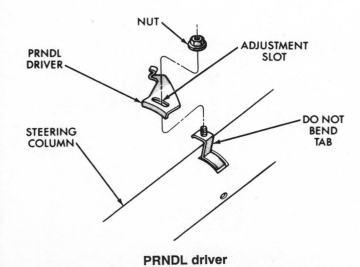

PRNDL driver

Do not remove retainer

10. Remove the 3 retaining screws and lift out the ignition switch.

To install:

11. Position the ignition switch in the center detent position (OFF).

12. Place the shift lever in PARK.

13. Feed the wires down through the space between the housing and jacket. Position the switch in the housing and install the 3 retaining screws.

14. Place the lock cylinder in the LOCK position and press it into place in the column. It will snap into position.

15. The remainder of assembly is the reverse of disassembly.

Tilt Columns

1. Remove the steering wheel.
2. Remove the tilt lever and turn signal lever.
3. If equipped, remove the turn signal lever.
4. Remove the turn signal switch.

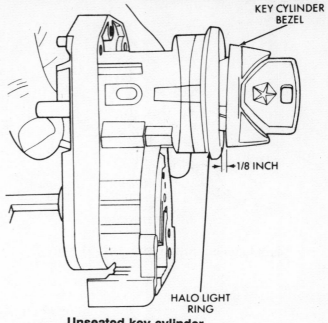

Unseated key cylinder

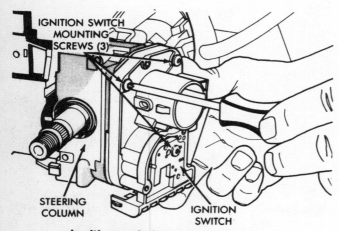

Ignition switch screw removal

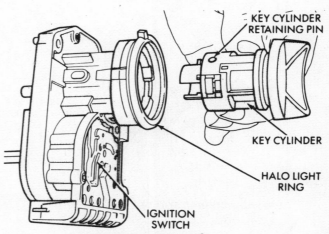

Key cylinder removal

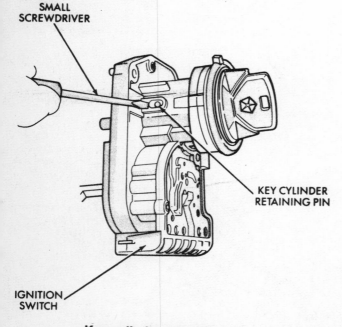

Key cylinder retaining pin

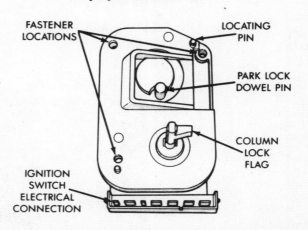

Ignition switch view from column

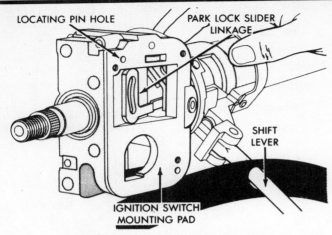

LOCATING PIN HOLE   PARK LOCK SLIDER LINKAGE

SHIFT LEVER

IGNITION SWITCH MOUNTING PAD

Ignition switch mounting pad

5. Using the key, place the lock cylinder in the **LOCK** position. Remove the key.

6. Insert a thin punch in the slot next to the switch mounting screw boss and depress the spring latch at the bottom of the slot. Hold the spring latch depressed and pull the lock cylinder out of the column.

7. Place the ignition switch in the ACCESSORY position and remove the mounting screws. Lift off the switch. The ACCESSORY position is the one opposite the spring-loaded end position.

To install:

8. First, install the lock cylinder. Place the cylinder in the LOCK position and push it into the housing. It will snap into place.

9. Rotate the lock cylinder to the ACCESSORY position.

10. Fit the actuator rod in the slider hole and position the switch on the column. Insert the mounting screws, but don't tighten them yet.

11. Push the switch gently down the column to remove all lash from the actuator rod. Tighten the mounting screws. Make sure that you didn't take the switch out of the ACCESSORY detent!

12. Complete the remainder of installation in reverse order.

Ignition Lock Cylinder

REMOVAL AND INSTALLATION

Standard Columns

1. Remove the steering wheel.
2. Remove the turn signal switch.
3. Remove the retaining screw and lift the ignition lock cylinder lamp out of the way.
4. Remove the bearing housing.
5. Remove the coil spring.
6. Remove the lock plate from the shaft.
7. Remove the 2 retaining screws and lift the lock lever guide plate to expose the lock cylinder release hole.
8. Insert the key and place the lock cylinder in the **LOCK** position. Remove the key.
9. Insert a thin punch into the lock cylinder release hole and push inward to release the spring-loaded lock retainer. At the same time, pull the lock cylinder out of the column.
10. When installing the lock cylinder, make sure it is in the **LOCK** position. Press it into the column. It will snap into place. The remainder of assembly is the reverse of disassembly.

Tilt Columns

1. Remove the steering wheel.

2. Remove the tilt lever and turn signal lever.
3. If equipped, remove the turn signal lever.
4. Remove the turn signal switch.
5. Using the key, place the lock cylinder in the **LOCK** position. Remove the key.
6. Insert a thin punch in the slot next to the switch mounting screw boss and depress the spring latch at the bottom of the slot. Hold the spring latch depressed and pull the lock cylinder out of the column.
7. Place the lock cylinder in the **LOCK** position and press it into place in the column. It will snap into position.
8. The remainder of assembly is the reverse of disassembly.

Steering Column

REMOVAL AND INSTALLATION

1989-90

1. Disconnect the battery ground cable.
2. On vans with a column shift, disconnect the link rod by prying the rod out of the grommet in the shift lever.
3. Remove the coupling roll pin.
4. Disconnect the wiring at the column.
5. Remove the steering wheel.
6. Remove the turn signal lever.
7. Remove the floor plate screws.
8. Remove the cluster bezel and panel lower reinforcement.
9. Disconnect the automatic shift indicator pointer from the shift housing.
10. Remove the column bracket-to-instrument panel nuts.
11. Carefully remove the coupling from the wormshaft and lift the column out of the vehicle.

To install:

12. Install a new shift grommet from the rod side of the lever using pliers and a back-up washer to snap the grommet into place. Coating the grommet with multi-purpose grease will aid in installation. A new grommet should be used whenever a rod is disconnected from the lever.
13. Position the column in the vehicle.
14. With the front wheels straight ahead and the master splines on the wormshaft and coupling aligned, engage the coupling and wormshaft and install the roll pin.

WARNING: Never force the shaft down into position!

15. Install, but don't tighten, the bracket nuts.
16. Make sure that both plastic spacers are fully seated in their slots in the column support bracket, then, tighten the bracket nuts to 110 inch lbs.
17. Install the floor plate screws.
18. Install the steering wheel.
19. Connect the wiring.
20. Connect and adjust the shift linkage.
21. Connect the shift indicator pointer to the operating bracket in its original location. Slowly move the shift lever from **L** to **P**, pausing briefly at each position. The pointer must align with each selector position. If necessary, loosen the bolt and readjust the linkage to align the pointer correctly.
22. Install the panel lower reinforcement and cluster bezel.

1991

1. Confirm that the wheels are in the straight ahead position. Disconnect the negative battery cable.
2. If equipped with a column shift, disconnect the link rod by prying it out of the shift lever grommet.
3. Remove the screws that attach the floor plate to the floor boards. Remove the steering column shaft to coupler roll pin. Remove the steering wheel center pad and disconnect the electrical components such as the horn leads and speed control.

4. Remove the steering wheel. Remove the lower instrument panel column cover.

5. Loosen, but do not remove the nuts that hold the column to the support bracket.

6. On column shift vehicles, make sure the shift lever is in the PARK position. Release the PRNDL cable from the driver on the column.

7. Remove the tilt lever. (Models equipped with a tilt wheel).

8. Remove the upper and lower lock housing shroud (torx screws). Remove the multi-function switch. Remove the electrical connections from the Key-in light, Main ignition switch, Horn, Clock spring and Speed control.

9. Remove the wiring harness from the column by prying out the plastic retainer buttons.

10. Remove the lower dash panel and support bracket standoff fasteners. Remove the column through the passenger compartment.

To install:

11. Align the master splines on the steering gear shaft and coupler. Slide the steering coupler up the steering column. Engage the coupler with the shaft and install the roll pin. Position the column with the bracket slots on the attaching mounting studs. Install the retaining nuts and bolts loosely. Connect all switches and wiring harnesses. Install the upper shroud.

12. After ensuring that both breakaway capsules are fully seated in the slots in the column support bracket, tighten the upper bracket nuts to 105 inch lbs. (12 N.m).

13. Tighten the floor plate to floor pan attaching screws. Complete connecting the wiring harnesses.

14. Install the lower shroud. Connect the PRNDL cable. Install the lock housing shrouds and tilt lever.

15. Install the lower dash panel cover. Install the steering wheel after all of the removed components have been connected.

Steering Linkage

REMOVAL AND INSTALLATION

1. Raise and safely support the front of the vehicle. Removal of the front wheels will make access easier, so remove them.

2. Remove the cotter pins and retaining nuts from the tie rod end, idler arm and pitman arm pivot ball studs.

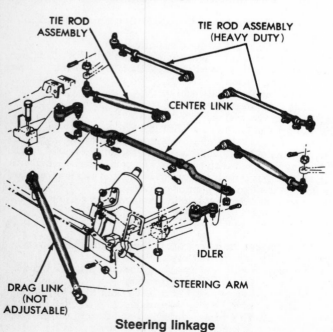

Steering linkage

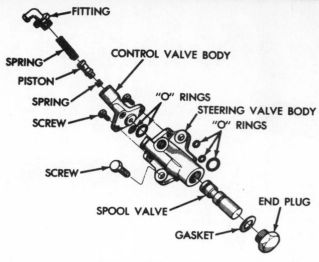

Steering gear valve

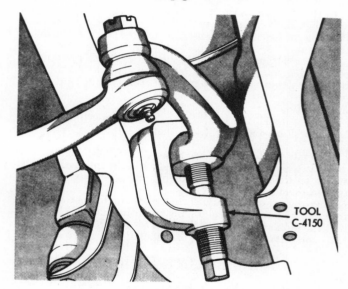

Pitman arm removal

3. Remove the outer tie rod ends from the steering knuckles, using the proper puller. Remove the inner tie rod ball studs from the center link. Once again using the proper puller.

4. Remove the idler arm ball studs from the center link. Remove the idler arm retaining nuts from the frame bracket bolts.

5. Remove the drag link ball studs from the pitman arm and the center link bore. Remove the retaining nut and the pitman arm from the steering gear, use puller C-4150, or the equivalent.

6. Examine all the steering linkage components. Replace as necessary.

To install:

7. Position the idler arms at the frame brackets and install the retaining nuts. Tighten them to 70 ft. lbs. (95 N.m).

8. Place the center link on the idler arm ball studs. Install and tighten the retaining nuts to 47 ft. lbs. (64 N.m). Install new cotter pins.

9. Connect the tie end ball studs to the steering knuckles and to the center link. Tighten the nuts to : $^9/_{16}$–18 — 55 ft. lbs. (75 N.m). $^5/_8$–18 — 75 ft. lbs. (102 N.m). Install new cotter pins.

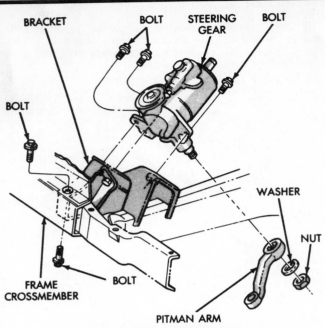

Steering gear removal/installation

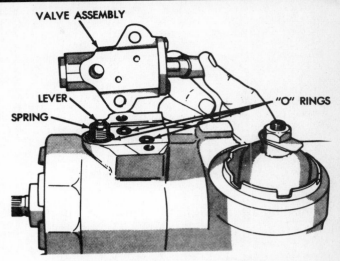

Steering gear valve removal/installation

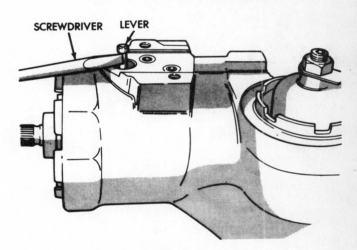

Pivot lever removal

10. Install the pitman arm on the steering gear and tighten the nut to 175 ft. lbs. (237 N.m.)

11. Connect the drag link ball studs to the pitman arm and to the center link. Tighten to 55 ft. lbs. (75 N.m). Install new cotter pins.

12. Install the front wheels and lower the vehicle. Have the front end alignment checked.

Tie Rod Ends

1. Raise and support the front end on jackstands.
2. Remove the cotter pin and nut from the tie rod end.
3. Using a separator, free the tie rod end from the knuckle arm or center link.
4. Count the exact number of threads visible between the tie rod end and the sleeve. Loosen the clamp bolts and unscrew the tie rod end.

To install:

5. Screw the end into the sleeve until the *exact* number of threads originally noted is showing.

6. Tighten the clamp bolts. Tighten the clamp bolts to 150-175 inch lbs. on 100 and 200 series or 20-30 ft. lbs. on 300 series; the tie rod end nut to 55 ft. lbs. (75 N.m), on HD (350) tighten to 75 ft. lbs. (102 N.m). Have the vehicle toe-in checked.

Power Steering Gear

ADJUSTMENTS

On Vehicle

NOTE: Removal of the drag link ball stud from the pitman arm by methods other than using Tool C-3894A, or the equivalent, will damage the ball stud seal.

PITMAN (SECTOR) SHAFT BACKLASH

1. Disconnect the drag link ball stud from the pitman arm.
2. Start the engine and allow it to idle.
3. Turn the steering wheel gently from one stop to the other and count the number of turns. Turn the steering wheel back exactly halfway, to the center position.

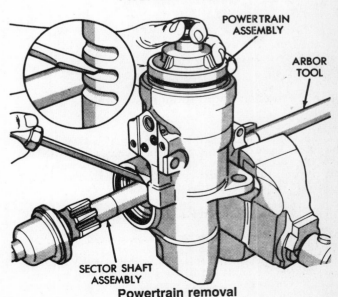

Powertrain removal

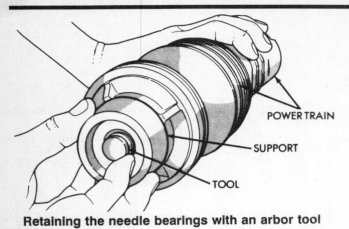

Retaining the needle bearings with an arbor tool

POWER TRAIN
SUPPORT
TOOL

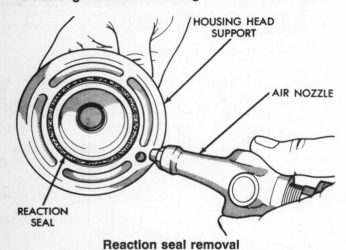

HOUSING HEAD SUPPORT
AIR NOZZLE
REACTION SEAL

Reaction seal removal

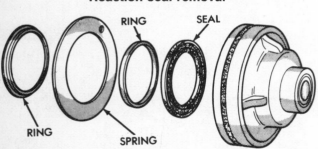

RING
SEAL
RING
SPRING

Reaction rings, springs and seals

4. Loosen the adjustment screw until the backlash exists in the pitman shaft. Tighten the adjustment screw until the backlash begins to disappear. Continue to tighten the adjustment screw ⅜ to ½ turn from this position. Tighten the locknut to 28 ft. lbs. (38 N.m). Connect the drag link.

REMOVAL AND INSTALLATION

1. Disconnect the battery negative ground cable.
2. Remove the steering column.
3. Place a drain pan under the gear and disconnect the hoses.
4. Remove the pitman arm, using a suitable puller..
5. Remove the gear-to-frame mounting bolts and remove the gear.

To install:

6. Place the gear assembly in position on the frame and install the mounting bolts. Torque the mounting bolts to 100 ft. lbs. (136 N.m). Make sure that the steering gear shaft is centered before connecting the linkage.

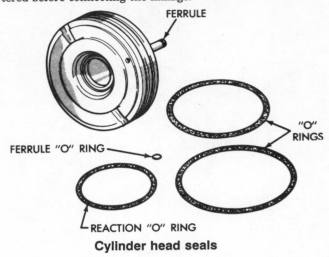

FERRULE
"O" RINGS
FERRULE "O" RING
REACTION "O" RING

Cylinder head seals

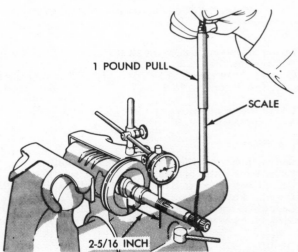

1 POUND PULL
SCALE
2-5/16 INCH

Wormshaft side play measurement

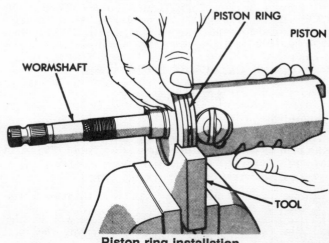

PISTON RING
PISTON
WORMSHAFT
TOOL

Piston ring installation

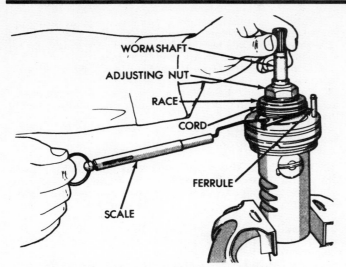

Thrust bearing preload torque adjustment

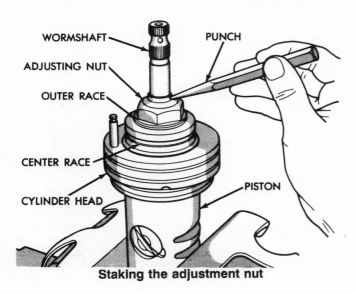

Staking the adjustment nut

OVERHAUL

1. Remove the steering gear assembly from the vehicle.
2. Clean the steering gear housing with a safe solvent, after draining all of the fluid by turning the wormshaft from one extreme of travel to the other.
3. Remove the steering gear valve, mounted on the side of the housing, and three O-rings.
4. Remove the pivot lever and spring, which was covered by the gear valve, from the housing.
5. Loosen the backlash adjustment screw locknut. Remove the shaft cover using a spanner nut wrench (Tool C-3988, or equivalent). Rotate the wormshaft to position the sector (pitman) shaft teeth at the midpoint of piston travel.
6. Position Arbor Tool C-3786 on the threaded end of the sector shaft. Slide the tool into the gear housing until the tool and shaft are engaged with the bearings.
6. Turn the wormshaft to the full left position to compress the powertrain components. Remove the powertrain retaining nut with the spanner wrench. Remove the housing head tang washer.

NOTE: It is important the cylinder head, center race, spacer, and head remain together during removal. This will eliminate the possibility of the reaction rings becoming disengaged from the grooves in both the cylinder head and the housing head. It will also prevent the center spacer from separating from the center race and becoming cocked in the housing. This can cause it to be impossible to remove the powertrain without damaging the spacer and/or the housing.

7. While the powertrain is fully compressed, pry on the piston teeth with a small prybar and remove the complete powertrain.
8. With the wormshaft facing upward, clamp the piston in a soft-jawed vise.
9. Raise the housing head until it clears the top of the wormshaft. Remove the housing head from the powertrain.

NOTE: If the needle bearings become detached from the wormshaft, insert and then retain them with wheel bearing grease. If the wormshaft bearing seal must be replace, replace it when the housing head is installed on the steering gear.

10. Remove the large O-ring from the groove in the housing head. Remove the reaction seal from the groove in the inner face of the head with compressed air directed into the ferrule rod chamber bore.
11. Inspect all of the seal groove passages for burrs. Ensure the fluid passage is not restricted.
12. Remove the outer reaction spring, reaction ring, wormshaft balancing ring and spacer.
13. Prevent the wormshaft from rotating. Turn the adjustment nut with sufficient torque to release it from the knurled section and remove the nut.
14. Clean the knurled section of the wormshaft with a wire brush. Use compressed air to remove any metal particles from the nut and wormshaft.
15. Remove the outer thrust bearing and (thin) race. Remove the thrust bearing center race. Remove the inner thrust bearing and (thick) race.
16. Remove the inner reaction ring and spring. Remove the cylinder head from the wormshaft.
17. Remove the O-ring seals from the outer perimeter grooves in the cylinder head.
18. Remove the reaction ring O-ring seal from the groove with compressed air. Remove the snapring and seal.
19. Remove the snap retaining ring, sleeve and rectangular-shaped ring seal from the cylinder head counterbore.
20. Test the wormshaft and piston for smooth operation. The torque required to rotate the wormshaft in or out of the piston must not exceed 1.5 inch lbs. (0.17 N.m).

NOTE: The wormshaft and piston are replaceable as a complete assembly only and should not be disassembled.

21. Test for excessive side play with the piston rack held firmly in a vise. Be sure the rack teeth are up and the wormshaft at the midpoint of travel. Lift the end of the wormshaft with a force equivalent to one pound. The vertical side play should not exceed 0.008 in. (0.2mm).
22. Inspect the Teflon® piston ring for excessive wear and cuts. If service is necessary, replace with a cast iron piston ring. Replace the ring and rectangular-shaped, rubber ring seal as follows:
• Remove the Teflon® piston ring and the rectangular-shaped seal from the piston groove
• Position the replacement rubber seal and cast iron ring in the piston groove
• Position the piston and ring on special tool C-3676, or equivalent, with the piston ring resting on the tool land
• Press down on the piston to seat the ring in the piston groove. Force the open ends of the ring out of the groove to aid in seating the ring

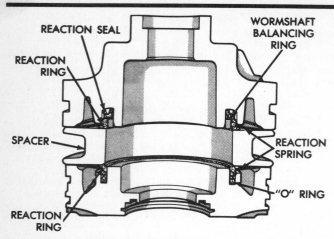

Reaction rings, springs and seals

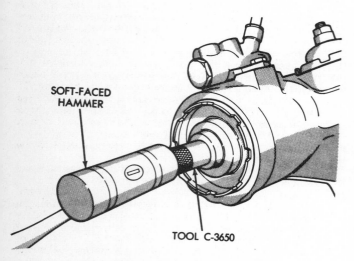

Wormshaft seal installation

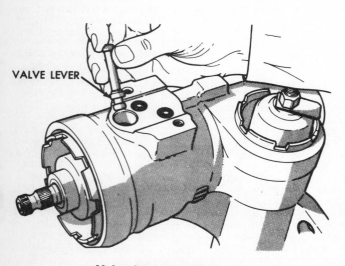

Valve lever installation

With the piston in a vertical position (wormshaft upwards), clamp the piston in a soft-jawed vise.

23. Inspect the cylinder head ferrule fluid passage bore for restrictions and the seal groove lands for burrs.

24. Lubricate the two large O-rings and install them in the cylinder head grooves.

25. Install the wormshaft sleeve seal, sleeve, and snap retaining ring. Ensure that the snap retaining ring is fully seated in the groove.

26. Install the inner reaction ring O-ring seal in the cylinder head groove.

NOTE: Check the wormshaft seal ring making sure the gap is closed. This will avoid damaging the ring when the cylinder head is positioned against the piston flange.

27. Slide the cylinder head assembly (ferrule up) on the wormshaft. Lubricate with power steering fluid and install the components in the following order:
- Inner thrust bearing (thick) race
- Inner thrust bearing
- Inner reaction spring with small hole positioned over the ferrule rod)
- Inner reaction ring with the flange up so it will protrude through the reaction ring spring and contact the reaction ring O-ring in the cylinder head
- Thrust bearing center race
- Outer thrust bearing
- Outer thrust bearing (thin) race
- Replacement wormshaft thrust bearing adjustment nut (finger tight)

28. Install a splined nut. Rotate the wormshaft clockwise (½ turn) and hold in this position with nut C-3637 and a socket wrench. Retain the wormshaft in this position. Retain the wormshaft in this position during the following Step. Tighten the thrust bearing adjustment nut to 50 ft. lbs. (68 N.m) to pre-stretch the threads.

29. Loosen the adjustment nut. Wrap several layers of cord around the thrust bearing center race. Tie a loop at the loose end of the cord and connect a spring scale hook to the loop. Pulling the cord outward will cause the thrust bearing center race to rotate. Tighten the wormshaft thrust bearing adjustment nut while pulling the cord with the scale. The scale should indicate 16 to 24 ounces (20 ounces is preferred) while the race is rotating.

30. Stake the upper part of the wormshaft adjustment nut to the knurled area of the shaft by positioning a ¼ in. (6mm) diameter flat-nose punch with a slight upward angle to the nut flange. Use a hammer to strike the end of the punch to stake the nut.

31. Measure the thrust bearing preload torque again. (Cord and scale procedure). If the adjustment nut moved during the staking, it can be corrected by striking the nut a glancing blow in the direction required.

32. After re-testing the preload, stake the adjustment nut at three additional locations around the upper part of the nut. Test the staking by applying 20 ft. lbs. (27 N.m) torque in each direction. If the nut does not move, the stakes are satisfactory. Check the pre-load again to ensure that the nut did not move.

33. Position the spacer over the center race while inserting the dowel pin into the slot. Insert the cylinder head ferrule rod into the spacer slot. This will align the hole in the race with the hole in the spacer. The small O-ring should not be installed until the upper reaction spring and spacer have been installed.

34. Install the outer reaction ring on the spacer with the flange facing down and against the spacer. Install the spring over the reaction ring with the cylinder head ferrule through the hole in the spring.

35. Install the wormshaft balancing ring (without flange) in-

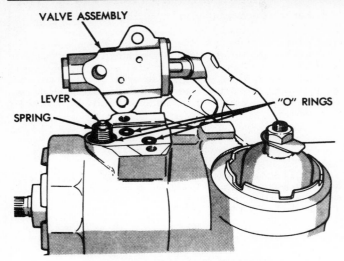

Steering gear valve installation

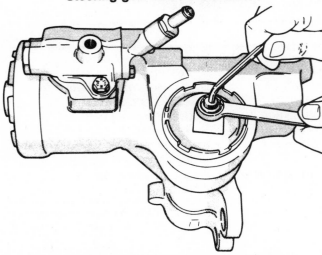

Pitman shaft backlash adjustment

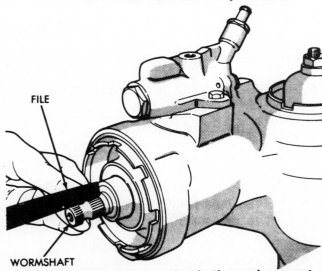

Creating a master serration in the replacement wormshaft

side the outer reaction ring. Lubricate the O-ring and install in the ferrule rod groove.

36. If the wormshaft seal was removed, install a replacement seal. Lubricate and install the reaction ring seal in the groove with the flat side of the seal facing outward.

37. Install the large O-ring seal in the groove on the housing head. Slide the housing head down on the wormshaft carefully engaging the ferrule rod and O-ring seal. Make sure the reaction rings enter the circular groove in the housing head face.

38. The powertrain is now ready for installation in the housing.

39. Service the sector (pitman) shaft next. The sector shaft rotates directly within the housing bore. Use care to avoid scoring the seal bore in the steering gear housing during seal removal.

40. Do not replace the sector (pitman) shaft seal unless it is leaking. Remove the seal cover by prying it out with a small pry bar. Remove the snapring that retains the seal in position. Remove the seal backup washer. Remove the seal by prying it out with a small pry bar.

41. Lubricate the sector shaft cover replacement O-ring and install the cover. Position the head of the backlash adjustment screw in the T slot in the end of the sector shaft. Position the cover over the adjustment screw. Insert an Allen wrench through the threaded hole in the cover. Turn the screw counterclockwise to thread the shaft completely into the cover. Install the combination aluminum gasket, date tag and locknut. Do not tighten at this time.

42. Proceed with final assembly as follows; Lubricate the powertrain bore in the steering gear housing with power steering fluid. Carefully insert the powertrain in the housing. To keep the reaction rings from separating from the grooves, rotate and retain the wormshaft fully counterclockwise. The piston teeth must be facing toward the right. The valve lever hole in the center race must be in the top position.

43. Ensure the cylinder head is in complete contact with the steering gear housing shoulder.

44. Align the valve lever bore in the bearing center race and spacer with the valve lever bore in the housing.

45. Insert the valve lever into the housing bore, double end first, parallel to the wormshaft. Tap lightly on the end of the lever with a hammer handle. Seat the inner end of the lever in the bearing center race/spacer.

46. Center the valve lever in the housing bore. Tap on a reinforcement rib with a hammer and drift to slightly rotate the housing head. Align the housing head tang washer with the groove in the housing and install it. Install the spanner nut and tighten it to 162 ft. lbs. (220 N.m). Make sure the valve lever remains centered in the steering gear housing bore. Turn the wormshaft until the piston bottoms in both directions and observe the action of the lever. It must be in the center of the bore and snap back to its centered position when the worm torque is relieved.

47. Install the valve lever spring, small end first. Position the piston at the midpoint of travel. Install the sector shaft and cover. The sector teeth must be correctly aligned with the piston rack teeth. Make sure the O-ring is correctly positioned on the sector shaft cover.

48. Install the sector shaft cover spanner nut and tighten to 155 ft. lbs. (210 N.m).

49. Install the valve body on the housing. Make sure the valve enters the spool valve bore. Make sure that the O-ring seals are correctly in place. Tighten the mounting screws to 7 ft. lbs. (9 N.m).

50. Lubricate the replacement sector shaft seal lips and install over the sector shaft with the lip of the seal facing inward toward the steering gear housing. Take care not to damage the seal lips. Push the seal into position using a seal driver. Install the seal backup washer and the retaining snapring. Fill the area outward from the snapring with multi-mileage grease. Install the seal cover in the housing bore.

51. The wormshaft and piston are replaceable as a unit only.

The master serration is machined on the original shaft, replacement shaft do not have a machined master serration. It is necessary to create the master serration with a file. The steering gear must be completely assembled and the wormshaft in the midpoint of travel. Remove on tooth of serration closest to the twelve o'clock position with a file.

52. Install the steering gear in the vehicle, fill and bleed the system. Check the backlash adjustment.

Power Steering Pump

REMOVAL AND INSTALLATION

1. If the pump is to be replaced or disassembled, remove the pulley nut before removing the belt.
2. Remove the belt.
3. Place a drain pan under the pump and disconnect both hoses from the pump.

4. Remove the mounting and adjusting bolts and lift out the pump.

To install:

5. Position the pump and install the bolts. Torque the bolts to 30 ft. lbs.
6. Connect the fluid lines. Make sure that the hoses are routed at least 1 in. (25mm) from all surfaces and at least 2 in. (51mm) from the exhaust manifold. Install and adjust the drive belt tension.

BLEEDING

After filling the power steering reservoir to the correct level. Start the engine and turn the steering wheel from lock to lock several times. Do not force against the stops. Stop the engine and check the fluid level. Add fluid if necessary.

Troubleshooting the Manual Steering Gear

Problem	Cause	Solution
Hard or erratic steering	• Incorrect tire pressure	• Inflate tires to recommended pressures
	• Insufficient or incorrect lubrication	• Lubricate as required (refer to Maintenance Section)
	• Suspension, or steering linkage parts damaged or misaligned	• Repair or replace parts as necessary
	• Improper front wheel alignment	• Adjust incorrect wheel alignment angles
	• Incorrect steering gear adjustment	• Adjust steering gear
	• Sagging springs	• Replace springs
Play or looseness in steering	• Steering wheel loose	• Inspect shaft spines and repair as necessary. Tighten attaching nut and stake in place.
	• Steering linkage or attaching parts loose or worn	• Tighten, adjust, or replace faulty components
	• Pitman arm loose	• Inspect shaft splines and repair as necessary. Tighten attaching nut and stake in place
	• Steering gear attaching bolts loose	• Tighten bolts
	• Loose or worn wheel bearings	• Adjust or replace bearings
	• Steering gear adjustment incorrect or parts badly worn	• Adjust gear or replace defective parts
Wheel shimmy or tramp	• Improper tire pressure	• Inflate tires to recommended pressures
	• Wheels, tires, or brake rotors out-of-balance or out-of-round	• Inspect and replace or balance parts
	• Inoperative, worn, or loose shock absorbers or mounting parts	• Repair or replace shocks or mountings
	• Loose or worn steering or suspension parts	• Tighten or replace as necessary
	• Loose or worn wheel bearings	• Adjust or replace bearings
	• Incorrect steering gear adjustments	• Adjust steering gear
	• Incorrect front wheel alignment	• Correct front wheel alignment

Troubleshooting the Manual Steering Gear

Problem	Cause	Solution
Tire wear	• Improper tire pressure	• Inflate tires to recommended pressures
	• Failure to rotate tires	• Rotate tires
	• Brakes grabbing	• Adjust or repair brakes
	• Incorrect front wheel alignment	• Align incorrect angles
	• Broken or damaged steering and suspension parts	• Repair or replace defective parts
	• Wheel runout	• Replace faulty wheel
	• Excessive speed on turns	• Make driver aware of conditions
Vehicle leads to one side	• Improper tire pressures	• Inflate tires to recommended pressures
	• Front tires with uneven tread depth, wear pattern, or different cord design (i.e., one bias ply and one belted or radial tire on front wheels)	• Install tires of same cord construction and reasonably even tread depth, design, and wear pattern
	• Incorrect front wheel alignment	• Align incorrect angles
	• Brakes dragging	• Adjust or repair brakes
	• Pulling due to uneven tire construction	• Replace faulty tire

Troubleshooting the Power Steering Gear

Problem	Cause	Solution
Hissing noise in steering gear	• There is some noise in all power steering systems. One of the most common is a hissing sound most evident at standstill parking. There is no relationship between this noise and performance of the steering. Hiss may be expected when steering wheel is at end of travel or when slowly turning at standstill.	• Slight hiss is normal and in no way affects steering. Do not replace valve unless hiss is extremely objectionable. A replacement valve will also exhibit slight noise and is not always a cure. Investigate clearance around flexible coupling rivets. Be sure steering shaft and gear are aligned so flexible coupling rotates in a flat plane and is not distorted as shaft rotates. Any metal-to-metal contacts through flexible coupling will transmit valve hiss into passenger compartment through the steering column.

Troubleshooting the Power Steering Gear (cont.)

Problem	Cause	Solution
Rattle or chuckle noise in steering gear	• Gear loose on frame	• Check gear-to-frame mounting screws.
	• Steering linkage looseness	• Check linkage pivot points for wear. Replace if necessary.
	• Pressure hose touching other parts of car	• Adjust hose position. Do not bend tubing by hand.
	• Loose pitman shaft over center adjustment	• Adjust to specifications
	NOTE: A slight rattle may occur on turns because of increased clearance off the "high point." This is normal and clearance must not be reduced below specified limits to eliminate this slight rattle.	
	• Loose pitman arm	• Tighten pitman arm nut to specifications
Squawk noise in steering gear when turning or recovering from a turn	• Damper O-ring on valve spool cut	• Replace damper O-ring
Poor return of steering wheel to center	• Tires not properly inflated	• Inflate to specified pressure
	• Lack of lubrication in linkage and ball joints	• Lube linkage and ball joints
	• Lower coupling flange rubbing against steering gear adjuster plug	• Loosen pinch bolt and assemble properly
	• Steering gear to column misalignment	• Align steering column
	• Improper front wheel alignment	• Check and adjust as necessary
	• Steering linkage binding	• Replace pivots
	• Ball joints binding	• Replace ball joints
	• Steering wheel rubbing against housing	• Align housing
	• Tight or frozen steering shaft bearings	• Replace bearings
	• Sticking or plugged valve spool	• Remove and clean or replace valve
	• Steering gear adjustments over specifications	• Check adjustment with gear out of car. Adjust as required.
	• Kink in return hose	• Replace hose
Car leads to one side or the other (keep in mind road condition and wind. Test car in both directions on flat road)	• Front end misaligned	• Adjust to specifications
	• Unbalanced steering gear valve	• Replace valve
	NOTE: If this is cause, steering effort will be very light in direction of lead and normal or heavier in opposite direction	
Momentary increase in effort when turning wheel fast to right or left	• Low oil level	• Add power steering fluid as required
	• Pump belt slipping	• Tighten or replace belt
	• High internal leakage	• Check pump pressure. (See pressure test)

Troubleshooting the Power Steering Gear (cont.)

Problem	Cause	Solution
Steering wheel surges or jerks when turning with engine running especially during parking	• Low oil level • Loose pump belt • Steering linkage hitting engine oil pan at full turn • Insufficient pump pressure • Pump flow control valve sticking	• Fill as required • Adjust tension to specification • Correct clearance • Check pump pressure. (See pressure test). Replace relief valve if defective. • Inspect for varnish or damage, replace if necessary
Excessive wheel kickback or loose steering	• Air in system • Steering gear loose on frame • Steering linkage joints worn enough to be loose • Worn poppet valve • Loose thrust bearing preload adjustment • Excessive overcenter lash	• Add oil to pump reservoir and bleed by operating steering. Check hose connectors for proper torque and adjust as required. • Tighten attaching screws to specified torque • Replace loose pivots • Replace poppet valve • Adjust to specification with gear out of vehicle • Adjust to specification with gear out of car
Hard steering or lack of assist	• Loose pump belt • Low oil level **NOTE:** Low oil level will also result in excessive pump noise • Steering gear to column misalignment • Lower coupling flange rubbing against steering gear adjuster plug • Tires not properly inflated	• Adjust belt tension to specification • Fill to proper level. If excessively low, check all lines and joints for evidence of external leakage. Tighten loose connectors. • Align steering column • Loosen pinch bolt and assemble properly • Inflate to recommended pressure
Foamy milky power steering fluid, low fluid level and possible low pressure	• Air in the fluid, and loss of fluid due to internal pump leakage causing overflow	• Check for leak and correct. Bleed system. Extremely cold temperatures will cause system aeration should the oil level be low. If oil level is correct and pump still foams, remove pump from vehicle and separate reservoir from housing. Check welsh plug and housing for cracks. If plug is loose or housing is cracked, replace housing.
Low pressure due to steering pump	• Flow control valve stuck or inoperative • Pressure plate not flat against cam ring	• Remove burrs or dirt or replace. Flush system. • Correct

Troubleshooting the Power Steering Gear (cont.)

Problem	Cause	Solution
Low pressure due to steering gear	• Pressure loss in cylinder due to worn piston ring or badly worn housing bore • Leakage at valve rings, valve body-to-worm seal	• Remove gear from car for disassembly and inspection of ring and housing bore • Remove gear from car for disassembly and replace seals

Troubleshooting the Power Steering Pump

Problem	Cause	Solution
Chirp noise in steering pump	• Loose belt	• Adjust belt tension to specification
Belt squeal (particularly noticeable at full wheel travel and stand still parking)	• Loose belt	• Adjust belt tension to specification
Growl noise in steering pump	• Excessive back pressure in hoses or steering gear caused by restriction	• Locate restriction and correct. Replace part if necessary.
Growl noise in steering pump (particularly noticeable at stand still parking)	• Scored pressure plates, thrust plate or rotor • Extreme wear of cam ring	• Replace parts and flush system • Replace parts
Groan noise in steering pump	• Low oil level • Air in the oil. Poor pressure hose connection.	• Fill reservoir to proper level • Tighten connector to specified torque. Bleed system by operating steering from right to left—full turn.
Rattle noise in steering pump	• Vanes not installed properly • Vanes sticking in rotor slots	• Install properly • Free up by removing burrs, varnish, or dirt
Swish noise in steering pump	• Defective flow control valve	• Replace part
Whine noise in steering pump	• Pump shaft bearing scored	• Replace housing and shaft. Flush system.
Hard steering or lack of assist	• Loose pump belt • Low oil level in reservoir **NOTE:** Low oil level will also result in excessive pump noise • Steering gear to column misalignment • Lower coupling flange rubbing against steering gear adjuster plug • Tires not properly inflated	• Adjust belt tension to specification • Fill to proper level. If excessively low, check all lines and joints for evidence of external leakage. Tighten loose connectors. • Align steering column • Loosen pinch bolt and assemble properly • Inflate to recommended pressure

Troubleshooting the Power Steering Pump (cont.)

Problem	Cause	Solution
Foaming milky power steering fluid, low fluid level and possible low pressure	• Air in the fluid, and loss of fluid due to internal pump leakage causing overflow	• Check for leaks and correct. Bleed system. Extremely cold temperatures will cause system aeriation should the oil level be low. If oil level is correct and pump still foams, remove pump from vehicle and separate reservoir from body. Check welsh plug and body for cracks. If plug is loose or body is cracked, replace body.
Low pump pressure	• Flow control valve stuck or inoperative • Pressure plate not flat against cam ring	• Remove burrs or dirt or replace. Flush system. • Correct
Momentary increase in effort when turning wheel fast to right or left	• Low oil level in pump • Pump belt slipping • High internal leakage	• Add power steering fluid as required • Tighten or replace belt • Check pump pressure. (See pressure test)
Steering wheel surges or jerks when turning with engine running especially during parking	• Low oil level • Loose pump belt • Steering linkage hitting engine oil pan at full turn • Insufficient pump pressure	• Fill as required • Adjust tension to specification • Correct clearance • Check pump pressure. (See pressure test). Replace flow control valve if defective.
Steering wheel surges or jerks when turning with engine running especially during parking (cont.)	• Sticking flow control valve	• Inspect for varnish or damage, replace if necessary
Excessive wheel kickback or loose steering	• Air in system	• Add oil to pump reservoir and bleed by operating steering. Check hose connectors for proper torque and adjust as required.
Low pump pressure	• Extreme wear of cam ring • Scored pressure plate, thrust plate, or rotor • Vanes not installed properly • Vanes sticking in rotor slots • Cracked or broken thrust or pressure plate	• Replace parts. Flush system. • Replace parts. Flush system. • Install properly • Freeup by removing burrs, varnish, or dirt • Replace part

TORQUE SPECIFICATIONS

Component	English	Metric
Front Suspension		
Idler arm-to-frame bracket nuts:	70 ft. lbs.	95 Nm
Lower control arm-to-crossmember bolts:	175 ft. lbs.	237 Nm
Lower ball joints nuts		
$^{11}/_{16}$ in.–18:	135 ft. lbs.	183 Nm
$^{3}/_{4}$ in.–16:	175 ft. lbs.	237 Nm
Pivot bar retaining bolt:	195 ft. lbs.	264 Nm
Shock absorber lower bolt:	17 ft. lbs.	23 Nm
Shock absorber upper nut:	25 ft. lbs.	34 Nm
Pitman arm-to-pitman shaft nut:	175 ft. lbs.	237 Nm
Strut front mounting bolt:	100 ft. lbs.	135 Nm
Strut rear mounting nut:	52 ft. lbs.	70 Nm
Drag link nuts:	55 ft. lbs.	75 Nm
Tie rod end nuts		
$^{9}/_{16}$ in.–20:	55 ft. lbs.	75 Nm
$^{5}/_{8}$ in.–18:	75 ft. lbs.	102 Nm
Upper ball joint nut:	135 ft. lbs.	183 Nm
Upper ball joint-to-control arm:	125 ft. lbs.	170 Nm
Tie rod sleeve clamp bolts		
Heavy duty:	26 ft. lbs.	35 Nm
Exc. heavy duty:	17 ft. lbs.	23 Nm
Rear Suspension		
Center bolt nuts:	15 ft. lbs.	20 Nm
Front eye bolt nuts:	100 ft. lbs.	135 Nm
U-bolt nuts		
150/250:	45 ft. lbs.	61 Nm
350:	110 ft. lbs.	149 Nm
Rear shackle bolt nuts		
Exc. 9000 lb. GVW:	35 ft. lbs.	47 Nm
9000 lb. GVW:	155 ft. lbs.	210 Nm
Rear spring hanger bolts:	50 ft. lbs.	68 Nm
Shock absober upper & lower nuts:	60 ft. lbs.	82 Nm
Steering Gear		
Gear housing-to-frame bolts:	100 ft. lbs.	136 Nm
Sector shaft adjusting screw locknut:	28 ft. lbs.	38 Nm
Sector shaft cover spanner nut:	155 ft. lbs.	210 Nm
Steering arm nut:	15 ft. lbs.	20 Nm
Housing head spanner nut:	162 ft. lbs.	220 Nm
Valve body attaching bolts:	17 ft. lbs.	23 Nm
Valve body end plug:	50 ft. lbs.	68 Nm
High pressure hose fittings		
Gear end:	19 ft. lbs.	26 Nm
Pump end:	32 ft. lbs.	43 Nm
Steering pump bracket bolts:	30 ft. lbs.	41 Nm
Steering Column		
Steering wheel nut:	45 ft. lbs.	61 Nm
Column bracket-to-support:	20 inch lbs.	2 Nm
Column clamp stud nut:	110 inch lbs.	12 Nm
Support plate bolts:	17 ft. lbs.	23 Nm
Wheels		
150/250, all:	85-110 ft. lbs.	115-149 Nm
350, w/½ in.–20 studs:	85-110 ft. lbs.	115-149 Nm
350 H.D., 90° cone:	175-225 ft. lbs.	217-305 Nm
350 H.D., flanged nut:	300-350 ft. lbs.	407-450 Nm

Brakes

Troubleshooting the Brake System

Problem	Cause	Solution
Low brake pedal (excessive pedal travel required for braking action.)	• Excessive clearance between rear linings and drums caused by inoperative automatic adjusters	• Make 10 to 15 alternate forward and reverse brake stops to adjust brakes. If brake pedal does not come up, repair or replace adjuster parts as necessary.
	• Worn rear brakelining	• Inspect and replace lining if worn beyond minimum thickness specification
	• Bent, distorted brakeshoes, front or rear	• Replace brakeshoes in axle sets
	• Air in hydraulic system	• Remove air from system. Refer to Brake Bleeding.
Low brake pedal (pedal may go to floor with steady pressure applied.)	• Fluid leak in hydraulic system	• Fill master cylinder to fill line; have helper apply brakes and check calipers, wheel cylinders, differential valve tubes, hoses and fittings for leaks. Repair or replace as necessary.
	• Air in hydraulic system	• Remove air from system. Refer to Brake Bleeding.
	• Incorrect or non-recommended brake fluid (fluid evaporates at below normal temp).	• Flush hydraulic system with clean brake fluid. Refill with correct-type fluid.
	• Master cylinder piston seals worn, or master cylinder bore is scored, worn or corroded	• Repair or replace master cylinder
Low brake pedal (pedal goes to floor on first application—o.k. on subsequent applications.)	• Disc brake pads sticking on abutment surfaces of anchor plate. Caused by a build-up of dirt, rust, or corrosion on abutment surfaces	• Clean abutment surfaces
Fading brake pedal (pedal height decreases with steady pressure applied.)	• Fluid leak in hydraulic system	• Fill master cylinder reservoirs to fill mark, have helper apply brakes, check calipers, wheel cylinders, differential valve, tubes, hoses, and fittings for fluid leaks. Repair or replace parts as necessary.
	• Master cylinder piston seals worn, or master cylinder bore is scored, worn or corroded	• Repair or replace master cylinder

Troubleshooting the Brake System (cont.)

Problem	Cause	Solution

Decreasing brake pedal travel (pedal travel required for braking action decreases and may be accompanied by a hard pedal.)	• Caliper or wheel cylinder pistons sticking or seized
	• Master cylinder compensator ports blocked (preventing fluid return to reservoirs) or pistons sticking or seized in master cylinder bore
	
	• Power brake unit binding internally
		• Repair or replace the calipers, or wheel cylinders
		• Repair or replace the master cylinder
		
		• Test unit according to the following procedure:
		  (a) Shift transmission into neutral and start engine
		  (b) Increase engine speed to 1500 rpm, close throttle and fully depress brake pedal
		  (c) Slow release brake pedal and stop engine
		  (d) Have helper remove vacuum check valve and hose from power unit. Observe for backward movement of brake pedal.
		  (e) If the pedal moves backward, the power unit has an internal bind—replace power unit

Grabbing brakes (severe reaction to brake pedal pressure.)	• Brakelining(s) contaminated by grease or brake fluid • Parking brake cables incorrectly adjusted or seized • Incorrect brakelining or lining loose on brakeshoes • Caliper anchor plate bolts loose • Rear brakeshoes binding on support plate ledges • Incorrect or missing power brake reaction disc • Rear brake support plates loose	• Determine and correct cause of contamination and replace brakeshoes in axle sets • Adjust cables. Replace seized cables. • Replace brakeshoes in axle sets • Tighten bolts • Clean and lubricate ledges. Replace support plate(s) if ledges are deeply grooved. Do not attempt to smooth ledges by grinding. • Install correct disc • Tighten mounting bolts
Spongy brake pedal (pedal has abnormally soft, springy, spongy feel when depressed.)	• Air in hydraulic system • Brakeshoes bent or distorted • Brakelining not yet seated with drums and rotors • Rear drum brakes not properly adjusted	• Remove air from system. Refer to Brake Bleeding. • Replace brakeshoes • Burnish brakes • Adjust brakes

Troubleshooting the Brake System (cont.)

Problem	Cause	Solution
Hard brake pedal (excessive pedal pressure required to stop vehicle. May be accompanied by brake fade.)	• Loose or leaking power brake unit vacuum hose • Incorrect or poor quality brake-lining • Bent, broken, distorted brakeshoes • Calipers binding or dragging on mounting pins. Rear brakeshoes dragging on support plate.	• Tighten connections or replace leaking hose • Replace with lining in axle sets • Replace brakeshoes • Replace mounting pins and bushings. Clean rust or burrs from rear brake support plate ledges and lubricate ledges with molydisulfide grease. **NOTE:** If ledges are deeply grooved or scored, do not attempt to sand or grind them smooth—replace support plate.
	• Caliper, wheel cylinder, or master cylinder pistons sticking or seized • Power brake unit vacuum check valve malfunction	• Repair or replace parts as necessary • Test valve according to the following procedure: (a) Start engine, increase engine speed to 1500 rpm, close throttle and immediately stop engine (b) Wait at least 90 seconds then depress brake pedal (c) If brakes are not vacuum assisted for 2 or more applications, check valve is faulty
	• Power brake unit has internal bind	• Test unit according to the following procedure: (a) With engine stopped, apply brakes several times to exhaust all vacuum in system (b) Shift transmission into neutral, depress brake pedal and start engine (c) If pedal height decreases with foot pressure and less pressure is required to hold pedal in applied position, power unit vacuum system is operating normally. Test power unit. If power unit exhibits a bind condition, replace the power unit.
	• Master cylinder compensator ports (at bottom of reservoirs) blocked by dirt, scale, rust, or have small burrs (blocked ports prevent fluid return to reservoirs). • Brake hoses, tubes, fittings clogged or restricted	• Repair or replace master cylinder **CAUTION:** Do not attempt to clean blocked ports with wire, pencils, or similar implements. Use compressed air only. • Use compressed air to check or unclog parts. Replace any damaged parts.

Troubleshooting the Brake System (cont.)

Problem	Cause	Solution
Hard brake pedal (excessive pedal pressure required to stop vehicle. May be accompanied by brake fade.)	• Brake fluid contaminated with improper fluids (motor oil, transmission fluid, causing rubber components to swell and stick in bores	• Replace all rubber components, combination valve and hoses. Flush entire brake system with DOT 3 brake fluid or equivalent.
	• Low engine vacuum	• Adjust or repair engine
Dragging brakes (slow or incomplete release of brakes)	• Brake pedal binding at pivot	• Loosen and lubricate
	• Power brake unit has internal bind	• Inspect for internal bind. Replace unit if internal bind exists.
	• Parking brake cables incorrrectly adjusted or seized	• Adjust cables. Replace seized cables.
	• Rear brakeshoe return springs weak or broken	• Replace return springs. Replace brakeshoe if necessary in axle sets.
	• Automatic adjusters malfunctioning	• Repair or replace adjuster parts as required
	• Caliper, wheel cylinder or master cylinder pistons sticking or seized	• Repair or replace parts as necessary
	• Master cylinder compensating ports blocked (fluid does not return to reservoirs).	• Use compressed air to clear ports. Do not use wire, pencils, or similar objects to open blocked ports.
Vehicle moves to one side when brakes are applied	• Incorrect front tire pressure	• Inflate to recommended cold (reduced load) inflation pressure
	• Worn or damaged wheel bearings	• Replace worn or damaged bearings
	• Brakelining on one side contaminated	• Determine and correct cause of contamination and replace brakelining in axle sets
	• Brakeshoes on one side bent, distorted, or lining loose on shoe	• Replace brakeshoes in axle sets
	• Support plate bent or loose on one side	• Tighten or replace support plate
	• Brakelining not yet seated with drums or rotors	• Burnish brakelining
	• Caliper anchor plate loose on one side	• Tighten anchor plate bolts
	• Caliper piston sticking or seized	• Repair or replace caliper
	• Brakelinings water soaked	• Drive vehicle with brakes lightly applied to dry linings
	• Loose suspension component attaching or mounting bolts	• Tighten suspension bolts. Replace worn suspension components.
	• Brake combination valve failure	• Replace combination valve
Chatter or shudder when brakes are applied (pedal pulsation and roughness may also occur.)	• Brakeshoes distorted, bent, contaminated, or worn	• Replace brakeshoes in axle sets
	• Caliper anchor plate or support plate loose	• Tighten mounting bolts
	• Excessive thickness variation of rotor(s)	• Refinish or replace rotors in axle sets

Troubleshooting the Brake System (cont.)

Problem	Cause	Solution
Noisy brakes (squealing, clicking, scraping sound when brakes are applied.)	• Bent, broken, distorted brakeshoes • Excessive rust on outer edge of rotor braking surface	• Replace brakeshoes in axle sets • Remove rust
Noisy brakes (squealing, clicking, scraping sound when brakes are applied.) (cont.)	• Brakelining worn out—shoes contacting drum of rotor • Broken or loose holdown or return springs • Rough or dry drum brake support plate ledges • Cracked, grooved, or scored rotor(s) or drum(s) • Incorrect brakelining and/or shoes (front or rear).	• Replace brakeshoes and lining in axle sets. Refinish or replace drums or rotors. • Replace parts as necessary • Lubricate support plate ledges • Replace rotor(s) or drum(s). Replace brakeshoes and lining in axle sets if necessary. • Install specified shoe and lining assemblies
Pulsating brake pedal	• Out of round drums or excessive lateral runout in disc brake rotor(s)	• Refinish or replace drums, re-index rotors or replace

BASIC OPERATING PRINCIPLES

Hydraulic systems are used to actuate the brakes of all automobiles. The system transports the power required to force the frictional surfaces of the braking system together from the pedal to the individual brake units at each wheel. A hydraulic system is used for two reasons.

First, fluid under pressure can be carried to all parts of an automobile by small pipes and flexible hoses without taking up a significant amount of room or posing routing problems.

Second, a great mechanical advantage can be given to the brake pedal end of the system, and the foot pressure required to actuate the brakes can be reduced by making the surface area of the master cylinder pistons smaller than that of any of the pistons in the wheel cylinders or calipers.

The master cylinder consists of a fluid reservoir and a double cylinder and piston assembly. Double type master cylinders are designed to separate the front and rear braking systems hydraulically in case of a leak.

Steel lines carry the brake fluid to a point on the vehicle's frame near each of the vehicle's wheels. The fluid is then carried to the calipers and wheel cylinders by flexible tubes in order to allow for suspension and steering movements.

In drum brake systems, each wheel cylinder contains two pistons, one at either end, which push outward in opposite directions.

In disc brake systems, the cylinders are part of the calipers. One cylinder in each caliper is used to force the brake pads against the disc.

All pistons employ some type of seal, usually made of rubber, to minimize fluid leakage. A rubber dust boot seals the outer end of the cylinder against dust and dirt. The boot fits around the outer end of the piston on disc brake calipers, and around the brake actuating rod on wheel cylinders.

The hydraulic system operates as follows: When at rest, the entire system, from the piston(s) in the master cylinder to those in the wheel cylinders or calipers, is full of brake fluid. Upon application of the brake pedal, fluid trapped in front of the master cylinder piston(s) is forced through the lines to the wheel cylinders. Here, it forces the pistons outward, in the case of drum brakes, and inward toward the disc, in the case of disc brakes. The motion of the pistons is opposed by return springs mounted outside the cylinders in drum brakes, and by spring seals, in disc brakes.

Upon release of the brake pedal, a spring located inside the master cylinder immediately returns the master cylinder pistons to the normal position. The pistons contain check valves and the master cylinder has compensating ports drilled in it. These are uncovered as the pistons reach their normal position. The piston check valves allow fluid to flow toward the wheel cylinders or calipers as the pistons withdraw. Then, as the return springs force the brake pads or shoes into the released position, the excess fluid reservoir through the compensating ports. It is during the time the pedal is in the released position that any fluid that has leaked out of the system will be replaced through the compensating ports.

Dual circuit master cylinders employ two pistons, located one behind the other, in the same cylinder. The primary piston is actuated directly by mechanical linkage from the brake pedal through the power booster. The secondary piston is actuated by fluid trapped between the two pistons. If a leak develops in front of the secondary piston, it moves forward until it bottoms against the front of the master cylinder, and the fluid trapped between the pistons will operate the rear brakes. If the rear brakes develop a leak, the primary piston will move forward until direct contact with the secondary piston takes place, and it will force the secondary piston to actuate the front brakes. In either case, the brake pedal moves farther when the brakes are applied, and less braking power is available.

All dual circuit systems use a switch to warn the driver when only half of the brake system is operational. This switch is located in a valve body which is mounted on the firewall or the frame below the master cylinder. A hydraulic piston receives pressure from both circuits, each circuit's pressure being applied to one end of the piston. When the pressures are in balance, the piston remains stationary. When one circuit has a leak, however, the greater pressure in that circuit during application of the brakes will push the piston to one side, closing the switch and activating the brake warning light.

In disc brake systems, this valve body also contains a metering valve and, in some cases, a proportioning valve. The metering valve keeps pressure from traveling to the disc brakes on the front wheels until the brake shoes on the rear wheels have contacted the drums, ensuring that the front brakes will never be used alone. The proportioning valve controls the pressure to the rear brakes to lessen the chance of rear wheel lock-up during very hard braking.

Warning lights may be tested by depressing the brake pedal and holding it while opening one of the wheel cylinder bleeder screws. If this does not cause the light to go on, substitute a new lamp, make continuity checks, and, finally, replace the switch as necessary.

The hydraulic system may be checked for leaks by applying pressure to the pedal gradually and steadily. If the pedal sinks very slowly to the floor, the system has a leak. This is not to be confused with a springy or spongy feel due to the compression of air within the lines. If the system leaks, there will be a gradual change in the position of the pedal with a constant pressure.

Check for leaks along all lines and at wheel cylinders. If no external leaks are apparent, the problem is inside the master cylinder.

Disc Brakes

BASIC OPERATING PRINCIPLES

Instead of the traditional expanding brakes that press outward against a circular drum, disc brake systems utilize a disc (rotor) with brake pads positioned on either side of it. Braking effect is achieved in a manner similar to the way you would squeeze a spinning phonograph record between your fingers. The disc (rotor) is a casting with cooling fins between the two braking surfaces. This enables air to circulate between the braking surfaces making them less sensitive to heat buildup and more resistant to fade. Dirt and water do not affect braking action since contaminants are thrown off by the centrifugal action

of the rotor or scraped off the by the pads. Also, the equal clamping action of the two brake pads tends to ensure uniform, straight line stops. Disc brakes are inherently self-adjusting. There are three general types of disc brake:

1. A fixed caliper.
2. A floating caliper.
3. A sliding caliper.

The fixed caliper design uses two pistons mounted on either side of the rotor (in each side of the caliper). The caliper is mounted rigidly and does not move.

The sliding and floating designs are quite similar. In fact, these two types are often lumped together. In both designs, the pad on the inside of the rotor is moved into contact with the rotor by hydraulic force. The caliper, which is not held in a fixed position, moves slightly, bringing the outside pad into contact with the rotor. There are various methods of attaching floating calipers. Some pivot at the bottom or top, and some slide on mounting bolts. In any event, the end result is the same.

All the vehicles covered in this book employ the sliding caliper design.

Drum Brakes

BASIC OPERATING PRINCIPLES

Drum brakes employ two brake shoes mounted on a stationary backing plate. These shoes are positioned inside a circular drum which rotates with the wheel assembly. The shoes are held in place by springs. This allows them to slide toward the drums (when they are applied) while keeping the linings and drums in alignment. The shoes are actuated by a wheel cylinder which is mounted at the top of the backing plate. When the brakes are applied, hydraulic pressure forces the wheel cylinder's actuating links outward. Since these links bear directly against the top of the brake shoes, the tops of the shoes are then forced against the inner side of the drum. This action forces the bottoms of the two shoes to contact the brake drum by rotating the entire assembly slightly (known as servo action). When pressure within the wheel cylinder is relaxed, return springs pull the shoes back away from the drum.

Most modern drum brakes are designed to self-adjust themselves during application when the vehicle is moving in reverse. This motion causes both shoes to rotate very slightly with the drum, rocking an adjusting lever, thereby causing rotation of the adjusting screw.

Power Boosters

Power brakes operate just as non-power brake systems except in the actuation of the master cylinder pistons. A vacuum diaphragm is located on the front of the master cylinder and assists the driver in applying the brakes, reducing both the effort and travel he must put into moving the brake pedal.

The vacuum diaphragm housing is connected to the intake manifold by a vacuum hose. A check valve is placed at the point where the hose enters the diaphragm housing, so that during periods of low manifold vacuum brake assist vacuum will not be lost.

Depressing the brake pedal closes off the vacuum source and allows atmospheric pressure to enter on one side of the diaphragm. This causes the master cylinder pistons to move and apply the brakes. When the brake pedal is released, vacuum is applied to both sides of the diaphragm, and return springs return the diaphragm and master cylinder pistons to the released position. If the vacuum fails, the brake pedal rod will butt against the end of the master cylinder actuating rod, and direct mechanical application will occur as the pedal is depressed.

The hydraulic and mechanical problems that apply to conventional brake systems also apply to power brakes, and should be checked for if the tests below do not reveal the problem.

Test for a system vacuum leak as described below:

1. Operate the engine at idle without touching the brake pedal for at least one minute.

2. Turn off the engine, and wait one minute.

3. Test for the presence of assist vacuum by depressing the brake pedal and releasing it several times. Light application will produce less and less pedal travel, if vacuum was present. If there is no vacuum, air is leaking into the system somewhere.

Test for system operation as follows:

1. Pump the brake pedal (with engine off) until the supply vacuum is entirely gone.

2. Put a light, steady pressure on the pedal.

3. Start the engine, and operate it at idle. If the system is operating, the brake pedal should fall toward the floor if constant pressure is maintained on the pedal.

Power brake systems may be tested for hydraulic leaks just as ordinary systems are tested.

BRAKE SYSTEM

WARNING: Clean, high quality brake fluid is essential to the safe and proper operation of the brake system. You should always buy the highest quality brake fluid that is available. If the brake fluid becomes contaminated, drain and flush the system and fill the master cylinder with new fluid. Never reuse any brake fluid. Any brake fluid that is removed from the system should be discarded.

Adjustments
DRUM BRAKES

The drum brakes are self-adjusting and require a manual adjustment only after the brake shoes have been replaced, or when the length of the adjusting screw has been changed while performing some other service operation, as i.e., taking off brake drums.

To adjust the brakes, follow the procedures given below:

Drum Installed

1. Raise and support the rear end on jackstands.

2. Remove the rubber plug from the adjusting slot on the backing plate.

3. Insert a brake adjusting spoon into the slot and engage the lowest possible tooth on the starwheel. Move the end of the brake spoon downward to move the starwheel upward and expand the adjusting screw. Repeat this operation until the brakes lock the wheels.

4. Insert a small screwdriver or piece of firm wire (coat hanger wire) into the adjusting slot and push the automatic adjusting lever out and free of the starwheel on the adjusting screw and hold it there.

5. Engage the topmost tooth possible on the starwheel with the brake adjusting spoon. Move the end of the adjusting spoon upward to move the adjusting screw starwheel downward and contract the adjusting screw. Back off the adjusting screw starwheel until the wheel spins freely with a minimum of drag. Keep track of the number of turns that the starwheel is backed off, or the number of strokes taken with the brake adjusting spoon.

6. Repeat this operation for the other side. When backing off the brakes on the other side, the starwheel adjuster must be

backed off the same number of turns to prevent side-to-side brake pull.

7. When the brakes are adjusted make several stops while backing the vehicle, to equalize the brakes at both of the wheels.

8. Remove the safety stands and lower the vehicle. Road test the vehicle.

Drum Removed

CAUTION

Brake shoes contain asbestos, which has been determined to be a cancer causing agent. Never clean the brake surfaces with compressed air! Avoid inhaling any dust from any brake surface! When cleaning brake surfaces, use a commercially available brake cleaning fluid.
---

1. Make sure that the shoe-to-contact pad areas are clean and properly lubricated.

2. Using an inside caliper check the inside diameter of the drum. Measure across the diameter of the assembled brake shoes, at their widest point.

3. Turn the adjusting screw so that the diameter of the shoes is 0.030 in. (0.76mm) less than the brake drum inner diameter.

4. Install the drum.

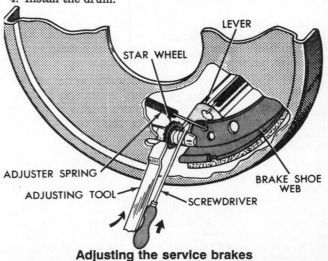

Adjusting the service brakes

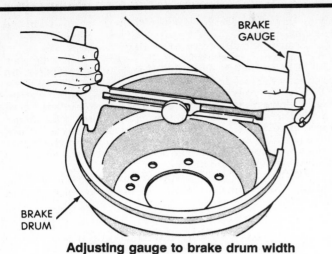

Adjusting gauge to brake drum width

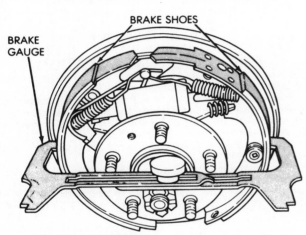

Adjusting the brake shoes to the gauge width

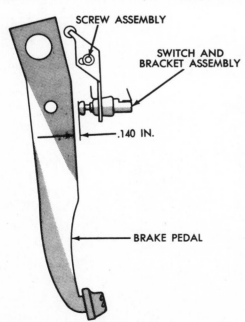

Stoplight adjustment, without speed control

Brake Light Switch

REMOVAL AND INSTALLATION

The switch is mounted on a bracket above the brake pedal and is contacted by the brake pedal arm.

ADJUSTMENT

Without Speed Control

1. Loosen the switch-to-pedal bracket screw and slide the switch away from the pedal arm.

2. Push the pedal down by hand and allow it to return on its own to the free-hanging position. Do not pull it back!

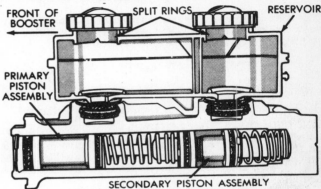

Master cylinder components

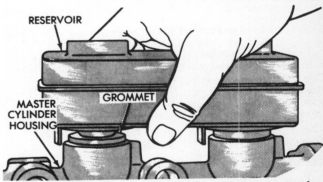

Removing/installing the master cylinder reservoir

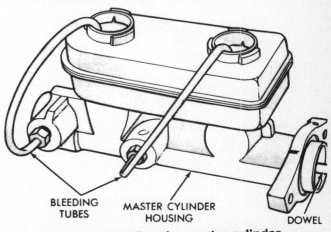

Bench bleeding the master cylinder

3. Slide the switch towards the pedal until there is a gap of 0.140 in. (3.5mm) between the plunger and pedal arm.

NOTE: The pedal must not move when measuring the gap.

4. Tighten the switch screw to 82 inch lbs. Recheck the gap.

With Speed Control

1. Push the switch through the clip in the mounting bracket until it is seated against the bracket. The pedal will move forward slightly.
2. Gently pull back on the pedal as far as it will go. The switch will ratchet backwards to the correct position.

Master Cylinder

The body of the master cylinder is made of aluminum and the reservoir is made of nylon. The body of the master cylinder is not a repairable unit, servicing is by replacement.

REMOVAL AND INSTALLATION

Master Cylinder

1. Disconnect the brake lines from the master cylinder.
2. Remove the master cylinder mounting nuts.
3. Slide the master cylinder off of the power brake booster.
4. Service as required. Place the master cylinder in position on the brake booster and secure the mounting nuts. Tighten the mounting nuts to 16 ft. lbs. Connect the brake lines. Bleed the master cylinder and brake system.

Master Cylinder Reservoir

1. Remove the master cylinder and reservoir from the power brake booster.
2. Drain the reservoir and install the reservoir caps. Clean the cylinder and reservoir with a safe solvent.
3. Position the master cylinder body in a soft jawed vise. Grasp the reservoir and firmly rock it from side to side while ap-plying upward pressure. Disengage the reservoir from the cylinder.
4. Remove and discard the mounting grommets.
5. Lubricate the master cylinder reservoir mounting area with clean brake fluid. Place the new mounting grommets into position in the master cylinder.
6. Place the reservoir into position over the mounting grommets. Using a rocking motion, push the reservoir into the master cylinder. Be sure that the reservoir is mounted in the proper direction. All lettering should be properly read from the left side of the master cylinder. Be sure the reservoir is completely seated on the master cylinder. The bottom of the reservoir should touch the mounting grommets.
7. Fill the reservoir with new brake fluid and bench bleed. Install the master cylinder assembly on the brake booster. Bleed the system.

OVERHAUL

Overhaul is not recommended by the factory on the aluminum bodied master cylinder. Service is by replacement only.

Power Booster

REMOVAL AND INSTALLATION

1. Remove the master cylinder.
2. Disconnect the vacuum hose from the power booster.
3. Disconnect the pushrod from the brake pedal. On transverse mounted boosters, remove the pivot bolt, O-ring and nut that attach the booster rod to the bellcrank.
4. Remove the power booster attaching nuts and remove the booster.
5. When installing the mounting nuts, tighten them to 23 ft. lbs. on in-lined mounted boosters and 30 ft. lbs for transverse mounts. Connect the pushrod to the pedal, or connect the bellcrank assembly. Coat the eyelet with Lubriplate® or equivalent.

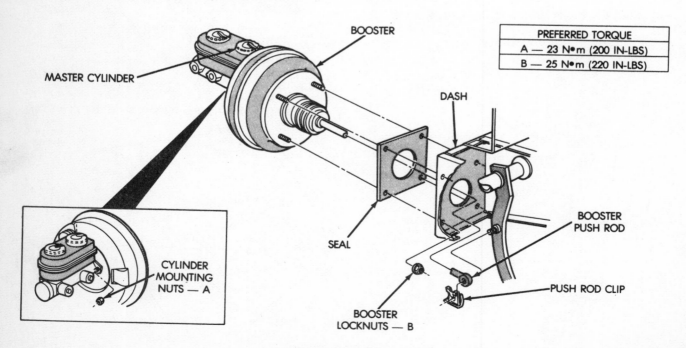

PREFERRED TORQUE	
A — 23 N•m (200 IN-LBS)	
B — 25 N•m (220 IN-LBS)	

Inline brake booster

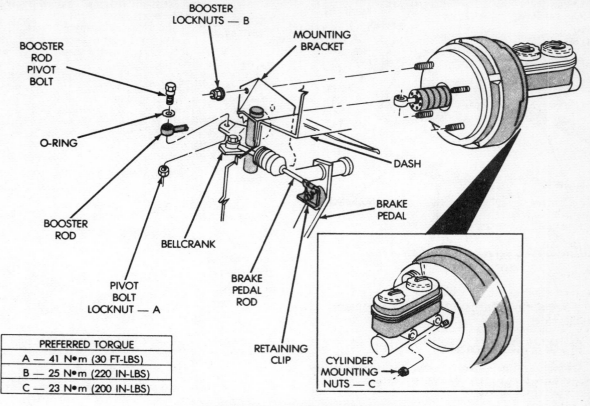

BOOSTER
LOCKNUTS — B

BOOSTER
ROD
PIVOT
BOLT

MOUNTING
BRACKET

O-RING

BOOSTER
ROD

BELLCRANK

DASH

BRAKE
PEDAL

PIVOT
BOLT
LOCKNUT — A

BRAKE
PEDAL
ROD

RETAINING
CLIP

CYLINDER
MOUNTING
NUTS — C

PREFERRED TORQUE	
A — 41 N•m (30 FT-LBS)	
B — 25 N•m (220 IN-LBS)	
C — 23 N•m (200 IN-LBS)	

Transverse mounted brake booster

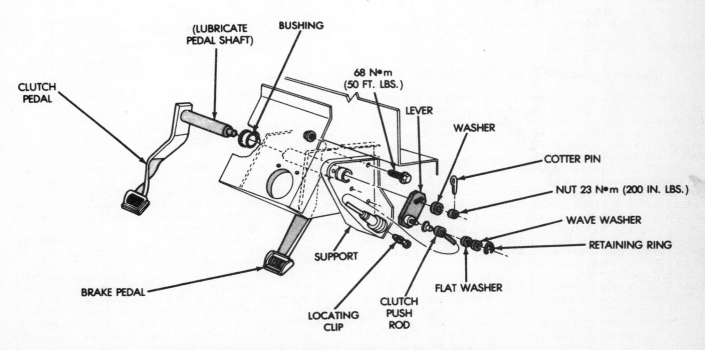

(LUBRICATE
PEDAL SHAFT)

BUSHING

68 N•m
(50 FT. LBS.)

LEVER

CLUTCH
PEDAL

WASHER

COTTER PIN

NUT 23 N•m (200 IN. LBS.)

WAVE WASHER

RETAINING RING

BRAKE PEDAL

SUPPORT

LOCATING
CLIP

CLUTCH
PUSH
ROD

FLAT WASHER

Brake pedal mounting, models with manual transmission

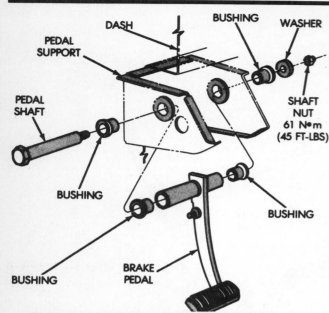

Brake pedal mounting, models with automatic transmission

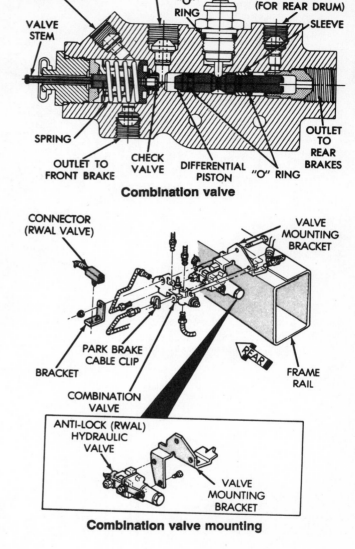

Combination valve

Combination valve mounting

Hydraulic Control Valves

NOTE: All models incorporate the brake safety switch into either the metering valve or combination metering/proportioning valve

REMOVAL AND INSTALLATION

1. Raise and support the front end on jackstands.
2. Remove the splash shield.
3. Disconnect the wiring at the warning light switch.
4. Disconnect the brake lines at the valve.
5. Unbolt and remove the valve from the frame.
6. Install the new valve and connect the lines and wiring.
7. Bleed the brake system.
8. Install the splash shield.

Brake System Control Valves

All models have a hydraulic system control valve in the brake system. The valve is usually mounted on the frame rail below the master cylinder. The control valve assembly combines a brake warning switch with a hold-off and proportioning valve assembly. On some models, such as the B350, the brake warning switch and hold-off valve assembly are combined.

Hold-off and Proportioning valves are used because of different braking characteristics between disc and drum brakes. The hold-off (metering) valve is used to balance brake action between the front discs and rear drum brakes. The metering valve is usually located in the front portion of the "combination valve". The metering valve hold off full apply pressure until the rear brake shoes are in full contact with the drums. At this point, the metering valve opens completely permitting full fluid apply pressure to the front disc brakes.

A brake warning valve is used to warn of a decrease or loss of fluid in either side (front/rear) of the hydraulic circuit. The switch valve will shuttle back and forth in response to the pressure differential. If the valve travels too far in one direction (indicating a pressure loss), the dash warning bulb will light.

The valve will remain in the warning position until system repairs have been completed an normal pressure differential is restored.

TESTING

To test the metering valve, have a helper apply the brakes. Observe the metering valve stem while the brakes are being applied and released. If the valve is operating correctly, the stem will extend slightly when the brakes are applied and retract when the brakes are released. If the valve is faulty, replace the entire combination valve assembly.

To test the pressure differential switch, raise and safely support the rear of the vehicle. Attach a bleeder hose to one of the rear wheel cylinders and immerse the hose end in a container of containing brake fluid. Have a helper maintain pressure on the brake pedal. Open the bleeder valve on the wheel cylinder slowly. While pressure is being applied to the brake pedal, fluid will escape through the bleeder hose. At a certain point, the loss of brake fluid should cause the warning bulb to light. If the light fails to illuminate, the switch is bad. If the light goes on, the switch is working properly. Close the bleeder valve. Repair or replace the necessary parts. Fill and bleed the brake system.

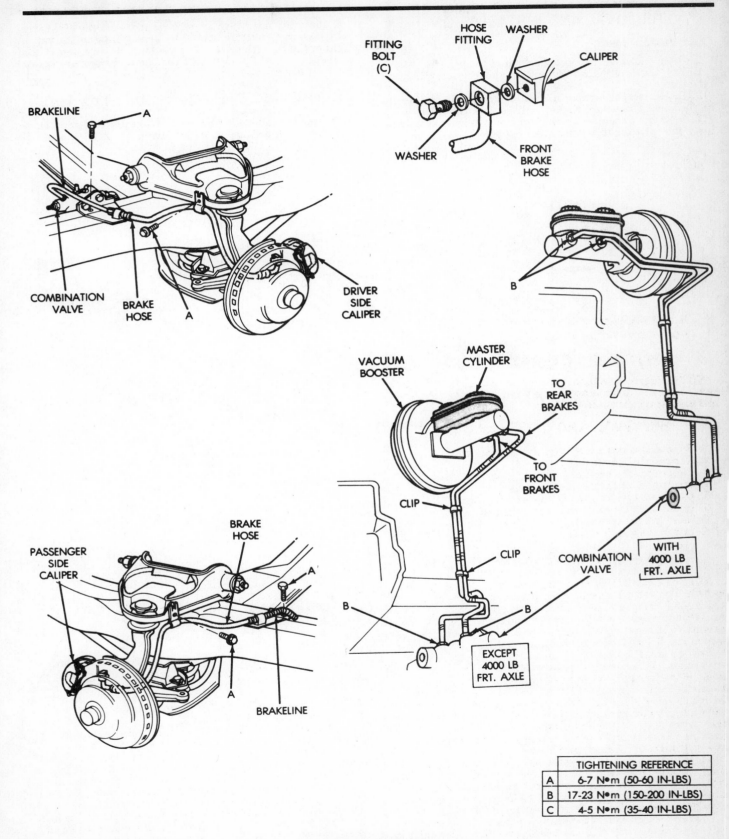

FITTING
BOLT
(C)

HOSE
FITTING

WASHER

CALIPER

WASHER

FRONT
BRAKE
HOSE

BRAKELINE

A

COMBINATION
VALVE

BRAKE
HOSE

A

DRIVER
SIDE
CALIPER

B

VACUUM
BOOSTER

MASTER
CYLINDER

TO
REAR
BRAKES

TO
FRONT
BRAKES

CLIP

CLIP

COMBINATION
VALVE

WITH
4000 LB
FRT. AXLE

PASSENGER
SIDE
CALIPER

BRAKE
HOSE

A

A

B

B

BRAKELINE

EXCEPT
4000 LB
FRT. AXLE

	TIGHTENING REFERENCE
A	6-7 N•m (50-60 IN-LBS)
B	17-23 N•m (150-200 IN-LBS)
C	4-5 N•m (35-40 IN-LBS)

Front brake lines

9 BRAKES

REMOVAL AND INSTALLATION

Combination Valve

1. Raise and safely support the front of the vehicle.
2. Tag the brake lines going into the valve for location reference.
3. Disconnect the parking brake cable from the mounting clip on the valve. Disconnect the brake lines from the valve.
4. Disconnect the wiring connector from the valve switch. remove the valve mounting bolts and the valve.

5. Place the new valve in position and start the mounting bolts loosely. Connected the brake lines loosely to the valve. Tighten the mounting bolts and the brake lines. Connect the wiring connector. Fill the and bleed the brake system. Lower and road test the vehicle.

CENTERING THE BRAKE SAFETY VALVE

The valve is self-centering after the required repairs have been done on the vehicle, and the system is filled and bled.

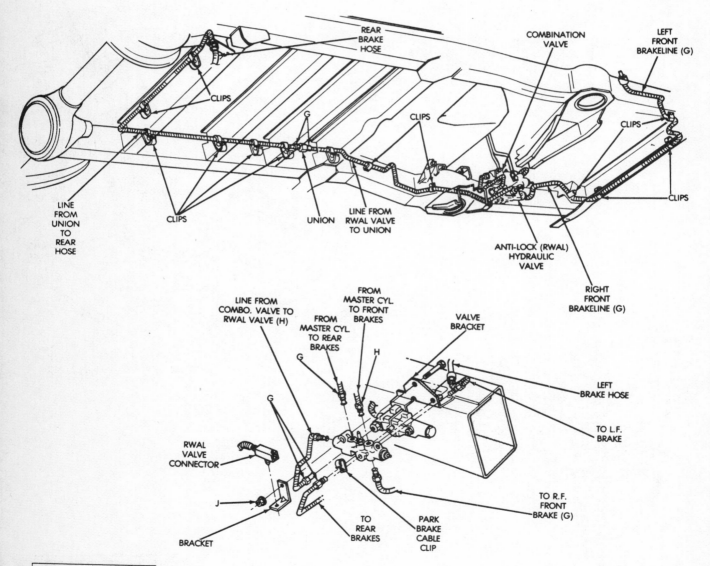

	TIGHTENING REFERENCE
G	16-20 N•m (145-175 IN-LBS)
H	16-23 N•m (140-200 IN-LBS)
J	19-38 N•m (168-252 IN-LBS)

Center brake lines and valves

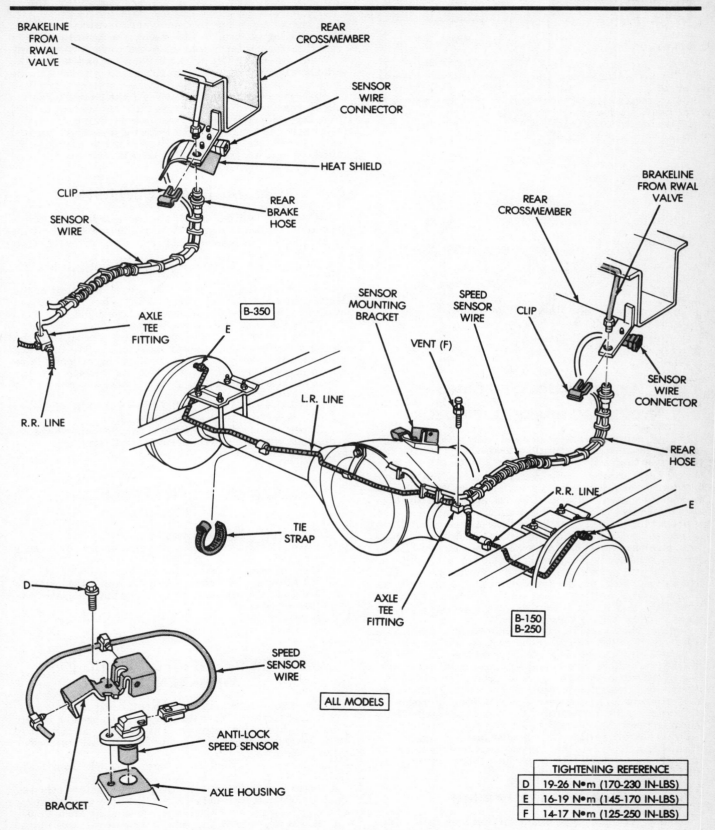

BRAKELINE FROM RWAL VALVE

REAR CROSSMEMBER

SENSOR WIRE CONNECTOR

HEAT SHIELD

CLIP

SENSOR WIRE

REAR BRAKE HOSE

AXLE TEE FITTING

B-350

R.R. LINE

E

SENSOR MOUNTING BRACKET

VENT (F)

SPEED SENSOR WIRE

REAR CROSSMEMBER

CLIP

BRAKELINE FROM RWAL VALVE

SENSOR WIRE CONNECTOR

L.R. LINE

REAR HOSE

R.R. LINE

E

TIE STRAP

AXLE TEE FITTING

B-150 B-250

D

SPEED SENSOR WIRE

ALL MODELS

ANTI-LOCK SPEED SENSOR

AXLE HOUSING

BRACKET

TIGHTENING REFERENCE		
D	19-26 N•m	(170-230 IN-LBS)
E	16-19 N•m	(145-170 IN-LBS)
F	14-17 N•m	(125-250 IN-LBS)

Rear brake lines and speed sensor

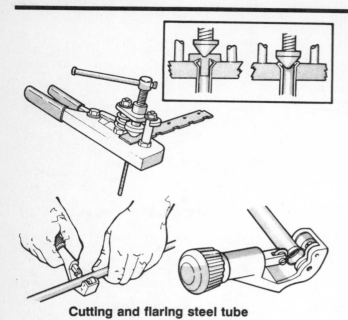

Cutting and flaring steel tube

Brake Hoses and Lines

HYDRAULIC BRAKE LINE CHECK

The hydraulic brake lines, hoses and brake pads and shoe lining should be inspected at regular intervals. Follow the steel tubing from the master cylinder to the combination valve, to the flexible hose fitting at each wheel. If a section of the tubing is found to be damaged, replace the entire section with tubing of the same type (steel, not copper), size, shape, and length. When installing a new section (available in different standard lengths) of brake tubing, flush clean brake fluid or denatured alcohol through to remove any dirt or foreign material from the line. If a standard replacement line is not available, be sure to use the required, special double flaring tool and flare both ends of the cut steel tubing to provide sound, leak-proof connections. When bending the tubing to fit the underbody contours, be careful not to kink or crack the line. Check the flexible brake hoses that connect the steel tubing to each wheel cylinder or caliper. Replace the hose if it shows any signs of softening, cracking, or other damage. When installing a new front brake hose, position the hose to avoid contact with other chassis parts. Place a new copper gasket over the hose fitting and thread the hose assembly into the front caliper. A new rear brake hose must be positioned clear of the exhaust pipe or shock absorber. Thread the hose into the rear brake tube connector. When installing either a new front or rear brake hose, engage the opposite end of the hose to the bracket on the frame. Install the horseshoe type retaining clip and connect the tube to the hose with the tube fitting nut.

Always bleed the system after hose or line replacement. Before bleeding, make sure that the master cylinder is topped up with high temperature, extra heavy duty fluid of at least SAE 70R3 quality.

Bleeding the Brakes

When any part of the hydraulic system has been disconnected for repair or replacement, air may get into the lines and cause spongy pedal action (because air can be compressed and brake fluid cannot). To correct this condition, it is necessary to bleed the hydraulic system after it has been properly connected to be sure that all air is expelled from the brake cylinders and lines.

When bleeding the brake system, bleed one brake cylinder/caliper at a time, beginning at the cylinder with the longest hydraulic line (farthest from the master cylinder) first. Keep the master cylinder reservoir filled with brake fluid during bleeding operation. Never use brake fluid that has been drained from the hydraulic system, no matter how clean it is.

It will not be necessary to centralize the pressure differential valve after a brake system failure has been corrected and the hydraulic system has been bled. The valve is self-centering.

MANUAL BRAKE BLEEDING

1. Clean all dirt from around the master cylinder reservoir filler caps, remove the cap and fill the master cylinder with brake fluid until the level is within ¼ in. (6mm) of the top of the edge of the reservoir.
2. Clean off the bleeder screws at the wheel cylinders and calipers.
3. Attach the length of rubber hose over the nozzle of the bleeder screw at the wheel to be done first. Place the other end of the hose in a glass jar, submerged in brake fluid.
4. Open the bleed screw valve ½-¾ turn.
5. Have an assistant slowly depress the brake pedal. Close the bleeder screw valve and tell your assistant to allow the brake pedal to return slowly. Continue this pumping action to force any air out of the system. When bubbles cease to appear at the end of the bleeder hose, close the bleed valve and remove the hose.
6. Check the master cylinder fluid level and add fluid accordingly. Do this after bleeding each wheel.
7. Repeat the bleeding operation at the remaining 3 wheels, ending with the one closest to the master cylinder. Fill the master cylinder reservoir.

MASTER CYLINDER BLEEDING (ON VEHICLE)

1. Fill the master cylinder reservoirs.
2. Place absorbent rags under the fluid lines at the master cylinder.
3. Have an assistant depress and hold the brake pedal.
4. With the pedal held down, slowly crack open the hydraulic line fitting, allowing the air to escape. Close the fitting and have the pedal released.
5. Repeat as necessary until all the air is released.

MASTER CYLINDER BLEEDING (ON WORKBENCH)

1. Secure the master cylinder in a soft jawed vise.
2. Attach bleeder tubes to the cylinder outlet ports and insert tubes into the reservoir compartments. The bleeder tubes may be made up from brake line equipped with the correct master cylinder end fitting.
3. Fill the master cylinder with the proper grade of new brake fluid.
4. Insert a wooden dowel into the piston end of the cylinder. Push the piston inward and release it. Continue this "pumping operation" until no air bubbles are visible in the brake fluid.
5. Remove the bleeder lines, install the reservoir caps and install the master cylinder.
6. Bleed the entire brake system.

FRONT DISC BRAKES

CAUTION

Brake shoes/pads contain asbestos, which has been determined to be a cancer causing agent. Never clean the brake surfaces with compressed air! Avoid inhaling any dust from any brake surface! When cleaning brake surfaces, use a commercially available brake cleaning fluid.

Application

The Dodge van uses a Chrysler sliding type caliper with 11.75 in. (298.5mm) rotor on models up to the 350. 350 models use the Chrysler sliding type caliper with a 12.82 in. (325.6mm) rotor.

Disc Brake Pads

INSPECTION

Remove the brake pads as described below and measure the thickness of the lining. If the lining at any point on the pad assembly is less 0.0625 in. ($\frac{1}{16}$ in.; 1.5mm) for LD brakes or 0.03125 in. ($\frac{1}{32}$ in.; 0.794mm) for HD brakes, thick (above the backing plate or rivets), or there is evidence of the lining being contaminated by brake fluid or oil, replace the brake pad.

REMOVAL AND INSTALLATION

NOTE: NEVER REPLACE THE PADS ON ONE SIDE ONLY! ALWAYS REPLACE PADS ON BOTH WHEELS AS A SET!

Chrysler Sliding Caliper

1. Raise and safely support the front of the vehicle on jackstands.

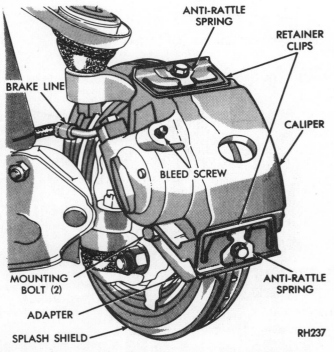

Fig. 1 Disc Brake Caliper Mounting—Rear View

Brake disc caliper mounting (rear view)

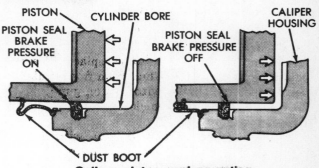

Caliper piston seal operation

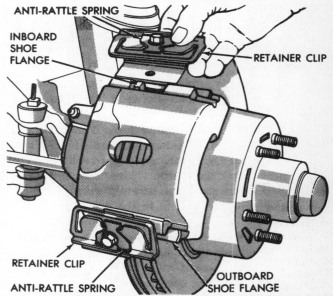

Remove/install the retainer clips and anti-rattle springs

2. Remove the wheels.
3. Remove the caliper retaining clips and anti-rattle springs.
4. Remove the caliper from the disc by slowly sliding the caliper and brake pad assembly out and away from the disc. Suspend the caliper with wire to avoid damage to the flexible brake hose.
5. Drain some of the fluid from the master cylinder.
6. Remove the outboard pad from the caliper by prying between the pad and the caliper fingers. Remove the inboard pad from the caliper support by the same method. DO NOT depress the brake pedal with the pads removed!
7. Push the caliper piston to the bottom of its bore. This may be done with a large C-clamp or a pair of large pliers by placing a flat metal bar against the piston and depressing the piston with a steady force. This operation will displace some of the fluid in the master cylinder.
8. Slide the new pads into the caliper and caliper support. The ears of the pad should rest on the bridges of the caliper.
9. Install the caliper on the disc and install the caliper retaining clips, pins and anti-rattle springs. Bolts are tightened to 20 ft. lbs. Pump the brake pedal until it is firm.
10. Check the fluid level in the master cylinder and add fluid as needed.
11. Install the wheels and lower the vehicle.
12. Check the brake pedal for a firm feel. Road test the vehicle.

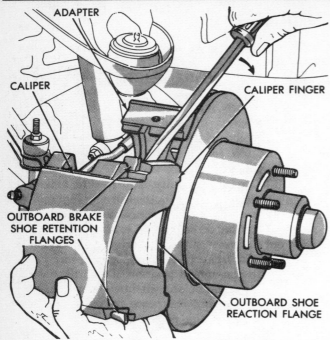

Remove/install the outboard brake pad

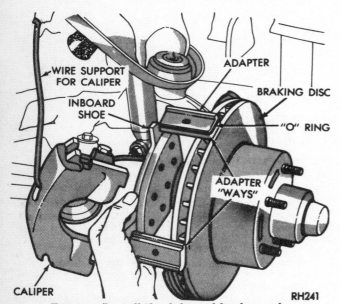

Remove/install the inboard brake pad

Disc Brake Calipers

REMOVAL AND INSTALLATION

Chrysler Sliding Caliper Type

1. Raise and support the front end on jackstands.
2. Remove the wheels.
3. Disconnect the rubber brake hose from the tubing at the frame mount. If the piston is to be removed from the caliper, leave the brake hose connected to the caliper. Check the rubber hose for cracks or chafed spots.

4. Plug the brake line to prevent loss of fluid.
5. Remove the retaining screw, clip and anti-rattle spring that attach the caliper to the adaptor.
6. Carefully slide the caliper out and away from the disc. Check the pads to be sure that they are reinstalled in the same position.

To install:

7. Position the outboard shoe in the caliper. The shoe should not rattle in the caliper. If it does, or if any movement is obvious, bend the shoe tabs over the caliper to tighten the fit.
8. Install the inboard shoe.
9. Slide the caliper into position on the adaptor and over the rotor.

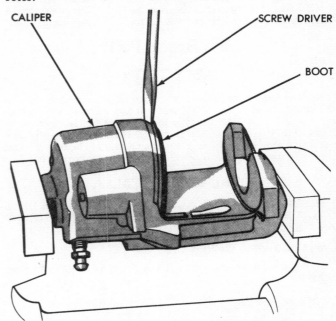

Removing the caliper piston dust boot

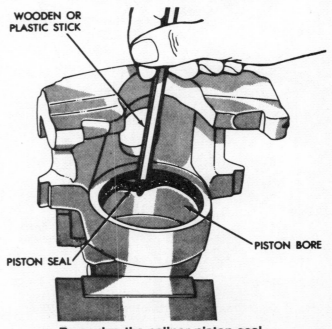

Removing the caliper piston seal

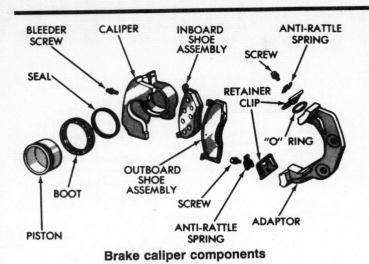

Brake caliper components

Installing the caliper piston

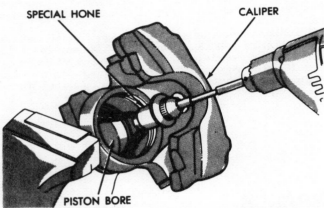

Lightly polish the caliper borer with a hone

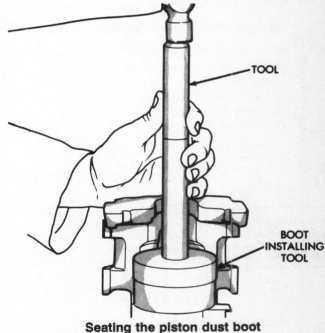

Seating the piston dust boot

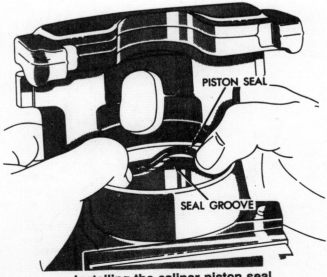

Installing the caliper piston seal

WARNING: Take great care to avoid dislodging the piston dust boot!

10. Install the anti-rattle springs and retaining clips and torque the retainer bolts to 20 ft. lbs.

NOTE: The inboard shoe must always be installed on top of the retainer spring plate.

11. Fill the system with fresh fluid and bleed the brakes.

OVERHAUL

Chrysler Sliding Caliper

1. Remove the caliper from the rotor.

2. If the piston is to be removed, support the caliper on the upper control arm and surround it with towels to absorb the hydraulic fluid that will be lost.

3. Carefully depress the brake pedal to hydraulically push the piston from its bore. Prop the pedal at any position below the first 1 in. (25mm) of travel to prevent further loss of fluid.

4. Disconnect and plug the brake line after the piston is removed.

5. Mount the caliper in a vise equipped with soft jaws.

6. Remove the dust boot.

7. Using a small wooden of plastic pointed tool, work the piston seal out of its groove in the piston bore.

8. Clean all parts in alcohol and allow to air dry.

9. Dip a new piston seal in the lubricant supplied with the factory rebuilding kit or in clean brake fluid. Install the seal in the groove in the bore.

10. Coat a new piston boot with lubricant and leave a generous amount inside the boot.

11. Install the piston seal into the caliper by working it into the groove with your fingers only. At first sight, the seal will appear larger than the bore, but will snap into place when properly seated.

12. Plug the high pressure inlet to the caliper and the bleeder screw hole and coat the piston with lubricant. Spread the boot with your fingers and work the piston into the boot, pressing down evenly.

13. Remove the plug and carefully push the piston down until it bottom.

14. Install the caliper on the vehicle, using the reverse order of removal.

15. Be sure to install the inboard pad anti-rattle spring on top

of the retainer spring plate. Bleed the brakes. Road test the vehicle making several tops to wear any foreign material off of the pads.

Brake Disc (Rotor)

REMOVAL AND INSTALLATION

1. Jack up the front of the vehicle and support it with jackstands. Remove the front wheel.

2. Remove the caliper assembly and support it to the frame with a piece of wire without disconnecting the brake fluid hose.

3. Remove the hub and rotor assembly as described in Section 1.

4. Install the rotor in the reverse order of removal, and adjust the wheel bearing as outlined in Section 1.

INSPECTION

If the rotor is scarred (grooved) or has shallow cracks, it may be refinished on a disc brake rotor lathe. If the lateral run-out exceeds 0.004 in. (0.1mm) within a 6 in. (152mm) radius when measured with a dial indicator, with the stylus 1 in. (25mm) in from the edge of the rotor, the rotor should be refinished or replaced.

Minimum thickness for an 11 in. (279.4mm) rotor is 1.080 in. (27.4mm); 1.125 in. (28.5mm) for a 12 in. (304.8mm) rotor. Material may be removed equally from each friction surface of the rotor. If the damage cannot be corrected when the rotor has been machined to the minimum thickness, it should be replaced.

REAR DRUM BRAKES

CAUTION

Brake shoes/pads contain asbestos, which has been determined to be a cancer causing agent. Never clean the brake surfaces with compressed air! Avoid inhaling any dust from any brake surface! When cleaning brake surfaces, use a commercially available brake cleaning fluid.

Brake Drum

REMOVAL AND INSTALLATION

11 in. (279.4mm) Brakes

1. Jack and support the vehicle.
2. Remove the plug from the brake adjustment access hole.
3. Insert a thin bladed screwdriver through the adjusting hole and hold the adjusting lever away from the star wheel.
4. Release the brake by prying down against the star wheel with a brake spoon.
5. Remove the rear wheel and clips from the wheel studs. Remove the brake drum.
6. Installation is the reverse of removal. Adjust the brakes.

12 in. (304.8mm) Brakes

1. Raise and support the vehicle.
2. Remove the rear wheel and tire.

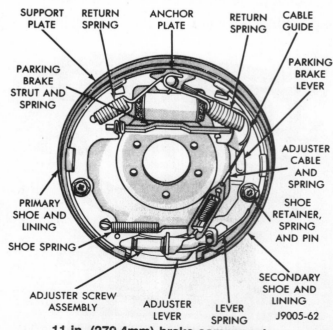

11 in. (279.4mm) brake components

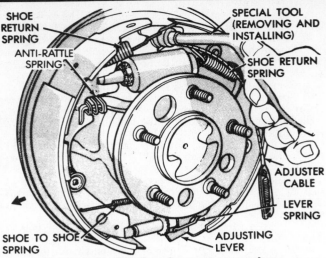

Remove/install the shoe return springs

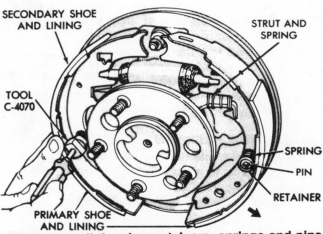

Remove/install the shoe retainers, springs and pins

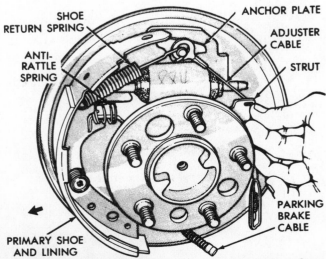

Remove/install the parking brake strut and spring

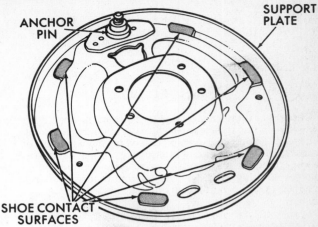

Shoe to backing plate contact surfaces

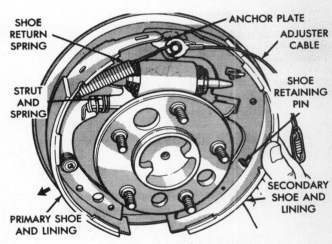

Brake shoe installation

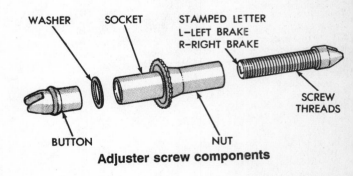

Adjuster screw components

3. Remove the axle shaft nuts, washers and cones. If the cones do not readily release, rap the axle shaft sharply in the center.

4. Remove the axle shaft.

5. Remove the outer hub nut.

6. Straighten the lockwasher tab and remove it along with the inner nut and bearing.

7. Carefully remove the drum.

8. Position the drum on the axle housing.

9. Install the bearing and inner nut. While rotating the wheel and tire, tighten the adjusting nut until a slight drag is felt.

10. Back off the adjusting nut $\frac{1}{6}$ turn so that the wheel rotates freely without excessive endplay.

11. Install the lockrings and nut. Place a new gasket on the hub and install the axle shaft, cones, lockwashers and nuts.

12. Install the wheel and tire.

13. Road test the vehicle.

INSPECTION

Check that there are no cracks or chips in the braking surface. Excessive bluing indicates overheating and a replacement drum is needed. The drum can be machined to remove minor damage and to establish a rounded braking surface on a warped drum. Never exceed the maximum oversize of the drum when machining the braking surface. The maximum inside diameter is stamped on the rim of the drum.

Brake Shoes

REMOVAL AND INSTALLATION

11 in. (279.4mm) Brakes

1. Raise and support the vehicle.

2. Remove the rear wheel, drum retaining clips and the brake drum.

3. Remove the brake shoe return springs, noting how the secondary spring overlaps the primary spring.

4. Remove the brake shoe retainer, springs and nails.

5. Disconnect the automatic adjuster cable from the anchor and unhook it from the lever. Remove the cable, cable guide, and anchor plate.

6. Remove the spring and lever from the shoe web.

7. Spread the anchor ends of the primary and secondary shoes and remove the parking brake spring and strut.

8. Disconnect the parking brake cable and remove the brake assembly.

9. Remove the primary and secondary brake shoe assemblies and the star adjuster as an assembly. Block the wheel cylinders to retain the pistons.

10. Inspect the brake drum. Service as required.

11. Apply a thin coat of lubricant to the support platforms.

12. Attach the parking brake lever to the back side of the secondary shoe.

13. Place the primary and secondary shoes in their relative positions on a workbench.

14. Lubricate the adjuster screw threads. Install it between the primary and secondary shoes with the star wheel next to the secondary shoe. The star wheels are stamped with an **L** (left) and **R** (right).

15. Overlap the ends of the primary and secondary brake shoes and install the adjusting spring and lever at the anchor end.

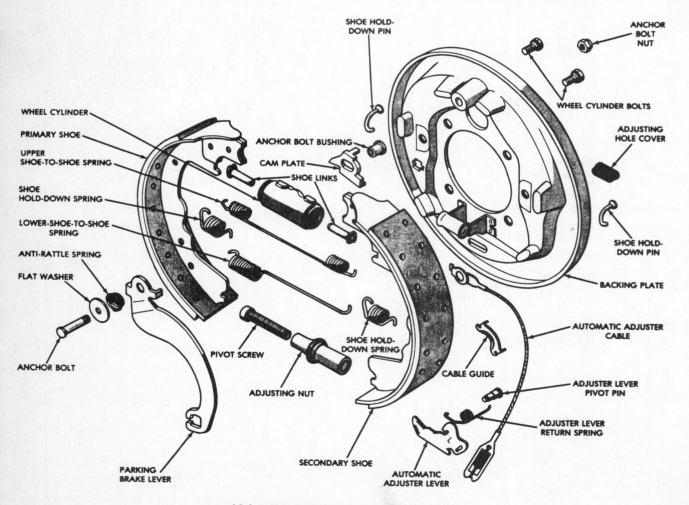

12 in. (304.8mm) brake components

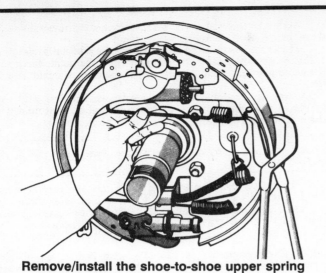

Remove/install the shoe-to-shoe upper spring

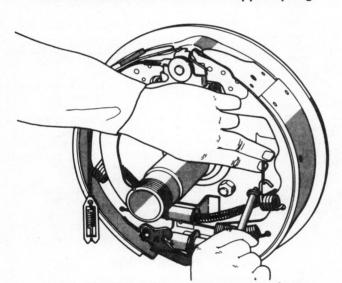

Remove/install the shoe holddown springs

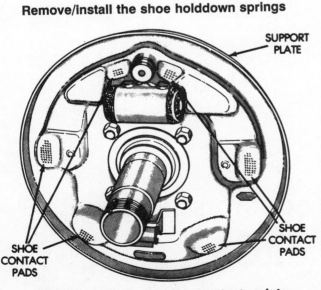

Support plate brake shoe contact points

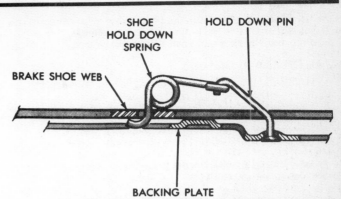

Holddown spring and pin attachment

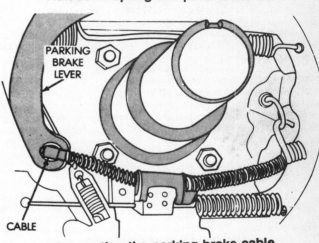

Connecting the parking brake cable

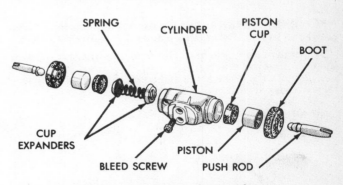

Wheel cylinder components

16. Hold the shoes in position and install the parking brake cable into the lever.

17. Install the parking brake strut and spring between the parking brake lever and primary shoe.

18. Place the brake shoes on the support and install the retainer nails and springs.

19. Install the anchor pin plate.

20. Install the eye of the adjusting cable over the anchor pin and install the return spring between the anchor pin and primary shoe.

21. Install the cable guide in the secondary shoe and install the secondary return spring. Be sure that the primary spring overlaps the secondary spring.

22. Position the adjusting cable in the groove of the cable guide and engage the hook of the cable in the adjusting lever.

23. Install the brake drum and retaining clips. Install the wheel and tire.

24. Adjust the brakes and road test the vehicle.

12 in. (304.8mm) Brakes

1. Unhook and remove the adjusting lever return spring.
2. Remove the lever from the lever pivot pin.
3. Unhook the adjuster lever from the adjuster cable.
4. Unhook the upper shoe-to-shoe spring.
5. Unhook and remove the shoe holddown springs.
6. Disconnect the parking brake cable from the parking brake lever.
7. Remove the shoes with the lower shoe-to-shoe spring and star wheel as an assembly.

To install the shoes:

8. The pivot screw and adjusting nut on the left side have left-hand threads and right hand threads on the right side.
9. Lubricate and assemble the star wheel assembly. Lubricate the guide pads on the support plates.
10. Assemble the star wheel, lower shoe-to-shoe spring, and the primary and secondary shoes. Position this assembly on the support plate.
11. Install and hook the holddown springs.
12. Install the upper shoe-to-shoe spring.
13. Install the cable and retaining clip.
14. Position the adjuster lever return spring on the pivot (green springs on left brakes and red springs on right brakes).
15. Install the adjuster lever. Route the adjuster cable and connect it to the adjuster.
16. Install the brake drum and adjust the brakes.

Wheel Cylinders

REMOVAL, INSTALLATION AND OVERHAUL

When the brake drums are removed, carry out an inspection of the wheel cylinder boots for cuts, tears, cracks, or leaks. If any of these are present, the wheel cylinder should have a complete overhaul performed.

NOTE: Preservative fluid is used during assembly; its presence in small quantities does not indicate a leak.

To remove and overhaul the wheel cylinders, proceed in the following manner:

1. Remove the brake shoes and check them. Replace them if they are soaked with grease or brake fluid.
2. Detach the brake hose.

3. Unfasten the wheel cylinder attachment bolts and slide the wheel cylinder off its support.
4. Pry the boots off from either end of the wheel cylinder and withdraw the push rods. Push in one of the pistons, to force out the other piston, its cup, the spring, the spring cup, and he piston, itself.
5. Wash the pistons, the wheel cylinder housing, and the spring in fresh brake fluid, or in denatured alcohol, and dry them off using compressed air.

CAUTION

Do not use a rag to dry them since the lint from it will stick to the surfaces.

---

6. Inspect the cylinder bore wall for signs of pitting, scoring, etc. If it is badly scored or pitted, the entire cylinder should be replaced. Light scratches or corrosion should be cleaned up with crocus cloth, or using a cylinder hone with fine grade stones. Do not over hone.

NOTE: Disregard the black stains from the piston cups that appear on the cylinder wall; they will do no damage.

Assembly and installation are performed in the following manner:

1. Dip the pistons and the cups in clean brake fluid. Replace the boots with new ones, if they show wear or deterioration. Coat the wall of the cylinder bore with clean brake fluid.
2. Place the spring in the cylinder bore. Position the cups in either end of the cylinder with the open end of the cups facing inward (toward each other).
3. Place the pistons in either end of the cylinder bore with the recessed ends facing outward (away from each other).
4. Install the boots over the ends of the cylinder and push down until each boot is seated, being careful not to damage either boot.
5. Install the wheel cylinder on its support.
6. Attach the jumper tube to the wheel cylinder. Install the brake hose on the frame bracket. Connect the brake line to the hose. Connect the end of the brake hose. Connect the end of the brake hose through the end of the stand-off. Attach the jumper tube to the brake hose and attach the hose to the stand-off.

Wheel Bearings

Rear wheel bearing adjustment (required only on Spicer 60 full-floating rear axles) is covered in Section 7 under "Axle Shafts and Bearings".

REAR WHEEL ANTI-LOCK SYSTEM

The purpose of Rear Wheel Anti-Lock (RWAL) is to prevent rear wheel lockup under heavy braking conditions on all types of road surfaces. The RWAL system uses a standard master cylinder and booster arrangement with a vertical split hydraulic circuit. An electronic control module, rear wheel speed sensor and a hydraulic pressure valve are the major components of the system.

There are some minor changes in brake lines, hoses and elec-

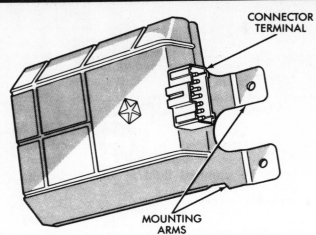

Anti-lock electronic control module

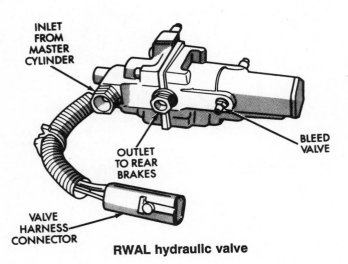

RWAL hydraulic valve

FAULT CODES

If a system fault is detected during the self-test, or at any other time, the control module will illuminate the antilock indicator lamp and store the fault code in the microprocessor memory. If a fault code is generated, the module will remember the code after the ignition is switched OFF. The microprocessor memory will store and display only one fault code at a time. The stored code can be displayed by grounding the RWAL diagnostic connector and counting the number of flashes on the indicator lamp. To clear the fault code, disconnect the control module connector or disconnect the battery for at least 5 seconds. During system retest, wait 30 seconds to make sure the fault code does not reappear.

FAULT CODE NUMBER	TYPICAL FAILURE DETECTED
1	Not used.
2	Open isolation valve wiring or bad control module.
3	Open dump valve wiring or bad control module.
4	Closed RWAL valve switch.
5	Over 16 dump pulses generated in 2WD vehicles (disabled for 4WD).
6	Erratic speed sensor reading while rolling.
7	Electronic control module fuse pellet open, isolation output missing, or valve wiring shorted to ground.
8	Dump output missing or valve wiring shorted to ground.
9	Speed sensor wiring/resistance (usually high reading).
10	Sensor wiring/resistance (usually low reading).
11	Brake switch always on. RWAL light comes on when speed exceeds 40 mph.
12	Not used.
13	Electronic control module phase lock loop failure.
14	Electronic control module program check failure.
15	Electronic control module RAM failure.

RWAL system fault codes

trical circuits to accommodate RWAL on models so equipped. No hydraulic pumps are used; brake pressure comes directly from the pedal application. The system provides stability by allowing at least one rear wheel to remain unlocked. It's still possible to lock the front wheels, since this system works on the rear wheels only.

An amber antilock warning light is installed on the instrument panel along with the standard red brake warning lights. Looking for this amber warning light is the quickest way of determining whether or not the vehicle is equipped with RWAL.

A speed sensor is mounted to the top of the rear differential housing. A toothed exciter ring is press fit onto the differential case next to the differential ring gear and provides the signal for the sensor.

The electronic control module is located on the lower right side cowl panel. The control module monitors the rear wheel speed and controls the dual solenoid valve. The module also performs a system self-check every time the ignition switch is turned from the OFF to the ON position.

The hydraulic pressure valve allows brake fluid to flow freely between the master cylinder and the rear brakes under normal operating conditions. Once antilock braking begins, the control module triggers the valve to either isolate or reduce pressure to the rear wheels.

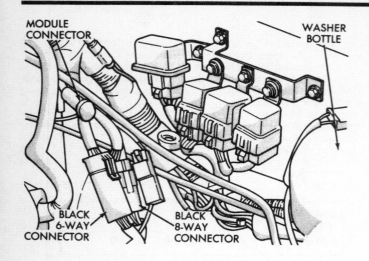

RWAL connector locations

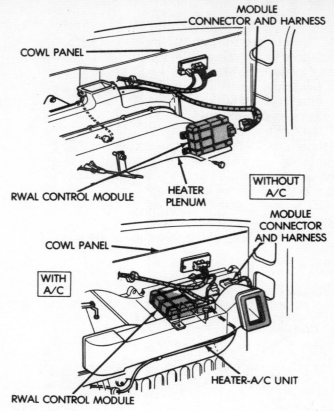

RWAL control module mounting

RWAL Control Module

REMOVAL AND INSTALLATION

The control module is on the right side of the vehicle. On models equipped with air conditioning, the module is attached to the top of the heater-A/C unit and accessed through the glovebox opening after removing the box.. On models without air conditioning, the module is attached to the face of the heater plenum, and is accessible from under the dash.

Disconnect the wiring to the module and remove the mounting screws. Replace the module by placing the module in position and securing the mounting screws (21 inch lbs.) and connecting the wiring.

RWAL Speed Sensor

REMOVAL AND INSTALLATION

1. Raise the vehicle and support it safely.
2. Remove the sensor holddown bolt.
3. Remove the sensor shield and sensor from the differential by pulling the sensor straight out.
4. Disconnect the wiring and remove the sensor.
   To install:
5. Connect the wiring to the sensor. Make sure the seal is in place between the sensor and wiring connector.
6. Install the sensor into the differential housing, making sure the new O-ring is in place.
7. Install the sensor shield.

8. Install the sensor holddown bolt and tighten it to 170-230 inch lbs. (19-29 Nm).
9. Lower the vehicle.

Hydraulic Valve

REMOVAL AND INSTALLATION

1. Raise the vehicle and support it safely.
2. Remove the brake lines from the hydraulic valve.
3. Remove the nuts attaching the valve to the frame rail.
4. Remove the valve from the frame and disconnect the wiring.
   To install:
5. Connect the wiring to the control valve.
6. Install the valve on the frame. Tighten the mounting bolts to 16-25 ft. lbs. (21-34 Nm).
7. Install the brake lines.
8. Bleed the brake system as previously described.
9. Lower the vehicle.

REAR WHEEL ANTILOCK DIAGNOSIS

The electronic control module has the capability of generating and storing fault codes. Only one code can be stored and shown at any one time. Also, if a fault code is generated the electronic control module will retain the code even after a key "off" condition.

If a problem is detected the electronic control module will illuminate the amber antilock brake warning lamp and set a fault code. When a fault code is set the red brake warning lamp will also be lit. To determine what the fault code is, you must momentarily ground the RWAL diagnostic connector and count the flashes of the amber antilock warning lamp. The initial flash will be a long flash followed by a number of short flashes. The long flash indicates the beginning of the fault number sequence and the short flashes are a continuation of that sequence. **You must count the long flash with the short flashes to have an accurate fault code count.**

To clear a fault code disconnect the control module connector from module or remove battery voltage for five seconds. During system retest wait 30 seconds to ensure the fault code does not reappear.

Test Step	What to do	Condition	Yes	No
TEST 1 VISUAL INSPECTION				
1.1	Inspect RWAL connectors and ground for defects and good connections (Figs. 1, 2 and 3).	Is connector free of defects and connected properly?	Go to Test 2	Repair or connect terminals as required.
TEST 2 SYSTEM SELF CHECK				
2.1	Turn ignition switch to run position.	Both lamps illuminate for 2 seconds then go out as system performs self check.	Go to Test 3	Choose another condition.
		Antilock and brake lamp stay on.	Go to Test 6	Choose another condition.
		Antilock lamp off and will not self check, Brake lamp checked OK.	Go to Test 7	Choose another condition.
		Brake lamp on, Antilock lamp off and does self check.	Go to Test 8	Choose another condition.
		Antilock and Brake lamps flashing.	Go to Test 9	Choose another condition.
		Brake lamp off, Antilock lamp on, Antilock lamp does self check.	Go to Test 10	Choose another condition.
TEST 3 CHECKING SENSOR OUTPUT AND PHYSICAL CONDITION				
3.1	Apply service brakes and check stop lights.	Stop lamps illuminate.	Go to Test 3.2	Repair stop lamp circuit.
3.2	Remove sensor from differential and inspect exciter ring for damage.	Exciter ring in good condition.	Reinstall sensor. Go to Test 3.3	Replace exciter ring and retest system.
3.3	Lift rear wheels, start engine, run wheels at 5 mph. Make sure vehicle is properly positioned on hoist or jack stands. **WARNING: STAY CLEAR OF ROTATING WHEELS.** Using a voltmeter set on 2 volt AC scale and connect between B01 PK and B02 LG/OR* wires of the sensor connector (Fig. 1).	Is voltage 650 MV (RMS) or greater.	Go to Test 5	Go to Test 3.4
3.4		Has sensor been replaced?	Go to Test 3.5	Replace sensor and go to Test 3.3.

Test Step	What to do	Condition	Yes	No
3.5	Disconnect 14 way module connector. Disconnect sensor connector and connect an ohmmeter between B01 RD/VT* wire in module connector and B01 PK wire in sensor connector.	Is there continuity?	Go to Test 3.6	Repair open circuit.
3.6	Connect an ohmmeter between B01 RD/VT* wire in module connector and ground.	Is there continuity?	Repair circuit for a short to ground.	Go to Test 3.7
3.7	Connect an ohmmeter between B02 WT/VT* wire in module connector and B02 LG/OR wire in sensor connector.	Is there continuity?	Repair circuit for an open circuit.	Go to Test 3.8
3.8	Connect an ohmmeter between B02 WT/VT* wire in module connector and ground.	Is there a short to ground?	Repair circuit for short to ground.	Go to Test 4

TEST 4 CHECKING SENSOR GAP

Test Step	What to do	Condition	Yes	No
4.1	Remove sensor from differential. Measure height of sensor pole piece from mounting face of sensor (should be 1.07"-1.08"). Measure top of exciter ring teeth from sensor mounting face on differential (should be 1.085"-1.12"). Subtract measurements as shown in Fig. 3 to obtain sensor gap. Gap must be a minimum of 0.005" and a maximum of 0.05".	Was gap within specifications?	Go to Test 5	Go to Test 4.2

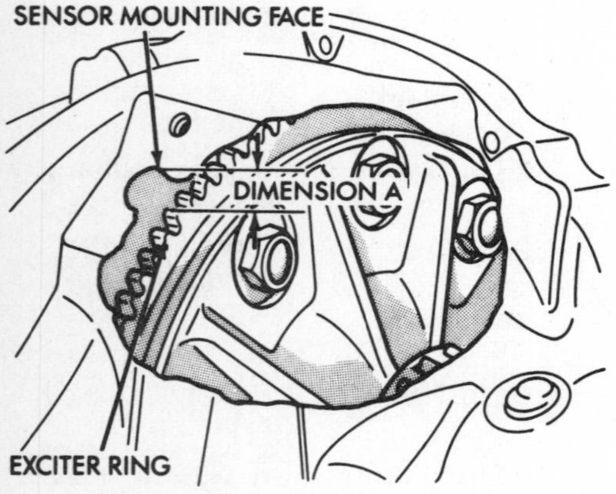

SENSOR MOUNTING FACE

DIMENSION A

EXCITER RING

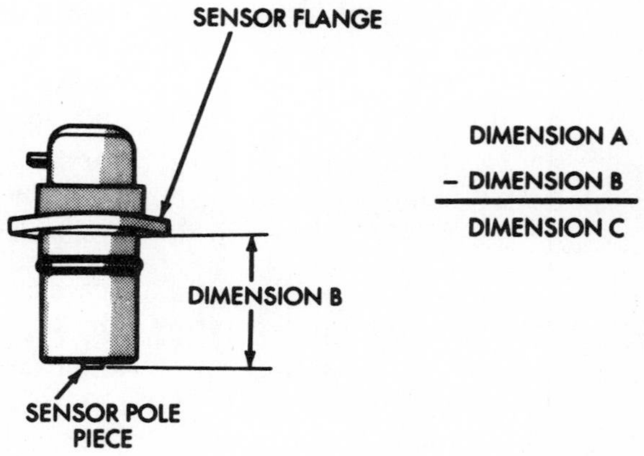

SENSOR FLANGE

DIMENSION B

SENSOR POLE PIECE

DIMENSION A
− DIMENSION B
———————
DIMENSION C

Test Step	What to do	Condition	Yes	No
4.2	Look at sensor measurement from Test 4.1 (should be 1.07"-1.08").	Was sensor measurement within specification?	Repair differential and retest system.	Replace sensor and retest system.

TEST 5 CHECKING FOR BRAKE MECHANICAL PROBLEMS

Test Step	What to do	Condition	Yes	No
5.1	Check rear brakes for mechanical problems such as grabbing, locking or pulling.	Are the rear brakes functioning properly?	Replace module and retest system.	Repair mechanical problem and retest system.

Test Step	What to do	Condition	Yes	No
TEST 6 CHECKING THE DIAGNOSTIC CONNECTOR GROUND				
6.1	Locate the black 2 way diagnostic connector below RWAL module. Connect a jumper wire between the diagnostic connector and ground.	Is there a flashout code?	Go to Test 11	Go to 6.2
6.2	Turn ignition off. Disconnect 14-way connector from module and connect an ohmmeter between the BK* in the 14 way connector and the BK* in the 2 way diagnostic connector.	Is there continuity?	Go to Test 6.3	Repair open circuit and retest system.
6.3	Check brake fluid level in master cylinder reservoir.	Is brake fluid level correct?	Go to Test 6.4	Find and repair leak and retest system.
6.4	Reconnect 14 way module connector. Disconnect connector from the pressure differential switch. Turn ignition to run position.	Do both antilock lamp and brake lamp stay on?	Go to Test 6.5	Check brake system for air in lines or mechanical damage.
6.5	Disconnect 14 way module connector. Turn ignition switch to the run position.	Are both antilock lamp and brake lamp off.	Go to Test 6.6	Choose another condition
		Antilock lamp on, brake lamp off.	Repair AT1 orange wire for short to ground between module and antilock lamp.	Repair short to ground in differential switch sensor wiring, B01 PK and B02 LG/OR.
6.6	Remove and inspect antilock fuse.	Is fuse open?	Check and repair all circuits fuse is protecting. Replace fuse.	Go to Test 6.7
6.7	Connect a voltmeter between pin 3 RD/YL wire of 14 way module connector and ground.	Is voltage greater than 9 volts?	Go to Test 6.8	Repair the D1 RD/YL wire for an open.
6.8	Remove and inspect the stop lamp fuse.	Is the fuse open?	Check and repair all circuits fuse is protecting for shorts and replace fuse.	Go to Test 6.9
6.9	Connect a voltmeter between pin 9 of the 14 way module connector and ground.	Is voltage greater than 9 volts?	Replace module and retest system.	Repair the D3B PK/DB* wire for open circuit.
TEST 7 CHECKING MODULE GROUND AND POWER				
7.1	Make sure module connector is fully plugged into module.	Is connector plugged in?	Go to Test 7.2	Plug connector in and retest system.
7.2	Disconnect battery and 14 way module connector. With an ohmmeter set on 200 ohm scale check resistance between pin 10 BK/LG wire on the 14 way harness connector and ground.	Is resistance less than 1 ohm?	Go to Test 7.3	Repair H40 BK/LG for an open or damaged circuit.

Test Step	What to do	Condition	Yes	No
7.3	Remove and inspect antilock lamp fuse.	Is fuse open?	Check and repair all circuits fuse is protecting for shorts and replace fuse.	Go to Test 7.4
7.4	Connect battery and turn ignition to run position. With a voltmeter set on 20 volt DC scale check voltage between Pin 2 Orange wire of the 14 way module connector and ground.	Is voltage less than 9 volts?	Go to Test 7.5	Replace the module and retest system.
7.5	Check antilock bulb.	Is bulb open?	Replace bulb and retest system.	Repair AT1 orange wire for open circuit between pin 2 and fuse.

TEST 8 CHECKING PARKING BRAKE SYSTEM AND MODULE

Test Step	What to do	Condition	Yes	No
8.1	Turn ignition to run position. Pull lever to release parking brake.	Does the brake lamp go off?	Disregard failure, retest antilock and brake lamp for 2 second self check.	Go to Test 8.2
8.2	Pull park brake release lever with one hand and pull pedal up with other hand.	Did brake lamp go off?	Repair park brake mechanism or switch.	Go to Test 8.3
8.3	Disconnect black 1 way park brake switch connector.	Did brake lamp go off?	Adjust or replace park brake switch and retest system.	Go to Test 8.4
8.4	Disconnect 14 way module connector.	Did brake lamp go off?	Replace module and retest system.	Repair P5 BK/GY wire for a short to ground (If brake light does not go out at this point, refer to "Basic Diagnosis Guide" in this section.

TEST 9 CHECKING FOR INTERMITTENT PROBLEMS

Test Step	What to do	Condition	Yes	No
9.1	Disconnect 14 way module connector. With a voltmeter set on 20 volt DC scale check voltage between pin 3 RD/YL wire on the 14 way module connector and ground. Turn ignition to run position and shake the instrument panel harness.	Is voltage steady at 9 volts?	Go to Test 9.2	Repair D1 RD/YL for open circuit.
9.2	Disconnect battery, set ohmmeter on 200 scale and connect between pin 12 BK wire of the 14 way module connector and ground, then shake the instrument panel harness.	Is resistance 100K or greater and steady?	Go to Test 9.3	Repair DK BK/* wire for a short to ground.

Test Step	What to do	Condition	Yes	No
9.3	With ohmmeter set on 200 ohm scale connect between Pin 10 BK/LG wire of the 14 way module connector and ground, then shake instrument panel harness.	Is resistance steady at 1 ohm?	Replace module and retest system.	Repair open circuit in H40 BK/LG.

TEST 10 CHECKING FOR OPEN OR DISCONNECTED PARK BRAKE SWITCH CONNECTOR

10.1	Make sure the P5 BK/GY wire at the park brake switch is connected.	Is P5 wire connected to park brake switch?	Go to Test 10.2	Connect and retest system.
10.2	Disconnect park brake switch connector. Disconnect instrument cluster 11 way red connector. Connect an ohmmeter between P5 BK/GY wire in both connectors.	Is there continuity?	Go to Test 10.3	Repair P5 BK/GY wire for open circuit.
10.3	Inspect instrument cluster printed circuit board for damage.	Is printed circuit board damaged?	Replace printed circuit board.	Replace brake warning lamp bulb.

TEST 11 FLASHCODES

11.1	Connect a jumper wire between the diagnostic connector and ground. Count the flashes including the long flash that starts the flash code count. Choose the proper condition.	Antilock lamp and brake lamp flash 1 time.	Go to Test 12
		Antilock lamp and brake lamp flash 2 times.	Go to Test 13
		Antilock lamp and brake lamp flash 3 times.	Go to Test 14
		Antilock lamp and brake lamp flash 4 times.	Go to Test 15
		Antilock lamp and brake lamp flash 5 times.	Go to Test 16
		Antilock lamp and brake lamp flash 6 times.	Go to Test 17
		Antilock lamp and brake lamp flash 7 times.	Go to Test 18
		Antilock lamp and brake lamp flash 8 times.	Go to Test 19
		Antilock lamp and brake lamp flash 9 times.	Go to Test 20
		Antilock lamp and brake lamp flash 10 times.	Go to Test 21
		Antilock lamp and brake lamp flash 11 times.	Go to Test 22
		Antilock lamp and brake lamp flash 12 times.	Go to Test 23
		Antilock lamp and brake lamp flash 13 times.	Go to Test 24
		Antilock lamp and brake lamp flash 14 times.	Go to Test 25
		Antilock lamp and brake lamp flash 15 times.	Go to Test 26

Test Step	What to do	Condition	Yes	No
		Antilock lamp and brake lamp flash 16 times or more.	Go to Test 27	

TEST 12 ONE FLASH

| 12.1 | One flashcode should not occur. Perform flashcode procedure several times. | Are you still getting code 1? | Go to Test 5 | Go to Test 11 |

TEST 13 TWO FLASHES

13.1	Disconnect battery and 14 way module connector. Set an ohmmeter on 200 ohm scale and connect between pin 1 LG wire of the 14 way module connector and ground.	Does the circuit have over 6 ohms?	Go to Test 13.2	Replace the module and retest system.
13.2	Disconnect valve harness connector from valve connector. Connect an ohmmeter between B09 GY/WT of harness connector and ground.	Is resistance greater than 1 ohm?	Repair B09 GY/WT wire for an open circuit or high resistance. Check for contaminated or loose connector pins and retest system.	Go to Test 13.3
13.3	Connect an ohmmeter between IS1 LG/* and B09 GY/WT wires in the 4-way black valve connector.	Does circuit have over 6 ohms?	Replace antilock valve and retest system.	Repair the IS1 LG/* for open circuit from valve to computer module and retest system.

TEST 14 THREE FLASHES

| 14.1 | Disconnect battery. Remove 14 way module harness connector from module. Set ohmmeter on 200 ohm scale and connect to pin 8 DS1 WT/BR and ground. | Does circuit have over 3 ohms? | Go to Test 14.2 | Replace module and retest system. |
| 14.2 | Disconnect the 4 way valve harness connector. Connect an ohmmeter between DS1 WT and B09 GY/NT wires in valve connector. | Does circuit have over 3 ohms? | Replace antilock valve and retest system. | Repair DS1 WT wire for open between module connector and valve connector. |

TEST 15 FOUR FLASHES

15.1	Disconnect the 4 way valve harness connector from valve connector. Set ohmmeter on 20K scale and connect between VS1 LB wire in valve body and ground.	Is resistance greater than 10K ohms?	Go to Test 15.2	Replace the antilock valve and retest system.
15.2	Connect an ohmmeter between VS1 LB and B09 GY/WY wires in the valve connector.	Is resistance greater than 10K ohms?	Go to Test 15.3	Replace the antilock valve and retest system.
15.3	Disconnect battery. Disconnect the 14 way module harness connector from module. Set ohmmeter on 200K scale and connect to pin 11, VS1 LB of 14 way connector and ground.	Is resistance greater than 100K ohms?	Replace module and retest system.	Repair VS1 LB wire for a short to ground between valve and module. Retest system.

Test Step	What to do	Condition	Yes	No
TEST 16 FIVE FLASHES				
16.1	Did the failure occur in 2 wheel drive mode?		Go to Test 16.2	Go to Test 16.3
16.2	Disconnect 14 way module connector from module to deactivate antilock system. Drive the vehicle in 2 wheel drive mode and make normal and safe stops to determine the condition of the rear brakes.	Are the brakes functioning normally?	Replace the antilock valve and retest system.	Repair rear brakes and retest system.
16.3	Disconnect 14 way module connector from module. Turn ignition key to run position. Shift into 4 wheel drive. Set a voltmeter to 20 vdc and connect between pin 4, X4 LG/BR wire and ground.	Is voltage greater than 1 volt?	Repair X4 wire for an open or 4 wheel drive indicator switch. Retest system.	Replace antilock valve and retest system.
TEST 17 SIX FLASHES				
17.1	Recheck flashcode after driving vehicle.	Antilock light and brake light flash 6 times.	Go to Test 17.2	Go to Test 11
17.2	Disconnect battery. Disconnect 14 way module connector. Set ohmmeter on 200 ohm scale and connect between pin 13, B02 WT/VT and pin 14, B01 RD/VT of harness connector. Shake antilock wiring harness from differential to module.	Is resistance constant at 1000-2000 ohms?	Go to Test 17.3	Repair circuit. Retest system.
17.3	Remove sensor from the differential and inspect for build up of metal chips on sensor pole piece.	Are metal chips present?	Drain and clean differential. Check exciter ring for broken or chipped teeth. Retest system.	Go to Test 17.4
17.4	Look into sensor hole in differential and rotate exciter ring and check for damage (missing or bent teeth)	Is exciter ring intact?	Go to 17.5	Replace exciter ring. Retest system.
17.5	Reinstall sensor. Disconnect 2 way RWAL sensor connector. With a voltmeter on 2 volt scale connect between B01 PK and B02 LG/OR wires of sensor connector. Raise rear wheels off floor and run at 5 mph. **WARNING: STAY CLEAR OF ROTATING WHEELS.**	Is voltage greater than 650 MV and steady?	Replace module. Retest system.	Replace sensor. Recheck sensor output. Retest system.
TEST 18 SEVEN FLASHES				
18.1	Disconnect 4 way valve harness connector from valve connector. Connect an ohmmeter between IS1 LG/* and B09 GY/WT wire in valve connector.	Is resistance less than 3 ohms?	Replace antilock valve. Retest system.	Go to Test 18.2

Test Step	What to do	Condition	Yes	No
18.2	Disconnect battery. Disconnect 4 way valve harness connector from valve connector. Disconnect 14 way module harness connector from module. Set ohmmeter on 20K ohms scale and connect between Pin 1, IS1 LG/* wire in harness connector and ground.	Is resistance greater than 20K ohms?	Replace module. Retest system.	Repair IS1 LG/* for a short between antilock valve and module. Retest system.

TEST 19 EIGHT FLASHES

19.1	Disconnect 4 way valve harness from valve connector. Set ohmmeter on 200 ohm scale and connect between DS1 WT and B09 GY/WT wires in valve connector.	Is resistance less than 1 ohm?	Replace antilock valve. Retest system.	Go to Test 19.2
19.2	Disconnect battery. Disconnect 4 way valve connector. Disconnect 14 way module connector. Set ohmmeter on 20K ohm scale and connect between pin 8, DS1 WT/BR and ground.	Is resistance greater than 20K ohms?	Replace module.	Repair DS1 WT/BR for a short to ground between antilock valve and module. Retest system.

TEST 20 NINE FLASHES

20.1	Disconnect 2 way sensor harness connector from sensor on differential housing. Set ohmmeter on 20K scale and connect to B01 PK and B02 LG/OR wires on sensor.	Is resistance greater than 2500 ohms?	Replace sensor. Recheck resistance. **Make sure seal is in place between sensor and connector.** Retest system.	Go to Test 20.2
20.2	Reconnect sensor harness **making sure seal is in place.** Disconnect battery. Disconnect 14 way module connector. Connect an ohmmeter between pin 13, B02 WT/VT and pin 14, B01 RD/VT wires in module harness connector.	Is resistance greater than 2500 ohms?	Repair B02 WT/VT and B01 RD/VT for open circuits between the module and sensor. Retest system.	Replace computer module. When reconnecting sensor **make sure seal is in place.** Retest system.

TEST 21 TEN FLASHES

21.1	Set ohmmeter on 20K ohm scale. Disconnect 2 way sensor connector from sensor on differential. Connect an ohmmeter between B01 PK and B02 LG/OR wires on sensor.	Is resistance less than 1000 ohms?	Replace sensor. Recheck resistance. **Make sure seal is in place between sensor and connector.** Retest system.	Go to Test 21.2
21.2	Disconnect battery. Disconnect 14 way module. Connect an ohmmeter between pin 14, B01 RD/VT and ground.	Is resistance greater than 20K ohms?	Go to Test 21.3	Repair B01 RD/VT circuit between the module and sensor. When reconnecting sensor **make sure seal is in place.** Retest system.

Test Step	What to do	Condition	Yes	No
21.3	Connect an ohmmeter between pin 13, B02 WT/VT and pin 14 B01 RD/VT wires in module harness connector.	Is resistance greater than 20K ohms?	Replace module. Retest system.	Repair B01 RD/VT and B02 WT/VT circuit. When reconnecting sensor **make sure seal is in place.** Retest system.

TEST 22 ELEVEN FLASHES

22.1	Recheck flash code after driving vehicle at 35 mph or greater.	Antilock light and brake light flash 11 times.	Go to Test 22.1	Go to Test 11
22.2	Apply vehicle service brakes and check vehicle stop lights.	Are stop lights operating correctly?	Go to Test 22.3	Repair the stop lamp circuit. Retest system.
22.3	Turn ignition switch off. Disconnect 14 way module connector from module. Connect a voltmeter between pin 7, D4 WT of harness connector and ground while stepping on brake pedal.	Is voltage less than 9 volts?	Repair D4 WT for open circuit between stop light switch and module. Retest system.	Check 4 way flasher, directional wiring, and feedback through stop light circuit. Also check for proper operation of cruise control. Recheck antilock and brake lights for proper 2 second bulb check.

TEST 23 TWELVE FLASHES

23.1	This code should not occur. Read flashcodes several times.	Are 12 flashes still present?	Replace module. Retest system.	Go to Test 11

TEST 24 THIRTEEN FLASHES

24.1		Are thirteen flashes present?	Replace module. Retest system.	Go to Test 11

TEST 25 FOURTEEN FLASHES

25.1		Are fourteen flashes present?	Replace module. Retest system.	Go to Test 11

TEST 26 FIFTEEN FLASHES

26.1		Are fifteen flashes present?	Replace module. Retest system.	Go to Test 11

TEST 27 SIXTEEN OR MORE FLASHES

27.1		Are sixteen or more flashes present?	Replace module. Retest system.	Go to Test 11

PARKING BRAKE

Cables

REMOVAL AND INSTALLATION

Front Cable

1. Raise and support the rear end on jackstands.
2. Disconnect the cable spring. Remove the adjusting nut at the equalizer.
3. Disengage the cable and housing from the frame and brake valve retaining clips.
4. Unseat the cable grommet from the floorboard inside the driver's compartment. Compress and remove the cable retainer from the pedal bracket. Loosen the cable retaining clamp screw and push the ball end of the cable out of the pedal assembly and slide the cable out.
5. Remove the front cable assembly from the vehicle. Place the new cable in position and install it in the reverse order. Adjust the parking brake.

Rear Cable

1. Raise and safely support the rear of the vehicle.
2. Release the parking brake.
3. Loosen the rear cable adjuster nut and disconnect the rear cable from the equalizer.
4. Remove the clip that secures the rear cable to the frame bracket.
5. Remove the rear wheel and brake drum, or axle shaft and then brake drum (350 models) for the side of the vehicle requiring cable replacement.
6. Remove the brake shoe that the parking brake cable is attached to. On 100-250 models, the secondary (rear) shoe. On 350 models the front or primary shoe. Disconnect the cable from the shoe lever and cable bracket (models equipped).
7. Compress the cable fingers and remove the cable from the brake backing plate. Use a small (mini) hose clamp to compress the fingers on the cable retainer.
8. Install the cable in the mounting bracket and connect the brake shoe lever. Install the brake shoe, brake drum (and axle).

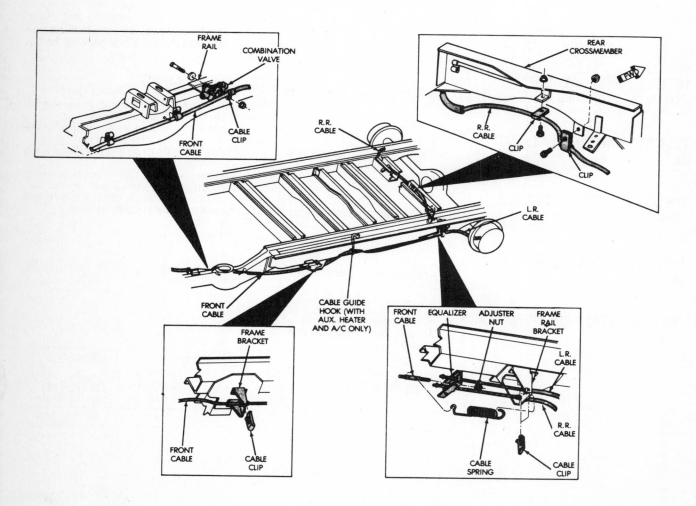

Parking brake cable routing and attachment

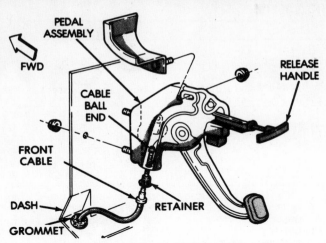

Parking brake pedal and front cable

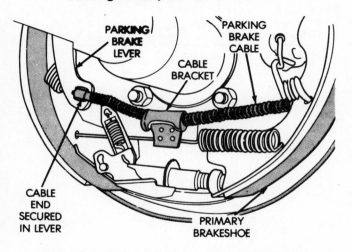

Rear cable attachment to lever and bracket

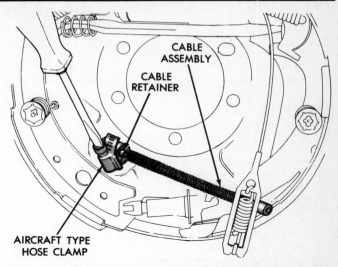

Compressing the cable retainer fingers with a mini hose clamp

Connect the cable to the equalizer. Mount the cable in the frame retaining clips. Install the wheel. Adjust the parking brake. Lower the vehicle.

ADJUSTMENT

1. Adjust the service brakes by making a few stops in reverse.
2. Raise and support the rear end on jackstands.
3. Release the parking brake lever and loosen the cable adjusting nut to be sure that the cable is slack.
4. Tighten the cable adjusting nut until a slight drag is felt while rotating the wheels.
5. Loosen the cable adjusting nut until the wheels can be rotated freely, then back off the cable adjusting nut two turns.
6. Apply the parking brake several times, then release it and check to be sure that the rear wheels rotate freely.

BRAKE SPECIFICATIONS
(All specifications in inches)

Years	Model	Master Cyl. Bore	Brake Disc			Brake Drum			Wheel Cyl. or Caliper Bore	
			Original Thickness	Minimum Thickness	Maximum Run-out	Orig. Inside Dia.	Max. Wear Limit	Maximum Machine O/S	Front	Rear
1989–91	B100–B350	1.125	①	②	0.004	11③	0.060	0.060	3.10	0.938

① Replace brake pad at 4.76mm (3/16 inch)
 Wheel lug torque
 85–110 ft. lbs. except below
 1/2 × 20 cone lug nut—105 ft. lbs.
 5/8 × 18 cone lug nut—200 ft. lbs.
 5/8 × 18 flanged lug nut—325 ft. lbs.
② 11 inch: 1.180
 12 inch: 1.125
③ B350: 12

9 BRAKES

TORQUE SPECIFICATIONS

Component	English	Metric
Power Booster		
Bellcrank pushrod nuts:	20-40 ft. lbs.	27-54 Nm
Booster bracket mounting nuts:	20-40 ft. lbs.	27-54 Nm
Booster-to-bracket nuts:	95-120 ft. lbs.	128-163 Nm
Master cylinder-to-booster nuts:	17 ft. lbs.	23 Nm
Disc Brakes		
Hose-to-caliper:	30-40 ft. lbs.	40-53 Nm
Caliper retaining plate bolts:	14-22 ft. lbs.	19-29 Nm
Caliper adapter-to-knuckle		
150/250/350, ½ in. diameter:	95-125	128-169 Nm
350, ⅝ in. diameter:	140-180 ft. lbs.	190-244 Nm
Drum Brakes		
Backing plate nuts		
150/250:	25-60 ft. lbs.	34-81 Nm
350 w/single wheels:	40-80 ft. lbs.	54-108 Nm
350 w/dual wheels:	65-105 ft. lbs.	88-142 Nm
Brake line-to-wheel cylinder		
⅜ in. or ⁷⁄₁₆ in.:	10-15 ft. lbs.	13-20 Nm
½ in. pr ⁹⁄₁₆ in.:	12-17 ft. lbs.	16-23 Nm
Wheel cylinder attaching nuts:	11-19 ft. lbs.	15-26 Nm

10

Body

QUICK REFERENCE INDEX

GENERAL INDEX

Hood, Trunk Lid, Hatch Lid, Glass and Doors

Problem	Possible Cause	Correction
HOOD/TRUNK/HATCH LID		
Improper closure.	• Striker and latch not properly aligned.	• Adjust the alignment.
Difficulty locking and unlocking.	• Striker and latch not properly aligned.	• Adjust the alignment.
Uneven clearance with body panels.	• Incorrectly installed hood or trunk lid.	• Adjust the alignment.
WINDOW/WINDSHIELD GLASS		
Water leak through windshield	• Defective seal. • Defective body flange.	• Fill sealant • Correct.
Water leak through door window glass.	• Incorrect window glass installation. • Gap at upper window frame.	• Adjust position. • Adjust position.
Water leak through quarter window.	• Defective seal. • Defective body flange.	• Replace seal. • Correct.
Water leak through rear window.	• Defective seal. • Defective body flange.	• Replace seal. • Correct.
FRONT/REAR DOORS		
Door window malfunction.	• Incorrect window glass installation. • Damaged or faulty regulator.	• Adjust position. • Correct or replace.
Water leak through door edge. Water leak from door center.	• Cracked or faulty weatherstrip. • Drain hole clogged. • Inadequate waterproof skeet contact or damage.	• Replace. • Remove foreign objects. • Correct or replace.
Door hard to open.	• Incorrect latch or striker adjustment.	• Adjust.
Door does not open or close completely.	• Incorrect door installation. • Defective door check strap. • Door check strap and hinge require grease.	• Adjust position. • Correct or replace. • Apply grease.
Uneven gap between door and body.	• Incorrect door installation.	• Adjust position.
Wind noise around door.	• Improperly installed weatherstrip. • Improper clearance between door glass and door weatherstrip. • Deformed door.	• Repair or replace. • Adjust. • Repair or replace.

EXTERIOR

Front Doors

REMOVAL AND INSTALLATION

NOTE: On doors equipped with power accessories such as radio speakers, power windows, etc. remove the door trim panel and disconnect the wiring harness to the accessory. Remove the harness from the door.

1. Matchmark the hinge-to-body locations. Support the door either on jackstands, a padded floor jack, or have somebody hold it for you.
2. Remove the lower hinge-to-frame bolts.
3. Remove the upper hinge-to-frame bolts and lift the door off of the body.
4. Install the door and hinges with the bolts finger tight.
5. Adjust the door and torque the hinge bolts to 26 ft. lbs. (35 Nm)

ADJUSTMENT

NOTE: Loosen the hinge-to-door bolts for lateral adjustment only. Loosen the hinge-to-body bolts for both lateral and vertical adjustment.

1. Determine which hinge bolts are to be loosened and back them out just enough to allow movement.
2. To move the door safely, use a padded pry bar. When the door is in the proper position, tighten the bolts to 24 ft. lbs. and check the door operation. There should be no binding or interference when the door is closed and opened.
3. Door closing adjustment can also be affected by the position of the lock striker plate. Loosen the striker plate bolts and move the striker plate just enough to permit proper closing and locking of the door.

Sliding Side Doors

REMOVAL AND INSTALLATION

1. Remove the lower rear screw from the roller track cover, body sheet metal side.
2. Remove the 2 bolts securing the upper hinge assembly mounting plate and remove the plate.
3. Remove the 2 screws securing the lower roller bracket to the door lower roller bracket support assembly.
4. Now, you'll need an assistant. Slide the door rearward, guiding the upper front and rear rollers out of the rails.
5. Lubricate both roller tracks with multi purpose lubricant. Insert the front and rear rollers into the tracks, slid the door forward to the front of each track. Position the rear, upper roller hinge plate at the hinge and align the screw holes. Install the retaining screws and tighten them to 200 inch lbs. (23 Nm). Install the upper hinge cover. Position the roller bracket under the support bracket with the screw holes aligned. Install the retaining screws in the support bracket finger tight. Adjust the door as required. Tighten the roller bracket screws to 30 ft. lbs. (41 Nm).

ADJUSTMENT

Fore and Aft

1. Remove the outer covers.
2. Loosen the hinge-roller screws and move the assembly to the desired position.
3. Tighten the screws and check the fit.

4. If necessary, shim the front striker to make sure that the pin enters the striker by 12–15mm.
5. If it was necessary to shim the striker, loosen the striker screws so they are just snug, close the door and open it carefully so as not to move the striker. Tighten the screws.

Vertical

1. Front edge fit:
   a. Loosen the 3 upper roller bracket screws.
   b. Loosen the 4 lower attaching screws.
   c. Move the door to the correct position and tighten the lower bolts.
   d. Make sure that there is 1.5mm clearance between the top roller and the top of the track.
2. Rear edge fit:
   a. Loosen the rear center striker and move it up or down the amount the door is to be raised or lowered, and tighten the screws.
   b. Close the door so that the rear latch is caught, but not fully closed.
   c. Loosen the 3 hinge screws and fully close the door.
   d. Push the front edge of the hinge up as far as it will go and snug down the front screw.
   e. Pull down firmly on the swing arm and full tighten all 3 screws.
   f. Make sure that there is 0.8mm clearance between the latch pawl and striker bar.

In-and-Out

1. Front upper corner adjustments are made by loosening the roller stud nut and moving the door to the correct position. Maintain a 1.5mm clearance between the rollers and the top of the track. Tighten the screws.
2. Front lower corner adjustments are made by loosening the 2 lower roller support bracket screws and moving the door to the desired position. Tighten the screws.
3. Rear adjustments are made by loosening the rear center striker and moving it to the desired position. Tighten the screws.

Side Doors/Rear Doors/Single Cargo Door

REMOVAL AND INSTALLATION

1. Remove the interior trim panel (models equipped) to gain access to the door pillar to hinge screws. Matchmark the hinge-to-body locations. Support the door either on jackstands, a padded floor jack, or have somebody hold it for you.
2. Remove the lower hinge-to-frame bolts.
3. Remove the upper hinge-to-frame bolts and lift the door off of the body.
4. Install the door and hinges with the bolts finger tight.
5. Adjust the door and torque the hinge bolts to 200 inch lbs. (23 Nm). Install the trim panel, if equipped.

ADJUSTMENT

NOTE: Loosen the hinge-to-door bolts for lateral adjustment only. Loosen the hinge-to-body bolts for both lateral and vertical adjustment.

1. Determine which hinge bolts are to be loosened and back them out just enough to allow movement.
2. To move the door safely, use a padded pry bar. When the door is in the proper position, tighten the bolts to 24 ft. lbs. and

check the door operation. There should be no binding or interference when the door is closed and opened.

3. Door closing adjustment can also be affected by the position of the lock striker plate. Loosen the striker plate bolts and move the striker plate just enough to permit proper closing and locking of the door.

Hood

REMOVAL AND INSTALLATION

1. Open and prop up the hood. Disconnect the underhood light, if equipped. Remove the harness from the fasteners.

2. Scribe the hood hinge outline on the hood for installation reference. Place a protective cover over the cowl. Have an assistant support the hood. Remove the bolts from each hinge and lift off the hood.

3. Place the hood in position and line up the hinge retaining bolt holes. Insert the bolts and tighten them enough to support the hood, but allow enough "slack" for hood alignment. Align the hood and tighten the mounting bolts to 200 inch lbs. (23 Nm). Install the under hood harness and connect the lamp.

ALIGNMENT

The hood can be adjusted fore-aft and up-and-down to obtain a proper fit.
1. Loosen the hood-to-hinge bolts until the are finger-tight.
2. Reposition the hood as required.
3. Tighten the bolts.

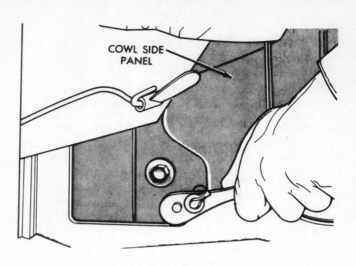

Front door hinge to body attachments

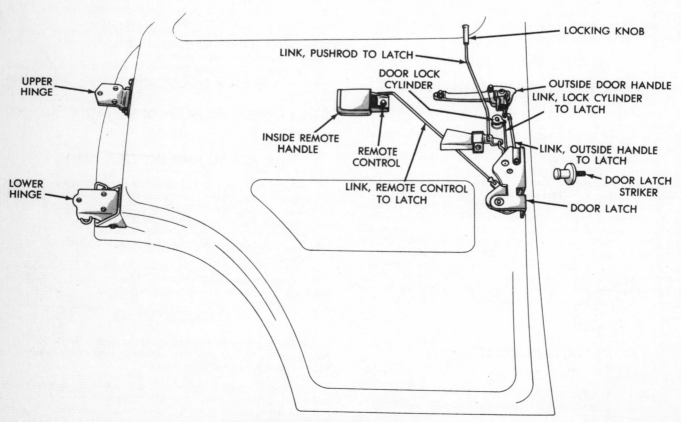

Front door lock and latch system

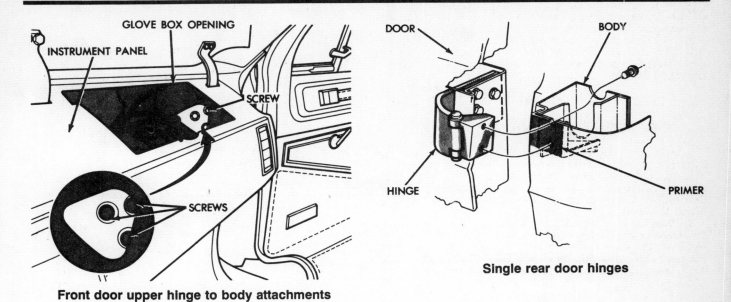

GLOVE BOX OPENING

INSTRUMENT PANEL

SCREW

SCREWS

Front door upper hinge to body attachments

DOOR

BODY

HINGE

PRIMER

Single rear door hinges

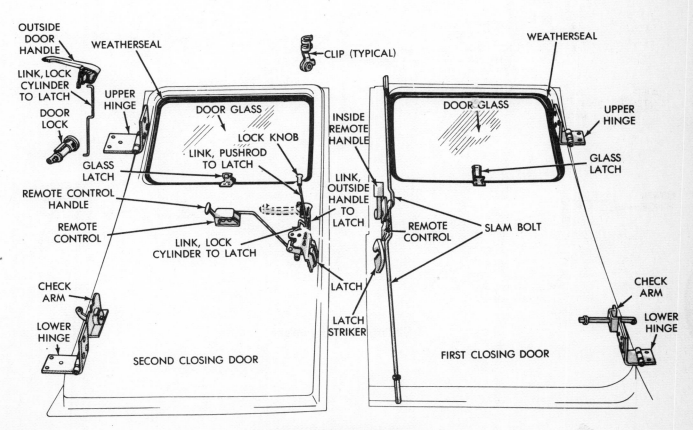

OUTSIDE DOOR HANDLE

WEATHERSEAL

CLIP (TYPICAL)

WEATHERSEAL

LINK, LOCK CYLINDER TO LATCH

UPPER HINGE

DOOR GLASS

INSIDE REMOTE HANDLE

DOOR GLASS

UPPER HINGE

DOOR LOCK

LOCK KNOB

GLASS LATCH

GLASS LATCH

LINK, PUSHROD TO LATCH

LINK, OUTSIDE HANDLE TO LATCH

REMOTE CONTROL HANDLE

REMOTE CONTROL

SLAM BOLT

REMOTE CONTROL

LINK, LOCK CYLINDER TO LATCH

CHECK ARM

LATCH

CHECK ARM

LOWER HINGE

LATCH STRIKER

LOWER HINGE

SECOND CLOSING DOOR

FIRST CLOSING DOOR

Side and rear cargo doors

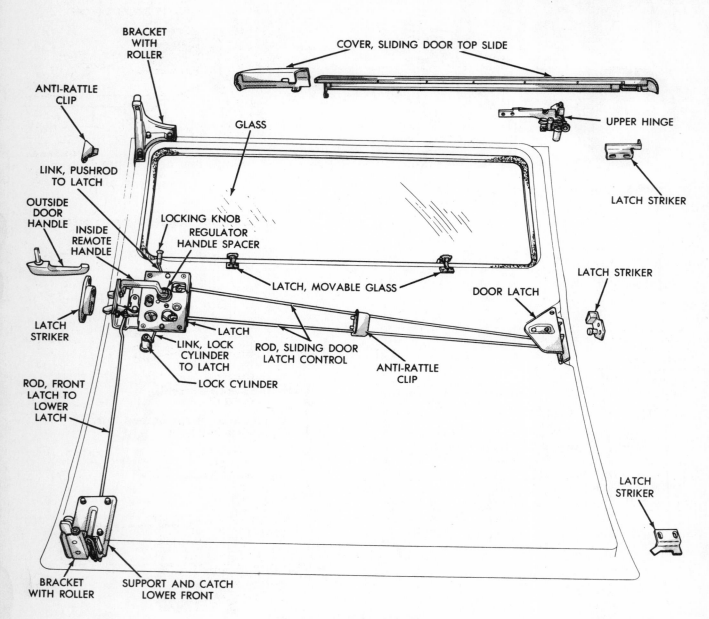

BRACKET
WITH
ROLLER

COVER, SLIDING DOOR TOP SLIDE

ANTI-RATTLE
CLIP

GLASS

UPPER HINGE

LATCH STRIKER

LINK, PUSHROD
TO LATCH

OUTSIDE
DOOR
HANDLE

INSIDE
REMOTE
HANDLE

LOCKING KNOB
REGULATOR
HANDLE SPACER

LATCH STRIKER

LATCH STRIKER

LATCH, MOVABLE GLASS

DOOR LATCH

LATCH

LINK, LOCK
CYLINDER
TO LATCH

ROD, SLIDING DOOR
LATCH CONTROL

ANTI-RATTLE
CLIP

LOCK CYLINDER

ROD, FRONT
LATCH TO
LOWER
LATCH

LATCH
STRIKER

BRACKET
WITH ROLLER

SUPPORT AND CATCH
LOWER FRONT

Sliding door assembly

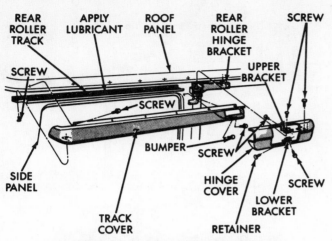

Sliding door hinge and track covers

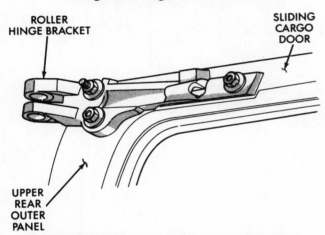

Sliding door rear, upper roller hinge bracket

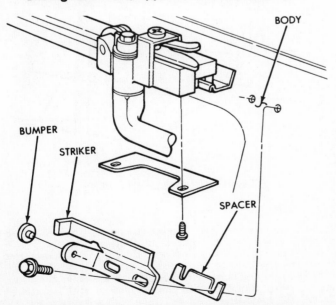

Sliding door closed position catch striker

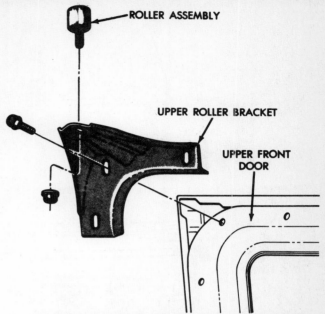

Sliding door front, upper support bracket

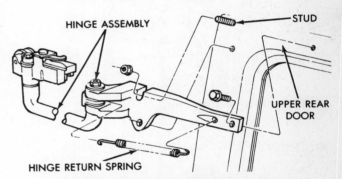

Sliding door hinge support bracket removal/installation

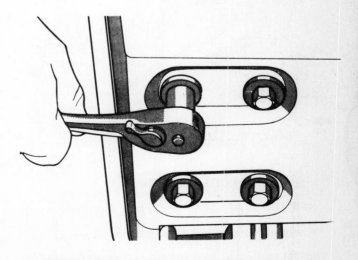

Sliding door loosening the lower attaching screws

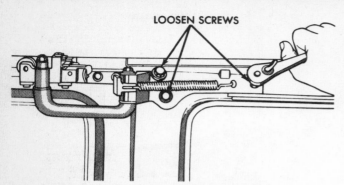

Sliding door loosening the hinge screws

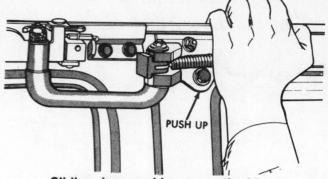

Sliding door pushing up on the hinge

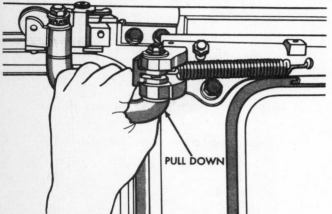

Sliding door pulling down on the swing arm

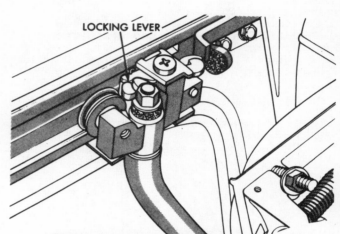

Sliding door locking lever engaged

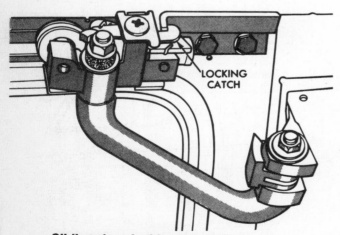

Sliding door locking catch engaged

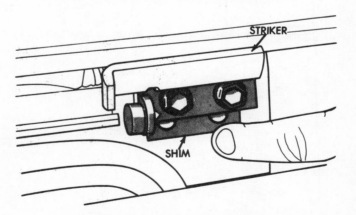

Sliding door striker shimming

Front Bumpers

REMOVAL AND INSTALLATION

1. Remove the bumper guards, if necessary to gain access to the bumper mounting bolts.

2. Support the bumper with jackstands. Remove the nuts and bolts that attach the upper and lower support brackets to the frame front crossmember. The upper support brackets are accessible via the headlamp panels, on some models.

3. Remove the front bumper with the support brackets from the frame crossmember.

4. Position the bumper and support brackets at the frame crossmember. Support the bumper in position with jackstands. Install the mounting bolts and nuts. Align the bumper and tighten the nuts and bolts to 30 ft. lbs. (41 Nm).

Rear Bumpers

REMOVAL AND INSTALLATION

1. Support the bumper.
2. Disconnect all of the wire harnesses to the various lights. Remove the nuts and bolts attaching the bumper and support brackets to the frame and/or bumper arms.
3. Position the bumper and mounting brackets at the frame mounts and support on jackstands. Install the mounting bolts and nuts. Align the bumper and tighten the mounting bolts and nuts to 30 ft. lbs. (41 Nm). Connect the various lamps.

Grille

REMOVAL AND INSTALLATION

1. Remove the headlamp trim, on models that it will interfere with grille removal. Remove the parking lamps.
2. Remove the grille mounting screws and the grille.
3. Place the grille in place and install the mounting screws. Install the parking lamps. Install the headlamp trim if removed.

Outside Mirrors

REMOVAL AND INSTALLATION

All mirrors are remove by removing the mounting screws and lifting off the mirror and gasket. On 6 in. × 9 in. (152mm × 229mm) wide "swing away" mirrors, remove the mounting cover from the base. Remove the mirror base mounting screws, the mirror and reinforcement base. Place the mounting bracket (if separate) and mirror in position and install and secure the mounting screws. Install the base cover, if equipped.

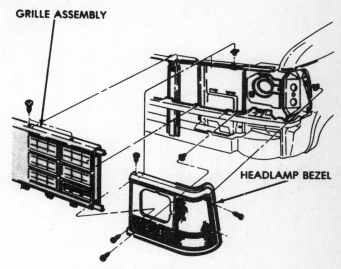

Grille and headlamp bezel

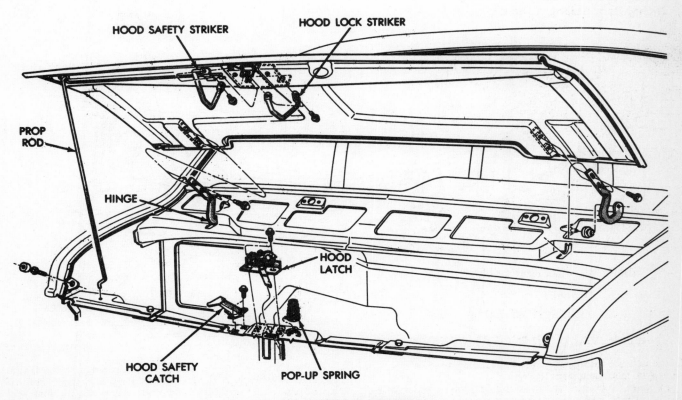

Hood and components

BOLT TORQUE
1/4-20 95 IN. LBS. (10.7 N•m)
5/16-18 200 IN. LB. (22.6 N•m)

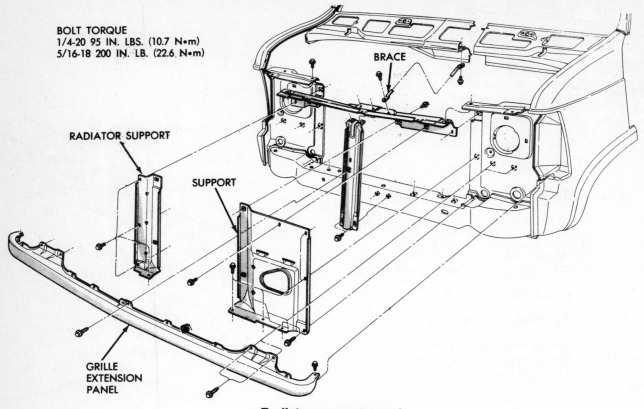

Radiator support panels

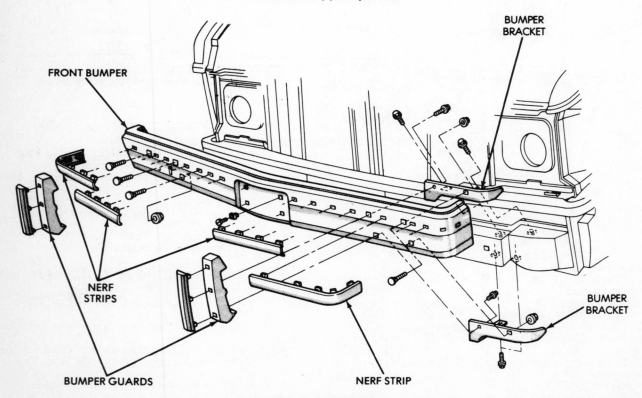

Front bumper assembly

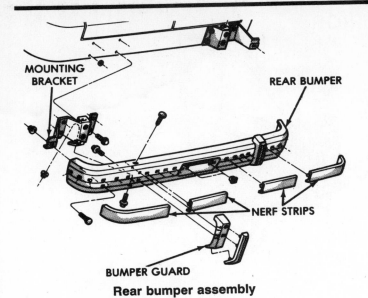

Rear bumper assembly

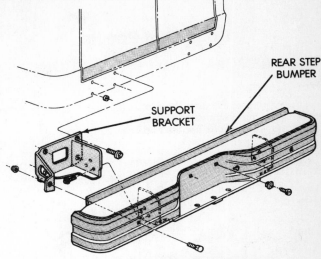

Rear step bumper assembly

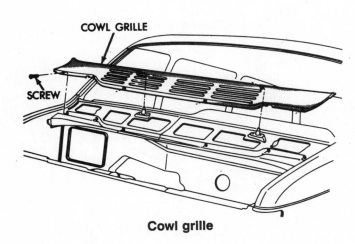

Cowl grille

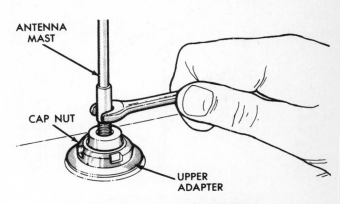

Antenna mast removal/installation

Antenna

REPLACEMENT

1. Disconnect the negative battery ground cable.
2. Remove the windshield wiper arms and disconnect the windshield washer hoses.
3. Remove the cowl grille. Snap out the glove box.
6. On vehicles equipped with air conditioning, remove the right air duct.
7. Reaching through the glove box opening, unplug the antenna cable from the radio.
8. Working through the cowl opening, remove the antenna cable and mounting grommet.
9. Unscrew the antenna mast from the mounting adapter.
10. Reach through the cowl opening and hold the antenna body.
11. Remove the adapter cap nut and pull the antenna from the sheet metal.
To install:
12. Place the antenna in the sheet metal and install the adapter cap nut. Torque the capnut to 100–150 inch lbs.

13. Install the mast in the adapter.
14. Install the cable and grommet through the cowl.
15. Plug the antenna cable into the radio.
16. On vehicles equipped with air conditioning, install the right air duct.
17. Install the glove box.
18. Connect the washer hoses at the nozzles.
19. Install the cowl grille.
20. Install the windshield wiper arms.
21. Connect the battery ground cable.

Cowl Grille and Screen

REMOVAL AND INSTALLATION

1. Remove the windshield wiper arms. Raise and support the hood.
2. Remove the cowl mounting screws. Disconnect the windshield washer hoses. Separate the grille from the cowl and remove.
3. Place the cowl in position. Connect the windshield washer hose. Secure the cowl with the mounting screws. Close the hood. Install the windshield wiper arms.

INTERIOR

Door Trim Panels

REMOVAL AND INSTALLATION

Front Doors

1. If equipped, remove the mounting screw and the power switch housing. Remove the armrest.
2. Remove the door handle and trim cup.

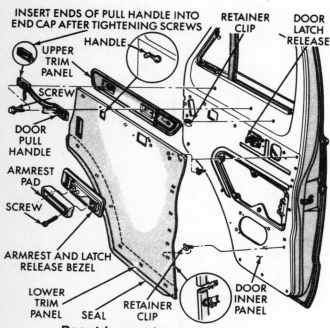

INSERT ENDS OF PULL HANDLE INTO END CAP AFTER TIGHTENING SCREWS

HANDLE

UPPER TRIM PANEL

SCREW

DOOR PULL HANDLE

ARMREST PAD

SCREW

ARMREST AND LATCH RELEASE BEZEL

LOWER TRIM PANEL

SEAL

RETAINER CLIP

RETAINER CLIP

DOOR LATCH RELEASE

DOOR INNER PANEL

Door trim panel and components

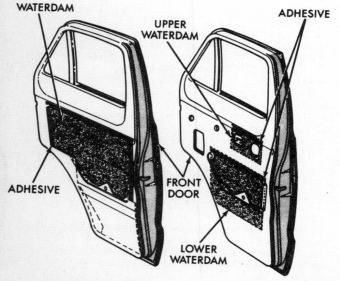

WATERDAM

UPPER WATERDAM

ADHESIVE

ADHESIVE

FRONT DOOR

LOWER WATERDAM

HIGH TRIM MODELS

LOW TRIM MODELS

Door waterdam

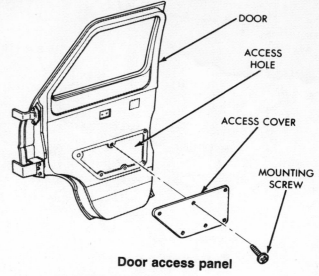

DOOR

ACCESS HOLE

ACCESS COVER

MOUNTING SCREW

Door access panel

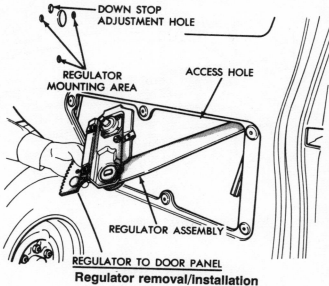

DOWN STOP ADJUSTMENT HOLE

REGULATOR MOUNTING AREA

ACCESS HOLE

REGULATOR ASSEMBLY

REGULATOR TO DOOR PANEL

Regulator removal/installation

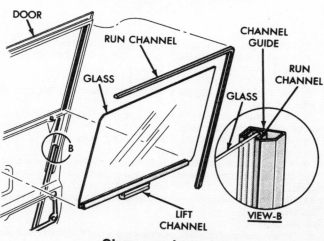

DOOR

RUN CHANNEL

CHANNEL GUIDE

RUN CHANNEL

GLASS

GLASS

B

LIFT CHANNEL

VIEW-B

Glass run channel

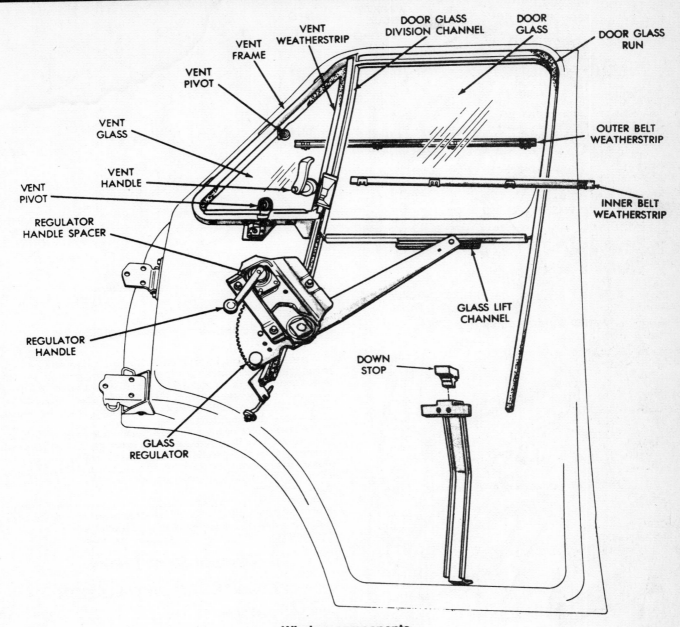

VENT
WEATHERSTRIP

DOOR GLASS
DIVISION CHANNEL

DOOR
GLASS

DOOR GLASS
RUN

VENT
FRAME

VENT
PIVOT

VENT
GLASS

VENT
HANDLE

VENT
PIVOT

REGULATOR
HANDLE SPACER

REGULATOR
HANDLE

GLASS
REGULATOR

OUTER BELT
WEATHERSTRIP

INNER BELT
WEATHERSTRIP

GLASS LIFT
CHANNEL

DOWN
STOP

Window components

3. If the vehicle is equipped with a stereo radio, remove the speaker grille.

4. Remove the setscrew and remove the window crank handle. On models with power windows, remove the window switch trim cup and switch (if not previously removed).

5. Using a flat wood spatula, insert it carefully behind the panel and slide it along to find the push-pins. When you encounter a pin, pry the pin outward. Do this until all the pins are out. NEVER PULL ON THE PANEL TO REMOVE THE PINS! If necessary, remove the door pull handle and upper trim panel. Carefully remove the inner paper waterdam and service the door as required.

6. Install the waterdam. Install the upper trim panel and door pull. Place the trim panel in position and carefully pound the pins into place with the palm of your hand. Be VERY careful to avoid missing the holes and breaking the pins or tearing the panel!

Sliding Doors

1. Carefully pry the pull-strap endcaps off.

2. Remove the retaining screws and remove the strap.

3. Using a flat wood spatula, insert it carefully behind the panel and slide it along to find the push-pins. When you encounter a pin, pry the pin outward. Do this until all the pins are out. NEVER PULL ON THE PANEL TO REMOVE THE PINS!

4. Place the panel in position and carefully pound the pins into place with the palm of your hand. Be VERY careful to avoid missing the holes and breaking the pins or tearing the panel!

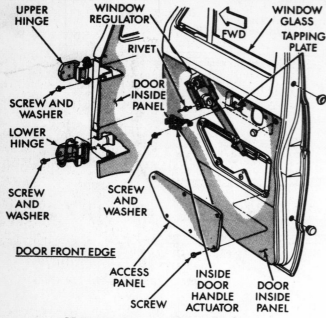

Manual window glass regulator

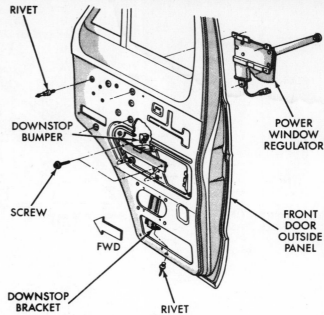

Power window glass regulator

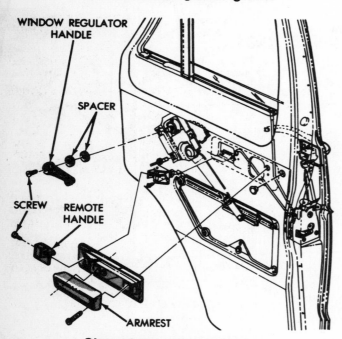

Glass channel and regulator

Interior Trim Panels

REMOVAL AND INSTALLATION

The rear door panels are removed and installed in a similar manner as the front door and sliding door panels. Remove the various handles and bezels. Pry the panel away from the door at the retaining clips with a flat wooden spatula.

Interior trim panels such as the front cowl side trim are secured by retaining screws. Unscrew the mounting screws and remove the panel.

When removing an interior panel that is overlapped by an adjacent trim panel, remove the nearest attaching screw in the adjacent trim panel to avoid possible damage to both panels. To ensure correct alignment when installing interior trim panels, first install the attaching screws finger tight, align the panels then secure the mounting screws.

The lower interior panels are serviced after seats and upper trim panels have been removed. The lower panels are retained by spring clips and possibly by a few screws. The panel is removed first by removing the mounting screws and then by using a flat wooden spatula to disengage the spring clips from their mounting holes.

On installation, make sure the spring clips are properly located over their mounting holes before attempting to push the clips into the holes. Do not overtighten retaining screws to avoid stripping the sheet metal they are screwed into.

Manual Door Locks

REMOVAL AND INSTALLATION

Door Lock Cylinder

1. Raise the window all the way.
2. Remove the door trim panel.
3. Disconnect the lock actuating rod from the lock control clip.

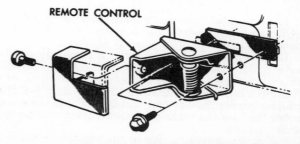

Remote control assembly

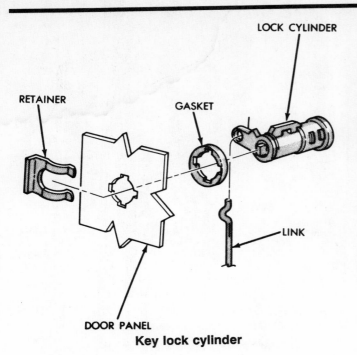

Key lock cylinder

LOCK CYLINDER

RETAINER

GASKET

LINK

DOOR PANEL

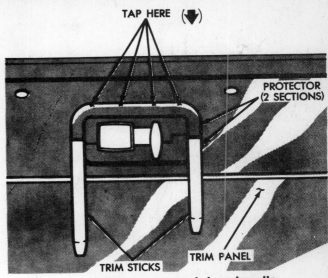

TAP HERE (⬇)

PROTECTOR (2 SECTIONS)

TRIM PANEL

TRIM STICKS

Protector removal door handle

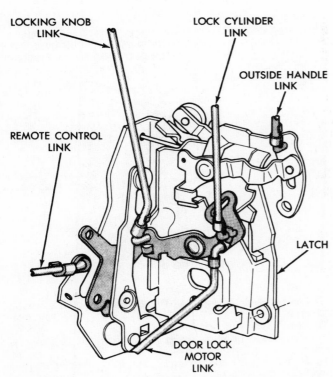

LOCKING KNOB LINK

LOCK CYLINDER LINK

OUTSIDE HANDLE LINK

REMOTE CONTROL LINK

LATCH

DOOR LOCK MOTOR LINK

Link attachment to latch

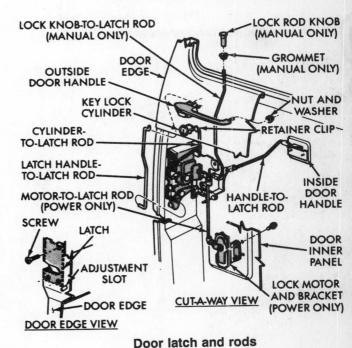

LOCK KNOB-TO-LATCH ROD (MANUAL ONLY)

LOCK ROD KNOB (MANUAL ONLY)

DOOR EDGE

GROMMET (MANUAL ONLY)

OUTSIDE DOOR HANDLE

KEY LOCK CYLINDER

NUT AND WASHER

CYLINDER-TO-LATCH ROD

RETAINER CLIP

LATCH HANDLE-TO-LATCH ROD

INSIDE DOOR HANDLE

MOTOR-TO-LATCH ROD (POWER ONLY)

HANDLE-TO-LATCH ROD

SCREW

LATCH

DOOR INNER PANEL

ADJUSTMENT SLOT

LOCK MOTOR AND BRACKET (POWER ONLY)

CUT-A-WAY VIEW

DOOR EDGE

DOOR EDGE VIEW

Door latch and rods

Power Door Locks

REMOVAL AND INSTALLATION

Lock Solenoid

1. Raise the glass to the full UP position.
2. Remove the door trim panel.
3. Unplug the wiring from the solenoid.
4. Disconnect the linkage from the solenoid.
5. Remove the mounting screws and lift out the solenoid.
6. Secure the solenoid with the mounting screws. Connect the linkage and wiring. Install the door panel.

4. Remove the lock cylinder retaining clip and pull the lock cylinder from the door. On the side doors, it will be necessary to loosen the inside lock control knob set screw and remove the knob.

5. Place the cylinder in position and install the retainer clip. Install the rods, controls and panel.

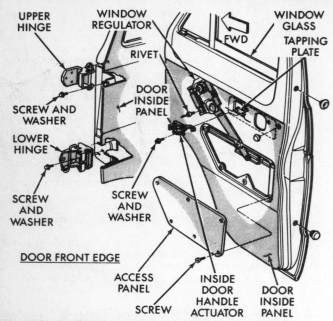

Inside handle actuator removal/installation

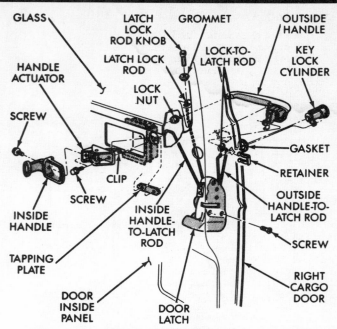

Right cargo door handle and latch assembly

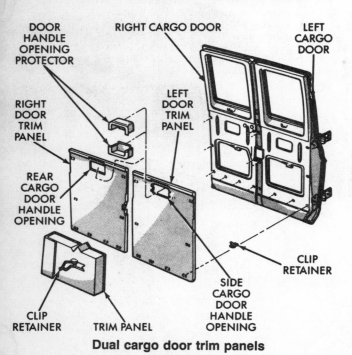

Dual cargo door trim panels

Door Glass and Regulator

REMOVAL AND INSTALLATION

Manual Windows

1. Remove the access or trim panel and watershield.
2. Lower the glass all the way.
3. Remove the lower vent window support bolt.
4. Locate the vent window retaining clip screw through the weatherstripping and remove the screw.

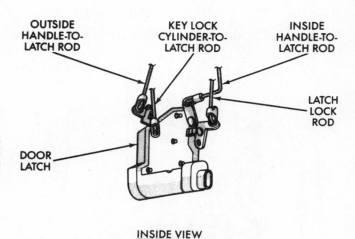

INSIDE VIEW

Right cargo door latch

5. Lower the door glass and tilt the vent window assembly rearward.
6. Remove the vent window.
7. Slide the door glass forward to disconnect it from the regulator.
8. Remove the inside weatherstripping from the glass opening.
9. Lift the glass from the door.
10. Unbolt and remove the regulator. Later models have the regulator secured with rivets. Drill these out.
11. Install the regulator. On later models, replace the rivets with 1/4–20 × 1/2 in. bolts and nuts.
12. Lower the glass into the door.
13. Install the inside weatherstripping in the glass opening.
14. Slide the door glass rearward to connect it to the regulator.
15. Install the vent window.
16. Install the access panel and watershield.

Power Windows

1. Raise the glass to the full UP position.
2. Remove the trim panel and watershield.
3. Remove the down-stop bumper bracket.
4. Lower the glass all the way.
5. Remove the lower vent window support bolt.
6. Locate the vent window retaining clip screw through the weatherstripping and remove the screw.
7. Lower the door glass and tilt the vent window assembly rearward.
8. Remove the vent window.
9. Disconnect the regulator wiring from the harness.
10. Slide the door glass forward to disconnect it from the regulator.
11. Remove the inside weatherstripping from the glass opening.
12. Lift the glass from the door.
13. Drill out the regulator mounting rivets.

To install:

14. Install the regulator. Replace the rivets with ¼–20 × ½ in. bolts and nuts. torqued to 90 inch lbs.
15. Lower the glass into the door.
16. Install the inside weatherstripping in the glass opening.
17. Slide the door glass rearward to connect it to the regulator.
18. Connect the wiring.
19. Install the vent window.
20. Install the down-stop bumper.
21. Install the access panel and watershield.

Windshield
REMOVAL AND INSTALLATION

1. Cover the cowl to protect the paint.

2. Remove the wiper arms. Remove the inside rear view mirror.
3. Remove the windshield retainer strip from the weatherstripping. To remove the strip, pry up one end and carefully pull it out. Later models have a bright cap over the retainer which must be pried out first.
4. Have a helper support the windshield from the outside while you push one corner of the glass out of the weatherstripping.
5. Once one corner is free, continue pushing *carefully* around the glass until the entire windshield is free.
6. The weatherstripping can then be removed from the frame. Replace any defective weatherstripping.

NOTE: Use only mineral spirits as a lubricant when installing the windshield!

7. Install the weatherstripping in the frame. Make sure it's completely seated.
8. With your helper's assistance, slide one corner of the glass into the lower groove in the weatherstripping.
9. Move the glass into the groove as far as possible.
10. Using a fiberglass spatula, force the lip of the weatherstripping over the glass, working around the entire circumference.
11. Starting at one of the lower corners, force the retaining strip into its groove in the weatherstripping. DON'T STRETCH THE STRIP WHEN INSTALLING IT!
12. On later models, install the bright cap.
13. Using a hose, water-test the windshield.
14. Install the inside rear view mirror. A new mirror mount will have to be attached to the windshield. Follow the adhesive manufacturer's directions for installation. Install the wiper arms.

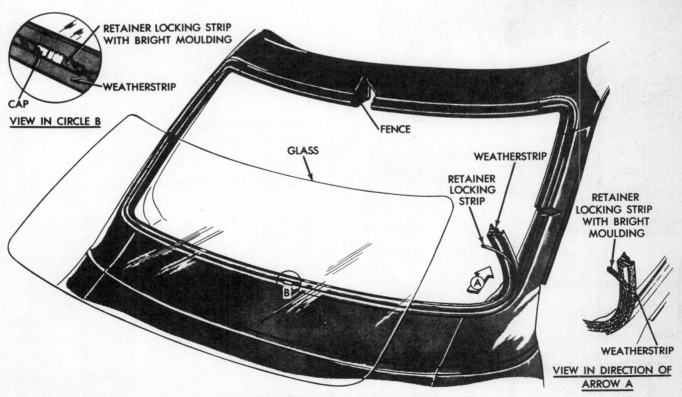

Windshield glass removal/installation

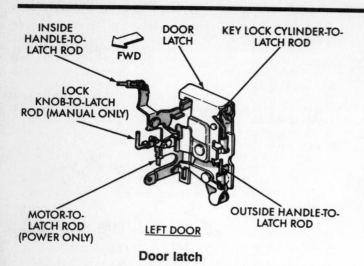

INSIDE HANDLE-TO-LATCH ROD

FWD

DOOR LATCH

KEY LOCK CYLINDER-TO-LATCH ROD

LOCK KNOB-TO-LATCH ROD (MANUAL ONLY)

MOTOR-TO-LATCH ROD (POWER ONLY)

LEFT DOOR

OUTSIDE HANDLE-TO-LATCH ROD

Door latch

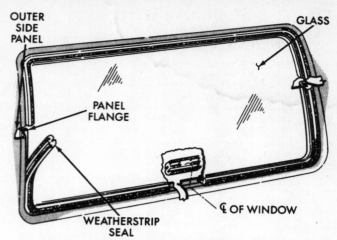

OUTER SIDE PANEL

PANEL FLANGE

GLASS

WEATHERSTRIP SEAL

Ȼ OF WINDOW

Weatherstrip seal with locking strip

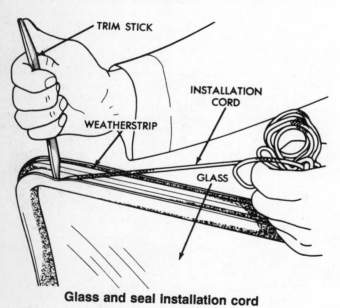

TRIM STICK

INSTALLATION CORD

WEATHERSTRIP

GLASS

Glass and seal installation cord

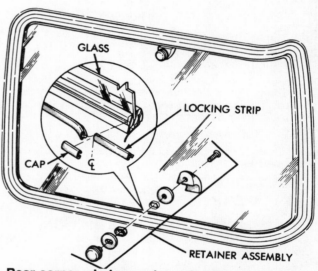

GLASS

LOCKING STRIP

CAP

Ȼ

RETAINER ASSEMBLY

Rear corner window and weatherstrip seal with locking strip

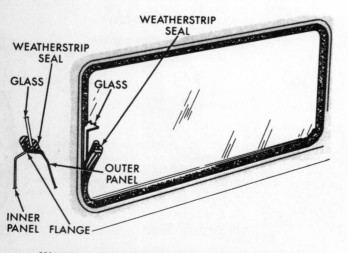

WEATHERSTRIP SEAL

WEATHERSTRIP SEAL

GLASS

GLASS

OUTER PANEL

INNER PANEL

FLANGE

Weatherstrip seal without locking strip

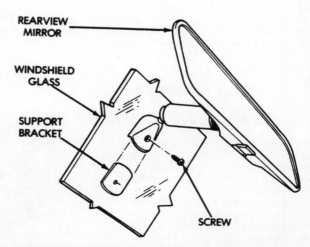

REARVIEW MIRROR

WINDSHIELD GLASS

SUPPORT BRACKET

SCREW

Inside rearview mirror removal/installation

Stationary Windows

REMOVAL AND INSTALLATION

NOTE: You'll need an assistant for this job.

1. Have your assistant stand outside and support the glass.
2. Some windows, such as the rear corner, are equipped with a retainer assembly, a cap and a locking strip. Remove them first. Then, working from the inside vehicle, start at one upper corner and work the weatherstripping across the top of the glass, pulling the weatherstripping down and pushing outward on the glass until your assistant can grab the glass and lift it out.
3. Remove the moldings.
4. Remove the weatherstripping from the glass.

To install:

5. Clean the weatherstripping, glass and glass opening with solvent to remove all old sealer.
6. Apply liquid butyl sealer in the glass channel of the weatherstripping and install the weatherstripping on the glass.
7. Install the moldings.
8. Apply a bead of sealer to the opening flange and in the inner flange crevice of the weatherstripping lip.
9. Place a length of strong cord, such as butcher's twine, in the flange crevice of the weatherstripping. The cord should go all the way around the weatherstripping with the ends, about 18 in. (457mm) long each, hanging down together at the bottom center of the window.
10. Apply soapy water to the weatherstripping lip.
11. Have your assistant position the window assembly in the channel from the outside, applying firm inward pressure.
12. From inside, you guide the lip of the weatherstripping into place using the cord, working each end alternately, until the window is locked in place.
13. Remove the cord, clean the glass and weatherstripping of excess sealer. If equipped, install the locking strip, cap and retainer. Leak test the window.

Inside Rear View Mirror

REMOVAL AND INSTALLATION

The inside rear view mirror is mounted, by a base, to the front windshield. The mirror and arm assembly slide over the base and are held in position by a set screw. Loosen the set screw and slide the mirror and arm up and off of the base.

The base is mounted to the windshield by adhesive. Mark the location of the base on the outside of the windshield with a china marker. Follow the directions that come with the adhesive for application, mount the base to the windshield. After the adhesive has cured, install the mirror assembly and carefully secure the set screw.

Engine Cover and Console

REMOVAL AND INSTALLATION

The console is attached to the engine cover with screws that are accessible through the ash tray housing. Open the ash tray and remove the console mounting screws and the console. Place the console in position on the engine cover and secure the mounting screws.

The inside engine cover is attached to the floor with retaining screws at the rear and latches at the dash. Remove the rear screws and unfasten the latches at the front for cover removal. Place the cover in position. Install the rear screws finger tight. Fasten the front latches, then tighten the mounting screws. Tighten the screws to 24 inch lbs. (3 Nm).

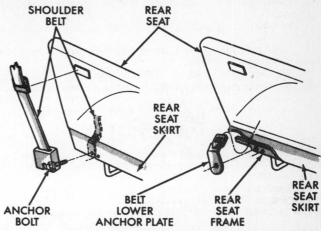

Shoulder belt lower anchor bolt removal/installation

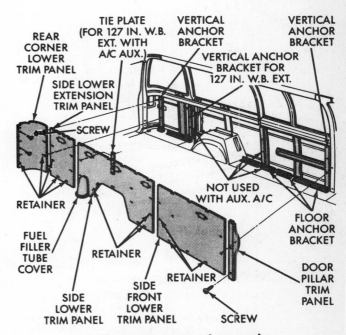

Left side lower trim panels

Seats

REMOVAL AND INSTALLATION

Front Bucket Seat

Remove the anchor bolts and shoulder belt from the seat platform. Remove the seat track-to-floor nuts and lift out the seat. Separate the seat from the platform if necessary. Install the seat to the platform. Place the seat in position inside the vehicle and secure the mounting bolts. Install the shoulder belt and anchors.

Center and Rear Seats

The 3 passenger seat is attached to the floor with hooks and anchors at the rear, and with latches and anchors at the front.

The 4 passenger seat is attached to the floor with brackets, nuts and studs.

1. On 3 passenger seats: Lift up the release lever and push up on the seat.

2. While holding up the front of the seat, push rearward to unlatch the rear anchors.

3. Lift the seat up and out of the vehicle.

4. On 4-passenger seats: Remove the holddown bolts and remove the seat.

5. Place the seat in position and mount it as required.

Travel Seat

A travel seat option is offered on some models. The package can be converted from a 6 passenger forward facing seating arrangement to:

- a lounge/full size sleeper
- a 6 place dinette with table

The seat/cushions, and table are arranged by various lowering and platform placement. Removal is accomplished by simultaneously inserting small prybars in the opening at each side of the seat platform and raising the travel latch pin to disengage the latches from their stops.

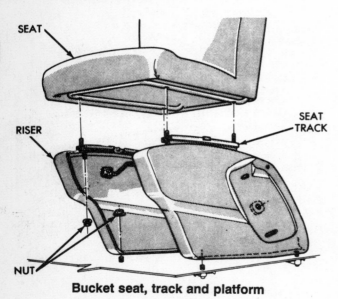

Bucket seat, track and platform

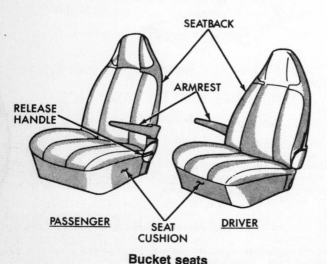

Bucket seats

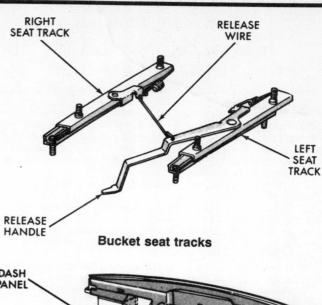

Bucket seat tracks

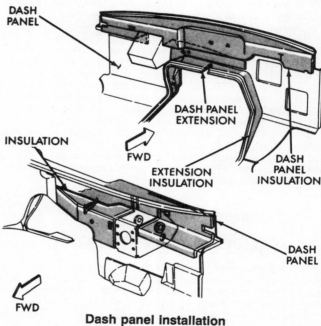

Dash panel installation

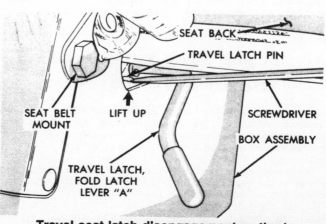

Travel seat latch disengagement method

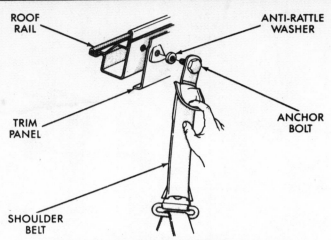

Front seat shoulder belt upper anchor bolt

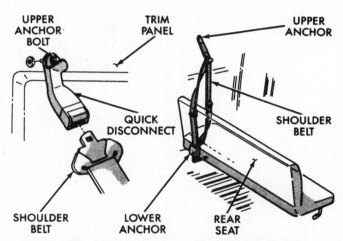

Left shoulder belt and quick disconnect coupler

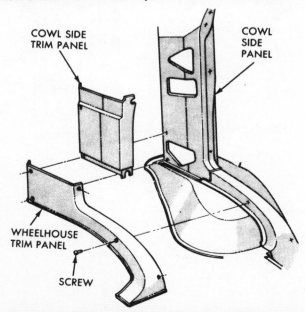

Right cowl side and front wheelhouse trim panels

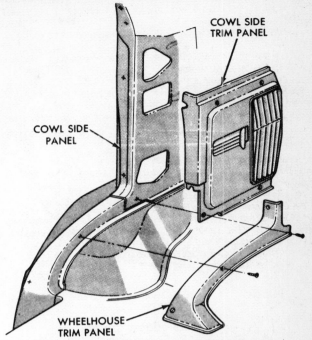

Left cowl side and front wheelhouse trim panels

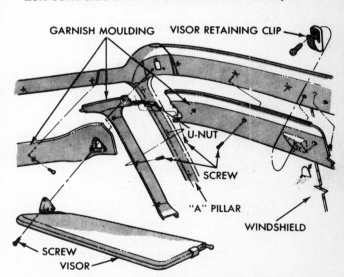

Windshield header, A pillar and front door header trim panels

Seat Belts

Shoulder belts and lap belts having end release buckles are provided.

When installing shoulder belt, lap belt and anchor bolts, they must be tightened to 350 inch lbs. (40 Nm) for safe anchoring to the body panels and seat frames.

1. Move the front seat forward to gain access to the lower belt anchors. Disconnect the shoulder belt warning wiring harness.

2. Remove the exposed cover retaining the shoulder bolts and disengage the cover upper, rear slot from the upper rear concealed shoulder bolt. Remove the cover for access to the belt/retractor anchor bolts.

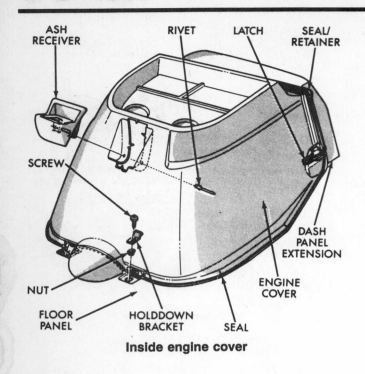

Inside engine cover

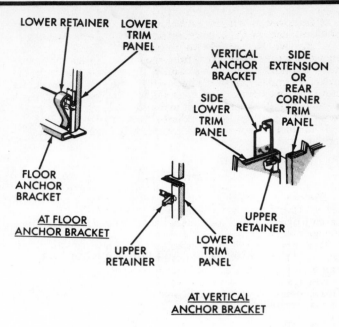

Lower trim panel retainers

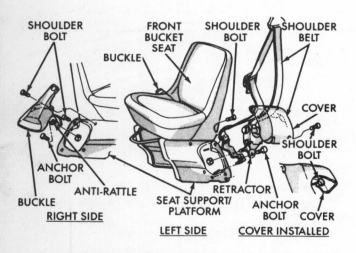

Left front seat shoulder belt/retractor and buckle

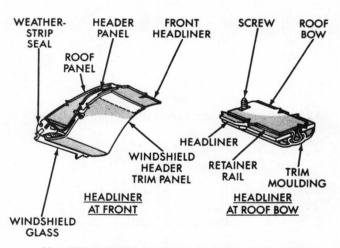

Headliner to header panel/roof attachments

3. Remove the belt and retractor anchor bolts from the seat platform. Remove the belt and retractor from the seat platform.

4. Remove the exposed cover retaining shoulder bolts and disengage the cover lower slots from the lower shoulder bolts. Remove the buckle anchor bolt and anti-rattle washer from the seat platform. Separate the buckle from the seat platform. Remove the cover concealing the shoulder belt upper anchor bolt. Remove the upper anchor bolt and anti-rattle washer from the roof panel. Remove the shoulder belt/retractor and buckle.

5. Replace as necessary and install the belt assemblies reversing the removal procedure.

Headliner

The headliner sections and the trim mouldings/panels are attached to the front/rear header panels and the roof bows with screws. The headliner sections are attached to the side roof panels with clips.

To remove a headliner section, all of the overlapping and interfering trim/mouldings/panels must be removed along with any lamps, coat hooks, and other accessories first.

Headliner sections are molded hardboard. They have limited flexibility and damage will occur if the are bent during removal/installation.

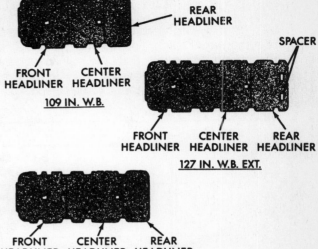

Headliners

TORQUE SPECIFICATIONS

Component	English	Metric
Grade 5 Bolts		
¼–20:	95 inch lbs.	11 Nm
¼–28:	95 inch lbs.	11 Nm
5/16–18:	200 inch lbs.	23 Nrn
5/16–24:	20 ft. lbs.	27 Nm
3/8–16:	30 ft. lbs.	41 Nm
3/8–24:	35 ft. lbs.	48 Nm
7/16–14:	50 ft. lbs.	68 Nm
7/16–20:	55 ft. lbs.	75 Nm
½–13:	75 ft. lbs.	102 Nm
½–20:	85 ft. lbs.	115 Nm
9/16–12:	105 ft. lbs.	142 Nm
9/16–18:	115 ft. lbs.	156 Nm
5/8–11:	150 ft. lbs.	203 Nm
5/8–18:	160 ft. lbs.	217 Nm
¾–16:	175 ft. lbs.	237 Nm
Grade 8 Bolts		
¼–20:	125 inch lbs.	14 Nm
¼–28:	150 inch lbs.	17 Nm
5/16–18:	270 inch lbs.	31 Nm
5/16–24:	25 ft. lbs.	34 Nm
3/8–16:	40 ft. lbs.	54 Nm
3/8–24:	45 ft. lbs.	61 Nm
7/16–14:	65 ft. lbs.	88 Nm
7/16–20:	70 ft. lbs.	95 Nm
½–13:	100 ft. lbs.	136 Nm
½–20:	110 ft. lbs.	149 Nm
9/16–12:	135 ft. lbs.	183 Nm
9/16–18:	150 ft. lbs.	203 Nm
5/8–11:	195 ft. lbs.	264 Nm
5/8–18:	210 ft. lbs.	285 Nm
¾–16:	225 ft. lbs.	305 Nm

How to Remove Stains from Fabric Interior

For best results, spots and stains should be removed as soon as possible. Never use gasoline, lacquer thinner, acetone, nail polish remover or bleach. Use a 3' x 3" piece of cheesecloth. Squeeze most of the liquid from the fabric and wipe the stained fabric from the outside of the stain toward the center with a lifting motion. Turn the cheesecloth as soon as one side becomes soiled. When using water to remove a stain, be sure to wash the entire section after the spot has been removed to avoid water stains. Encrusted spots can be broken up with a dull knife and vacuumed before removing the stain.

Type of Stain	How to Remove It
Surface spots	Brush the spots out with a small hand brush or use a commercial preparation such as K2R to lift the stain.
Mildew	Clean around the mildew with warm suds. Rinse in cold water and soak the mildew area in a solution of 1 part table salt and 2 parts water. Wash with upholstery cleaner.
Water stains	Water stains in fabric materials can be removed with a solution made from 1 cup of table salt dissolved in 1 quart of water. Vigorously scrub the solution into the stain and rinse with clear water. Water stains in nylon or other synthetic fabrics should be removed with a commercial type spot remover.
Chewing gum, tar, crayons, shoe polish (greasy stains)	Do not use a cleaner that will soften gum or tar. Harden the deposit with an ice cube and scrape away as much as possible with a dull knife. Moisten the remainder with cleaning fluid and scrub clean.
Ice cream, candy	Most candy has a sugar base and can be removed with a cloth wrung out in warm water. Oily candy, after cleaning with warm water, should be cleaned with upholstery cleaner. Rinse with warm water and clean the remainder with cleaning fluid.
Wine, alcohol, egg, milk, soft drink (non-greasy stains)	Do not use soap. Scrub the stain with a cloth wrung out in warm water. Remove the remainder with cleaning fluid.
Grease, oil, lipstick, butter and related stains	Use a spot remover to avoid leaving a ring. Work from the outisde of the stain to the center and dry with a clean cloth when the spot is gone.
Headliners (cloth)	Mix a solution of warm water and foam upholstery cleaner to give thick suds. Use only foam—liquid may streak or spot. Clean the entire headliner in one operation using a circular motion with a natural sponge.
Headliner (vinyl)	Use a vinyl cleaner with a sponge and wipe clean with a dry cloth.
Seats and door panels	Mix 1 pint upholstery cleaner in 1 gallon of water. Do not soak the fabric around the buttons.
Leather or vinyl fabric	Use a multi-purpose cleaner full strength and a stiff brush. Let stand 2 minutes and scrub thoroughly. Wipe with a clean, soft rag.
Nylon or synthetic fabrics	For normal stains, use the same procedures you would for washing cloth upholstery. If the fabric is extremely dirty, use a multi-purpose cleaner full strength with a stiff scrub brush. Scrub thoroughly in all directions and wipe with a cotton towel or soft rag.

Glossary

AIR/FUEL RATIO: The ratio of air to gasoline by weight in the fuel mixture drawn into the engine.

AIR INJECTION: One method of reducing harmful exhaust emissions by injecting air into each of the exhaust ports of an engine. The fresh air entering the hot exhaust manifold causes any remaining fuel to be burned before it can exit the tailpipe.

ALTERNATOR: A device used for converting mechanical energy into electrical energy.

AMMETER: An instrument, calibrated in amperes, used to measure the flow of an electrical current in a circuit. Ammeters are always connected in series with the circuit being tested.

AMPERE: The rate of flow of electrical current present when one volt of electrical pressure is applied against one ohm of electrical resistance.

ANALOG COMPUTER: Any microprocessor that uses similar (analogous) electrical signals to make its calculations.

ARMATURE: A laminated, soft iron core wrapped by a wire that converts electrical energy to mechanical energy as in a motor or relay. When rotated in a magnetic field, it changes mechanical energy into electrical energy as in a generator.

ATMOSPHERIC PRESSURE: The pressure on the Earth's surface caused by the weight of the air in the atmosphere. At sea level, this pressure is 14.7 psi at 32°F (101 kPa at 0°C).

ATOMIZATION: The breaking down of a liquid into a fine mist that can be suspended in air.

AXIAL PLAY: Movement parallel to a shaft or bearing bore.

BACKFIRE: The sudden combustion of gases in the intake or exhaust system that results in a loud explosion.

BACKLASH: The clearance or play between two parts, such as meshed gears.

BACKPRESSURE: Restrictions in the exhaust system that slow the exit of exhaust gases from the combustion chamber.

BAKELITE: A heat resistant, plastic insulator material commonly used in printed circuit boards and transistorized components.

BALL BEARING: A bearing made up of hardened inner and outer races between which hardened steel balls roll.

BALLAST RESISTOR: A resistor in the primary ignition circuit that lowers voltage after the engine is started to reduce wear on ignition components.

BEARING: A friction reducing, supportive device usually located between a stationary part and a moving part.

BIMETAL TEMPERATURE SENSOR: Any sensor or switch made of two dissimilar types of metal that bend when heated or cooled due to the different expansion rates of the alloys. These types of sensors usually function as an on/off switch.

BLOWBY: Combustion gases, composed of water vapor and unburned fuel, that leak past the piston rings into the crankcase during normal engine operation. These gases are removed by the PCV system to prevent the buildup of harmful acids in the crankcase.

BRAKE PAD: A brake shoe and lining assembly used with disc brakes.

BRAKE SHOE: The backing for the brake lining. The term is, however, usually applied to the assembly of the brake backing and lining.

BUSHING: A liner, usually removable, for a bearing; an anti-friction liner used in place of a bearing.

BYPASS: System used to bypass ballast resistor during engine cranking to increase voltage supplied to the coil.

CALIPER: A hydraulically activated device in a disc brake system, which is mounted straddling the brake rotor (disc). The caliper contains at least one piston and two brake pads. Hydraulic pressure on the piston(s) forces the pads against the rotor.

CAMSHAFT: A shaft in the engine on which are the lobes (cams) which operate the valves. The camshaft is driven by the crankshaft, via a belt, chain or gears, at one half the crankshaft speed.

CAPACITOR: A device which stores an electrical charge.

CARBON MONOXIDE (CO): A colorless, odorless gas given off as a normal byproduct of combustion. It is poisonous and extremely dangerous in confined areas, building up slowly to toxic levels without warning if adequate ventilation is not available.

CARBURETOR: A device, usually mounted on the intake manifold of an engine, which mixes the air and fuel in the proper proportion to allow even combustion.

CATALYTIC CONVERTER: A device installed in the exhaust system, like a muffler, that converts harmful byproducts of combustion into carbon dioxide and water vapor by means of a heat-producing chemical reaction.

CENTRIFUGAL ADVANCE: A mechanical method of advancing the spark timing by using flyweights in the distributor that react to centrifugal force generated by the distributor shaft rotation.

CHECK VALVE: Any one-way valve installed to permit the flow of air, fuel or vacuum in one direction only.

GLOSSARY

CHOKE: A device, usually a moveable valve, placed in the intake path of a carburetor to restrict the flow of air.

CIRCUIT: Any unbroken path through which an electrical current can flow. Also used to describe fuel flow in some instances.

CIRCUIT BREAKER: A switch which protects an electrical circuit from overload by opening the circuit when the current flow exceeds a predetermined level. Some circuit breakers must be reset manually, while most reset automatically

COIL (IGNITION): A transformer in the ignition circuit which steps up the voltage provided to the spark plugs.

COMBINATION MANIFOLD: An assembly which includes both the intake and exhaust manifolds in one casting.

COMBINATION VALVE: A device used in some fuel systems that routes fuel vapors to a charcoal storage canister instead of venting them into the atmosphere. The valve relieves fuel tank pressure and allows fresh air into the tank as the fuel level drops to prevent a vapor lock situation.

COMPRESSION RATIO: The comparison of the total volume of the cylinder and combustion chamber with the piston at BDC and the piston at TDC.

CONDENSER: 1. An electrical device which acts to store an electrical charge, preventing voltage surges.
  2. A radiator-like device in the air conditioning system in which refrigerant gas condenses into a liquid, giving off heat.

CONDUCTOR: Any material through which an electrical current can be transmitted easily.

CONTINUITY: Continuous or complete circuit. Can be checked with an ohmmeter.

COUNTERSHAFT: An intermediate shaft which is rotated by a mainshaft and transmits, in turn, that rotation to a working part.

CRANKCASE: The lower part of an engine in which the crankshaft and related parts operate.

CRANKSHAFT: The main driving shaft of an engine which receives reciprocating motion from the pistons and converts it to rotary motion.

CYLINDER: In an engine, the round hole in the engine block in which the piston(s) ride.

CYLINDER BLOCK: The main structural member of an engine in which is found the cylinders, crankshaft and other principal parts.

CYLINDER HEAD: The detachable portion of the engine, fastened, usually, to the top of the cylinder block, containing all or most of the combustion chambers. On overhead valve engines, it contains the valves and their operating parts. On overhead cam engines, it contains the camshaft as well.

DEAD CENTER: The extreme top or bottom of the piston stroke.

DETONATION: An unwanted explosion of the air/fuel mixture in the combustion chamber caused by excess heat and compression, advanced timing, or an overly lean mixture. Also referred to as "ping".

DIAPHRAGM: A thin, flexible wall separating two cavities, such as in a vacuum advance unit.

DIESELING: A condition in which hot spots in the combustion chamber cause the engine to run on after the key is turned off.

DIFFERENTIAL: A geared assembly which allows the transmission of motion between drive axles, giving one axle the ability to turn faster than the other.

DIODE: An electrical device that will allow current to flow in one direction only.

DISC BRAKE: A hydraulic braking assembly consisting of a brake disc, or rotor, mounted on an axle, and a caliper assembly containing, usually two brake pads which are activated by hydraulic pressure. The pads are forced against the sides of the disc, creating friction which slows the vehicle.

DISTRIBUTOR: A mechanically driven device on an engine which is responsible for electrically firing the spark plug at a predetermined point of the piston stroke.

DOWEL PIN: A pin, inserted in mating holes in two different parts allowing those parts to maintain a fixed relationship.

DRUM BRAKE: A braking system which consists of two brake shoes and one or two wheel cylinders, mounted on a fixed backing plate, and a brake drum, mounted on an axle, which revolves around the assembly. Hydraulic action applied to the wheel cylinders forces the shoes outward against the drum, creating friction, slowing the vehicle.

DWELL: The rate, measured in degrees of shaft rotation, at which an electrical circuit cycles on and off.

ELECTRONIC CONTROL UNIT (ECU): Ignition module, module, amplifier or igniter. See Module for definition.

ELECTRONIC IGNITION: A system in which the timing and firing of the spark plugs is controlled by an electronic control unit, usually called a module. These systems have no points or condenser.

ENDPLAY: The measured amount of axial movement in a shaft.

ENGINE: A device that converts heat into mechanical energy.

EXHAUST MANIFOLD: A set of cast passages or pipes which conduct exhaust gases from the engine.

FEELER GAUGE: A blade, usually metal, of precisely predetermined thickness, used to measure the clearance between two parts. These blades usually are available in sets of assorted thicknesses.

F-Head: An engine configuration in which the intake valves are in the cylinder head, while the camshaft and exhaust valves are located in the cylinder block. The camshaft operates the intake valves via lifters and pushrods, while it operates the exhaust valves directly.

FIRING ORDER: The order in which combustion occurs in the cylinders of an engine. Also the order in which spark is distributed to the plugs by the distributor.

FLATHEAD: An engine configuration in which the camshaft and all the valves are located in the cylinder block.

FLOODING: The presence of too much fuel in the intake manifold and combustion chamber which prevents the air/fuel mixture from firing, thereby causing a no-start situation.

FLYWHEEL: A disc shaped part bolted to the rear end of the crankshaft. Around the outer perimeter is affixed the ring gear. The starter drive engages the ring gear, turning the flywheel, which rotates the crankshaft, imparting the initial starting motion to the engine.

FOOT POUND (ft.lb. or sometimes, ft. lbs.): The amount of energy or work needed to raise an item weighing one pound, a distance of one foot.

FUSE: A protective device in a circuit which prevents circuit overload by breaking the circuit when a specific amperage is present. The device is constructed around a strip or wire of a lower amperage rating than the circuit it is designed to protect. When an amperage higher than that stamped on the fuse is present in the circuit, the strip or wire melts, opening the circuit.

GEAR RATIO: The ratio between the number of teeth on meshing gears.

GENERATOR: A device which converts mechanical energy into electrical energy.

HEAT RANGE: The measure of a spark plug's ability to dissipate heat from its firing end. The higher the heat range, the hotter the plug fires.

HUB: The center part of a wheel or gear.

HYDROCARBON (HC): Any chemical compound made up of hydrogen and carbon. A major pollutant formed by the engine as a byproduct of combustion.

HYDROMETER: An instrument used to measure the specific gravity of a solution.

INCH POUND (in.lb. or sometimes, in. lbs.): One twelfth of a foot pound.

INDUCTION: A means of transferring electrical energy in the form of a magnetic field. Principle used in the ignition coil to increase voltage.

INJECTION PUMP: A device, usually mechanically operated, which meters and delivers fuel under pressure to the fuel injector.

INJECTOR: A device which receives metered fuel under relatively low pressure and is activated to inject the fuel into the engine under relatively high pressure at a predetermined time.

INPUT SHAFT: The shaft to which torque is applied, usually carrying the driving gear or gears.

INTAKE MANIFOLD: A casting of passages or pipes used to conduct air or a fuel/air mixture to the cylinders.

JOURNAL: The bearing surface within which a shaft operates.

KEY: A small block usually fitted in a notch between a shaft and a hub to prevent slippage of the two parts.

MANIFOLD: A casting of passages or set of pipes which connect the cylinders to an inlet or outlet source.

MANIFOLD VACUUM: Low pressure in an engine intake manifold formed just below the throttle plates. Manifold vacuum is highest at idle and drops under acceleration.

MASTER CYLINDER: The primary fluid pressurizing device in a hydraulic system. In automotive use, it is found in brake and hydraulic clutch systems and is pedal activated, either directly or, in a power brake system, through the power booster.

MODULE: Electronic control unit, amplifier or igniter of solid state or integrated design which controls the current flow in the ignition primary circuit based on input from the pick-up coil. When the module opens the primary circuit, the high secondary voltage is induced in the coil.

NEEDLE BEARING: A bearing which consists of a number (usually a large number) of long, thin rollers.

OHM: (Ω) The unit used to measure the resistance of conductor to electrical flow. One ohm is the amount of resistance that limits current flow to one ampere in a circuit with one volt of pressure.

OHMMETER: An instrument used for measuring the resistance, in ohms, in an electrical circuit.

OUTPUT SHAFT: The shaft which transmits torque from a device, such as a transmission.

OVERDRIVE: A gear assembly which produces more shaft revolutions than that transmitted to it.

OVERHEAD CAMSHAFT (OHC): An engine configuration in which the camshaft is mounted on top of the cylinder head and operates the valve either directly or by means of rocker arms.

OVERHEAD VALVE (OHV): An engine configuration in which all of the valves are located in the cylinder head and the camshaft is located in the cylinder block. The camshaft operates the valves via lifters and pushrods.

OXIDES OF NITROGEN (NOx): Chemical compounds of nitrogen produced as a byproduct of combustion. They combine with hydrocarbons to produce smog.

OXYGEN SENSOR: Used with the feedback system to sense the presence of oxygen in the exhaust gas and signal the computer which can reference the voltage signal to an air/fuel ratio.

PINION: The smaller of two meshing gears.

PISTON RING: An open ended ring which fits into a groove on the outer diameter of the piston. Its chief function is to form a seal between the piston and cylinder wall. Most automotive pistons have three rings: two for compression sealing; one for oil sealing.

PRELOAD: A predetermined load placed on a bearing during assembly or by adjustment.

PRIMARY CIRCUIT: Is the low voltage side of the ignition system which consists of the ignition switch, ballast resistor or resistance wire, bypass, coil, electronic control unit and pick-up coil as well as the connecting wires and harnesses.

PRESS FIT: The mating of two parts under pressure, due to the inner diameter of one being smaller than the outer diameter of the other, or vice versa; an interference fit.

GLOSSARY

RACE: The surface on the inner or outer ring of a bearing on which the balls, needles or rollers move.

REGULATOR: A device which maintains the amperage and/or voltage levels of a circuit at predetermined values.

RELAY: A switch which automatically opens and/or closes a circuit.

RESISTANCE: The opposition to the flow of current through a circuit or electrical device, and is measured in ohms. Resistance is equal to the voltage divided by the amperage.

RESISTOR: A device, usually made of wire, which offers a preset amount of resistance in an electrical circuit.

RING GEAR: The name given to a ring-shaped gear attached to a differential case, or affixed to a flywheel or as part a planetary gear set.

ROLLER BEARING: A bearing made up of hardened inner and outer races between which hardened steel rollers move.

ROTOR: 1. The disc-shaped part of a disc brake assembly, upon which the brake pads bear; also called, brake disc.
2. The device mounted atop the distributor shaft, which passes current to the distributor cap tower contacts.

SECONDARY CIRCUIT: The high voltage side of the ignition system, usually above 20,000 volts. The secondary includes the ignition coil, coil wire, distributor cap and rotor, spark plug wires and spark plugs.

SENDING UNIT: A mechanical, electrical, hydraulic or electromagnetic device which transmits information to a gauge.

SENSOR: Any device designed to measure engine operating conditions or ambient pressures and temperatures. Usually electronic in nature and designed to send a voltage signal to an on-board computer, some sensors may operate as a simple on/off switch or they may provide a variable voltage signal (like a potentiometer) as conditions or measured parameters change.

SHIM: Spacers of precise, predetermined thickness used between parts to establish a proper working relationship.

SLAVE CYLINDER: In automotive use, a device in the hydraulic clutch system which is activated by hydraulic force, disengaging the clutch.

SOLENOID: A coil used to produce a magnetic field, the effect of which is produce work.

SPARK PLUG: A device screwed into the combustion chamber of a spark ignition engine. The basic construction is a conductive core inside of a ceramic insulator, mounted in an outer conductive base. An electrical charge from the spark plug wire travels along the conductive core and jumps a preset air gap to a grounding point or points at the end of the conductive base. The resultant spark ignites the fuel/air mixture in the combustion chamber.

SPLINES: Ridges machined or cast onto the outer diameter of a shaft or inner diameter of a bore to enable parts to mate without rotation.

TACHOMETER: A device used to measure the rotary speed of an engine, shaft, gear, etc., usually in rotations per minute.

THERMOSTAT: A valve, located in the cooling system of an engine, which is closed when cold and opens gradually in response to engine heating, controlling the temperature of the coolant and rate of coolant flow.

TOP DEAD CENTER (TDC): The point at which the piston reaches the top of its travel on the compression stroke.

TORQUE: The twisting force applied to an object.

TORQUE CONVERTER: A turbine used to transmit power from a driving member to a driven member via hydraulic action, providing changes in drive ratio and torque. In automotive use, it links the driveplate at the rear of the engine to the automatic transmission.

TRANSDUCER: A device used to change a force into an electrical signal.

TRANSISTOR: A semi-conductor component which can be actuated by a small voltage to perform an electrical switching function.

TUNE-UP: A regular maintenance function, usually associated with the replacement and adjustment of parts and components in the electrical and fuel systems of a vehicle for the purpose of attaining optimum performance.

TURBOCHARGER: An exhaust driven pump which compresses intake air and forces it into the combustion chambers at higher than atmospheric pressures. The increased air pressure allows more fuel to be burned and results in increased horsepower being produced.

VACUUM ADVANCE: A device which advances the ignition timing in response to increased engine vacuum.

VACUUM GAUGE: An instrument used to measure the presence of vacuum in a chamber.

VALVE: A device which control the pressure, direction of flow or rate of flow of a liquid or gas.

VALVE CLEARANCE: The measured gap between the end of the valve stem and the rocker arm, cam lobe or follower that activates the valve.

VISCOSITY: The rating of a liquid's internal resistance to flow.

VOLTMETER: An instrument used for measuring electrical force in units called volts. Voltmeters are always connected parallel with the circuit being tested.

WHEEL CYLINDER: Found in the automotive drum brake assembly, it is a device, actuated by hydraulic pressure, which, through internal pistons, pushes the brake shoes outward against the drums.